CARBOHYDRATE NANOTECHNOLOGY

CARBOHYDRATE NANOTECHNOLOGY

Edited by
KEITH J. STINE

WILEY

Published by John Wiley & Sons, Inc., Hoboken, New Jersey
Published simultaneously in Canada

Library of Congress Cataloging-in-Publication Data:

Carbohydrate nanotechnology / edited by Keith J. Stine.
pages cm
Includes bibliographical references and index.
ISBN 978-1-118-86053-3 (cloth)
1. Nanomedicine. 2. Nanostructured materials. 3. Carbohydrates. 4. Proteins.
I. Stine, Keith J.
R857.N34C367 2016
610.28–dc23

2015021572

Set in 10/12pt Times by SPi Global, Pondicherry, India

Printed in the United States of America

10 9 8 7 6 5 4 3 2 1

1 2016

CONTENTS

CONTRIBUTORS

Shailesh G. Ambre, Glycoconjugate and NMR Section, Chemical Biology Laboratory, Center for Cancer Research, National Cancer Institute at Frederick, Frederick, MD, USA

Joseph J. Barchi Jr., Glycoconjugate and NMR Section, Chemical Biology Laboratory, Center for Cancer Research, National Cancer Institute at Frederick, Frederick, MD, USA

C. Remzi Becer, School of Engineering and Materials Science, Queen Mary, University of London, London, UK

Tomas Bertok, Institute of Chemistry, Slovak Academy of Sciences, Bratislava, Slovakia

Rabah Boukherroub, Institute of Electronics, Microelectronics and Nanotechnology (IEMN), UMR 8520 CNRS, Lille 1 University, Avenue Poincaré – BP 60069, 59652 Villeneuve d'Ascq, France

Juan Manuel Casas-Solvas, Department of Chemistry and Physics, University of Almería, Almería, Spain

Willem M. de Vos, Laboratory of Microbiology, Wageningen University, Wageningen, the Netherlands and Department of Bacteriology & Immunology and Department of Veterinary Biosciences, University of Helsinki, Helsinki, Finland

Aline Debrassi, Laboratory of Organic Chemistry, Wageningen University, Wageningen, the Netherlands

Jared Q. Gerlach, Glycoscience Group, National Centre for Biomedical Engineering Science, National University of Ireland Galway, Galway, Ireland

Xuefei Huang, Department of Chemistry, Michigan State University, East Lansing, MI, USA

Andras Hushegyi, Institute of Chemistry, Slovak Academy of Sciences, Bratislava, Slovakia

Knud J. Jensen, Department of Chemistry, Centre for Carbohydrate Recognition and Signalling, Faculty of Science, University of Copenhagen, Frederiksberg, Copenhagen, Denmark

Lokesh Joshi, Glycoscience Group, National Centre for Biomedical Engineering Science, National University of Ireland Galway, Galway, Ireland

Kagan Kerman, Department of Physical and Environmental Sciences, University of Toronto, Scarborough, Toronto, Ontario, Canada

Michelle Kilcoyne, Glycoscience Group, National Centre for Biomedical Engineering Science and Microbiology, School of Natural Sciences, National University of Ireland Galway, Galway, Ireland

Ludmila Klukova, Institute of Chemistry, Slovak Academy of Sciences, Bratislava, Slovakia

Hovig Kouyoumdjian, Department of Chemistry, Michigan State University, East Lansing, MI, USA

Nan Li, Department of Physical and Environmental Sciences, University of Toronto, Scarborough, Toronto, Ontario, Canada

Yinfa Ma, Department of Chemistry, Center for Single Nanoparticle, Single Cell, and Single Molecule Monitoring (CS^{3}M), Missouri University of Science and Technology, Rolla, MO, USA

Zachary P. Michael, Department of Chemistry, University of Pittsburgh, Pittsburgh, PA, USA

Ravin Narain, Chemical and Materials Engineering, University of Alberta, Edmonton, Alberta, Canada

Alexander Star, Department of Chemistry, University of Pittsburgh, Pittsburgh, PA, USA

Keith J. Stine, Department of Chemistry and Biochemistry and Center for Nanoscience, University of Missouri–St. Louis, St. Louis, MO, USA

Xue-Long Sun, Department of Chemistry, Chemical and Biomedical Engineering and Center for Gene Regulation in Health and Disease (GRHD), Cleveland State University, Cleveland, OH, USA

Rajesh Sunasee, Department of Chemistry, State University of New York at Plattsburgh, Plattsburgh, NY, USA

Sabine Szunerits, Institute of Electronics, Microelectronics and Nanotechnology (IEMN), UMR 8520 CNRS, Lille 1 University, Avenue Poincaré – BP 60069, 59652 Villeneuve d'Ascq, France

Yih Horng Tan, Department of Chemistry and Biochemistry and Center for Nanoscience, University of Missouri–St. Louis, St. Louis, MO, USA

Mikkel B. Thygesen, Department of Chemistry, Centre for Carbohydrate Recognition and Signalling, Faculty of Science, University of Copenhagen, Frederiksberg, Copenhagen, Denmark

Jan Tkac, Institute of Chemistry, Slovak Academy of Sciences, Bratislava, Slovakia

Pratima Vabbilisetty, Department of Chemistry, Chemical and Biomedical Engineering and Center for Gene Regulation in Health and Disease (GRHD), Cleveland State University, Cleveland, OH, USA

Antonio Vargas-Berenguel, Department of Chemistry and Physics, University of Almería, Almería, Spain

Sébastien Vidal, Institut de Chimie et Biochimie Moléculaires et Supramoléculaires, Laboratoire de Chimie Organique 2—Glycochimie, UMR 5246, Université Lyon 1 and CNRS, Villeurbanne, France

Jacob J. Weingart, Department of Chemistry, Chemical and Biomedical Engineering and Center for Gene Regulation in Health and Disease (GRHD), Cleveland State University, Cleveland, OH, USA

Tom Wennekes, Laboratory of Organic Chemistry, Wageningen University, Wageningen, the Netherlands

Qingbo Yang, Department of Chemistry, Center for Single Nanoparticle, Single Cell, and Single Molecule Monitoring (CS^3M), Missouri University of Science and Technology, Rolla, MO, USA

Gokhan Yilmaz, Department of Chemistry, University of Warwick, Coventry, UK and Department of Basic Sciences, Turkish Military Academy, Ankara, Turkey

Han Zuilhof, Laboratory of Organic Chemistry, Wageningen University, Wageningen, the Netherlands and Department of Chemical and Materials Engineering, King Abdulaziz University, Jeddah, Saudi Arabia

PREFACE

Glycoscience and nanoscience are two fields that have been growing significantly in interest and impact over the past decade or so, and thus, the emergence of a fertile intersection between these fields seems natural given the important biological role of carbohydrate-decorated structures and interactions on the nanoscale in biological systems. Carbohydrates are involved in fundamental biological processes including fertilization, viral infection, bacterial adhesion, immunity and immune response, immunodeficiency diseases, and neuroscience and in cancers where altered glycosylation is common. The fact that many proteins are glycoproteins and that the attached glycans are heterogeneous in structure and they play key roles in protein function and interaction provides a strong motivation to develop technologies to assay and ultimately exploit these interactions for diagnostic and therapeutic aims. Glycoscience has steadily reached into and become a new and integral part of many of the areas of nanoscience, including nanomaterials, supramolecular design, drug delivery, self-assembly, and others such that the two fields are now advancing together in synergistic ways. This book is meant to provide a range of chapters in some of the major fundamental areas that have emerged under the heading of "Carbohydrate Nanotechnology."

In Chapter 1 by Debrassi, de Vos, Zuilhof, and Wennekes, the presentation of carbohydrates at the surfaces of self-assembled monolayers (SAMs) is covered including direct modification of hydrogen-terminated silicon surfaces as an alternative to thiols on gold SAMs. Chemical and photochemical means of glycan conjugation, physical methods for characterization of the SAM structure, and biological applications to binding of bacteria, sensing of bacterial toxins, and multivalency effects on these surfaces are described.

In Chapter 2 by Szunerits and Boukherroub, the basic aspects of plasmonics that are the foundation of the traditional surface plasmon resonance (SPR) technologies widely used in label-free analysis of glycan interactions with proteins and other partners are reviewed. The advances in development of chips and arrays surface modified by various chemical strategies to present glycans suited for SPR analysis are reviewed.

In Chapter 3 by Thygesen and Jensen, the area of carbohydrate-modified gold nanoparticles is surveyed covering many chemical attachment methods. This is a core area for advancement of carbohydrate nanotechnology, with the unique physical behavior of metal nanoparticles and the multivalent nature of carbohydrate-binding converging.

In Chapter 4 by Li and Kerman, the field of quantum dot glycoconjugates is reviewed. Preparation, physical properties, and conjugation strategies are described for these nanoparticles that are finding valuable applications in imaging and in biosensor development involving glycans.

In Chapter 5 by Michael, Star, and Vidal, the conjugation of carbohydrates with carbon nanostructures including fullerenes, nanotubes, and graphene by both covalent and noncovalent means is reviewed. These conjugate structures are shown to have applications in biosensors, biofuel cells, and biomedical research.

In Chapter 6 by Yilmaz and Becer, glycopolymers and their synthesis by a range of controlled polymerization methods are reviewed. The elegant design of precisely structured glycopolymers has fueled studies of their multivalent binding by lectins and created new possibilities for their application in glycobiology, vaccine development, and other areas.

In Chapter 7 by Casas-Solvas and Vargas-Berenguel, the development of glycoclusters intended to function as inhibitors to viral entry and bacterial adhesion, as vaccine platforms, and as vehicles for drug or gene delivery is examined. The use of a wide range of scaffolds for building multivalent structures is a key aspect of this chapter.

In Chapter 8 by Weingart, Vabbilisetty, and Sun, the surface modification of liposomes to incorporate carbohydrate structures and also their direct assembly are surveyed. Methods for the characterization of glycoliposomes are described, and biomedical applications to drug, gene, or antigen delivery and as multivalent inhibitors of lectin binding are reviewed.

In Chapter 9 by Stine, applications of nanoporous or what are referred to also as mesoporous materials development to glycoscience are surveyed. Many of these applications are in the areas of affinity materials for glycan recognition and separation, with other aspects including controlled release and supported synthesis.

In Chapter 10 by Gerlach, Kilcoyne, and Joshi, advances in glycomic microarray technology that involves incorporating nanostructures are reviewed including both arrays supporting glycans and those supporting lectins. The microarrays provide affinity analysis of many interactions simultaneously and can be used for analysis of small quantities of sample and for cases where binding partners are not known.

In Chapter 11 by Tan, the application of atomic force microscopy (AFM) to gain information on carbohydrate nanostructures assembled on surfaces by imaging at the nearly molecular level is described. The procedure and subtleties of AFM analysis applied to protein binding to carbohydrate presenting SAMs, to glycolipid containing supported bilayers, and to analysis of carbohydrate–lectin interactions using modified tips are reviewed.

In Chapter 12 by Kouyoumdjian and Huang, it is described how sialic acids presented on the surfaces of cells facilitate aggregation of amyloid peptides (Aβ) that play a crucial role in Alzheimer's disease. Methods for creating sialic acid-modified nanoparticles and using them to detect aggregation of Aβ and possibly protect cells from the toxic effects of Aβ aggregates are reviewed.

In Chapter 13 by Ambre and Barchi, how glycan-modified nanoparticles of various kinds can be used to develop new cancer therapeutics that exploit specific features of tumor biology is described. It is also described how the glycan can serve as a therapeutic agent or as a targeting agent and how nanoparticles made of polysaccharides can serve as a basis for the design of these potential new treatments.

In Chapter 14 by Sunasee and Narain, vaccine development using synthetic glycopolymers or glyconanoparticles is the focus. The growing ability to precisely control the architecture of these particles leads to their application in delivery of antigens, adjuvants, and anticancer drugs, but much remains to be learned about their interaction with biological systems.

In Chapter 15 by Hushegyi, Klukova, Bertok, and Tkac, strategies for surface modification and conjugation of glycans onto surfaces are reviewed that are needed for the creation of glycan-based biosensors. Conjugation chemistry is reviewed in detail, along with properties of SAMs and label-free detection methods such as electrochemical impedance, surface plasmon, and field-effect transistor, among others.

In Chapter 16 by Ma and Yang, nanotoxicology aspects of carbohydrate-modified nanostructures are covered. In order for these nanostructures to advance further in their applications, understanding their unique toxicity issues and verifying their safety are areas that must be give detailed consideration.

It is hoped that this collection of chapters can provide an overview of a rapidly advancing multidisciplinary field. While many topics in carbohydrate nanotechnology are represented here, there are many that were not able to be included but are also of current interest or are emerging. Reviews of some of these topics can be found elsewhere, as the literature in this field is now growing steadily. It is also hoped that it can serve as a resource for those whose research enters this field either from the direction of being a glycoscientist seeking to integrate aspects of nanoscience into their work or from the direction of a nanoscientist seeking to collaborate or approach some of the many opportunities offered by glycoscience. All of the contributors are acknowledged for their most fascinating and valued contributions.

Keith J. Stine
Department of Chemistry and Biochemistry
Center for Nanoscience
University of Missouri–St. Louis
St. Louis, MO, USA

1

CARBOHYDRATE-PRESENTING SELF-ASSEMBLED MONOLAYERS: PREPARATION, ANALYSIS, AND APPLICATIONS IN MICROBIOLOGY

ALINE DEBRASSI[1], WILLEM M. DE VOS[2,3], HAN ZUILHOF[1,4], AND TOM WENNEKES[1]

[1] *Laboratory of Organic Chemistry, Wageningen University, Wageningen, the Netherlands*
[2] *Laboratory of Microbiology, Wageningen University, Wageningen, the Netherlands*
[3] *Department of Bacteriology & Immunology and Department of Veterinary Biosciences, University of Helsinki, Helsinki, Finland*
[4] *Department of Chemical and Materials Engineering, King Abdulaziz University, Jeddah, Saudi Arabia*

1.1 INTRODUCTION

Carbohydrates are a complex class of essential biomolecules that can be considered as the dark matter of the biological universe as they are greatly understudied yet omnipresent in all kingdoms of life and vital to fully understand biological processes. The structurally diverse carbohydrates are present both on the cell surface and inside cells. They decorate the cell surface to form the so-called glycocalyx, a dense and complex layer of carbohydrates unique for every type of cell or organism, and as such are key to many important biological recognition events by interacting with carbohydrate-binding proteins. Carbohydrate–protein interactions play an important role in various biological events occurring at the cell surface, such as bacterial and viral infections [1,2], cancer metastasis [3,4], and immune response [4]. The study of the interactions between carbohydrates and other biomolecules at biological surfaces

Carbohydrate Nanotechnology, First Edition. Edited by Keith J. Stine.

and interfaces is instrumental in the understanding of these processes and contributing to the development of novel diagnostic methods and medicines. The study of carbohydrates, compared to, for example, nucleic acids and proteins, however, poses unique challenges because their structure is nonlinear and their biosynthesis not template driven. The native glycocalyx is too complex, dense, and dynamic for studying these interactions individually, with the current techniques at our disposal. Therefore, a simplified version is often created by the placement of well-defined, synthetic carbohydrates on a surface, so-called glycoarrays or glycosurfaces, to study specific carbohydrate–protein interactions. These fabricated glycosurfaces can also be more readily incorporated in a sensor or a nanostructure and as such used to elicit, detect, or quantify binding events, for example, in diagnostic devices, molecular imaging, and drug delivery applications. Various approaches have been developed to prepare glycosurfaces, each of them with their advantages and drawbacks, and these approaches will be the main focus of this chapter.

We will start the chapter by presenting an overview of the different methods most commonly used to prepare glycosurfaces. These methods will be discussed divided over three sections that each reflect one of the three distinct approaches used to create glycosurfaces: (i) direct formation of carbohydrate-containing self-assembled monolayers (SAMs), (ii) use of secondary (or tertiary) reactions to install a carbohydrate on a preformed SAM, and (iii) noncovalent immobilization of carbohydrates on a surface. The discussion of the secondary reaction approach (ii) is subdivided into two subsections: one addressing the use of unmodified "natural" carbohydrates and the other, the use of synthetic carbohydrate derivatives with a special emphasis on attachment using so-called "click" chemistry. Next, we will focus on several key surface analysis techniques that can be used to characterize a prepared glycosurface and the type of information that can be obtained from each technique. As previously mentioned, carbohydrate–protein interactions are involved in bacterial pathogenesis and symbiosis. A famous example of carbohydrate-mediated bacterial adhesion is between the gut microbiota and the carbohydrates present on the surface of human intestinal cells. Glycosurfaces can be used for the binding, capture, and sensing of gut bacteria. A representative example of this from our own group is the study of interactions between the mannose-specific adhesin of *Lactobacillus plantarum* [5]—a lactic acid bacterium present in various probiotic products, fermented foods, and our gut—and fabricated mannose-terminated glycosurfaces (vide infra) [6]. At the end of this chapter, we will discuss several more applications of glycosurfaces in microbiology, focusing on binding, capture, and sensing of bacteria and bacterial toxins and on the multivalency effects that exert a large influence on the interaction between carbohydrates and proteins in biological systems and on fabricated glycosurfaces.

1.2 PREPARATION OF SAMs CONTAINING CARBOHYDRATES

SAMs are ordered molecular assemblies that spontaneously form on a substrate by chemisorption (or strong interaction) of molecules containing a chemical functionality with a strong affinity for the substrate surface. The chemical structure of molecules that are used to prepare a SAM is usually subdivided in its constituting parts; the part that adsorbs on the substrate surface can be called the attaching group, the part on the opposing end of the molecule that ends up at the top of the monolayer is called the end group or terminal group, and the intermediate part is called the chain or backbone [7,8]. In this section, we will present an overview of the recent scientific literature on the preparation and properties of SAMs containing carbohydrates as end groups (Table 1.1).

One of the most common combinations of substrate and attaching group is the formation of SAMs of thiols on gold (Table 1.1; entry a), and to our knowledge, this was also the first example of a carbohydrate-presenting SAM. In 1996, Spencer and coworkers reported the formation of SAMs on gold surfaces with a thiol-terminated hexasaccharide. The thiol-terminated hexasaccharide, a truncated amylose derivative consisting of six α-1,4-linked glucopyranosides, was assembled on gold surfaces in its protected (peracetylated) and deprotected form. Both protected and deprotected compounds readily formed SAMs on gold, although the kinetics of SAM formation varied, with the deprotected hexasaccharides achieving an SAM with higher density. The protected hexasaccharide was also successfully deprotected on the surface after the SAM formation: however, the degree of deprotection was slightly lower than when conducted in solution before SAM formation [24]. These early studies already indicate that thiol SAMs on gold are best prepared directly with deprotected carbohydrate derivatives in order to circumvent incomplete deprotection of carbohydrates on the surface and degradation of the unstable thiol on gold SAM itself.

Using a similar approach, Russell and coworkers [9] synthesized protected and deprotected thiol-terminated monosaccharides that were assembled as SAMs on gold-coated glass substrates and afterwards assessed for their interaction with a series of lectins. The SAM formed with a thiol-terminated mannose derivative was exposed to concanavalin A (Con A), a lectin known to bind strongly with mannose, and a lectin from *Tetragonolobus purpureas*, which specifically binds L-fucose. As expected, the mannose-terminated SAM showed selective interaction with Con A, demonstrating that carbohydrate-presenting SAMs can be used to study interactions between carbohydrates and proteins as a simplified version of natural cell surfaces [9].

Houseman and Mrksich [18] were the first to prepare mixed SAMs that consisted of various ratios of a carbohydrate and oligoethylene glycol end group, in which the latter was incorporated to minimize nonspecific interactions. The authors prepared SAMs using *N*-acetylglucosamine and tri(ethylene glycol) with thiol attaching groups and studied the effect of the concentration of *N*-acetylglucosamine in the monolayer on an enzymatic reaction [18]. Later in this chapter, we will further discuss the strategy of using molecules to "dilute" the amount of carbohydrate on a surface and thereby tune the carbohydrate presentation and concentration (multivalency effect and optimization of density; page 50).

The relatively easy preparation of thiol SAMs on gold and high tolerance for additional functional groups, including carbohydrate hydroxyls, have made it a popular method to immobilize also other carbohydrates with various levels of complexity: monosaccharides (mannose [10–14], glucose [15–17,32], galactose [13,16,17,37],

TABLE 1.1 Approaches Used For the Direct Preparation of Carbohydrate-Presenting SAMs

Substrate		Functional Group	Immobilized Product	Immobilized Carbohydrates
(a)		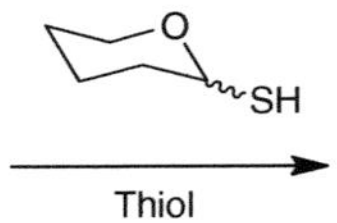	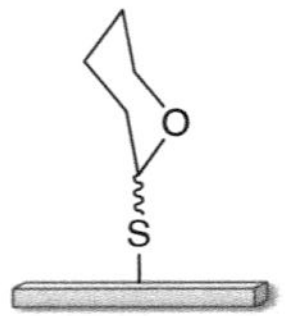	Mannose [9–14], glucose [9,15–17], galactose [13,16,17], *N*-acetylglucosamine [18], lactose [15], rhamnose [17], maltose [17,19], maltotriose [17], abequose [20], paratose [20], tyvelose [20], globotriose [21], xylose [17], dimethylated maltose [17], GM1 [15], other disaccharides [22,23], hexasaccharide [24]
(b)		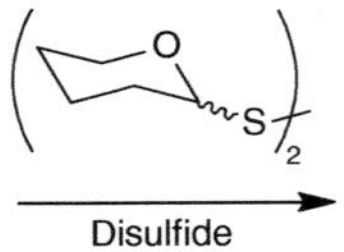	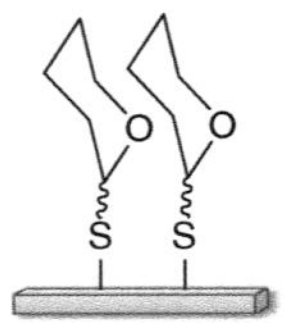	Globotriose [25–27], maltose [28], P^k trisaccharide [29], asialo-GM2 disaccharide [29]
(c)	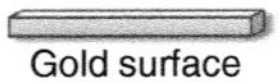	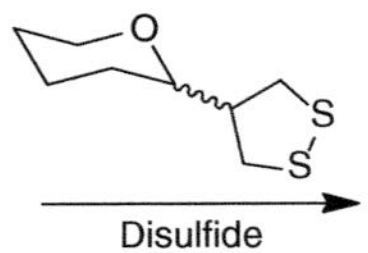	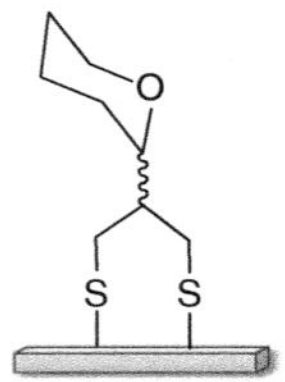	Mannose [30], glucose [30–32], fucose [30], galactose [30,31], *N*-acetylglucosamine [30], sialic acid [30], lactose [31]

(d)

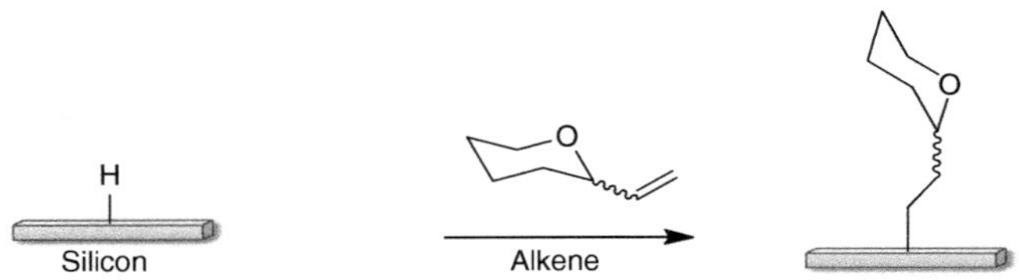

Lactose [33]

(e)

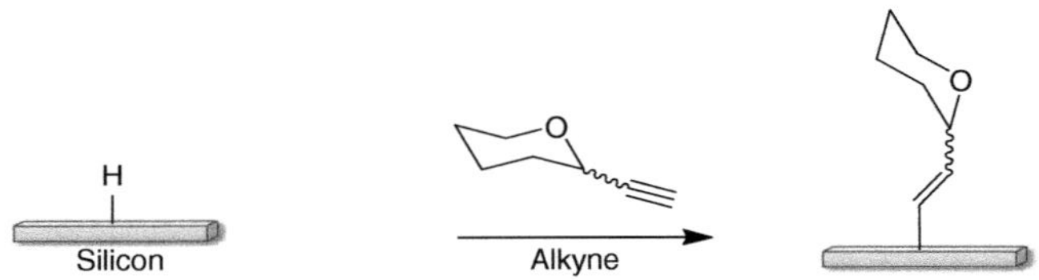

Mannose [34]

(f)

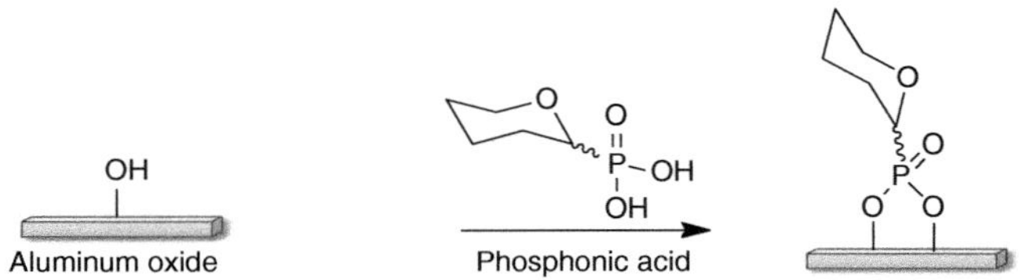

Mannose, Gb3, globo H [35]

(g)

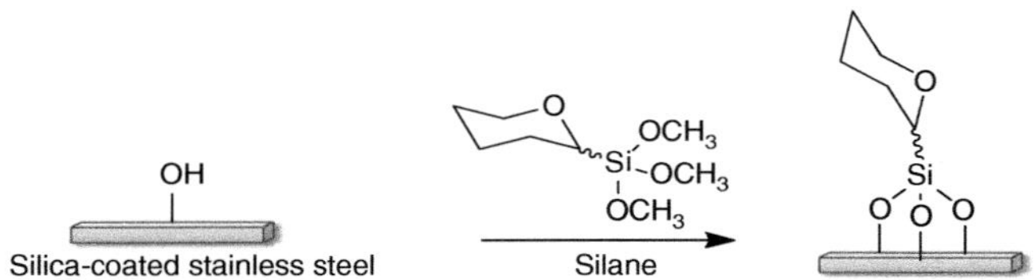

N-Acetylglucosamine, galactose [36]

(a) Thiol on gold, (b) disulfide on gold (monovalent binding), (c) disulfide on gold (multidentate binding), (d) alkene on silicon, (e) alkyne on silicon, (f) phosphonic acid on aluminum oxide, and (g) silane on silica.

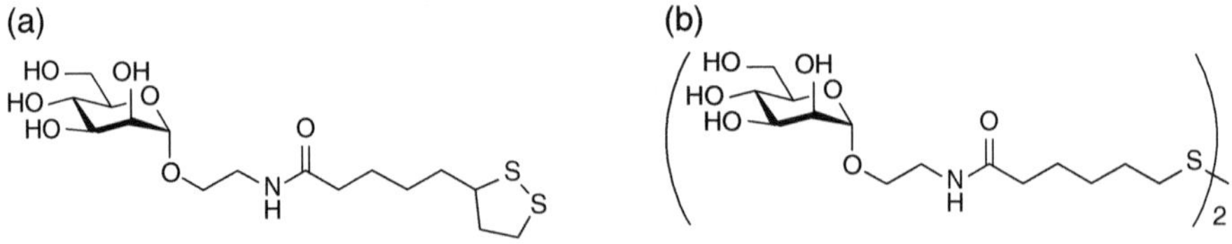

FIGURE 1.1 Mannose derivatives containing disulfides: (a) disulfide that can form multidentate binding on gold and (b) disulfide that results in monodentate binding on gold.

xylose [17], rhamnose [17]), disaccharides (lactose [15], maltose [17,19], dimethylated maltose [17]) [20,22,23], and oligosaccharides (GM1 pentasaccharide [15], globotriose [21], maltotriose [17]) [37].

A general drawback of SAMs created by the adsorption of thiols on gold is their relative limited stability. In order to increase the stability of SAMs on gold, some research groups have prepared SAMs with molecules that can form multiple bonding interactions with the substrate (multidentate adsorbates) (Table 1.1; entry c). The increased stability enables their use under conditions that are not compatible with the monodentated ones [38]. Disulfides can be used to generate more stable SAMs on gold (Fig. 1.1a), and this strategy has been applied to various carbohydrate derivatives: mannose [10,30], galactose [30,31], glucose [30,31], fucose [30], *N*-acetyl glucosamine [30], sialic acid [30], and lactose [31]. However, some carbohydrate derivatives containing disulfides are designed in a way that does not enable multidentate binding to the surface (Fig. 1.1b and Table 1.1; entry b). Although these molecules also form SAMs on gold, their binding mode and presentation of the carbohydrate are comparable to the binding of single thiol attaching groups [25–29].

As is clear from the previous paragraphs, carbohydrate-presenting SAMs have up till now been prepared mostly by thiol absorption on gold, but several alternative methods exist, which are discussed next. One of these is the formation of SAMs on hydrogen-terminated silicon surfaces using terminal alkenes as attaching group (Table 1.1; entry d). In this case, the SAMs can be obtained by thermal or photochemical radical reaction of the alkene, resulting in the formation of a Si–C bond. Acetyl-protected β-glucose, a mixture of β and α-sialic acid, and a sialic acid derivative were successfully immobilized on hydrogen-terminated silicon surfaces using either thermal or photochemical method, depending on the thermal stability of the carbohydrate [39,40].

Using a similar approach, lactose was immobilized as *p*-vinylbenzyllactonoamide on silicon (Fig. 1.2). Through a thermal radical reaction, a silicon-centered radical, which was formed by the activation of a Si–H bond, reacted with the terminal alkene of the *p*-vinylbenzyllactonoamide molecule in an anti-Markovnikov fashion. After SAM formation, the lactoside-covered surface was patterned by UV irradiation using a copper grid. The authors showed specific binding of a lactose-binding lectin (*Ricinus communis* agglutinin, RCA_{120}) on the regions that were not irradiated with UV light, without nonspecific adsorption of the protein on the SiO_x regions. Compared to the earlier SAMs on gold, this technique offers the advantage that an additional

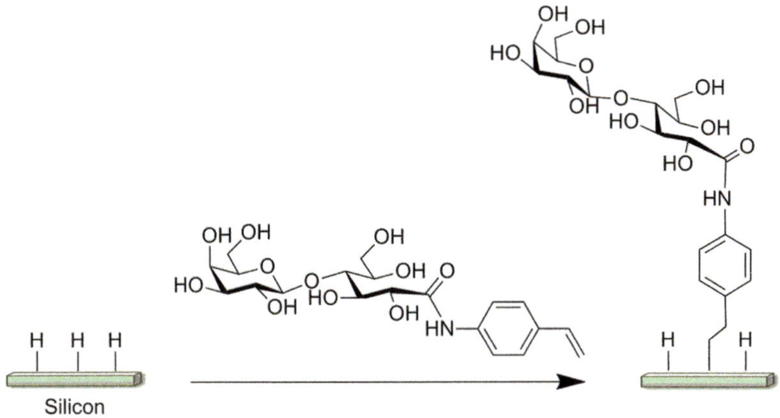

FIGURE 1.2 Immobilization of lactose as *p*-vinylbenzyllactonoamide on silicon.

resistant SAM, such as a polyethylene glycol chain, is not needed to prevent nonspecific adsorption of proteins on silicon surfaces [33].

In a similar approach, a mannose derivative containing a terminal alkyne group was used to form SAMs on hydrogen-terminated silicon surfaces by a photochemical radical reaction (Table 1.1; entry e). Hydrosilation of the Si–H surface was achieved by UV/visible light irradiation-generated radicals, which initiate the Si–C bond formation that over time generates the SAM. The mannose-presenting SAM was formed on a patterned substrate and displayed specific protein recognition of fluorescently labeled mannose-binding lectin (Con A) [34].

Another approach to generate covalent SAMs uses carbohydrate derivatives containing a phosphonic acid attaching group that is able to form SAMs on oxide surfaces (Table 1.1; entry f). Using this approach, Wong and coworkers [35] prepared phosphonic acid-presenting derivatives of simple monosaccharides, like mannose, and more complex carbohydrates, like the trisaccharide Gb3 and the hexasaccharide Globo H that were allowed to form SAMs on aluminum oxide-coated glass slides. The glycan arrays generated by this technique were successfully used to study several carbohydrate–protein interactions [35].

Although one of the most common methods to prepare SAMs in general is the modification of surface oxides with alkylsilanes [41], there are not many examples of carbohydrate derivatives containing alkylsilanes to form SAMs, probably due to the reactivity of silanes with the hydroxyls of unprotected carbohydrates and the consequently laborious synthesis routes required to circumvent this. One of the few existing examples is the synthesis of *N*-acetyl-D-glucosamine and galactose derivatives containing a trialkoxysilane attaching group and their use to form SAMs on silica-coated stainless steel surfaces (Table 1.1; entry g). The presence and availability for biological interactions of the carbohydrates were confirmed by the successful binding of *N*-acetyl-D-glucosamine- and galactose-binding lectins [36].

In general, there are not many methods for the direct formation of SAMs with carbohydrate derivatives. It is evident that the most well-known and frequently used method is the formation of SAMs of thiols or disulfides on gold surfaces. Although this is an easy and well-established technique for carbohydrate SAMs formation, the limited stability of the thiol SAMs on gold may hamper the scope of their potential applications [42]. However, the formation of thiol SAMs on gold is the most simple method to immobilize carbohydrates on a surface in only one step and is currently still being used successfully, especially to study carbohydrate–protein interactions by surface plasmon resonance (SPR) [14], electrochemical impedance spectroscopy (EIS) [12,13,21], cyclic voltammetry [16], quartz crystal microbalance (QCM) [30], and a cantilever sensor platform [37]. An alternative for the direct formation of SAMs with carbohydrate derivatives is to use a secondary reaction to attach the carbohydrates via the end groups of a previously formed SAM, an approach that is discussed in the following section.

1.3 PREPARATION OF GLYCOSURFACES VIA A SECONDARY REACTION ON SAMs

1.3.1 Glycosurfaces Obtained Stepwise Using Unmodified Carbohydrates

The attachment of unmodified carbohydrates to a reactive surface is the simplest method, because it does not require prior chemical modification of the carbohydrates, which is usually a time-consuming step. For the methods described in this section, in general a preformed SAM presents end groups that react with a functional group of a carbohydrate to form a covalent bond (Table 1.2).

Koberstein and coworkers [43] described a photochemical method for immobilization of a variety of unmodified mono-, oligo-, and polysaccharides on glass, quartz, and silicon substrates. The authors initially synthesized a phthalimide-derivatized silane, which was self-assembled on the substrates to generate phthalimide-terminated surfaces. Upon exposure to UV light, an excited n–π* state was produced that abstracts a hydrogen atom from a nearby molecule (Fig. 1.3a and Table 1.2; entry a). The resulting radicals then recombined and formed a covalent bond that in this case was with a nearby carbohydrate present in the reaction solution. Because of the photochemical nature of the process, this method can be used for direct chemical patterning of surfaces with carbohydrates via a photolithography process. Similar experiments were also successfully performed on benzophenone-terminated surfaces (Fig. 1.3b), which also contain aromatic carbonyls that can photochemically react with natural carbohydrates. However, the thickness of these carbohydrate layers was lower, and the water contact angle was higher than that of the carbohydrates immobilized on the phthalimide-terminated surfaces [43].

Another more recently reported application of a photochemical reaction to immobilize unmodified carbohydrates on surfaces employs perfluorophenylazide-terminated SAMs (Fig. 1.3c and Table 1.2; entry b). Initially, SAMs were formed on gold with perfluorophenylazide-containing thiol groups. Upon irradiation with UV light, the azide moiety yields perfluorophenylnitrene, which is able to insert into C–H bonds (Fig. 1.3c). A series of mono- and oligosaccharides was successfully immobilized in

TABLE 1.2 Immobilization of Unmodified Carbohydrates On Surfaces With Different End Group Terminations

	Surface Termination	Unmodified Carbohydrates	Immobilized Product	Immobilized Carbohydrates
(a)	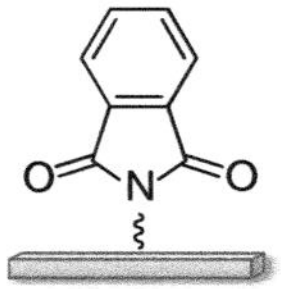 Phthalimide-terminated surface	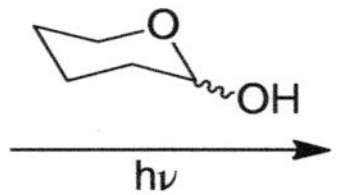	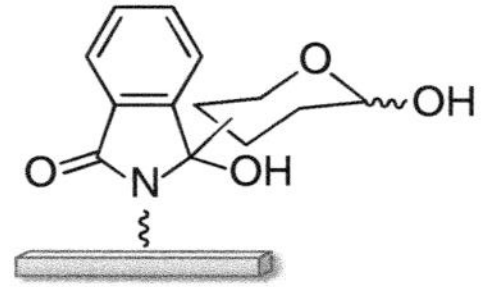	Galactose, *N*-acetylgalactosamine, arabinose, rhamnose, mannose, glucose, isomaltotriose, isomaltopentose, isomaltoheptaose [43]
(b)	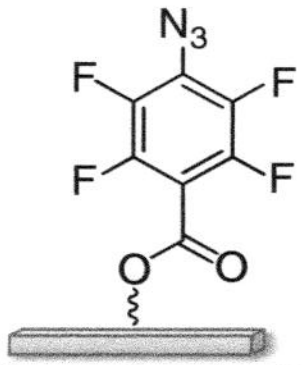 Perfluorophenyl azide-terminated surface	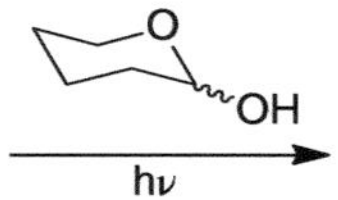	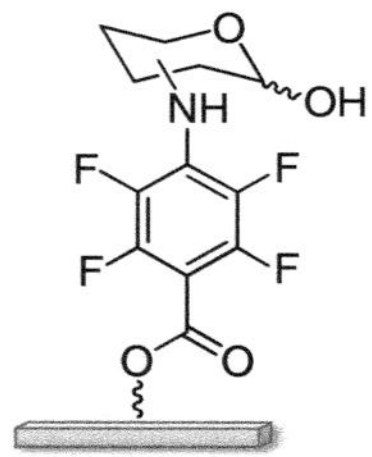	Mannose, glucose, galactose [44]
(c)	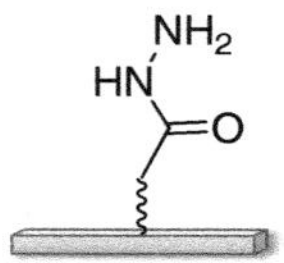 Hydrazide-terminated surface	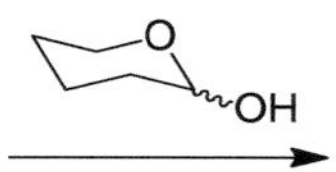	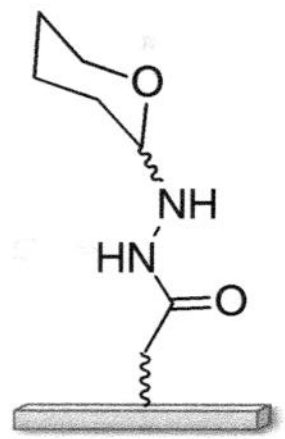	*N*-Acetylglucosamine, mannobiose, methyl mannopyranoside, mannan, sialy Lewis X, isomaltopentaose [45], mannose, heparin decasaccharides [46]

(continued)

TABLE 1.2 (Continued)

Surface Termination	Unmodified Carbohydrates	Immobilized Product	Immobilized Carbohydrates
(d) 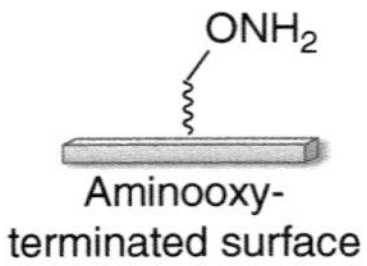	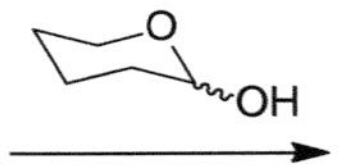	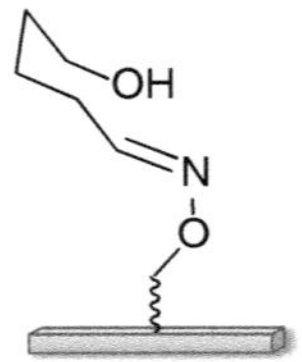	*N*-Acetylglucosamine, mannobiose, methyl mannopyranoside, mannan, sialyl Lewis X, isomaltopentaose [45]
(e) 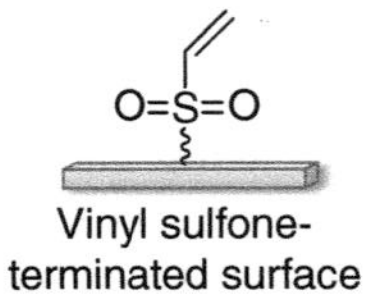	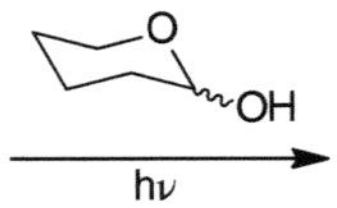	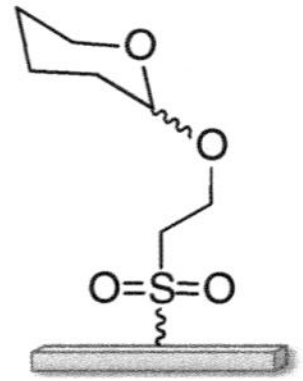	Mannose [47], various complex carbohydrates [48]

(a) Phthalimide, (b) perfluorophenyl azide, (c) hydrazide, (d) aminooxy, and (e) vinyl sulfone.

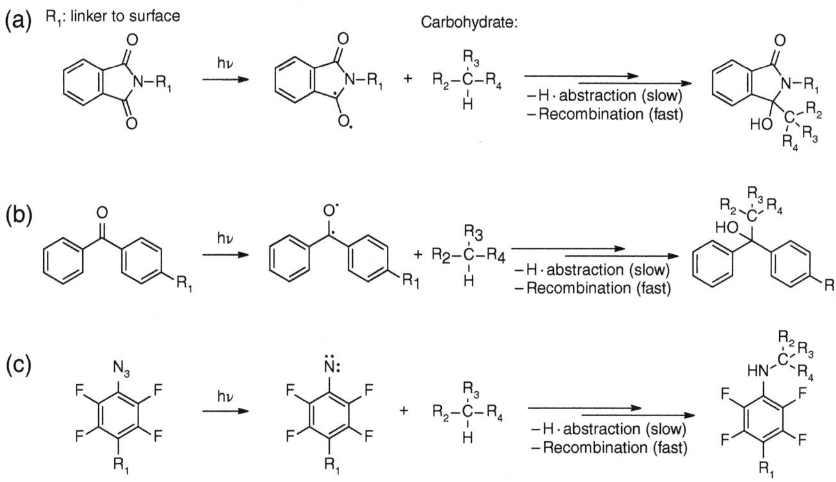

FIGURE 1.3 Photochemical reactions used to immobilize unmodified carbohydrates on surfaces with photoactive end groups: (a) phthalimide, (b) benzophenone, and (c) perfluorophenylazide.

this way onto SPR sensors and used for carbohydrate–protein binding studies. Through these binding studies, it was shown that the surface-bound carbohydrates retained their binding affinities and selectivity. Thus, this technique apparently enables the formation of robust and stable carbohydrate arrays, which can be repeatedly used to study carbohydrate–protein interactions [44]. These photochemical reactions form the basis for convenient methods to immobilize various unmodified carbohydrates onto surfaces, although a major drawback is that the carbohydrates are immobilized in an ill-defined way due to the many reactive sites in the same molecule.

A way to overcome this problem and still use unmodified carbohydrates is to use the anomeric hemiacetal present in reducing carbohydrates for the surface immobilization. In solution, this functional group is in equilibrium with the open chain form aldehyde that can undergo various selective reactions. Wang and coworkers [45] used this approach to prepare carbohydrate microarrays on glass slides. They initially immobilized a three-dimensional poly(amidoamine) starburst dendrimer on epoxy-terminated glass, followed by functionalization of the dendrimer with terminal hydrazide (Table 1.2; entry c) and aminooxy (Table 1.2; entry d) groups (Fig. 1.4). These functional groups react with the aldehyde of the reducing carbohydrates, leading to site-specific immobilization via oxime and hydrazine formation. Using these techniques, the authors immobilized various unmodified mono-, oligo-, and polysaccharides with preservation of their specific binding activity [45].

In a similar approach, Zhi and coworkers [46] prepared carbohydrate microarrays by reacting the aldehyde group of a reducing carbohydrate with hydrazide-terminated surfaces. The difference between this approach and the previous one is that the latter uses an additional reduction step of the oligosaccharides to form a reducing end aldehyde moiety, which reacts with the hydrazide groups present on the surface, forming

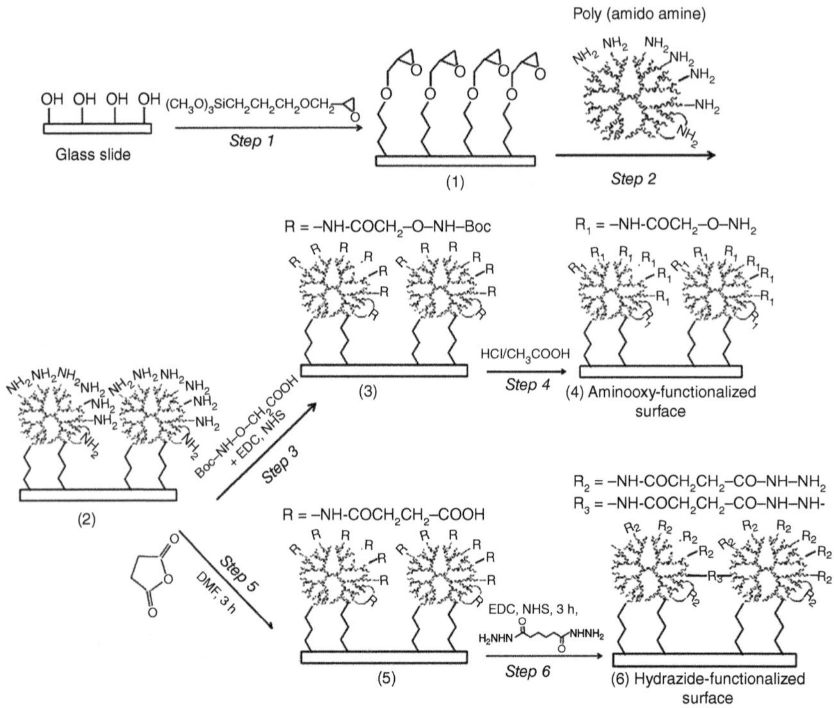

FIGURE 1.4 Chemical process for preparation of 3D aminooxy- and hydrazide functionalized glass slides. *Source*: Reprinted with permission from Ref. 45, Copyright 2009, American Chemical Society.

a hydrazone. This hydrazone is then mainly converted into the native β-pyranose form, immobilizing the carbohydrates in a site-specific way [46].

Another approach that leads to a certain degree of site-specific immobilization of unmodified carbohydrates on surfaces makes use of divinyl sulfone as a cross-linking agent between hydroxy-terminated surfaces and the hydroxyl groups of the carbohydrate (Table 1.2; entry e) [47,48]. In the first step, a hydroxy-terminated thiol-based SAM is generated on gold, followed by the immobilization of divinyl sulfone and the unmodified carbohydrate via a Michael addition. The increased nucleophilicity of the anomeric hydroxyl contributes to the immobilization of the carbohydrates mainly via their anomeric center. However, an important drawback of this method is that the carbohydrate may also be immobilized by any of its other multiple hydroxyl groups and can exist as a mixture of α and β anomers, which is difficult to characterize on a surface and can have an effect on subsequent biological assays. To overcome these problems and to improve the reactivity of the carbohydrates, mannose derivatives containing amine and thiol groups were synthesized and immobilized on these vinyl-terminated surfaces (Table 1.3; entry i). The results indeed showed that the aminated and thiolated mannose derivatives are more efficiently immobilized on the vinyl sulfone-terminated surfaces [47].

Although the approaches described in this section are easy and versatile as they can be applied to a variety of natural carbohydrates, their major drawback is the nonspecific attachment of carbohydrates onto the surface. The use of chemically modified carbohydrates derivatives for site-selective attachment on surfaces is therefore a more commonly used approach to ensure that all molecules present on the surface are immobilized in a well-defined manner and thus have the same orientation. The reactions that are most frequently used for site-selective attachment of carbohydrates on surfaces are discussed in the following section.

1.3.2 Glycosurfaces Obtained Stepwise Using Synthetic Carbohydrate Derivatives

The most extensively developed method to immobilize carbohydrates on surfaces involves the prior attachment of surface-reactive groups at the anomeric position of carbohydrates, resulting in site-specific immobilization (Table 1.3) [49]. Of course, if one invests the additional time and effort in synthesizing a tailor-made carbohydrate derivative, the subsequent SAM attachment reaction should proceed in a controlled and efficient fashion to allow for a well-defined glycosurface and under mild conditions to allow for a large scope of (complex) carbohydrates.

In view of these desired reaction characteristics, the most frequently used reactions to immobilize carbohydrates on surfaces via this approach belong to the popular so-called "click" reactions. The most used is the copper(I)-catalyzed azide–alkyne cycloaddition (CuAAC) reaction (Table 1.3; entries a and b), which can be performed in various solvents and tolerates most functionalities. One of the first examples of immobilization of carbohydrates on surfaces using a CuAAC reaction was reported by Wang and coworkers [43]. In their study, azide-containing carbohydrate derivatives (a mannoside, lactoside, and galactose-containing trisaccharide) were successfully immobilized on alkyne-terminated gold surfaces via the CuAAC reaction. The immobilized carbohydrates presented specific binding toward proteins, as analyzed by SPR and QCM [50]. Overall, two different approaches have been used to immobilize carbohydrates on surfaces via CuAAC: Either the alkyne functionality is present on the surface and reacts with azide-containing carbohydrate derivatives [6,51–53,55,100–102], or the azide group is on the surface and reacts with an alkyne-containing carbohydrate [56,57]. While the yield of CuAAC is typically high, a significant drawback of this reaction is the requirement of the toxic copper catalyst, which cannot always be completely removed and might limit the use of the resulting glycosurfaces for diagnostic and other biotechnological applications [103,104].

An interesting alternative to circumvent the toxicity of copper is the use of strained cyclic alkynes that are able to react with azides via a copper-free strain-promoted azide–alkyne cycloaddition (SPAAC) reaction (Table 1.3; entries c and d) [105]. The SPAAC reaction was first described by Bertozzi and coworkers [106] and has been used by our group to attach lactose derivatives containing azide groups on cyclooctyne-terminated Si_3N_4 surfaces. The bioactivity of the lactoside immobilized on Si_3N_4 was analyzed by binding studies with a fluorescently labeled lectin [59]. In the same year, Ravoo and coworkers immobilized a mannose derivative containing a

TABLE 1.3 Immobilization of Synthetic Carbohydrates Derivatives On Surfaces With Different End Group Terminations

	Surface Termination	Functionalized Carbohydrates	Immobilized Product	Immobilized Carbohydrates
(a)	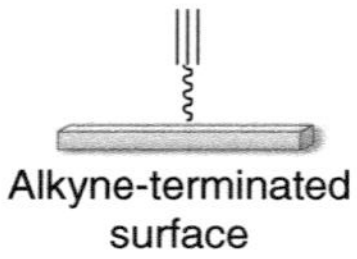	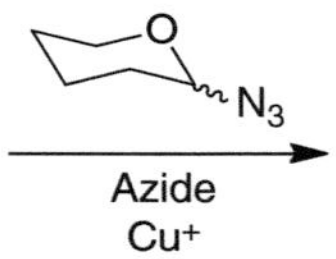	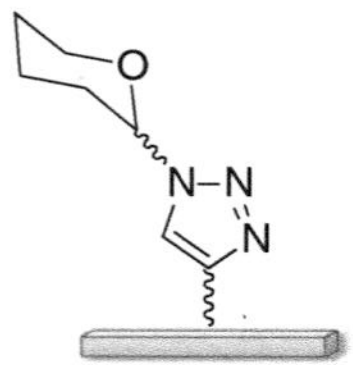	Mannose [6,50–54], galactose [52], glucose [52,55], *N*-acetylglucosamine [52], sulfo-*N*-acetylglucosamine [52], sialic acid [52], lactose [50,53], α-Gal trisaccharide [50]
(b)	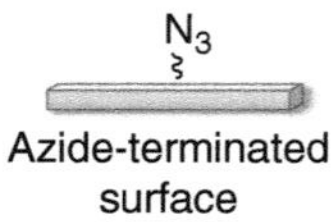	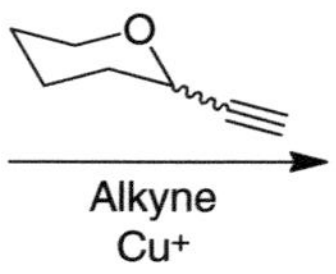	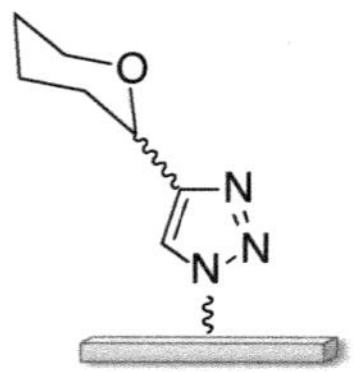	Mucin mimic glycopolymer [56], maltoheptaose [57]
(c)	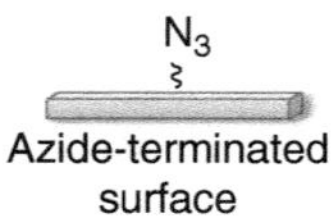	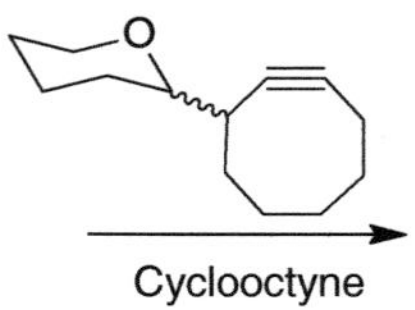	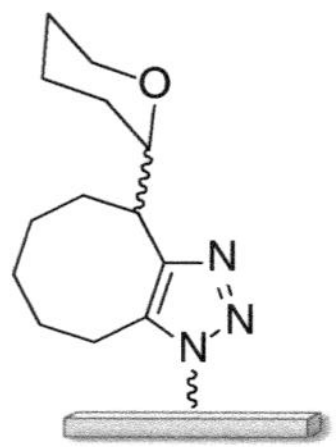	Mannose [58]

(d) 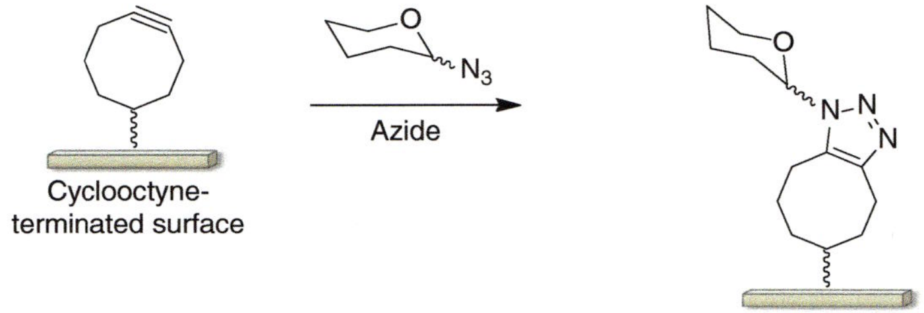

Lactose [59]

(e) 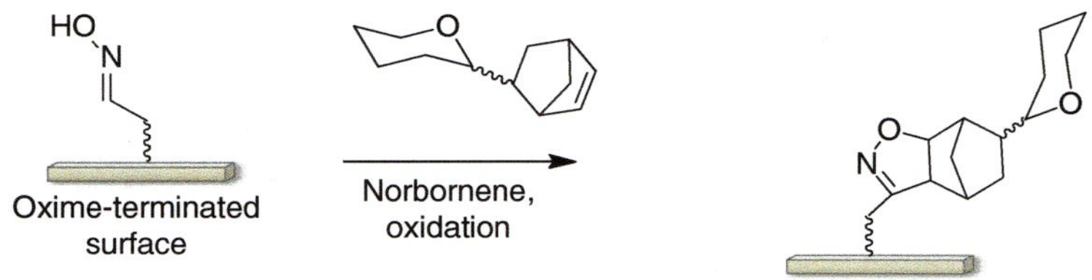

Galactose [58]

(f) 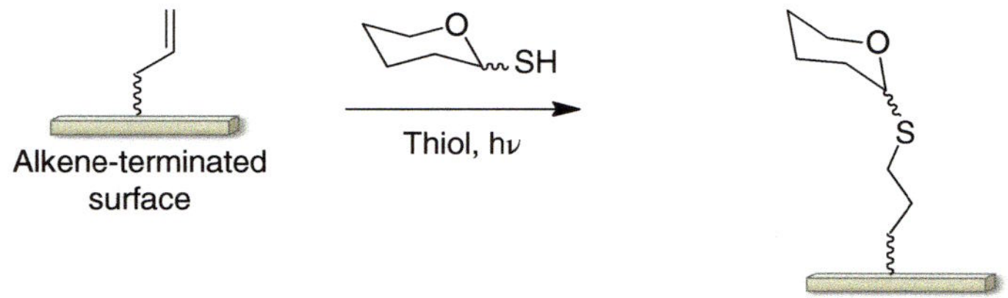

Mannose [60,61], glucose [62], galactose [61,62]

(g) 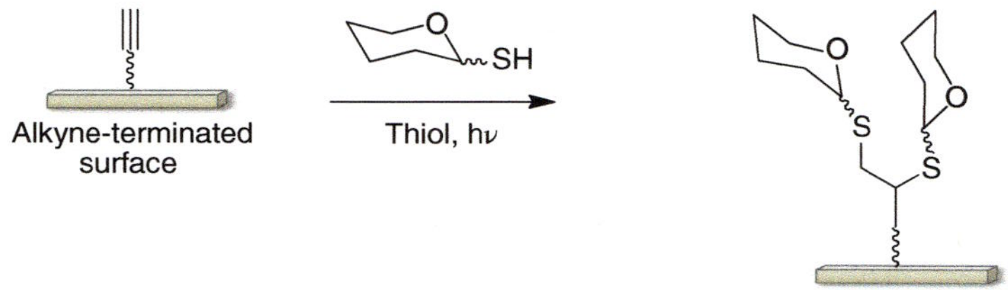

Mannose [61], galactose [61], glucose [63,64]

(continued)

TABLE 1.3 (Continued)

Surface Termination	Functionalized Carbohydrates	Immobilized Product	Immobilized Carbohydrates
(h) 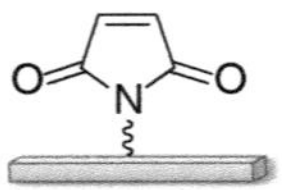 Maleimide-terminated surface	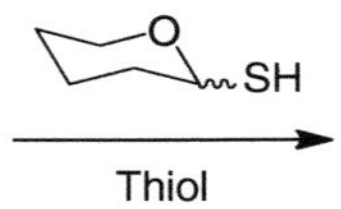	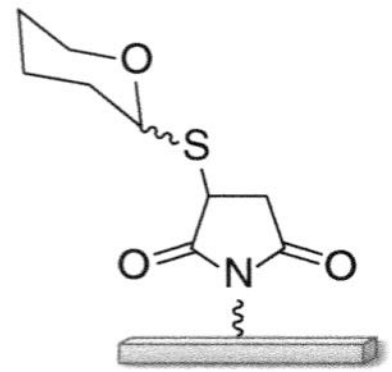	Mannose [65–67], galactose [65], glucose [65,68], *N*-acetylglucosamine [65]
(i) 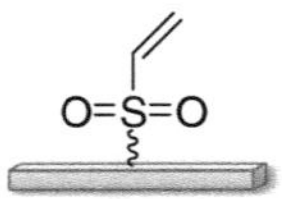 Vinyl sulfone-terminated surface	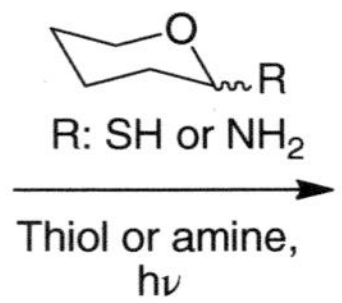	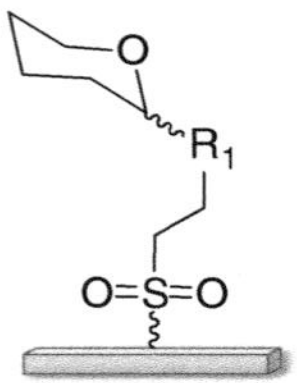 R_1: S or NH	Mannose [47,69], galactose [69], complex carbohydrates [69]
(j) 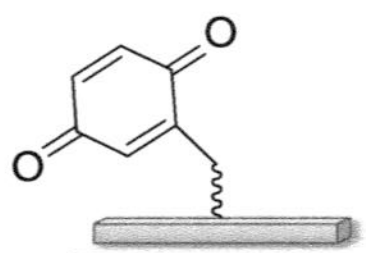 Benzoquinone-terminated surface	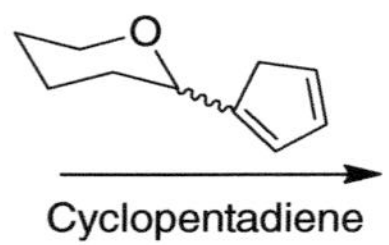	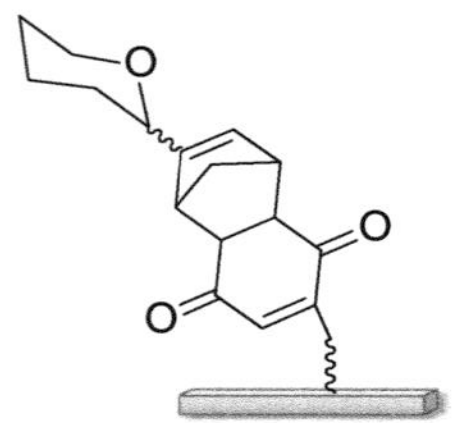	Mannose, galactose, glucose, fucose, rhamnose, *N*-acetylglucosamine [70]

(k)	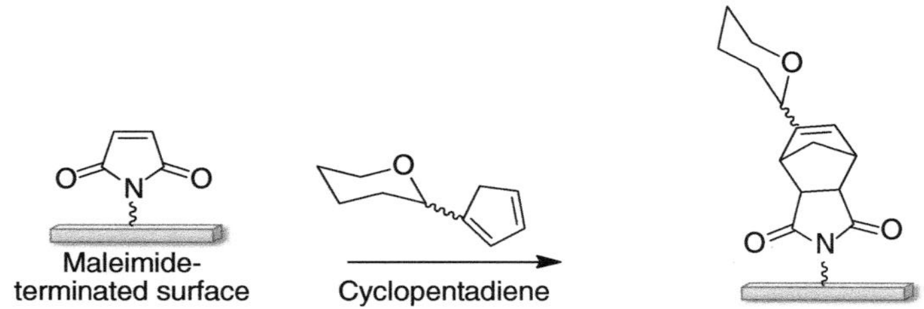	Mannose, galactose, glucose, lactose, maltose [71]
(l)	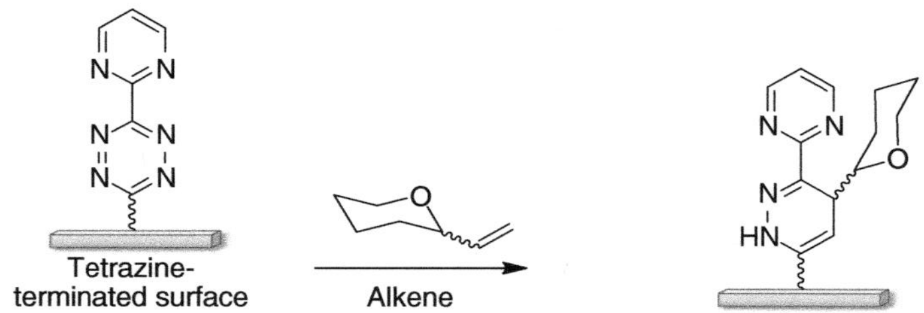	Mannose, lactose, *N*-acetylglucosamine [72]
(m)	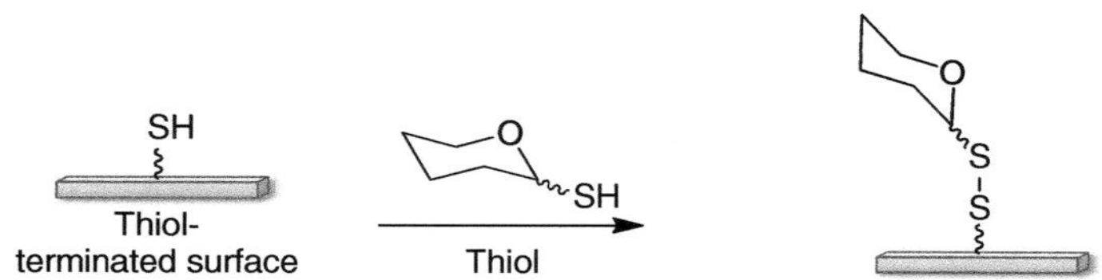	Mannose [73], galactose [73,74], maltose [74]

(continued)

TABLE 1.3 (Continued)

Surface Termination	Functionalized Carbohydrates	Immobilized Product	Immobilized Carbohydrates
(n) 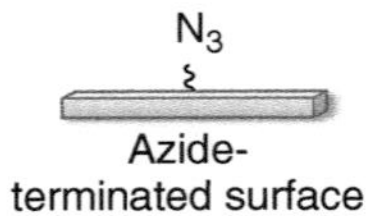	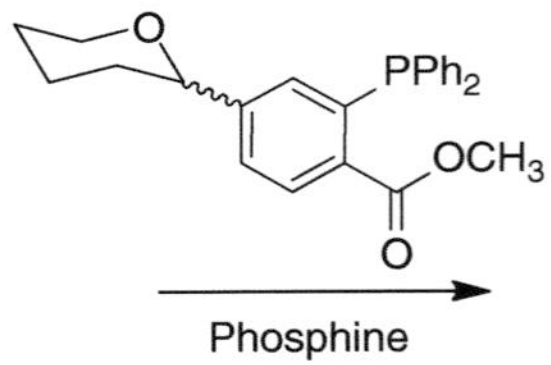	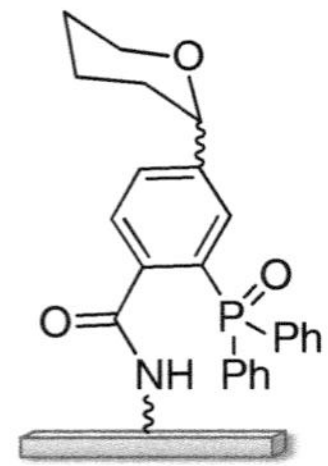	Galactose, lactose, *N*-acetyl-lactosamine [75]
(o) 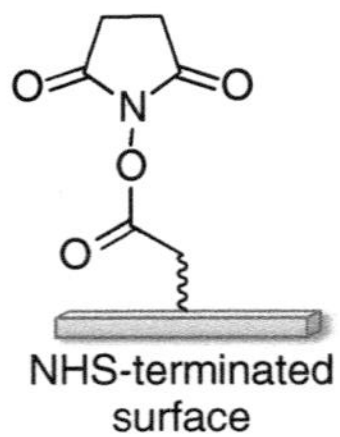	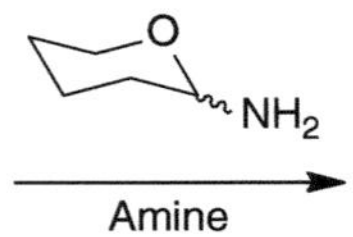	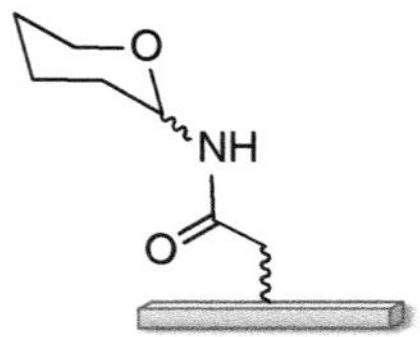	Galactose [76,77], glucose [76,78,79], mannose [76,80], *N*-acetylglucosamine [76,77], *N*-acetylgalactosamine [78], sialic acid [77,81], lactose [60,76], *N*-acetyl-lactosamine [79,81], mannobiose [79], heparin decasaccharide [76], Globo H [82], complex carbohydrates [76,80–84]
(p) 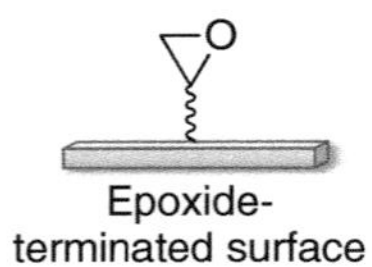	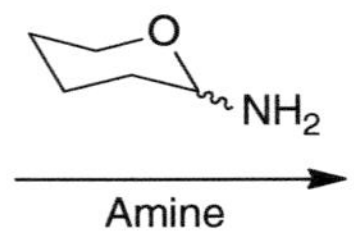	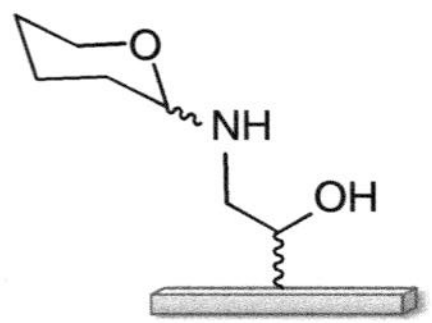	Complex carbohydrates [84,85]

(q)	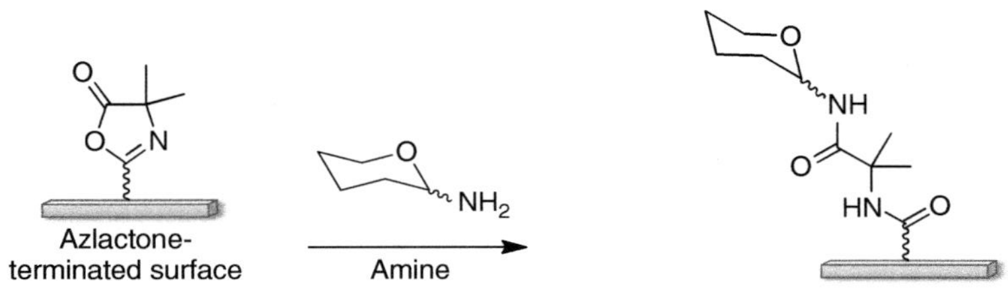	Glucamine [86]
(r)	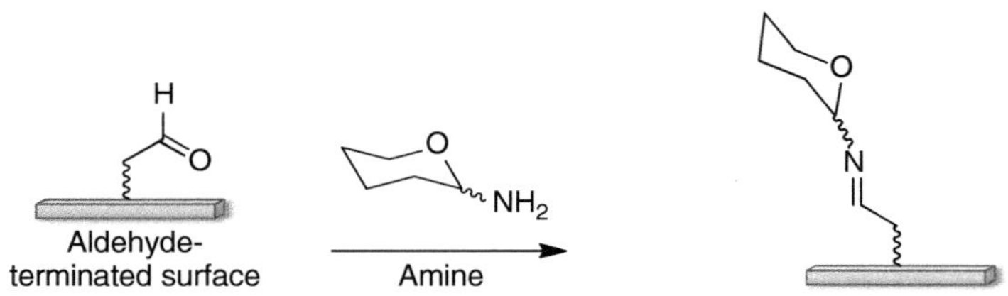	Mannose, galactose, glucose [87]
(s)	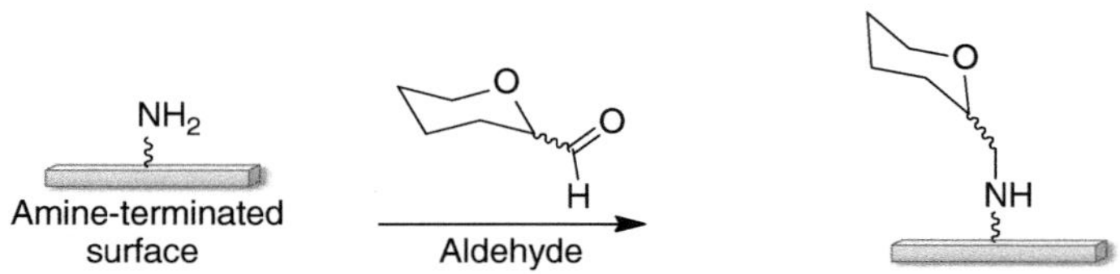	Mannose, galactose, glucose, glucosamine, cellobiose, lactose, lactosamine [88]
(t)	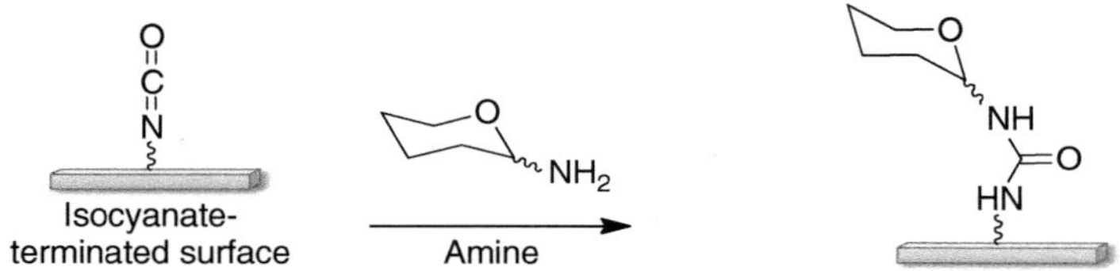	*N*-Acetylglucosamine, *N*-acetyl-lactosamine [89]

(continued)

TABLE 1.3 (Continued)

Surface Termination	Functionalized Carbohydrates	Immobilized Product	Immobilized Carbohydrates
(u) 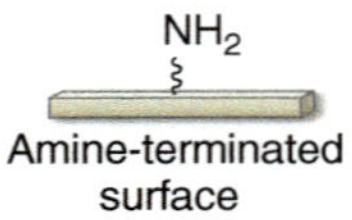	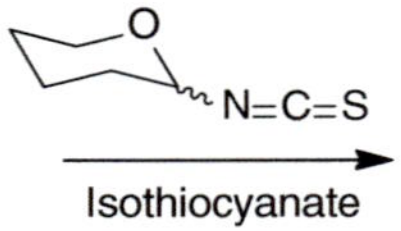	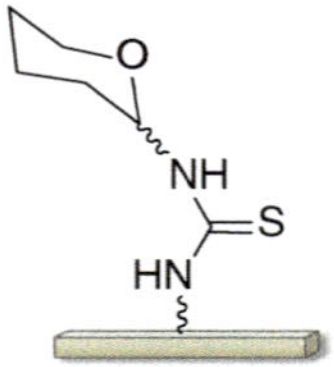	Mannose [90]
(v) 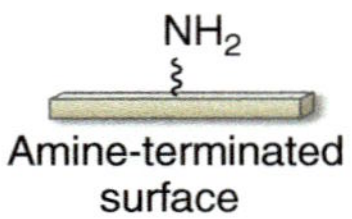	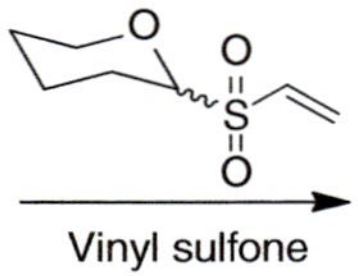	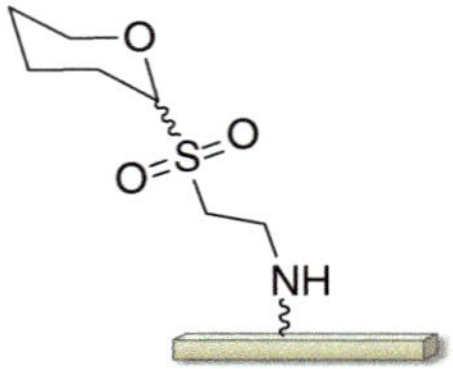	Mannose, glucose, *N*-acetylglucosamine, fucose, lactose, maltose, melibiose [91]
(w) 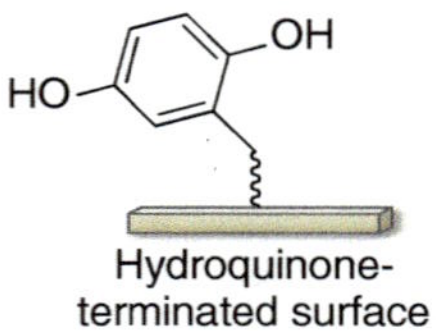	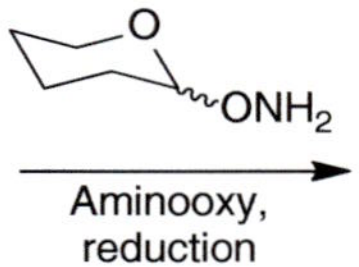	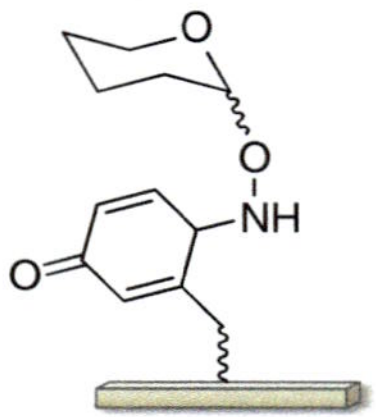	Galactose, glucose [92]

(x)	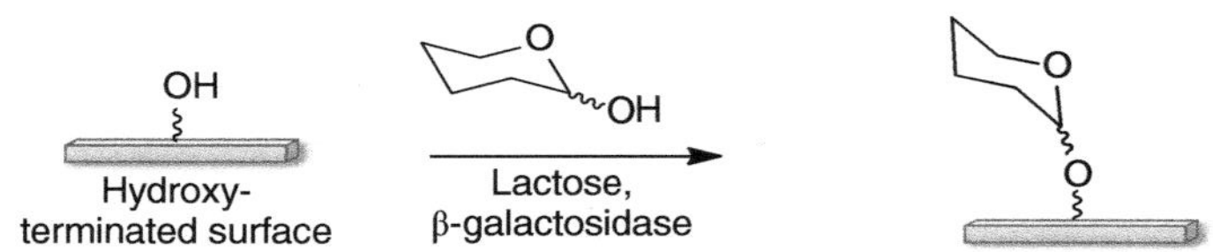	Galactose [93]
(y)	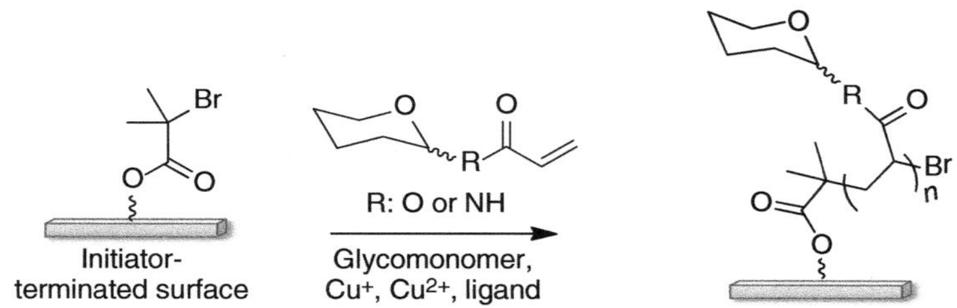	Glucose [94–97], sulfonated glucose [95], mannose [97,98], galactose [97]
(z)	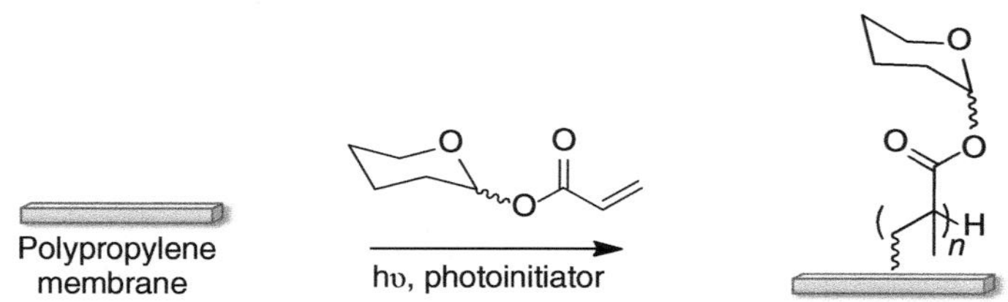	Galactose [99]

(a) CuAAC on alkyne-terminated surfaces, (b) CuAAC on azide-terminated surfaces, (c) SPAAC on azide-terminated surfaces, (d) SPAAC on cyclooctyne-terminated surfaces, (e) nitrile oxide–alkene cycloaddition, (f) thiol–ene, (g) thiol–yne, (h) thiol–maleimide coupling, (i) Michael addition, (j) Diels–Alder on benzoquinone, and (k) maleimide-terminated surfaces, (l) Diels–Alder with inverse electron demand, (m) disulfide formation, (n) Staudinger ligation, (o) amide coupling on NHS-terminated surfaces, (p) epoxide–amine addition, (q) azlactone–amine coupling, (r) imine formation on aldehyde-terminated surfaces, (s) reductive amination on amine-terminated surfaces, (t) isocyanate–amine coupling on isocyanate-terminated surfaces, (u) isothiocyanate–amine coupling on amine-terminated surfaces, (v) Michael addition on amine-terminated surfaces, (w) benzoquinone–aminooxy reaction, (x) enzymatic transglycosylation, (y) surface-initiated atom transfer radical polymerization (SI-ATRP), and (z) UV-induced graft polymerization on polypropylene membranes.

cyclooctyne group on azide-terminated surfaces [58]. The main disadvantage of the SPAAC reaction is the more challenging synthesis of carbohydrate derivatives containing cyclooctynes when compared with the ones containing linear terminal alkynes. The limited stability of cyclooctyne-terminated monolayers also needs to be considered when using the SPAAC reaction to immobilize biomolecules on surfaces.

Another reaction used by Ravoo and coworkers [58] to immobilize carbohydrates on surfaces involves *in situ* generation of nitrile oxide moieties on the surface, which react with alkenes or alkynes, forming isoxazolines (Table 1.3; entry e). The nitrile oxide groups on the surface were generated by oxidation of oxime-terminated surfaces using diacetoxy iodobenzene as the oxidizing agent. Using this approach, the authors immobilized a galactose derivative containing a norbornene group, and the galactose-terminated surfaces were shown to bind fluorescently labeled peanut agglutinin (PNA) [58].

Another approach immobilizes carbohydrates on surfaces using a "click" reaction between thiols and alkenes or alkynes, called the thiol–ene or thiol–yne reaction (Table 1.3; entries f and g), respectively. The radical thiol–ene and thiol–yne reactions are usually carried out under UV light using a photoinitiator such as 2,2-dimethoxy-2-phenylacetophenone (DMPA). Ravoo and coworkers [58] used the thiol–ene photochemical reaction to attach a thiol-containing mannose derivative on alkene-terminated surfaces [60]. The same type of reaction was also used by Ramström and coworkers [61] to attach mannose and galactose derivatives on polymer-coated quartz surfaces functionalized with alkenes or alkynes. The authors carried out the reactions at room temperature and in aqueous solutions of the thiol-containing carbohydrates, in contrast with the reported reaction conditions for both thiol–ene and thiol–yne reactions, that is, using nonpolar solvents. The addition of a radical initiator was not necessary for the reactions, and as expected, the thiol–ene reaction proceeded at a higher rate than the thiol–yne. The carbohydrate-terminated surfaces were then tested for their binding affinity of various lectins, showing good selectivity [61]. The thiol–yne reaction was also used to immobilize glucose derivatives on an alkyne-presenting microporous polypropylene membrane [63,64] and on an alkyne-terminated silicon surface [107]. In the latter example, the consecutive double addition of thiol-terminated carbohydrates onto the alkyne in the thiol–yne reaction resulted in a higher surface density of carbohydrates when compared with the thiol–ene reaction.

Thiol-containing carbohydrate derivatives can also be used for immobilization onto maleimide-terminated surfaces (Table 1.3; entry h). Mrksich and coworkers used thiol–maleimide coupling to attach mannose, galactose, glucose, and *N*-acetyl-glucosamine derivatives on mixed disulfide monolayers on gold [65]. The same approach was also used later to prepare mannose-terminated surfaces, which were used to compare the binding responses with Con A of propagating SPR and localized SPR sensors [66]. Glucose [68]- and mannose [67]-terminated surfaces were also prepared using the same reaction.

Another reaction used to immobilize carbohydrates on surfaces is the Diels–Alder reaction. Diels–Alder was used to attach various carbohydrate–diene conjugates on benzoquinone- (Table 1.3; entry j) [70] and maleimide- (Table 1.3; entry k) [71] terminated surfaces. The Diels–Alder reaction with inverse electron demand (Table 1.3; entry l) was also successfully used to immobilize alkene-containing carbohydrate derivatives on tetrazine-terminated surfaces [72].

Thiol-containing carbohydrate derivatives have also been used for immobilization on thiol-terminated monolayers by formation of disulfides (Table 1.3; entry m). Based on this method, Corn and coworkers attached mannose and galactose derivatives on gold surfaces to fabricate carbohydrate arrays for imaging SPR [73]. Badyal and coworkers also used the formation of disulfides as a method to immobilize galactose and maltose on thiol-terminated surfaces [74]. An advantage of this immobilization method is the possibility of reducing the disulfide bonds and reusing the same thiol-terminated surfaces to attach other carbohydrates, if the initial monolayer is stable enough to survive the reduction step.

Cairo and coworkers prepared galactose, lactose, and *N*-acetyl-lactosamine-terminated surfaces using the Staudinger ligation as immobilization method (Table 1.3; entry n). For this technique, the authors initially introduced an azide functionality to a carboxymethyldextran surface, which reacted with phosphane-containing carbohydrate derivatives. The modified surfaces were then used to analyze carbohydrate–lectin interactions by SPR [75].

Another functional group frequently used for attachment of carbohydrates on surfaces is the amine group, which can either be present on the surface or in the carbohydrate derivative. Amine-containing carbohydrate derivatives can be attached to carboxy-terminated surfaces using carbodiimide chemistry (e.g., dicyclohexylcarbodiimide, DCC) or *N*-hydroxysuccinimide (NHS) coupling chemistry (Table 1.3; entry o). Using this latter technique, a series of mono- and disaccharide derivatives containing an amine group was synthesized and attached on NHS-terminated monolayers on gold surfaces. The specific binding of proteins on the carbohydrate arrays was analyzed by SPR, matrix-assisted laser desorption ionization time-of-flight mass spectrometry (MALDI-ToF-MS), and on-chip enzymatic modification [76]. This NHS approach has also been used to immobilize a variety of monosaccharides [77,79,80], disaccharides, [60,77,81], and also more complex carbohydrates [80–84] and has been used by the Consortium for Functional Glycomics (CFG) for the surface immobilization of carbohydrates for years [108]. The resulting surfaces were used to prepare arrays used for various biotechnological applications such as profiling diverse glycan-binding proteins [83] and studying carbohydrate-specific cell adhesion [77].

Carbohydrate derivatives containing amine groups have also been immobilized on epoxy-terminated surfaces (Table 1.3; entry p). De Boer and coworkers used this technique to immobilize a series of complex carbohydrates and prepare an array to study carbohydrate–protein interactions [85]. Cummings and coworkers [84] compared the immobilization of amine-containing complex carbohydrate derivatives on epoxy and NHS-terminated surfaces. The authors observed that the minimum detectable printing concentrations are lower on the epoxy slides, but that this reaction is less specific when compared with printing on NHS-terminated surfaces [84].

Amine-containing molecules can also be immobilized on azlactone-terminated surfaces (Table 1.3; entry q). Using this approach, Lynn and coworkers immobilized D-glucamine on polymeric films with terminal azlactones, and the resulting surfaces prevented the adhesion and growth of mammalian cells *in vitro* [86]. Another technique to immobilize amine-containing carbohydrate derivatives is by reacting with aldehyde-terminated surfaces, forming an imine bond (Table 1.3; entry r). Wang and coworkers used this approach to attach monosaccharides onto a polyacrylamide hydrogel activated with glutaraldehyde and employed the resulting surfaces for bacterial capture [87]. The use of carbohydrate-terminated surfaces to capture microorganisms will be discussed in more detail at the end of this chapter. The formation of an imine bond can also be used to attach aldehyde-functionalized carbohydrates on amine-terminated surfaces. The imine bond can then be reduced with mild reducing agents such as sodium cyanoborohydride to obtain irreversible immobilization (Table 1.3; entry s) [88]. The reaction between amines and isocyanate or isothiocyanate groups (Table 1.3; entries t and u) has also been used to immobilize carbohydrate derivatives on surfaces. Amine-functionalized carbohydrates were successfully immobilized onto 96-well plates coated with an isocyanate-presenting polymer [89]. Using a similar reaction, a mannose derivative containing an isothiocyanate group was attached to amine-terminated surfaces, and the resulting surfaces were used for bacterial capture [90]. Another strategy that has been used to prepare glycosurfaces involves the reaction between amines and vinyl sulfones (Table 1.3; entry v). Ratner and coworkers used this reaction to immobilize carbohydrates on the surface of silicon photonic microring resonators, used for label-free detection of glycan–protein and glycan–virus interactions [69]. Another possibility is to synthesize carbohydrate derivatives with a vinyl sulfone group, which then reacts with amine-terminated surfaces such as commercially available amine-functionalized glass slides or microwell plates [91].

Another strategy used to prepare glycosurfaces is to synthesize carbohydrates presenting an oxyamine group, which reacts with the ketone group in electroactive quinone-terminated monolayers (Table 1.3; entry w). A remarkable advantage of this approach is the possibility to release the immobilized carbohydrate by applying a constant potential under reducing conditions to regenerate the original surface. Using this technique, glucose, galactose, and mannose derivatives were immobilized on gold surfaces, which were used to prepare microarrays to analyze binding of lectins [92].

Carbohydrates can also be attached on surfaces via enzymatic reactions. In a recent work, Xu and coworkers [93] immobilized lactose on QCM chips modified with hydroxy-terminated poly(ethylene glycol) (PEG) brushes via enzymatic transglycosylation (Table 1.3; entry x). The β-galactosidase enzyme catalyzed the transfer of the galactose unit from lactose to the hydroxyl acceptor, yielding the glycoside compound with a β-1,4-glycosidic bond. The modified surface was used to selectively bind a lectin, while it displayed minimal nonspecific adsorption of bovine serum albumin (BSA) [93].

Finally, another quite distinct method to prepare glycosurfaces is the use of polymerization reactions with carbohydrate-containing monomers. The most commonly used polymerization method is the surface-initiated atom transfer radical polymerization (SI-ATRP) (Table 1.3; entry y), a powerful and versatile controlled radical polymerization method that enables control over polymer thickness and functionality [109]. Using this method, Fukuda and coworkers [94] prepared glucose-terminated surfaces with protected glucose-functionalized monomers. The glucose units were subsequently deprotected quantitatively by removing the isopropylidenyl protecting groups, which did not result in degrafting or chain breaking of the glycopolymers [94]. The same monomer was later also used by Ayres and coworkers and Yoon and coworkers, but with different surface-bound initiators [95,96]. Additionally, Ayres and coworkers performed a sulfonation reaction to also obtain glycosurfaces that may mimic heparin, a naturally occurring sulfonated glycosaminoglycan [95]. Another example using SI-ATRP to functionalize surfaces with glycopolymers employs D-gluconamidoethyl methacrylate as the glycomonomer used for the polymerization. The capability of preventing the binding of nonspecific proteins, such as lysozyme, BSA, and fibrinogen, on this glycopolymer was shown by SPR studies [110]. The property of certain glycopolymers to prevent nonspecific protein adsorption on surfaces was also explored by Yu and Kizhakkedathu [97]. In their work, the authors prepared glycopolymer brushes by SI-ATRP using unprotected mannoside, glucoside, and galactoside *N*-substituted acrylamide derivatives as the monomers. The surfaces modified with the glycopolymers showed low BSA and fibrinogen adsorption, but also preserved the specific protein interaction as shown in the binding study with Con A [97]. Another example of a mannose-containing polymer was prepared by free radical polymerization on a gold surface, and binding with Con A was studied by QCM, atomic force microscopy (AFM), and SPR techniques [98]. A galactose-containing monomer was synthesized by Ulbricht and coworkers [99] and grafted on a polypropylene microfiltration membrane surface by UV-induced graft polymerization (Table 1.3; entry z). The galactose-presenting glycopolymer was successfully recognized by the bacterium *Enterococcus faecalis*, increasing the adhesion of bacteria (by a factor of 39) when compared with the unmodified polypropylene surfaces [99]. The main advantage of preparing glycopolymers on surfaces is that the orientations in which they are presented probably better resemble the complex natural presentation of carbohydrates when compared to SAMs. However, one current challenge is to limit the polydispersity in order to obtain well-defined glycosurfaces.

The chemoselective reactions discussed in this section are the most frequently used approach to immobilize carbohydrates on surfaces in a well-defined and controlled manner. In general, the main benefit of this approach is the possibility of obtaining a homogeneous carbohydrate layer, in which all the carbohydrate units are presented in the same orientation. They are usually bound to the surface via a linker connected to the anomeric carbon of the carbohydrate, yielding a carbohydrate layer with defined anomeric centers and thus hopefully similar properties compared to the glycosurfaces present in nature. However, this approach also has the disadvantage of requiring chemical modification of the carbohydrates for the immobilization on the previously prepared initial SAM. The synthetic methods required for the modification of the carbohydrates might limit the wider applicability of these chemoselective reactions in the carbohydrate nanotechnology scientific community.

1.3.3 Noncovalent Immobilization of Carbohydrates to Prepare Glycosurfaces

The preparation of glycosurfaces using secondary reactions, either using unmodified carbohydrates or in a chemoselective way, usually has the advantage of yielding stable modifications due to the covalent nature of the bond used for the immobilization of the carbohydrate. Noncovalent immobilization is a quick and easy alternative to immobilize carbohydrates on surfaces, although it usually yields less stable glycosurfaces.

An interesting approach to prepare glycosurfaces using noncovalent immobilization methods is using glycolipids (Table 1.4; entry a), which can interact with surfaces presenting hydrophobic monolayers or also with surfaces that are intrinsically hydrophobic [115,126]. Using this approach, Kiessling and coworkers [111] prepared a mannose-presenting glycolipid, which interacted with a previously prepared hydrophobic SAM. The carbohydrate density on the glycolipid bilayer was varied by using different percentages of the mannose glycolipid relative to phosphatidylcholine. The modified surfaces were then used to investigate the specific binding of Con A by SPR [111]. Mannose glycolipids with two hydrophobic alkyl chains were also immobilized on surfaces by a similar technique [113–115]. More complex carbohydrates such as gangliosides were also immobilized on surfaces using a similar approach and used to investigate the binding of cholera toxin [112,119,127]. The same technique was also used to prepare glycosurfaces with C-lactose and C-galactose glycolipids on a commercially available HPA hydrophobic chip (Biacore) [117]. Natural glycolipids secreted by *Pseudozyma* yeast, known as mannosylerythritol lipids, were also immobilized on the HPA hydrophobic chip [128]. Using a slightly different technique, but also with a carbohydrate derivative containing hydrophobic chains, Liu and coworkers [118] prepared *N*-acetylglucosamine-terminated surfaces. In their approach (Fig. 1.5), a glass slide was initially functionalized to present amino groups. The amino-terminated glass slide was then covered with gold nanoparticles, and these nanoparticles were then functionalized with thiols containing hydrophobic chains, which were used to interact with the glycolipids [118].

Noncovalent immobilization of carbohydrates by hydrophobic interactions is not restricted to long alkyl chains. Other groups that provide hydrophobic or fluorophase interactions can be bound to carbohydrates and used to prepare glycosurfaces, such as aromatic rings and perfluorinated hydrocarbon chains (Table 1.4; entry b). Lindhorst and coworkers prepared mannose-terminated surfaces using trityl-containing mannose derivatives, which were immobilized onto polystyrene microplates [67,116]. Complex carbohydrates were also immobilized on aluminum oxide-coated glass surfaces via interaction between perfluorinated hydrocarbon chains immobilized on the surface and the ones present on the carbohydrate derivative [35].

Another type of interaction that can be used to prepare glycosurfaces via noncovalent immobilization is ionic interaction between surfaces and molecules with opposite charges (Table 1.4; entries c and d). The anionic polysaccharide heparin was immobilized on a positively charged amine-terminated surface [120]. Coullerez and coworkers also prepared a glycosurface based on the polycationic graft copolymer poly(L-lysine)-*graft*-poly(ethylene glycol) (PLL-g-PEG), which was modified to present mannose units and immobilized via ionic interactions on a negatively charged surface [121].

TABLE 1.4 Methods of Noncovalent Immobilization of Carbohydrates On Surfaces

Substrate		Functional Group	Immobilized Product	Immobilized Carbohydrates
(a)	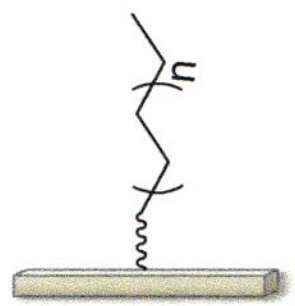 Glycolipids-terminated or other hydrophobic surface	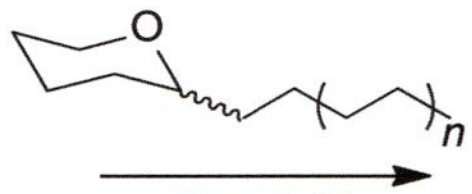 Glycolipid	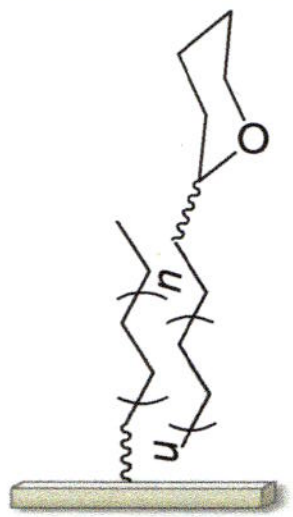	Mannose [111–116], galactose [117], *N*-acetylglucosamine [116,118], lactose [117], cellobiose [117], maltose [117], GM1 [112,119], other complex carbohydrates [112]
(b)	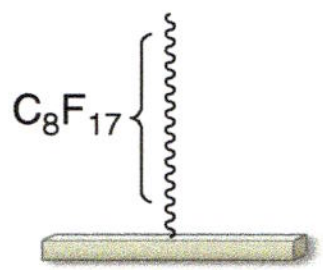 Fluoroalkyl-terminated surface	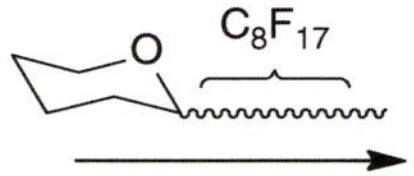 Fluorous-tagged carbohydrate	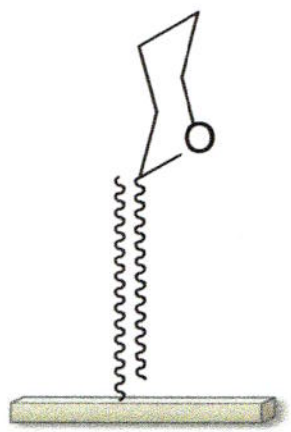	Mannose, lactose, cellobiose, cellotriose, cellotetraose, Gb5, Globo H [35]
(c)	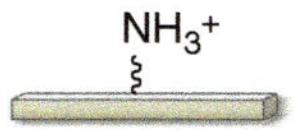 Amine-terminated surface	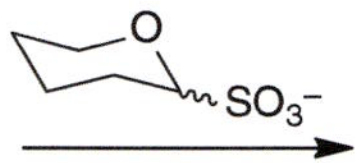 Heparin	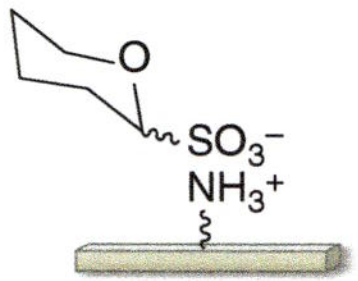	Heparin [120]
(d)	 Nb_2O_5-coated surface	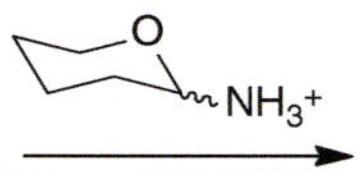 PLL-*g*-PEG/PEG-mannoside	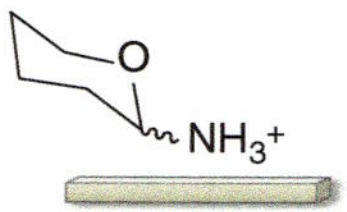	Mannose [121]

(continued)

TABLE 1.4 (Continued)

Substrate	Functional Group	Immobilized Product	Immobilized Carbohydrates
(e) 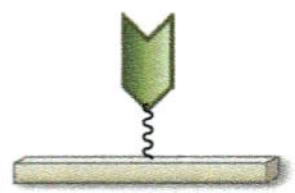 Streptavidin-terminated surface	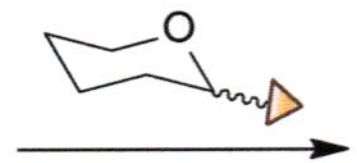 Biotin-conjugated carbohydrates		Complex carbohydrates [122–124]
(f) 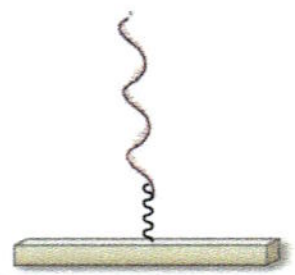 Oligonucleotide-terminated surface	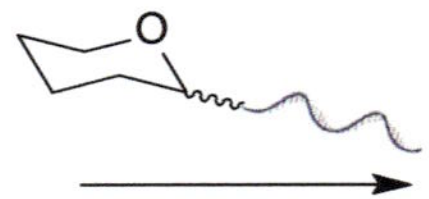 Oligonucleotide-conjugated carbohydrates	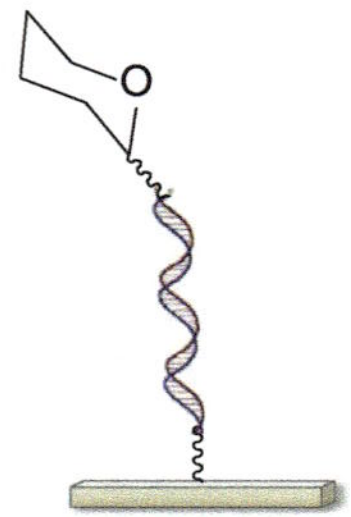	Mannose, galactose [125]

(a) Glycolipids, (b) perfluorinated hydrocarbon chains, ionic interactions on (c) positively and (d) negatively charged surfaces, (e) biotin–streptavidin interaction, and (f) DNA-directed immobilization.

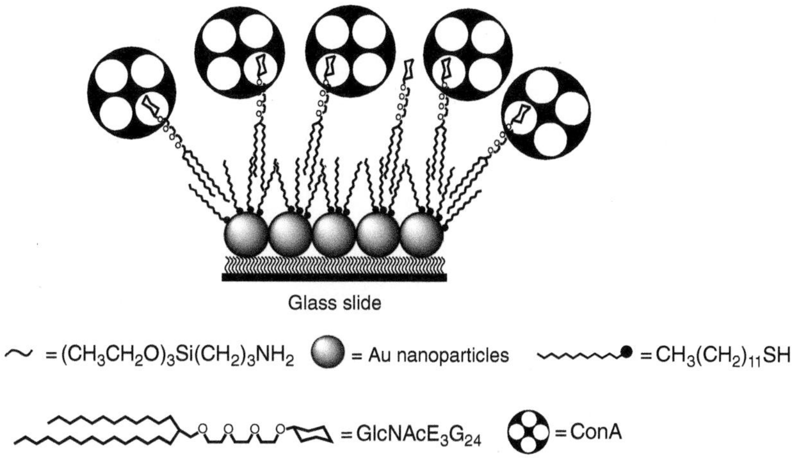

FIGURE 1.5 Schematic representation of the self-assembled multicomponent system for the investigation of biomolecular binding. *Source*: Reprinted with permission from Ref. 118, Copyright 2007, Elsevier.

More specific noncovalent interactions, as between biotin and its protein binding partners (avidin, neutravidin, or streptavidin), have also been used to prepare glycosurfaces for biotin-linked carbohydrates (Table 1.4; entry e) [129]. For the carbohydrate immobilization, biotin-terminated layers were initially prepared, followed by binding of neutravidin on the biotin-terminated surfaces (Fig. 1.6). Finally, the biotin group linked to sialoside derivatives could then interact with the remaining free binding sites of the neutravidin units. The surface-bound sialosides were used to study carbohydrate–protein interactions [124,130].

Additionally, DNA-directed immobilization can also be used to prepare glycosurfaces (Table 1.4; entry f). Chevolot and coworkers successfully immobilized galactose derivatives on borosilicate glass slides using this technique [125].

The carbohydrates present on the cell surface in nature are usually present in the fluid lipid bilayer of membranes as glycoconjugates of proteins or lipids and, consequently, are exposed to their interaction partner in a dynamic fashion. In order to better mimic this aspect of natural glycosurfaces, some noncovalent immobilization techniques can be used, such as solid-supported lipid bilayers [115,126]. In this case, the carbohydrates may be exposed in a more dynamic way and may behave more similarly to the natural glycosurfaces.

In general, various techniques to prepare glycosurfaces have been developed in the past few decades. These techniques can be divided in four main groups: SAMs of molecules containing carbohydrates, covalent immobilization of unmodified carbohydrates by secondary reactions, chemoselective covalent immobilization of carbohydrate derivatives, and noncovalent immobilization of carbohydrates. All the approaches discussed have their advantages and their drawbacks, and the optimal technique usually depends on the application of the

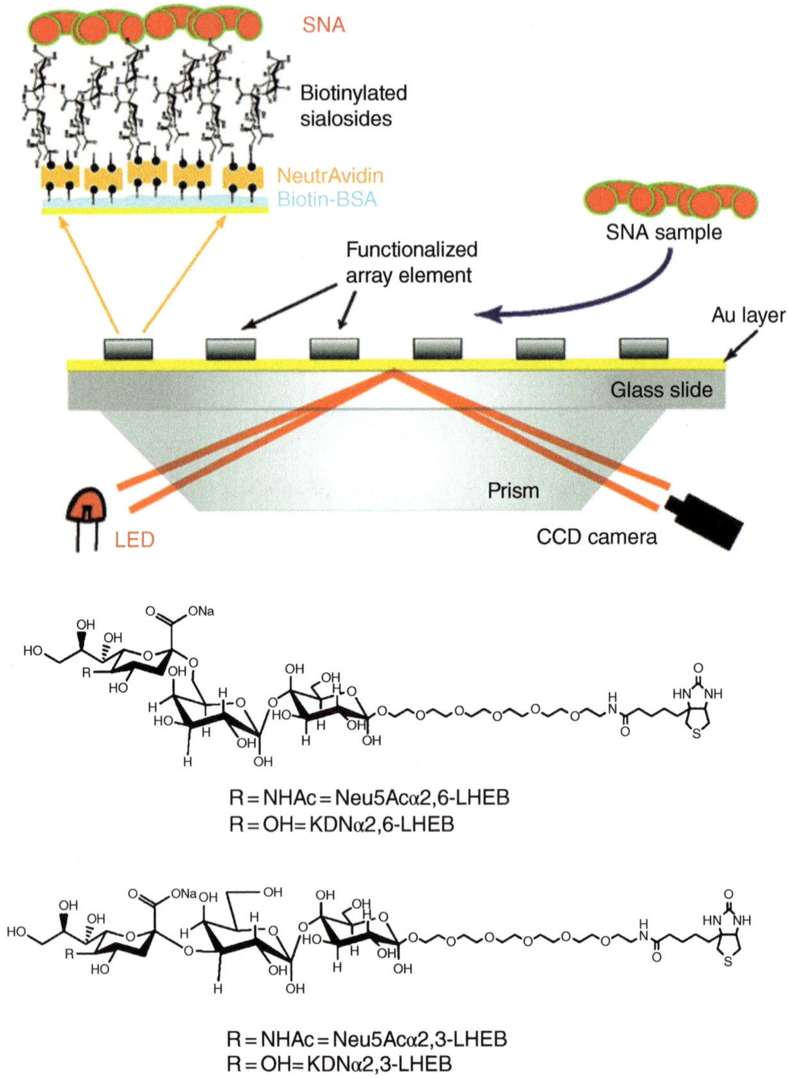

FIGURE 1.6 Schematic representation of the carbohydrate immobilization using the noncovalent interaction between NeutrAvidin and biotin. *Source*: Reprinted with permission from Ref. 124, Copyright 2009, American Chemical Society.

glycosurfaces and the expertise of the scientists responsible for their design and preparation. However, what all these techniques have in common is the need to comprehensively characterize the obtained glycosurfaces. The main techniques used for characterization of glycosurfaces, including a few examples, will be discussed in the following section.

1.4 CHARACTERIZATION OF GLYCOSURFACES

Various methods are reported in the literature for preparation of glycosurfaces, as described in the previous sections. Glycosurfaces generated via all these methods need to be characterized to prove the successful modification of the surface. In this section, we provide an overview of the most frequently used techniques used to characterize glycosurfaces (Fig. 1.7), including the type of information that can possibly be obtained with each of these techniques.

1.4.1 Contact Angle Measurements (Wettability)

One of the most often used techniques for characterization of modified surfaces, including glycosurfaces, is the measurement of the surface wettability. The surface wettability is measured statically by placing a small drop of liquid on a surface and measuring the angle formed between the liquid and the solid surface, named contact angle. Nowadays, these measurements are obtained using a camera to record the image of the drop and a software that provides different fitting models to calculate the contact angle [131,132]. The main solvent used for contact angle measurements is water, but more apolar solvents can also be used. In general, according to the values of the static water contact angle, the surfaces can be divided in hydrophilic (θ_{CA}<90°) or hydrophobic (θ_{CA}>90°) [132]. A surface after functionalization with carbohydrates typically displays a low water contact angle due to the hydrophilicity

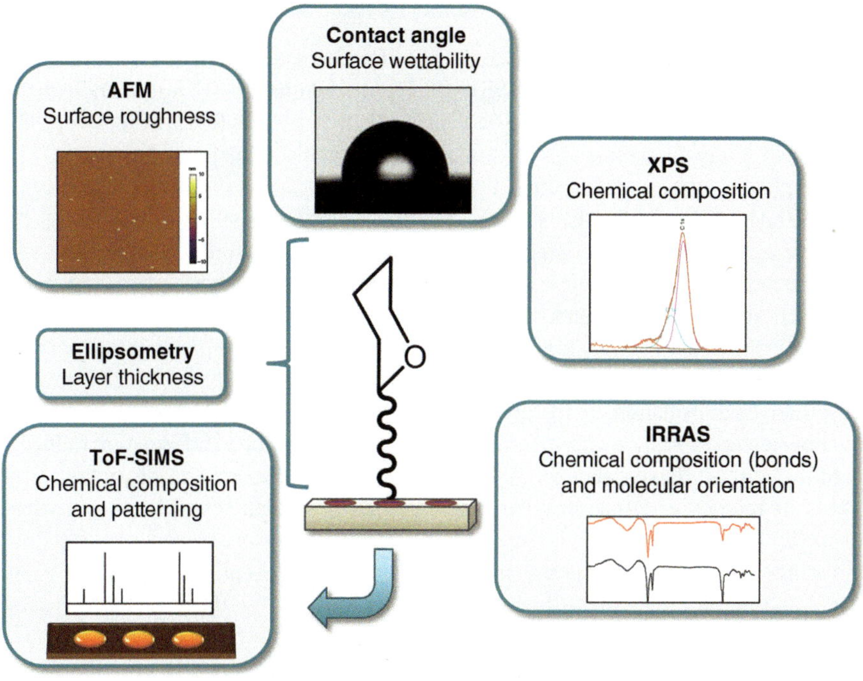

FIGURE 1.7 Main analysis techniques used to characterize glycosurfaces.

of the unprotected carbohydrates. The most frequently used contact angle measurement is the static contact angle; however, advancing and receding contact angles can also be measured. The difference between the advancing and receding contact angles, called contact angle hysteresis, represents the "nonideality" of a surface and is related to the adhesion of materials on the surface [132].

Water contact angle measurements have been used by many researchers as the first technique to quickly and easily follow reactions with carbohydrates on surfaces. Yu and Kizhakkedathu [97], for example, compared the water contact angles of glycopolymers on surface generated both with protected carbohydrate derivatives and deprotected ones. The surfaces generated with the protected carbohydrates presented a contact angle of 53.8°, whereas the ones generated with the deprotected carbohydrates had a contact angle of 7.2°. When the deprotection step was performed on the surface after generation of the glycopolymer, the contact angle of the resulting surfaces was 33.5°, showing that the deprotection on the surface was incomplete [97]. Xu and coworkers [64] used water contact angle measurements to evaluate the effect of glycosyl density on the hydrophilicity of a polypropylene membrane. The authors observed a significant decrease of the contact angle with increase of the glycosyl density [64].

Although other techniques need to be used to analyze the functionalization of a surface with carbohydrates at the molecular/structural level, the contact angle measurement is an easy and quick method that provides a good indication about whether a functionalization step succeeded and the overall surface wettability.

1.4.2 X-Ray Photoelectron Spectroscopy (XPS)

Another technique that has been frequently used for characterization of glycosurfaces is X-ray photoelectron spectroscopy (XPS), which provides quantitative information on the elemental and chemical composition of a surface to a depth of a few nanometers [133]. In XPS, an X-ray source induces the emission of photoelectrons from the surface. The emitted photoelectrons are collected, identified, and quantified according to their energy. The method is quantitative because the binding energies of the core levels that are analyzed are element specific, while the signal intensities are not influenced significantly by the precise nature of chemical bonding [134]. An important advantage of XPS when compared to other techniques that provide elemental and chemical composition information is that it is essentially a nondestructive technique [135]. Two different types of surface analysis can be performed in XPS: wide-scan analysis, which provides (quantitative) information on the elemental composition, and high-resolution analysis (yielding a narrow-scan spectrum), which provides chemical bond information [136].

Many authors have used XPS to characterize glycosurfaces made with mixed SAMs of molecules containing carbohydrates and oligoethylene glycol or hydrocarbon units. In this case, the elemental analysis (wide-scan spectrum) is not sufficient to distinguish between the carbon and oxygen atoms of the carbohydrates and those of the ethylene glycol or hydrocarbon units. However, in the C1s narrow scan spectrum, it is possible to obtain quantitative information about the chemical bonds that are present on the glycosurface. One can conveniently identify an XPS signal that is characteristic of carbohydrates, corresponding to the anomeric acetal carbon

(O–C–O), that is present around 288 eV [6,137–139]. Dhayal and Ratner, for example, used this signal to quantify the relative coverage of carbohydrate in mixed monolayers with ethylene glycol units [137].

Overall, XPS provides very useful information on the composition and chemical state of elements present on surfaces. Although it has been used for the characterization of glycosurfaces, its more restricted availability to laboratories specialized in surface chemistry has limited its widespread application.

1.4.3 Infrared Reflection–Absorption Spectroscopy (IRRAS)

Infrared reflection–absorption spectroscopy (IRRAS) is another technique that has been used for characterization of glycosurfaces. IRRAS is a truly nondestructive technique that provides information about the chemical composition of surfaces, based on frequencies and intensities of molecular vibrations [140]. IRRAS measurements can, in principle, also provide information about the molecular orientation within the layer, using polarized light [141,142], but the short-range disorder of glycosurfaces will typically hamper providing such detailed information.

An advantage of infrared spectroscopy is that it does not require ultrahigh vacuum conditions, as opposed to XPS. Most IRRAS analysis of SAMs can be performed in air or inert atmosphere [141]. Depending on the substrate, transmission infrared spectroscopy can be used instead of IRRAS, for example, for glass [143] and porous aluminum oxide [6,101].

The characteristic bands of carbohydrates are present at around 3300–3500 cm^{-1}, related to the stretching of intramolecular and intermolecular OH groups, and as three shoulders at around 1147, 1090, and 1045 cm^{-1}, related to the hemiacetal of carbohydrates (O–C–O) [55,141]. Some other IR bands that can often be found on prepared glycosurfaces are found at around 2850 and 2920 cm^{-1}. They relate to the symmetric and antisymmetric stretching vibrations of the CH_2 chain, respectively, which are present when the carbohydrates are attached to a hydrocarbon chain [55,141]. The position of these bands may give information about how ordered the monolayer is [141,144].

IRRAS is a useful and accessible technique for characterization of glycosurfaces. However, quantification of the carbohydrate density is challenging, and another technique is probably required for more comprehensive characterization of a glycosurface.

1.4.4 Mass Spectrometry

Mass spectrometry techniques, such as time-of-flight secondary ion mass spectrometry (ToF-SIMS) and direct analysis in real-time high-resolution mass spectrometry (DART-HRMS), have also been used to obtain information about the composition of glycosurfaces. ToF-SIMS is a highly sensitive technique that can be used for surface chemical mapping and also enables the elucidation of molecular orientation [145]. In SIMS, a pulsed ion beam removes molecular fragments from the surface, and the secondary ion fragments are accelerated by an electrical field. The ToF detector measures the velocity of the ions, which depends on the mass-to-charge ratio [146].

Some characteristic peaks that can be observed in ToF-SIMS analysis of glycosurfaces are related to oxygen-rich anions, namely, $C_3H_3O_2^-$, $C_2H_3O_2^-$, $C_2H_2O_2^-$, CHO_2^-, C_2HO^- [71], and $C_3H_5O_2^+$ [47]. Other signals mostly depend on the type of SAM present on the surface, such as CH_2AuS, C_2H_4AuS, and C_4H_8AuS for alkanethiol SAMs on gold, for example [138]. ToF-SIMS has also been used for characterization of patterned glycosurfaces, with detection of characteristic carbohydrate ions depending on the surface localization [71,141].

Another mass spectrometry technique, which has recently been studied in detail in our group, for the characterization of glycosurfaces is DART-HRMS. In DART-HRMS, excited He gas is impinged onto the surface under ambient conditions. This effects an interaction between such excited He and the surface or between atmospheric species obtained from ionization by excited He and the surface. Surface-bound groups are in this process removed from the surface and ionized [147]. We used DART-HRMS to characterize lactose-, mannose-, and fucose-functionalized porous aluminum oxide surfaces. Using the tremendous structural power of HRMS, it is thus easily possible to discriminate between the immobilized monosaccharides and between various disaccharides [100]. In this way, DART-HRMS provides structural information about glycosurfaces that is not easy to obtain otherwise.

Mass spectrometry techniques, such as ToF-SIMS and DART-HRMS, are useful to obtain information about chemical composition of glycosurfaces and also to prove the localization of carbohydrates on patterned surfaces. A drawback is that these techniques are destructive, so the glycosurfaces that are analyzed cannot be further used for biological tests.

1.4.5 Ellipsometry

Ellipsometry is a powerful optical tool to measure the thickness of thin films, including glycosurfaces. Ellipsometric measurements are usually performed with a light beam that propagates in air (or vacuum) and is reflected by (or transmitted through) the sample, eventually arriving at the detector. This technique can determine the thickness of monolayers or glycopolymers on surfaces, because the light reflected by the interface of an initial layer interferes with the light reflected by the interface of a second layer on top of it [148].

Ellipsometry has usually been used in combination with other techniques to characterize glycosurfaces, because it does not provide information on chemical composition. An interesting characteristic of ellipsometry is the possibility of measuring layer thickness in various conditions, including the presence of different solvents. This technique is useful to measure SAMs directly prepared with carbohydrate derivatives [17,25,34], glycosurfaces in which the carbohydrate is immobilized via a secondary (or tertiary reaction) [90], and also for glycopolymers on surfaces [96,97,110].

1.4.6 AFM

AFM is a scanning probe microscopy imaging technique used to investigate roughness of surfaces, size, shape, structure, dispersion, and aggregation of nanomaterials. The equipment consists of a micro-machined cantilever, usually made of silicon or silicon nitride, with a sharp tip at one end. This tip detects the deflection of the cantilever caused by repulsion or attraction between the tip and the analyzed surface. It presents different scanning modes: noncontact or static mode, contact mode, and intermittent sample contact mode (or dynamic and tapping mode) [149].

AFM can also be used as a highly sensitive force machine that can measure forces as small as 10–20 pN. This property enables the use of AFM to measure nano-adhesive properties of samples. By using a (bio)functionalized cantilever tip, it is even possible to measure the attraction or binding event, between a (bio)molecule and a (bio)functionalized surface [150,151].

For the characterization of glycosurfaces, AFM has mainly been used to investigate surface roughness, especially for glycopolymers on surfaces [28,32,97,124]. AFM is an interesting technique to investigate how the carbohydrates are distributed on a surface; however, it provides limited information on the nature of the carbohydrate molecule. One way to increase the information provided by AFM is the use of specific lectins at its cantilever tip. In addition, AFM can be combined with other characterization techniques, such as XPS, ToF-SIMS, and IR, for confirmation of the biofunctionalization of the surface with carbohydrates.

Various techniques have been used for characterization of glycosurfaces, and each of them provides different types of information. It is usually necessary to apply a combination of two or three techniques to properly characterize a glycosurface. The best selection of characterization methods often depends on the material of the modified substrate, the type of modification (SAMs, glycopolymers, or noncovalent immobilization), and the intended final application of the glycosurface.

1.5 APPLICATION OF GLYCOSURFACES IN MICROBIOLOGY

As mentioned in the introduction, carbohydrates are present on the surface of most multicellular and single cell organisms and their interaction with other biomolecules, especially receptor proteins are critical in many biological processes, but currently only poorly understood. Glycosylated surfaces are an emerging type of biomaterials that can be used in the investigation of interactions between carbohydrates and other biomolecules, mainly proteins, either in isolation or on microbial cells. The study of these interactions can contribute to the discovery of novel carbohydrate–protein (usually lectins)-binding partners, which can for instance be useful to identify disease biomarkers or to better characterize certain microorganisms [152]. The main application of prepared glycosurfaces is currently for carbohydrate microarrays [49,153], which will be discussed in details in Chapter 9 (carbohydrate nanotechnology and its application to microarray technology). When compared to antibodies or DNA-based sensors, the carbohydrate-based sensors have the advantages of being more stable under various conditions [154]. However, the main drawbacks are the lower selectivity and lower binding affinity of the carbohydrate-based sensors.

Glycosurfaces have also been used for binding, capture, and sensing of bacteria via the interaction with the lectins present on the surface of many bacterial species. In bacterial infections, a well-known carbohydrate–protein interaction is the one between mannosides and the type 1 fimbrial lectin (FimH) present on the cell surface of

Enterobacteriaceae species, including the uropathogenic *Escherichia coli* (UPEC). This interaction is responsible for the adhesion of UPEC on the luminal surface of the bladder epithelium, causing bladder infection and inflammation. In this case, mannose-presenting surfaces can be used to detect these pathogenic bacteria [1].

In this section, we will focus on some examples of applications that use glycosurfaces for binding, capture and sensing of bacteria and bacterial toxins, as well as the multivalency effect that plays an important part in carbohydrate–protein interactions on glycosurfaces.

1.5.1 Binding, Capture, and Sensing of Bacteria

The genetic model bacterium *E. coli* has been frequently used as a model for microbial capture using glycosurfaces due to the well-known interaction between their fimbrial adhesins and the glycocalyx of their host cells. The interaction between α-D-mannoside residues and the FimH adhesin present on the type 1 fimbriae of many strains of *E. coli* is the most investigated bacterial adhesion process so far [1]. Wang and coworkers [11] used two different approaches for *E. coli* detection via carbohydrate–adhesin interactions: direct detection and Con A-mediated detection (Fig. 1.8). The direct detection involved detecting the interaction between a mannose-terminated SAM and the adhesin units present on the fimbriae of the bacteria. The Con A-mediated detection, on the other hand, involved the initial binding of the lipopolysaccharide present on the bacterial surface to Con A that in turn binds to the mannose-terminated SAM. The signal response was around eight times larger when the Con A-mediated detection was used. This is attributed to the formation of bridges between *E. coli* and the mannose-presenting surface, leading to a relatively rigid and strong attachment [11]. Most studies using glycosurfaces for binding, capture, or sensing of *E. coli* however use the direct detection approach [13,90,121,155–157], as the use of the multivalent binding capacity of Con A as an intermediate adds additional complexity to interpreting the already complex carbohydrate–protein interaction of interest [87]. In general, the bacterial binding can be detected and/or quantified by, for example, (fluorescence) microscopy [90,121], faradaic EIS [13], metal mesh device (MMD) sensors [155], or quartz crystal microbalance with dissipation (QCM-D) techniques [156].

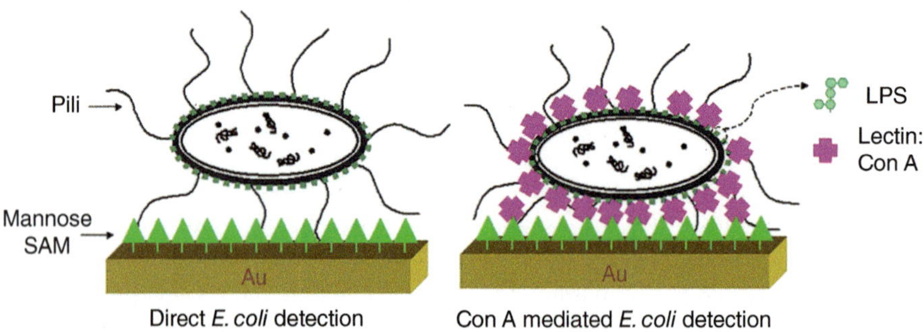

FIGURE 1.8 Schematic representation of direct *E. coli* detection and Con A-mediated *E. coli* detection. *Source*: Reprinted with permission from Ref. 11, Copyright 2007, American Chemical Society.

Different *E. coli* strains have been used as a model for investigating this binding, such as the ORN178 [13,115,155] and DH5α [157] strains—which both possess the wild-type FimH domain—or a fluorescently GFP-expressing strain (*E. coli* pPKL1162) (Fig. 1.9) that also overexpresses the type 1 fimbriae at its surface [90]. The interaction between bacteria and glycosurfaces is a complex process, and proper negative controls are therefore required. Some *E. coli* strains that do not present the type 1 fimbriae have been used as such a negative control, such as AAEC191A [121] and HB101 strains (nonfimbriated *E. coli*), a strain lacking the plasmid for FimH expression (but expressing the rest of the *fim* gene cluster) [121], and the ORN 208 strain (expresses abnormal type 1 pili that fail to mediate mannose-specific binding) [13,115,155]. The use of these strains as negative controls is important to show that the binding process is mostly due to the interaction between the carbohydrates and the adhesins, partially excluding nonspecific interactions as the source for any observed binding. Another option for a negative control is the addition of soluble carbohydrates to the suspension of bacteria before the binding to the glycosurface. If the soluble carbohydrate is the same as the one present on the glycosurface, it will saturate the carbohydrate-binding sites of any present bacterial adhesins and thereby strongly decreases the bacterial glycosurface binding when compared with the experiments in the absence of the soluble carbohydrate [6,87].

Another bacterium known to have adhesins that are related to its binding of natural glycosurfaces and the formation of biofilms is *Pseudomonas aeruginosa*. *P. aeruginosa* synthesizes two surface adhesins, named PA-IL (or LecA), which specifically binds galactose, and PA-IIL (or LecB), which specifically binds fucose [158,159]. Wei and coworkers [160] studied the binding of *P. aeruginosa* on glycosurfaces with trisaccharides containing β-D-*N*-acetylgalactosamine or α-L-fucose. The bacteria adhered to both glycosurfaces; however, they displayed a lower adhesion to the surface with the fucose-containing trisaccharide than on the one containing *N*-acetylgalactosamine, which naturally occurs on the epithelium of the pulmonary tract. This difference might be due to the fact that the fucose-binding adhesin PA-IIL is regulated in an environmentally dependent fashion, requiring certain compounds in the culture medium for proper expression [160]. More recently, Liu and coworkers [156] investigated the binding of *P. aeruginosa* by QCM-D using surfaces coated with glycopolymers generated with glucose and lactose-containing monomers. The bacteria showed an increased, and calcium-dependent, adhesion to the lactose-presenting surface, probably due to the interaction between PA-IL and the galactose unit [156].

Surfaces coated with glycopolymers containing lactose were also used to increase the binding of *E. faecalis* via interaction between an adhesin on its surface and the galactose unit [99]. It was shown that the presence of soluble galactose decreased the bacterial binding, whereas a soluble glucose negative control did not influence the binding. The binding of *E. faecalis* was also significantly affected by the blocking of the carbohydrates with peanut agglutinin (PNA), a lectin that specifically binds galactose. Prior exposure of the glycosurface to BSA or Con A had considerably less influence on the bacterial binding when compared with PNA [99].

Glycosurfaces were also recently used by us for increased binding of *L. plantarum*, a probiotic bacterium that presents a mannose-binding adhesin on the cell surface [161]. Mannose-terminated porous aluminum oxide (PAO) was used for binding

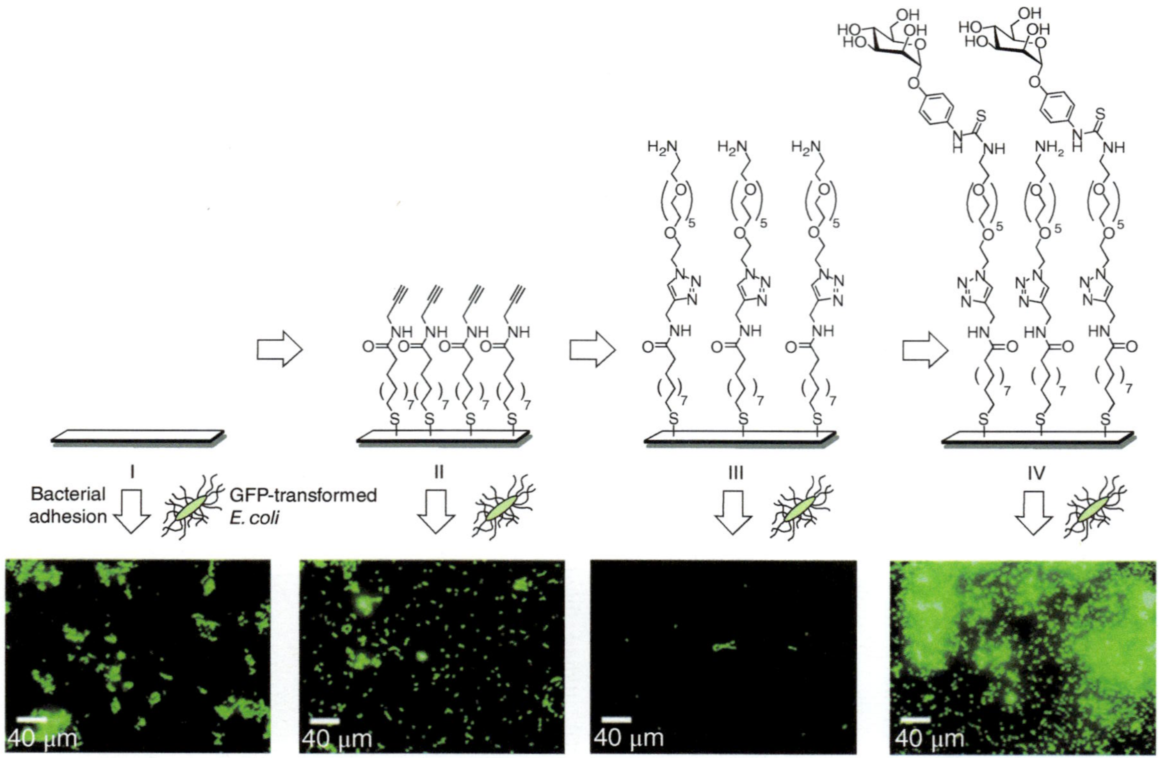

FIGURE 1.9 Adhesion of fluorescent bacteria to the different stages of the SAM during the 'dual click' approach. The GFP-transformed *E. coli* bacteria (pPKL1162) enable a fast, direct fluorescence readout to investigate bacterial adhesion on surfaces. The native gold surface (I) was used as reference in each of the other experiments. As can be seen in the epifluorescence micrographs, the (non-specific) adhesivity of the alkyne-terminated SAM II is comparable to the one of the native Au surface. Introduction of the OEG chain reduces the adhesion significantly, while the α-mannosyl-terminated SAM is effectively recognized by the *E. coli* leading to heavy adhesion. *Source*: Reproduced from Ref. 90 with permission from The Royal Society of Chemistry.

and subsequent growth of *L. plantarum* on the same surface. PAO is a cheap nanoporous surface that has been used for various biotechnological applications. Bacterial growth is possible on PAO because nutrients can diffuse from the medium on which PAO is placed to the top surface via the pores present on the surface [162]. We also showed in a control experiment that a soluble mannose derivative was able to reduce the binding of *L. plantarum* to the mannose-terminated glycosurface by blocking the mannose-binding adhesin [6].

1.5.2 Binding and Sensing of Bacterial Toxins

Although the use of glycosurfaces for detection of bacterial cells is still limited, there are various examples of detection of bacterial toxins using glycosurfaces. This is probably because the interaction between carbohydrates and whole cells is more complex and challenging to characterize than the interaction between carbohydrates and toxins. The detection of bacterial toxins using glycosurfaces is possible because a large number of bacterial toxins target carbohydrate derivatives on host cell surfaces to enter into the cells. Some examples of bacterial toxins that target host cell surface carbohydrates include cholera toxin, shiga and shiga-like toxins, tetanus toxin, botulinum toxin, and pertussis toxin [163].

Cholera toxin, secreted by *Vibrio cholerae*, is the cause of the pathology observed in cholera, mainly watery diarrhea and vomiting, that can lead to lethal dehydration and still is a common illness in many developing countries. The natural ligand of cholera toxin is the glycosphingolipid ganglioside GM1, which is present on the cellular membrane of the intestinal epithelial surface of the host. The major contributors to the binding are the galactose and the *N*-acetylneuraminic acid units of the GM1 ganglioside [164]. Glycosurfaces containing these carbohydrates may therefore be useful for detection of cholera toxin. GM1-terminated surfaces have been used for detection of cholera toxin [112,119,165–167] and also to study the inhibition of cholera toxin binding by soluble carbohydrate derivatives [127]. Ligler and coworkers [163] detected cholera toxin on both *N*-acetylneuraminic acid and *N*-acetylgalactosamine-terminated surfaces, with a limit of detection of 100 ng ml^{-1}. As expected, the binding of the toxin to the *N*-acetylneuraminic acid-terminated surface was more intense than to the *N*-acetylgalactosamine-terminated one [163]. Even more simple carbohydrates, such as galactose, showed interaction with cholera toxin when immobilized on a surface used to generate carbohydrate microarrays [168].

Shiga toxin and shiga-like toxins have also been detected using glycosurfaces. These toxins are known to cause gastrointestinal diseases after invading mammalian cells through binding to the carbohydrate portion of glycolipids on the cell surface. A shiga-like toxin produced by *E. coli* O157:H7 presents specificity toward globotriaosylceramide (Gb3), which contains the trisaccharide αGal(1→4)βGal(1→4)βGlc [169]. Consequently, Gb3 derivatives and more simple carbohydrates that are part of Gb3 have been immobilized on surfaces and used for the binding of shiga-like toxins as detected by QCM [26,170] and SPR [29,171–173].

Other glycosurfaces have been used to investigate the interactions between carbohydrates and other bacterial toxins [108,174], such as *E. coli* heat-labile enterotoxin [129], and tetanus toxin [163].

1.5.3 Multivalency Effect

Although single interactions between carbohydrates and proteins are in general very weak, multiple interactions can reinforce one another and together achieve avidities that result in strong binding. Almost all carbohydrate–protein interactions in nature are multivalent, including the interactions between carbohydrates and bacterial adhesins or toxins discussed in the previous sections. The avidity and even the selectivity of carbohydrate–protein interactions clearly depend on the density of the carbohydrate units on a studied surface [175]. In this case, for an optimal binding of a protein on a glycosurface, it is important that the carbohydrate is homogeneously distributed on the surface and present in a suitable density. Obviously, even when surfaces are functionalized with monovalent ligands, they can act as a multivalent system based only on the immobilization of the monovalent agents on the surface [176].

Kahne and coworkers [22] investigated the influence of carbohydrate surface density on protein-binding selectivity. They showed that the *Bauhinia purpurea* lectin switches its selectivity from one carbohydrate to another depending on the surface density of the ligands in a mixed SAM [22]. Houseman and Mrksich [18] investigated the influence of the surface density of carbohydrates on an enzymatic glycosylation reaction. In their system, the optimal density of carbohydrate for the enzymatic glycosylation was an intermediate density, around 70%, with significant decrease of glycosylation rates at higher densities of carbohydrates [18]. The influence of the carbohydrate density on antibody binding was investigated by SPR. When the carbohydrate was present on the SAM at a high density, the antibody binding was minimal. The carbohydrate density needed to be reduced to around 25% to obtain an increase of antibody binding [20]. Similar results were found for binding of Con A on maltose-terminated surfaces, with increase of the binding being observed when the carbohydrate density decreased (Fig. 1.10). The binding also changed from monovalent recognition to bivalent recognition when the carbohydrate was diluted to 10% [177]. One explanation for the low binding of proteins to glycosurfaces with a high carbohydrate density is attributed to the tight packing of the end groups in the SAM, resulting in limited access of the carbohydrates to the protein binding pocket [20].

Ratner and coworkers [137] showed that the effect of the carbohydrate density in the binding of proteins also depends on the type of molecule used for the dilution (Fig. 1.11). When a molecule with a long chain (longer than the carbohydrate) was used for dilution, the increase of carbohydrate density increased the lectin binding. On the other hand, when a short molecule was used for the dilution, the maximum lectin binding was found around 40–60% carbohydrate density, followed by a plateau and a slight decrease of binding [137].

Stine and coworkers [12] observed a difference in the optimal mannose density for Con A binding on flat gold and nanoporous gold. Although the dilution of the carbohydrate in the SAM increased the Con A binding for both surfaces, the optimal mannose density proved lower for nanoporous gold. Whereas the optimal mannose density for Con A binding was 1:9 ratio (mannoside/alkane thiol) for flat gold, the best Con A binding was found to be 1:19 for nanoporous gold [12].

Buriak and coworkers [36] did not find a decrease of protein binding with an increase of carbohydrate density, but rather an optimal protein binding when the

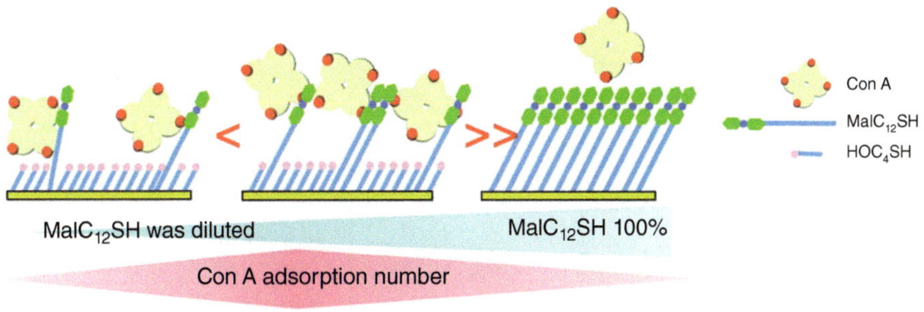

FIGURE 1.10 Model structure of three-dimensional carbohydrate positioning in the interface provides the best strategy for overcoming the weak binding detection. The maltoside–OH terminated hybrid monolayer represents multivalent binding for Con A when the maltoside was diluted. The affinity of Con A and maltoside on the surface was enhanced in ~10 mol % surfaces. *Source:* Reprinted with permission from Ref. 177, Copyright 2012, American Chemical Society.

carbohydrate is present at 30% density, followed by a plateau when the carbohydrate density was increased [36]. Szunerits and coworkers obtained similar results for binding of *Lens culinaris* lectin on mannose-terminated surfaces, with a maximum response for carbohydrate density of 60%, with a plateau of response when increasing the mannose density [53].

Another way of studying the effect of multivalency on the binding of lectins to glycosurfaces is using carbohydrate derivatives with different valencies. Pieters and coworkers [168] investigated the binding of a series of lectins to five different carbohydrates presenting valencies ranging from one to eight. Using a microarray, it was possible to investigate the lectin specificity and the multivalency effect for each carbohydrate in a single experiment. The authors found various profiles, depending on the carbohydrate and the lectin, but with maintenance of specificity [168].

Changes in the carbohydrate density of SAMs or in the way of presenting the carbohydrate units can also interfere in the bacterial binding to glycosurfaces. Carbohydrate derivatives with different valencies were immobilized on a surface, and the binding of *E. coli* was investigated. The bacterial binding increased when the amount of carbohydrate units per molecule increased from one to three. However, when the carbohydrate units further increased to six and nine, there was a decrease in bacterial binding [121]. The binding of *E. coli* to dynamic glycosurfaces with different densities showed a change in the avidity from monovalent to multivalent as the mannose density increased (Fig. 1.12) [115]. Recently, Lindhorst and coworkers [178] investigated the influence on bacterial binding to mannose-presenting surfaces of both the carbohydrate density and of different ways of presenting mannose. The authors showed that at high densities of carbohydrate, the simple monovalent mannoside provided a more adhesive surface for *E. coli* than the bivalent and trivalent ones. However, at lower densities, the binding of *E. coli* was higher on the surfaces with bivalent and trivalent mannosides when compared to the monovalent one. The highest binding affinity was found for the lowest concentration tested with the trivalent mannoside (Fig. 1.12) [178].

It is interesting to notice that a higher carbohydrate density on a SAM does not necessarily result in greater binding of lectins or bacteria. It can either not provide

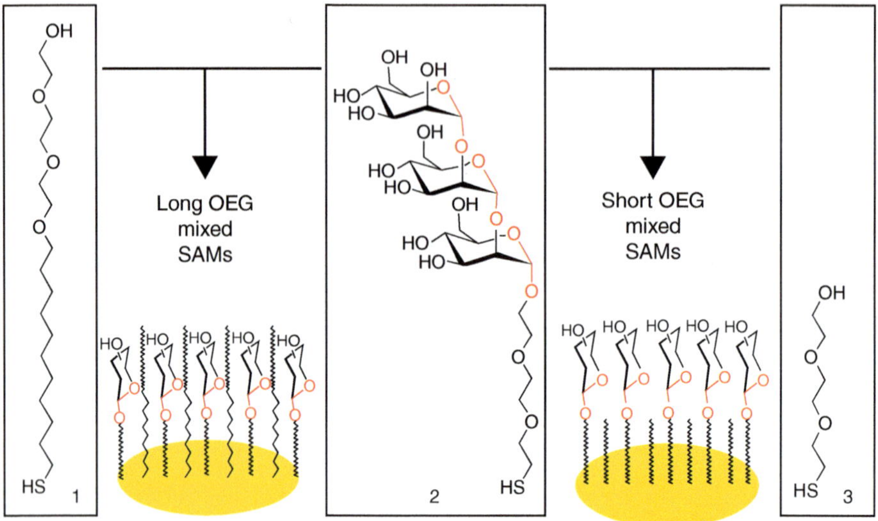

FIGURE 1.11 Assembly of mixed sugar/oligoethylene glycol (OEG) SAMs on gold. *Source*: Reprinted with permission from Ref. 137, Copyright 2009, American Chemical Society.

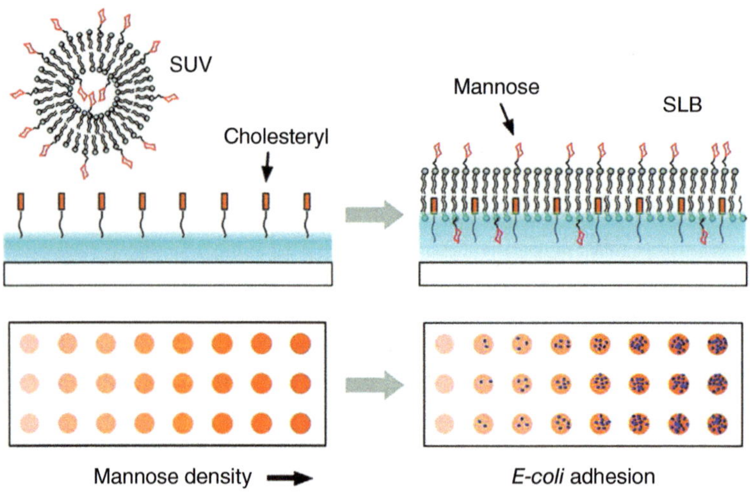

FIGURE 1.12 (Upper) Formation of a glycan presenting supported lipid bilayer (SLB) surface from a small unilamellar vesicle (SUV) solution. (Lower) Schematic illustration of a glycan density gradient microarray for pathogen adhesion. *Source*: Reprinted with permission from Ref. 115, Copyright 2009, American Chemical Society.

any improvement in binding, resulting in waste of the excess of carbohydrate or, in the worst case, decrease the binding [36]. For this reason, it is important to determine the optimal carbohydrate density in a glycosurface to maximize the biological effect and also to reduce the waste of carbohydrate derivative.

1.6 OUTLOOK

In the previous sections, we presented an overview of the methods frequently used to prepare glycosurfaces via three different general approaches: (i) direct formation of SAMs containing carbohydrates, (ii) secondary reactions (with unmodified and modified carbohydrates), and (iii) noncovalent immobilization. The main challenge in the modification of surfaces with carbohydrates is to develop an approach that yields well-defined and stable glycosurfaces using the minimal number of steps. These steps should ideally be reproducible, simple, cheap, and available to a great number of scientists. Additionally, the ideal glycosurface should be stable to a wide range of conditions, and for certain applications, it should be resistant to multiple cycles of regeneration, without loss of its original properties. For industrial applications, it would also be important to consider the possibilities of scaling up the production process. The most currently used approach for straightforward glycosurfaces is still the method that was first developed, formation of thiol-based SAMs of carbohydrate derivatives on gold, mainly because of their easy preparation. As the study and utilization of the biological role of carbohydrates has been rapidly increasing over the past decades, glycosurfaces are required that enable an ever increasing amount of sensitivity, complexity, and precise control over their composition. In general, the trend is therefore to explore other approaches that enable stable and controlled site-specific immobilization of unprotected carbohydrates via an attaching group on their anomeric position.

Besides the often applied copper(I)-catalyzed azide–alkyne cycloaddition (CuAAC), an increasing range of other "click" reactions [179,180] have become available also for use in surface attachment of carbohydrates. The full potential of these reactions for the preparation of glycosurfaces has not yet been reached, but over the coming years will allow for more advanced glycosurfaces and perhaps the ideal one. When preparing a glycosurface, it is of course essential to know what it looks like before using it. Various characterization techniques have been applied for glycosurfaces, namely, contact angle measurements, XPS, IRRAS, mass spectrometry, ellipsometry, and AFM. All these methods provide different types of information, and the main challenge is to select what is the best combination of techniques to characterize a glycosurface. This selection usually depends on the substrate used to prepare the glycosurface, the type of modification, and the application of the glycosurface. It should give the highest level of information in the shortest time possible, ideally without destroying the glycosurface.

Glycosurfaces have already proven to be very useful in increasing our understanding of the binding between carbohydrates and other biomolecules, mainly proteins, and have been successfully used for detection of known carbohydrate–protein interactions and helping the discovery of new binding partners and the development of novel therapies. Because of the enormous number and diversity of carbohydrate-containing biomolecules and the fact that most are still uncharted, there is a great need for reliable and sensitive methods for their high-throughput characterization, and it is evident that the next-generation glycosurfaces will be instrumental in this [181]. What is needed now is a focus on developing and optimizing immobilization techniques that yield well-defined and stable complex glycosurfaces and also on developing and optimizing automated (chemo-enzymatic) synthetic methods [182–184] to prepare the required complex carbohydrates with suitable reactive functional groups at the anomeric position for surface immobilization. Well-established automated synthetic methods may eventually make the synthesis and application of complex carbohydrate derivatives a routine procedure that is more accessible to non-specialists and thereby widen their application to slowly but surely illuminate this complex dark matter-world.

ACKNOWLEDGMENTS

Part of this research was supported by the NanoNext program, a micro and nanotechnology consortium of the Government of the Netherlands and 130 partners, the Netherlands Organization for Scientific Research (VENI grant 722.011.006 & Spinoza grant), and the European Research Council (ERC 250172).

REFERENCES

1. Hartmann, M.; Lindhorst, T. K., *Eur. J. Org. Chem.* **2011**, *2011* (20–21), 3583–3609.
2. Kamhi, E.; Joo, E. J.; Dordick, J. S.; Linhardt, R. J., *Biol. Rev. Camb. Philos. Soc.* **2013**, *88* (4), 928–943.
3. Berois, N.; Osinaga, E., *Front. Oncol.* **2014**, *4*, 114.
4. Krengel, U.; Bousquet, P. A., *Front. Immunol.* **2014**, *5*, 325.
5. Pretzer, G.; Snel, J.; Molenaar, D.; Wiersma, A.; Bron, P. A.; Lambert, J.; de Vos, W. M.; van der Meer, R.; Smits, M. A.; Kleerebezem, M., *J. Bacteriol.* **2005**, *187* (17), 6128–6136.
6. Debrassi, A.; Ribbera, A.; de Vos, W. M.; Wennekes, T.; Zuilhof, H., *Langmuir* **2014**, *30* (5), 1311–1320.
7. Schreiber, F., *Prog. Surf. Sci.* **2000**, *65* (5–8), 151–256.
8. Love, J. C.; Estroff, L. A.; Kriebel, J. K.; Nuzzo, R. G.; Whitesides, G. M., *Chem. Rev.* **2005**, *105* (4), 1103–1169.
9. Revell, D. J.; Knight, J. R.; Blyth, D. J.; Haines, A. H.; Russell, D. A., *Langmuir* **1998**, *14* (16), 4517–4524.
10. Karamanska, R.; Mukhopadhyay, B.; Russell, D. A.; Field, R. A., *Chem. Commun.* **2005**, (26), 3334–3336.
11. Shen, Z.; Huang, M.; Xiao, C.; Zhang, Y.; Zeng, X.; Wang, P. G., *Anal. Chem.* **2007**, *79* (6), 2312–2319.
12. Pandey, B.; Tan, Y. H.; Fujikawa, K.; Demchenko, A. V.; Stine, K. J., *J. Carbohyd. Chem.* **2012**, *31* (4–6), 466–503.
13. Guo, X.; Kulkarni, A.; Doepke, A.; Halsall, H. B.; Iyer, S.; Heineman, W. R., *Anal. Chem.* **2012**, *84* (1), 241–246.
14. Kaplan, J. M.; Shang, J.; Gobbo, P.; Antonello, S.; Armelao, L.; Chatare, V.; Ratner, D. M.; Andrade, R. B.; Maran, F., *Langmuir* **2013**, *29* (26), 8187–8192.
15. Seo, J. H.; Adachi, K.; Lee, B. K.; Kang, D. G.; Kim, Y. K.; Kim, K. R.; Lee, H. Y.; Kawai, T.; Cha, H. J., *Bioconjugate Chem.* **2007**, *18* (6), 2197–2201.
16. He, X.-P.; Wang, X.-W.; Jin, X.-P.; Zhou, H.; Shi, X.-X.; Chen, G.-R.; Long, Y.-T., *J. Am. Chem. Soc.* **2011**, *133* (10), 3649–3657.

17. Fyrner, T.; Lee, H.-H.; Mangone, A.; Ekblad, T.; Pettitt, M. E.; Callow, M. E.; Callow, J. A.; Conlan, S. L.; Mutton, R.; Clare, A. S.; Konradsson, P.; Liedberg, B.; Ederth, T., *Langmuir* **2011**, *27* (24), 15034–15047.
18. Houseman, B. T.; Mrksich, M., *Angew. Chem. Int. Ed.* **1999**, *38* (6), 782–785.
19. Sato, Y.; Yoshioka, K.; Tanaka, M.; Murakami, T.; Ishida, M. N.; Niwa, O., *Chem. Commun.* **2008**, (40), 4909–4911.
20. Grant, C. F.; Kanda, V.; Yu, H.; Bundle, D. R.; McDermott, M. T., *Langmuir* **2008**, *24* (24), 14125–14132.
21. Pandey, B.; Tan, Y. H.; Parameswar, A. R.; Pornsuriyasak, P.; Demchenko, A. V.; Stine, K. J., *Carbohydr. Res.* **2013**, *373*, 9–17.
22. Horan, N.; Yan, L.; Isobe, H.; Whitesides, G. M.; Kahne, D., *Proc. Natl. Acad. Sci. U. S. A.* **1999**, *96* (21), 11782–11786.
23. Hederos, M.; Konradsson, P.; Borgh, A.; Liedberg, B., *J. Phys. Chem. B* **2005**, *109* (33), 15849–15859.
24. Fritz, M. C.; Hahner, G.; Spencer, N. D.; Burli, R.; Vasella, A., *Langmuir* **1996**, *12* (25), 6074–6082.
25. Svedhem, S.; Ohberg, L.; Borrelli, S.; Valiokas, R.; Andersson, M.; Oscarson, S.; Svensson, S. C. T.; Liedberg, B.; Konradsson, P., *Langmuir* **2002**, *18* (7), 2848–2858.
26. Miura, Y.; Sasao, Y.; Dohi, H.; Nishida, Y.; Kobayashi, K., *Anal. Biochem.* **2002**, *310* (1), 27–35.
27. Miura, Y.; Sasao, Y.; Kamihira, M.; Sakaki, A.; Iijima, S.; Kobayashi, K., *BBA: Gen. Subj.* **2004**, *1673* (3), 131–138.
28. Tromas, C.; Eaton, P.; Mimault, J.; Rojo, J.; Penades, S., *Langmuir* **2005**, *21* (14), 6142–6144.
29. Kanda, V.; Kitov, P.; Bundle, D. R.; McDermott, M. T., *Anal. Chem.* **2005**, *77* (23), 7497–7504.
30. Wang, Y.; El-Boubbou, K.; Kouyoumdjian, H.; Sun, B.; Huang, X.; Zeng, X., *Langmuir* **2010**, *26* (6), 4119–4125.
31. Uzawa, H.; Ohga, K.; Shinozaki, Y.; Ohsawa, I.; Nagatsuka, T.; Seto, Y.; Nishida, Y., *Biosens. Bioelectron.* **2008**, *24* (4), 923–927.
32. Kadalbajoo, M.; Park, J.; Opdahl, A.; Suda, H.; Kitchens, C. A.; Garno, J. C.; Batteas, J. D.; Tarlov, M. J.; DeShong, P., *Langmuir* **2007**, *23* (2), 700–707.
33. Shirahata, N.; Hozumi, A.; Miura, Y.; Kobayashi, K.; Sakka, Y.; Yonezawa, T., *Thin Solid Films* **2006**, *499* (1–2), 213–218.
34. Funato, K.; Shirahata, N.; Miura, Y., *Thin Solid Films* **2009**, *518* (2), 699–702.
35. Chang, S.-H.; Han, J.-L.; Tseng, S. Y.; Lee, H.-Y.; Lin, C.-W.; Lin, Y.-C.; Jeng, W.-Y.; Wang, A. H. J.; Wu, C.-Y.; Wong, C.-H., *J. Am. Chem. Soc.* **2010**, *132* (38), 13371–13380.
36. Slaney, A. M.; Wright, V. A.; Meloncelli, P. J.; Harris, K. D.; West, L. J.; Lowary, T. L.; Buriak, J. M., *ACS Appl. Mater. Interfaces* **2011**, *3* (5), 1601–1612.
37. Gruber, K.; Horlacher, T.; Castelli, R.; Mader, A.; Seeberger, P. H.; Hermann, B. A., *ACS Nano* **2011**, *5* (5), 3670–3678.
38. Chinwangso, P.; Jamison, A. C.; Lee, T. R., *Acc. Chem. Res.* **2011**, *44* (7), 511–519.
39. de Smet, L.; Stork, G. A.; Hurenkamp, G. H. F.; Sun, Q. Y.; Topal, H.; Vronen, P. J. E.; Sieval, A. B.; Wright, A.; Visser, G. M.; Zuilhof, H.; Sudholter, E. J. R., *J. Am. Chem. Soc.* **2003**, *125* (46), 13916–13917.
40. de Smet, L.; Pukin, A. V.; Sun, Q. Y.; Eves, B. J.; Lopinski, G. P.; Visser, G. M.; Zuilhof, H.; Sudholter, E. J. R., *Appl. Surf. Sci.* **2005**, *252* (1), 24–30.
41. Pujari, S. P.; Scheres, L.; Marcelis, A. T. M.; Zuilhof, H., *Angew. Chem. Int. Ed.* **2014**, *53* (25), 6322–6356.

42. Bhairamadgi, N. S.; Pujari, S. P.; Trovela, F. G.; Debrassi, A.; Khamis, A. A.; Alonso, J. M.; Al Zahrani, A. A.; Wennekes, T.; Al-Turaif, H. A.; van Rijn, C.; Alhamed, Y. A.; Zuilhof, H., *Langmuir* **2014**, *30* (20), 5829–5839.

43. Carroll, G. T.; Wang, D. N.; Turro, N. J.; Koberstein, J. T., *Langmuir* **2006**, *22* (6), 2899–2905.

44. Tyagi, A.; Wang, X.; Deng, L.; Ramstrom, O.; Yan, M., *Biosens. Bioelectron.* **2010**, *26* (2), 344–350.

45. Zhou, X.; Turchi, C.; Wang, D., *J. Proteome Res.* **2009**, *8* (11), 5031–5040.

46. Zhi, Z.-L.; Powell, A. K.; Turnbull, J. E., *Anal. Chem.* **2006**, *78* (14), 4786–4793.

47. Cheng, F.; Shang, J.; Ratner, D. M., *Bioconjugate Chem.* **2011**, *22* (1), 50–57.

48. Shang, J.; Piskarev, V. E.; Xia, M.; Huang, P.; Jiang, X.; Likhosherstov, L. M.; Novikova, O. S.; Newburg, D. S.; Ratner, D. M., *Glycobiology* **2013**, *23* (12), 1491–1498.

49. Park, S.; Gildersleeve, J. C.; Blixt, O.; Shin, I., *Chem. Soc. Rev.* **2013**, *42* (10), 4310–4326.

50. Zhang, Y.; Luo, S. Z.; Tang, Y. J.; Yu, L.; Hou, K. Y.; Cheng, J. P.; Zeng, X. Q.; Wang, P. G., *Anal. Chem.* **2006**, *78* (6), 2001–2008.

51. Kleinert, M.; Winkler, T.; Terfort, A.; Lindhorst, T. K., *Org. Biomol. Chem.* **2008**, *6* (12), 2118–2132.

52. Matsumoto, E.; Yamauchi, T.; Fukuda, T.; Miura, Y., *Sci. Technol. Adv. Mat.* **2009**, *10* (3), 7, 034605.

53. Szunerits, S.; Niedziolka-Joensson, J.; Boukherroub, R.; Woisel, P.; Baumann, J.-S.; Siriwardena, A., *Anal. Chem.* **2010**, *82* (19), 8203–8210.

54. Yang, J.; Chazalviel, J.-N.; Siriwardena, A.; Boukherroub, R.; Ozanam, F.; Szunerits, S.; Gouget-Laemmel, A. C., *Anal. Chem.* **2014**, *86* (20), 10340–10349.

55. Leone, G.; Consumi, M.; Tognazzi, A.; Magnani, A., *Thin Solid Films* **2010**, *519* (1), 462–470.

56. Godula, K.; Rabuka, D.; Nam, K. T.; Bertozzi, C. R., *Angew. Chem. Int. Ed.* **2009**, *48* (27), 4973–4976.

57. Bouchet-Spinelli, A.; Reuillard, B.; Coche-Guerente, L.; Armand, S.; Labbe, P.; Fort, S., *Biosens. Bioelectron.* **2013**, *49*, 290–296.

58. Wendeln, C.; Singh, I.; Rinnen, S.; Schulz, C.; Arlinghaus, H. F.; Burley, G. A.; Ravoo, B. J., *Chem. Sci.* **2012**, *3* (8), 2479–2484.

59. Manova, R. K.; Pujari, S. P.; Weijers, C. A. G. M.; Zuilhof, H.; van Beek, T. A., *Langmuir* **2012**, *28* (23), 8651–8663.

60. Wendeln, C.; Rinnen, S.; Schulz, C.; Kaufmann, T.; Arlinghaus, H. F.; Ravoo, B. J., *Chem.-Eur. J.* **2012**, *18* (19), 5880–5888.

61. Norberg, O.; Lee, I. H.; Aastrup, T.; Yan, M.; Ramstrom, O., *Biosens. Bioelectron.* **2012**, *34* (1), 51–56.

62. Biggs, C. I.; Edmondson, S.; Gibson, M. I., *Biomater. Sci.* **2015**, *3* (1), 175–181.

63. Wang, C.; Ren, P.-F.; Huang, X.-J.; Wu, J.; Xu, Z.-K., *Chem. Commun.* **2011**, *47* (13), 3930–3932.

64. Wang, C.; Fan, Y.; Hu, M.-X.; Xu, W.; Wu, J.; Ren, P.-F.; Xu, Z.-K., *Colloid. Surface. B* **2013**, *110*, 105–112.

65. Houseman, B. T.; Gawalt, E. S.; Mrksich, M., *Langmuir* **2003**, *19* (5), 1522–1531.

66. Yonzon, C. R.; Jeoungf, E.; Zou, S. L.; Schatz, G. C.; Mrksich, M.; Van Duyne, R. P., *J. Am. Chem. Soc.* **2004**, *126* (39), 12669–12676.

67. Wehner, J. W.; Weissenborn, M. J.; Hartmann, M.; Gray, C. J.; Sardzik, R.; Eyers, C. E.; Flitsch, S. L.; Lindhorst, T. K., *Org. Biomol. Chem.* **2012**, *10* (44), 8919–8926.

68. Yatawara, A. K.; Tiruchinapally, G.; Bordenyuk, A. N.; Andreana, P. R.; Benderskii, A. V., *Langmuir* **2009**, *25* (4), 1901–1904.
69. Shang, J.; Cheng, F.; Dubey, M.; Kaplan, J. M.; Rawal, M.; Jiang, X.; Newburg, D. S.; Sullivan, P. A.; Andrade, R. B.; Ratner, D. M., *Langmuir* **2012**, *28* (6), 3338–3344.
70. Houseman, B. T.; Mrksich, M., *Chem. Biol.* **2002**, *9* (4), 443–454.
71. Wendeln, C.; Heile, A.; Arlinghaus, H. F.; Ravoo, B. J., *Langmuir* **2010**, *26* (7), 4933–4940.
72. Beckmann, H. S. G.; Niederwieser, A.; Wiessler, M.; Wittmann, V., *Chem.-Eur. J.* **2012**, *18* (21), 6548–6554.
73. Smith, E. A.; Thomas, W. D.; Kiessling, L. L.; Corn, R. M., *J. Am. Chem. Soc.* **2003**, *125* (20), 6140–6148.
74. Harris, L. G.; Schofield, W. C. E.; Doores, K. J.; Davis, B. G.; Badyal, J. P. S., *J. Am. Chem. Soc.* **2009**, *131* (22), 7755–7761.
75. Loka, R. S.; Cairo, C. W., *Carbohydr. Res.* **2010**, *345* (18), 2641–2647.
76. Zhi, Z.-L.; Laurent, N.; Powel, A. K.; Karamanska, R.; Fais, M.; Voglmeir, J.; Wright, A.; Blackburn, J. M.; Crocker, P. R.; Russell, D. A.; Flitsch, S.; Field, R. A.; Turnbull, J. E., *ChemBioChem* **2008**, *9* (10), 1568–1575.
77. Sardzik, R.; Sharma, R.; Kaloo, S.; Voglmeir, J.; Crocker, P. R.; Flitsch, S. L., *Chem. Commun.* **2011**, *47* (19), 5425–5427.
78. Javier Munoz, F.; Andre, S.; Gabius, H.-J.; Sinisterra, J. V.; Hernaiz, M. J.; Linhardt, R. J., *Green Chem.* **2009**, *11* (3), 373–379.
79. Clo, E.; Blixt, O.; Jensen, K. J., *Eur. J. Org. Chem.* **2010**, (3), 540–554.
80. Liang, P.-H.; Wang, S.-K.; Wong, C.-H., *J. Am. Chem. Soc.* **2007**, *129* (36), 11177–11184.
81. Bohorov, O.; Andersson-Sand, H.; Hoffmann, J.; Blixt, O., *Glycobiology* **2006**, *16* (12), 21C–27C.
82. Wang, C.-C.; Huang, Y.-L.; Ren, C.-T.; Lin, C.-W.; Hung, J.-T.; Yu, J.-C.; Yu, A. L.; Wu, C.-Y.; Wong, C.-H., *Proc. Natl. Acad. Sci. U. S. A.* **2008**, *105* (33), 11661–11666.
83. Blixt, O.; Head, S.; Mondala, T.; Scanlan, C.; Huflejt, M. E.; Alvarez, R.; Bryan, M. C.; Fazio, F.; Calarese, D.; Stevens, J.; Razi, N.; Stevens, D. J.; Skehel, J. J.; van Die, I.; Burton, D. R.; Wilson, I. A.; Cummings, R.; Bovin, N.; Wong, C. H.; Paulson, J. C., *Proc. Natl. Acad. Sci. U. S. A.* **2004**, *101* (49), 17033–17038.
84. Song, X.; Xia, B.; Stowell, S. R.; Lasanajak, Y.; Smith, D. F.; Cummings, R. D., *Chem. Biol.* **2009**, *16* (1), 36–47.
85. de Boer, A. R.; Hokke, C. H.; Deelder, A. M.; Wuhrer, M., *Anal. Chem.* **2007**, *79* (21), 8107–8113.
86. Buck, M. E.; Breitbach, A. S.; Belgrade, S. K.; Blackwell, H. E.; Lynn, D. M., *Biomacromolecules* **2009**, *10* (6), 1564–1574.
87. Liu, X.; Lei, Z.; Liu, F.; Liu, D.; Wang, Z., *Biosens. Bioelectron.* **2014**, *58*, 92–100.
88. Kopitzki, S.; Jensen, K. J.; Thiem, J., *Chem.-Eur. J.* **2010**, *16* (23), 7017–7029.
89. Beer, M. V.; Rech, C.; Diederichs, S.; Hahn, K.; Bruellhoff, K.; Moeller, M.; Elling, L.; Groll, J., *Anal. Bioanal. Chem.* **2012**, *403* (2), 517–526.
90. Grabosch, C.; Kind, M.; Gies, Y.; Schweighoefer, F.; Terfort, A.; Lindhorst, T. K., *Org. Biomol. Chem.* **2013**, *11* (24), 4006–4015.
91. Javier Lopez-Jaramillo, F.; Ortega-Munoz, M.; Megia-Fernandez, A.; Hernandez-Mateo, F.; Santoyo-Gonzalez, F., *Bioconjugate Chem.* **2012**, *23* (4), 846–855.

92. Pulsipher, A.; Yousaf, M. N., *Chem. Commun.* **2011**, *47* (1), 523–525.
93. Fang, Y.; Xu, W.; Wu, J.; Xu, Z.-K., *Chem. Commun.* **2012**, *48* (91), 11208–11210.
94. Ejaz, M.; Ohno, K.; Tsujii, Y.; Fukuda, T., *Macromolecules* **2000**, *33* (8), 2870–2874.
95. Ayres, N.; Holt, D. J.; Jones, C. F.; Corum, L. E.; Grainger, D. W., *J. Polym. Sci. A1* **2008**, *46* (23), 7713–7724.
96. Yoon, K. R.; Ramaraj, B.; Lee, S.; Yu, J.-S.; Choi, I. S., *J. Biomed. Mater. Res. A* **2009**, *88A* (3), 735–740.
97. Yu, K.; Kizhakkedathu, J. N., *Biomacromolecules* **2010**, *11* (11), 3073–3085.
98. Yu, L.; Huang, M.; Wang, P. G.; Zeng, X., *Anal. Chem.* **2007**, *79* (23), 8979–8986.
99. Yang, Q.; Strathmann, M.; Rumpf, A.; Schaule, G.; Ulbricht, M., *ACS Appl. Mater. Interfaces* **2010**, *2* (12), 3555–3562.
100. Manova, R. K.; Joshi, S.; Debrassi, A.; Bhairamadgi, N. S.; Roeven, E.; Gagnon, J.; Tahir, M. N.; Claassen, F. W.; Scheres, L. M. W.; Wennekes, T.; Schroen, K.; van Beek, T. A.; Zuilhof, H.; Nielen, M. W. F., *Anal. Chem.* **2014**, *86* (5), 2403–2411.
101. ter Maat, J.; Regeling, R.; Ingham, C. J.; Weijers, C. A. G. M.; Giesbers, M.; de Vos, W. M.; Zuilhof, H., *Langmuir* **2011**, *27* (22), 13606–13617.
102. Sun, X. L.; Stabler, C. L.; Cazalis, C. S.; Chaikof, E. L., *Bioconjugate Chem.* **2006**, *17* (1), 52–57.
103. Patterson, D. M.; Nazarova, L. A.; Prescher, J. A., *ACS Chem. Biol.* **2014**, *9* (3), 592–605.
104. King, M.; Wagner, A., *Bioconjugate Chem.* **2014**, *25* (5), 825–839.
105. Manova, R.; van Beek, T. A.; Zuilhof, H., *Angew. Chem. Int. Ed.* **2011**, *50* (24), 5428–5430.
106. Agard, N. J.; Prescher, J. A.; Bertozzi, C. R., *J. Am. Chem. Soc.* **2004**, *126* (46), 15046–15047.
107. Bhairamadgi, N. S.; Gangarapu, S.; Campos, M. A. C.; Paulusse, J. M. J.; van Rijn, C. J. M.; Zuilhof, H., *Langmuir* **2013**, *29* (14), 4535–4542.
108. Rillahan, C. D.; Paulson, J. C., Glycan Microarrays for Decoding the Glycome. *Annu. Rev. Biochem.*, **2011**, *80*, 797–823.
109. Siegwart, D. J.; Oh, J. K.; Matyjaszewski, K., *Prog. Polym. Sci.* **2012**, *37* (1), 18–37.
110. Yang, Q.; Kaul, C.; Ulbricht, M., *Langmuir* **2010**, *26* (8), 5746–5752.
111. Mann, D. A.; Kanai, M.; Maly, D. J.; Kiessling, L. L., *J. Am. Chem. Soc.* **1998**, *120* (41), 10575–10582.
112. Liebau, M.; Hildebrand, A.; Neubert, R. H. H., *Eur. Biophys. J. Biophys.* **2001**, *30* (1), 42–52.
113. Zheng, H.; Du, X., *J. Phys. Chem. B* **2009**, *113* (32), 11330–11337.
114. Zheng, H.; Du, X., *J. Phys. Chem. B* **2010**, *114* (1), 577–584.
115. Zhu, X. Y.; Holtz, B.; Wang, Y.; Wang, L.-X.; Orndorff, P. E.; Guo, A., *J. Am. Chem. Soc.* **2009**, *131* (38), 13646–13650.
116. Weissenborn, M. J.; Wehner, J. W.; Gray, C. J.; Sardzik, R.; Eyers, C. E.; Lindhorst, T. K.; Flitsch, S. L., Beilstein J. *Org. Chem.* **2012**, *8*, 753–762.
117. Critchley, P.; Clarkson, G. J., *Org. Biomol. Chem.* **2003**, *1* (23), 4148–4159.
118. Guo, C.; Boullanger, P.; Jiang, L.; Liu, T., *Biosens. Bioelectron.* **2007**, *22* (8), 1830–1834.

119. Parikh, A. N.; Beers, J. D.; Shreve, A. P.; Swanson, B. I., *Langmuir* **1999**, *15* (16), 5369–5381.

120. Sato, H.; Miura, Y.; Saito, N.; Kobayashi, K.; Takai, O., *Surf. Sci.* **2007**, *601* (18), 3871–3875.

121. Barth, K. A.; Coullerez, G.; Nilsson, L. M.; Castelli, R.; Seeberger, P. H.; Vogel, V.; Textor, M., *Adv. Funct. Mater.* **2008**, *18* (9), 1459–1469.

122. Guo, Y.; Feinberg, H.; Conroy, E.; Mitchell, D. A.; Alvarez, R.; Blixt, O.; Taylor, M. E.; Weis, W. I.; Drickamer, K., *Nat. Struct. Mol. Biol.* **2004**, *11* (7), 591–598.

123. Bochner, B. S.; Alvarez, R. A.; Mehta, P.; Bovin, N. V.; Blixt, O.; White, J. R.; Schnaar, R. L., *J. Biol. Chem.* **2005**, *280* (6), 4307–4312.

124. Linman, M. J.; Yu, H.; Chen, X.; Cheng, Q., *ACS Appl. Mater. Interfaces* **2009**, *1* (8), 1755–1762.

125. Zhang, J.; Pourceau, G.; Meyer, A.; Vidal, S.; Praly, J.-P.; Souteyrand, E.; Vasseur, J.-J.; Morvan, F.; Chevolot, Y., *Biosens. Bioelectron.* **2009**, *24* (8), 2515–2521.

126. Castellana, E. T.; Cremer, P. S., *Surf. Sci. Rep.* **2006**, *61* (10), 429–444.

127. Sinclair, H. R.; Kemp, F.; de Slegte, J.; Gibson, G. R.; Rastall, R. A., *Carbohydr. Res.* **2009**, *344* (15), 1968–1974.

128. Konishi, M.; Imura, T.; Fukuoka, T.; Morita, T.; Kitamoto, D., *Biotechnol. Lett.* **2007**, *29* (3), 473–480.

129. Pukin, A. V.; Florack, D. E. A.; Brochu, D.; van Lagen, B.; Visser, G. M.; Wennekes, T.; Gilbert, M.; Zuilhof, H., *Org. Biomol. Chem.* **2011**, *9* (16), 5809–5815.

130. Linman, M. J.; Taylor, J. D.; Yu, H.; Chen, X.; Cheng, Q., *Anal. Chem.* **2008**, *80* (11), 4007–4013.

131. Dorrer, C.; Ruehe, J., *Soft Matter* **2009**, *5* (1), 51–61.

132. Chu, Z.; Seeger, S., *Chem. Soc. Rev.* **2014**, *43* (8), 2784–2798.

133. Kingshott, P.; Andersson, G.; McArthur, S. L.; Griesser, H. J., *Curr. Opin. Chem. Biol.* **2011**, *15* (5), 667–676.

134. Papp, C.; Steinrueck, H.-P., *Surf. Sci. Rep.* **2013**, *68* (3–4), 446–487.

135. Hajati, S.; Tougaard, S., *Anal. Bioanal. Chem.* **2010**, *396* (8), 2741–2755.

136. Rabanel, J.-M.; Hildgen, P.; Banquy, X., *J. Control. Release* **2014**, *185*, 71–87.

137. Dhayal, M.; Ratner, D. A., *Langmuir* **2009**, *25* (4), 2181–2187.

138. Dietrich, P. M.; Horlacher, T.; Girard-Lauriault, P.-L.; Gross, T.; Lippitz, A.; Min, H.; Wirth, T.; Castelli, R.; Seeberger, P. H.; Unger, W. E. S., *Langmuir* **2011**, *27* (8), 4808–4815.

139. Chernyy, S.; Jensen, B. E. B.; Shimizu, K.; Ceccato, M.; Pedersen, S. U.; Zelikin, A. N.; Daasbjerg, K.; Iruthayaraj, J., *J. Colloid Interf. Sci.* **2013**, *404*, 207–214.

140. Mendelsohn, R.; Mao, G.; Flach, C. R., *BBA-Biomembranes* **2010**, *1798* (4), 788–800.

141. Leone, G.; Consumi, M.; Lamponi, S.; Magnani, A., *Appl. Surf. Sci.* **2012**, *258* (17), 6302–6315.

142. Ramin, M. A.; Le Bourdon, G.; Daugey, N.; Bennetau, B.; Vellutini, L.; Buffeteau, T., *Langmuir* **2011**, *27* (10), 6076–6084.

143. Dinh, D. H.; Vellutini, L.; Bennetau, B.; Dejous, C.; Rebiere, D.; Pascal, E.; Moynet, D.; Belin, C.; Desbat, B.; Labrugere, C.; Pillot, J.-P., *Langmuir* **2009**, *25* (10), 5526–5535.

144. Pujari, S. P.; Spruijt, E.; Stuart, M. A. C.; van Rijn, C. J. M.; Paulusse, J. M. J.; Zuilhof, H., *Langmuir* **2012**, *28* (51), 17690–17700.

145. Barnes, T. J.; Kempson, I. M.; Prestidge, C. A., *Int. J. Pharm.* **2011**, *417* (1–2), 61–69.
146. Wei, Q.; Becherer, T.; Angioletti-Uberti, S.; Dzubiella, J.; Wischke, C.; Neffe, A. T.; Lendlein, A.; Ballauff, M.; Haag, R., *Angew. Chem. Int. Ed.* **2014**, *53* (31), 8004–8031.
147. Manova, R. K.; Claassen, F. W.; Nielen, M. W. F.; Zuilhof, H.; van Beek, T. A., *Chem. Commun.* **2013**, *49* (9), 922–924.
148. Garcia-Caurel, E.; De Martino, A.; Gaston, J.-P.; Yan, L., *Appl. Spectrosc.* **2013**, *67* (1), 1–21.
149. Lin, P. C.; Lin, S.; Wang, P. C.; Sridhar, R., *Biotechnol. Adv.* **2014**, *32* (4), 711–726.
150. Pillet, F.; Chopinet, L.; Formosa, C.; Dague, E., *BBA: Gen. Subj.* **2014**, *1840* (3), 1028–1050.
151. Wang, C.; Yadavalli, V. K., *Micron* **2014**, *60*, 5–17.
152. Kennedy, D. C.; Gruenstein, D.; Lai, C.-H.; Seeberger, P. H., *Chem.-Eur. J.* **2013**, *19* (12), 3794–3800.
153. Paulson, J. C.; Blixt, O.; Collins, B. E., *Nat. Chem. Biol.* **2006**, *2* (5), 238–248.
154. Zeng, X.; Andrade, C. A. S.; Oliveira, M. D. L.; Sun, X.-L., *Anal. Bioanal. Chem.* **2012**, *402* (10), 3161–3176.
155. Seto, H.; Kamba, S.; Kondo, T.; Hasegawa, M.; Nashima, S.; Ehara, Y.; Ogawa, Y.; Hoshino, Y.; Miura, Y., *ACS Appl. Mater. Interfaces* **2014**, *6* (15), 13234–13241.
156. Wang, Y.; Narain, R.; Liu, Y., *Langmuir* **2014**, *30* (25), 7377–7387.
157. Dechtrirat, D.; Gajovic-Eichelmann, N.; Wojcik, F.; Hartmann, L.; Bier, F. F.; Scheller, F. W., *Biosens. Bioelectron.* **2014**, *58*, 1–8.
158. Winzer, K.; Falconer, C.; Garber, N. C.; Diggle, S. P.; Camara, M.; Williams, P., *J. Bacteriol.* **2000**, *182* (22), 6401–6411.
159. Chemani, C.; Imberty, A.; de Bentzmann, S.; Pierre, M.; Wimmerova, M.; Guery, B. P.; Faure, K., *Infect. Immun.* **2009**, *77* (5), 2065–2075.
160. Adak, A. K.; Leonov, A. P.; Ding, N.; Thundimadathil, J.; Kularatne, S.; Low, P. S.; Wei, A., *Bioconjugate Chem.* **2010**, *21* (11), 2065–2075.
161. Gross, G.; Snel, J.; Boekhorst, J.; Smits, M. A.; Kleerebezem, M., *Benef. Microbes* **2010**, *1* (1), 61–66.
162. Ingham, C. J.; ter Maat, J.; de Vos, W. M., *Biotechnol. Adv.* **2012**, *30* (5), 1089–1099.
163. Ngundi, M. M.; Taitt, C. R.; McMurry, S. A.; Kahne, D.; Ligler, F. S., *Biosens. Bioelectron.* **2006**, *21* (7), 1195–1201.
164. Garcia-Hartjes, J.; Bernardi, S.; Weijers, C. A. G. M.; Wennekes, T.; Gilbert, M.; Sansone, F.; Casnati, A.; Zuilhof, H., *Org. Biomol. Chem.* **2013**, *11* (26), 4340–4349.
165. Arigi, E.; Blixt, O.; Buschard, K.; Clausen, H.; Levery, S. B., *Glycoconjugate J.* **2012**, *29* (1), 1–12.
166. Seo, J. H.; Lee, H. Y.; Cha, H. J., *Analyst* **2012**, *137* (12), 2860–2865.
167. Seo, J. H.; Kim, C. S.; Cha, H. J., *Analyst* **2013**, *138* (22), 6924–6929.
168. Pera, N. P.; Branderhorst, H. M.; Kooij, R.; Maierhofer, C.; van der Kaaden, M.; Liskamp, R. M. J.; Wittmann, V.; Ruijtenbeek, R.; Pieters, R. J., *ChemBioChem* **2010**, *11* (13), 1896–1904.
169. Chien, Y. Y.; Jan, M. D.; Adak, A. K.; Tzeng, H. C.; Lin, Y. P.; Chen, Y. J.; Wang, K. T.; Chen, C. T.; Chen, C. C.; Lin, C. C., *ChemBioChem* **2008**, *9* (7), 1100–1109.
170. Mori, T.; Ohtsuka, T.; Okahata, Y., *Langmuir* **2010**, *26* (17), 14118–14125.

171. Uzawa, H.; Ito, H.; Neri, P.; Mori, H.; Nishida, Y., *ChemBioChem* **2007**, *8* (17), 2117–2124.

172. Toyoshima, M.; Oura, T.; Fukuda, T.; Matsumoto, E.; Miura, Y., *Polym. J.* **2010**, *42* (2), 172–178.

173. Nagatsuka, T.; Uzawa, H.; Sato, K.; Kondo, S.; Izumi, M.; Yokoyama, K.; Ohsawa, I.; Seto, Y.; Neri, P.; Mori, H.; Nishida, Y.; Saito, M.; Tamiya, E., *ACS Appl. Mater. Interfaces* **2013**, *5* (10), 4173–4180.

174. Haseley, S. R., *Anal. Chim. Acta* **2002**, *457* (1), 39–45.

175. Bernardi, A.; Jimenez-Barbero, J.; Casnati, A.; De Castro, C.; Darbre, T.; Fieschi, F.; Finne, J.; Funken, H.; Jaeger, K. E.; Lahmann, M.; Lindhorst, T. K.; Marradi, M.; Messner, P.; Molinaro, A.; Murphy, P. V.; Nativi, C.; Oscarson, S.; Penades, S.; Peri, F.; Pieters, R. J.; Renaudet, O.; Reymond, J. L.; Richichi, B.; Rojo, J.; Sansone, F.; Schaffer, C.; Turnbull, W. B.; Velasco-Torrijos, T.; Vidal, S.; Vincent, S.; Wennekes, T.; Zuilhof, H.; Imberty, A., *Chem. Soc. Rev.* **2013**, *42* (11), 4709–4727.

176. Huskens, J., *Curr. Opin. Chem. Biol.* **2006**, *10* (6), 537–543.

177. Sato, Y.; Yoshioka, K.; Murakami, T.; Yoshimoto, S.; Niwa, O., *Langmuir* **2012**, *28* (3), 1846–1851.

178. Wehner, J. W.; Hartmann, M.; Lindhorst, T. K., *Carbohydr. Res.* **2013**, *371*, 22–31.

179. Nebhani, L.; Barner-Kowollik, C., *Adv. Mater.* **2009**, *21* (34), 3442–3468.

180. Xi, W.; Scott, T. F.; Kloxin, C. J.; Bowman, C. N., *Adv. Funct. Mater.* **2014**, *24* (18), 2572–2590.

181. Wang, L. X.; Davis, B. G., *Chem. Sci.* **2013**, *4* (9), 3381–3394.

182. Seeberger, P. H., *Chem. Soc. Rev.* **2008**, *37* (1), 19–28.

183. Wang, Z.; Chinoy, Z. S.; Ambre, S. G.; Peng, W.; McBride, R.; de Vries, R. P.; Glushka, J.; Paulson, J. C.; Boons, G.-J., *Science* **2013**, *341* (6144), 379–383.

184. Bennett, C. S., *Org. Biomol. Chem.* **2014**, *12* (11), 1686–1698.

2

PLASMONIC METHODS FOR THE STUDY OF CARBOHYDRATE INTERACTIONS

Sabine Szunerits and Rabah Boukherroub

Institute of Electronics, Microelectronics and Nanotechnology (IEMN), UMR 8520 CNRS, Lille 1 University, Avenue Poincaré – BP 60069, 59652 Villeneuve d'Ascq, France

2.1 INTRODUCTION

Advances in carbohydrate synthesis are beginning to spin off new biophysical methods to study glycoconjugates and their many roles in biology. Surface plasmon resonance (SPR) is one of these analytical tools as it allows probing the bioactivity of carbohydrate-related interactions. The first documented observation of surface plasmons dates back to 1902 when R.W. Wood reported unexplained narrow dark bands in the diffracted spectrum of metallic gratings illuminated with polychromatic light [1]. This anomalous phenomenon, termed SPR [2], was experimentally confirmed several years later by Otto [3] and Kretschmann [4]. However, it was only in 1983, through the seminal work by Liedberg, Nylander, and Lunström, that SPR was proposed as a new optical approach for sensing of affinity reactions [5] and made commercially available in 1990 by Biacore® (GE Healthcare). SPR spectroscopy is now an accepted bioanalytical technique for routine characterization of molecular recognition events at a solid interface. It has in particular proven to be a powerful approach for the determination of kinetic parameters (association and dissociation rate constants) as well as thermodynamic parameters, including affinity constants of interacting partners without the requirement of tagging one of the two partners. While initially SPR was mainly used to characterize protein–protein [6,7] and DNA–DNA interactions [8],

Carbohydrate Nanotechnology, First Edition. Edited by Keith J. Stine.

with the advent of new methods in synthetic carbohydrate chemistry [9], SPR became one of the routine techniques employed to study glycan-based interactions. Indeed, with half or more of all proteins containing some carbohydrate motives, the use of a bioanalytical tool allowing glycosignatures to be scrutinized rapidly is without doubt one of the principal driving forces behind the rapid development of SPR-based glyconomics.

Carbohydrates are however by nature more complicated substrates than oligonucleotides, and the development of carbohydrate microarrays requires several key challenges to be addressed. As carbohydrate-mediated recognition events commonly involve low-affinity interactions, carbohydrate interfaces must display the glycan in a fashion that is readily accessible to interrogating biomolecules [10]. A careful consideration of the glycan source and the way glycan molecules are anchored onto the surface is one of the first requirements for an SPR interface.

We begin this chapter with a brief discussion of some physical fundamentals of surface plasmons and their optical generation scheme. The chemistries for the construction of SPR glycochips and arrays as well as their performance to tackle carbohydrate-related questions will be presented. The chapter will be concluded with a discussion about recent advances on plasmonic-based sensing concepts and their contributions to the growing field of glyconomics.

2.2 PHYSICS OF SPR

2.2.1 SPR in Kretschmann Configuration: The Bases of Biacore

Although the basics of SPR theory are well documented in several books and reviews [11–13], a short description will be given here. Surface plasmons, also often known in the literature as surface plasmon polaritons (SPPs) or surface plasma waves (SPWs), are longitudinal charge density distributions generated on the metal interface (typically gold or silver) when light propagates through the metal. Such SPPS are transversal magnetic waves that propagate parallel along the interface between the metal and a dielectric layer (e.g., air, water, etc). According to the Drude model, the dispersion relation of an SPP, which essentially correlates the relationship between the wave vector along the interface and the angular frequency ω, can be described by

$$k_{SP} = \frac{\omega}{c}\left[\frac{\varepsilon_m \varepsilon_d}{\varepsilon_m + \varepsilon_d}\right]^{1/2} \tag{2.1}$$

where k_{SP} is the propagation constant of the surface plasmons, c is the speed of light in vacuum, and ε_m and ε_d are the dielectric constants of the metal and dielectric, respectively. This dispersion relation is sketched in Figure 2.1a, where k_x is the component along the x-axis of the incident light wave vector. One can see from the curves in Figure 2.1a that the wave vector of surface plasmons at a metal–dielectric

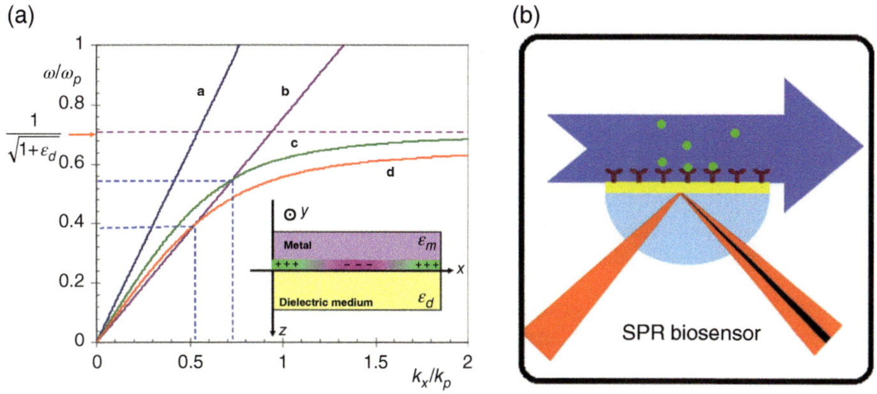

FIGURE 2.1 (a) Schematic illustration of surface plasmon dispersion relation in free space (line b) and inside the dielectric (line a). Curves c and d show the dispersion of SP at the interface between metal and dielectric. Line c corresponds to higher RI for the dielectric medium than case d. The frequency is normalized to the plasma frequency ω_p, and the x component of the light wave vector is normalized to the plasma wave number, which is defined in the text. The inset in the right bottom shows metal–dielectric interface. (b) Excitation of surface plasmons by the attenuated total reflection (ATR) method in the so-called Kretschmann configuration.

interface is by nature larger than that of the light wave in the dielectric. This means that the surface plasmons cannot be excited directly by a light wave incident on the interface.

To allow optical excitation of surface plasmons, the wave vector of the incident light must be increased. Indeed, if for a given ω, the magnitude of kx (or in case that $\varepsilon_{mr} \to -\varepsilon_d$) of the light incident at the metal–air boundary can be increased, the dispersion curves will intersect as the value of the plasmon frequency approaches $\omega_p / \sqrt{1+\varepsilon_d}$. Physically, this can be achieved by passing the light wave through an optically denser medium in the attenuated total reflection (ATR) method (Figure 2.1b). The glass prism has the effect of multiplying the wave vector by the refractive index of glass (n_{glass}). The sufficiently thin metal film allows some of the incident light to be transmitted through the metal–air boundary, where the surface plasmon is excited. Prism-based angular modulation with a conventional SPR structure consisting of a high refractive index glass prism and a thin metallic layer (typically around 50 nm of gold) is the most commonly used configuration in SPR sensing technology and illustrated in Figure 2.1b.

The excitation of surface plasmons is characterized by a dip in the angular spectrum of the reflected light at the angle of resonance (Figure 2.2a). The promising potential of SPR sensors lies in the very high sensitivity of the position and shape of the SPR dip to a change in the refractive index of the dielectric or formation of a nanoscale-thick film. SPR sensing thus relies on the principle that any change on the dielectric sensing surface will cause a shift in the angle of reflectivity in order to satisfy the resonance conditions (Figure 2.2b).

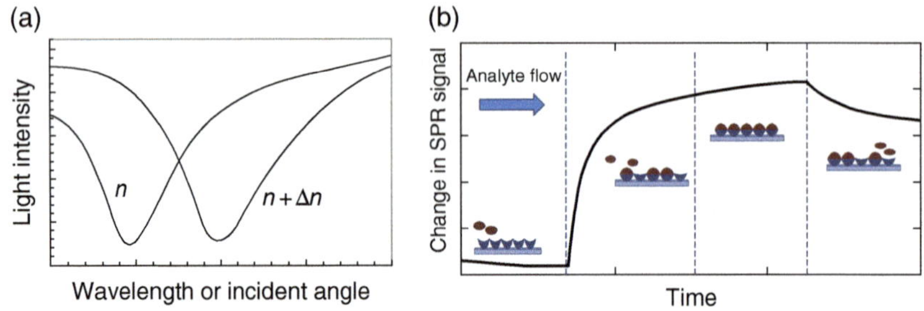

FIGURE 2.2 (a) Illustration of angular modulation as the refractive index of the sensed dielectric changes from n (solid line) to $n + \Delta n$ (dashed line). (b) SPR sensorgram angle change can be monitored as a function of time.

To make such changes specific to a certain analyte, suitable surface functionalization has to be performed on the SPR interface (Figure 2.1b). In the case of glyconomics, this implies linking of the glycan itself or its interaction species (protein, enzyme, bacteria, etc.). The first step of any SPR experiments is thus concerned with efficient surface immobilization of the respective ligand. Some of the most widely used immobilization strategies will be shown in section 3. In most cases, the glycan is immobilized to the SPR surface rather than the protein or bacteria. This is partly due to SPR-linked sensitivity issues. The quality of a SPR sensor such as sensitivity and selectivity is defined by the chemical and optical stability of the SPR chip as well as by the dip depth and width of the sensor signal [11,14,15]. The sensor's sensitivity S is defined as the ratio of the change of the SPR angle dY to the change of the refractive index of the sensed medium dn (RIU for refractive index unit):

$$S = \frac{dY}{dn} \tag{2.2}$$

To take into account the width of the resonance, sensor sensitivity is also often evaluated in terms of a figure of merit (FOM):

$$FOM = \frac{S}{FWHM} \tag{2.3}$$

where FWHM is the full width at half maximum of the resonance.

Occasionally, the expression "sensitivity" is used in place of limit of detection (LOD). While the sensor sensitivity is solely determined by the physical characteristics of the device and refers to the smallest change in refractive index of the sensed medium that produces a detectable change in the sensor output (determined by the noise in the sensor instrumentation), the LOD is in addition influenced by the size and mass of the analyte to be studied. Low molecular weight analytes (<1000 Da) are more difficult to quantify by SPR than larger ones [16]. The immobilization of the low molecular weight carbohydrates rather than lectins makes affinity studies more reliable. This is why the immobilization of carbohydrates has been of uttermost importance for the use of SPR in glyconomics. Since Mrksich's group showed specific binding of lectin concanavalin A (*Con A*) with certain surface-linked carbohydrates by SPR [17], many examples of carbohydrate-based SPR experiments have been reported and will be discussed in the next section.

2.2.2 Analysis of SPR Data

The rates of changes of the SPR signal can be analyzed to yield apparent rate constants for the association and dissociation phases of the reaction [18,19]. Practical considerations about the use of SPR technique for reliable determination of binding affinity and kinetics have been given in several articles [20]. A complete discussion is out of the scope of this article. In most cases of carbohydrate-based interactions, a classical biomolecular association model can be considered, and the equilibrium dissociation constant (K_D) is defined as

$$K_D = \frac{1}{K_A} = \frac{k_{diss}}{k_{ass}} = \frac{[L]*[A]}{[LA]} \tag{2.4}$$

where k_{diss} (s^{-1}) is the dissociation rate constant, k_{ass} ($l\ mol^{-1}\ s^{-1}$) is the association and rate constant, [L] is the ligand concentration (in our case mostly the glycan), [A] is the analyte concentration (in our case mostly the lectin, bacteria, etc.), [LA] is the concentration of the ligand–analyte complex, and K_A ($l\ mol^{-1}$) is the affinity constant. From the SPR sensorgrams (Fig. 2.3a), association and dissociation constants can be derived using linear transformation of the SPR sensorgram data according to Equation 4:

$$k_s = k_{ass}c + k_{diss} \tag{2.5}$$

where k_s is the slope of refractive unit change *d*RU/*dt* versus RU plot, *R* being the change in the SPR response [19,21].

A range of concentrations are analyzed, and a plot of k_s versus concentration of the lectin has a slope of k_{ass} and a *Y*-intercept of k_{diss} (Figure 2.3b). In the case of interaction of PNA lectin with lactose-modified interfaces [22], a $k_{ass} = 2.45 \times 10^3$ $M^{-1}\ s^{-1}$ and $k_{diss} = 1.09 \times 10^{-3}\ s^{-1}$ were determined, resulting in $K_D = 4.45 \times 10^{-7}$ M. Scatchard analysis (Fig. 2.3c) allows also determining the equilibrium dissociation constant found to be in this case $K_D = 5.8 \times 10^{-7}$ M.

The rate constants can also be calculated by nonlinear fitting of the sensorgram data (Fig. 2.4) according to

$$R_t = \frac{k_{ass} * R_{max} * c_A}{k_{ass} * c_A + k_{diss}}\left(1 - e^{-(k_{ass} * c_A + (k_{diss})t}\right) \tag{2.6a}$$

$$R_t = R_a e^{-(t-t_0)k_{diss}} \tag{2.6b}$$

where R_t is the SPR response at a given time *t*, R_{max} is the maximal response, c_A is the analyte concentration in solution ($mol\ l^{-1}$), and R_a is the binding response at the beginning of the dissociation.

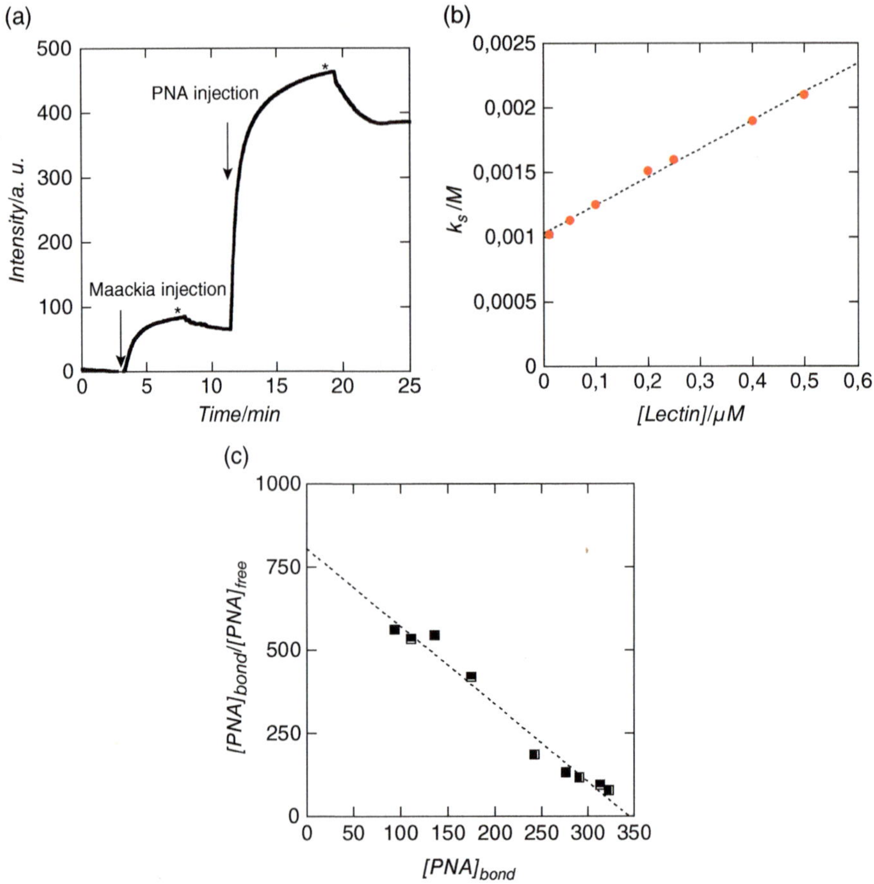

FIGURE 2.3 (a) Interaction on *PNA* lectin (8.3 μmol l^{-1}) with lactose-modified SPR interface. (b) Plot of k_s of PNA/lactose interaction versus lectin concentration. The values of k_s were obtained from analysis of SPR sensorgrams in (a). (c) Scatchard plots of lactose-modified SPR interfaces with increasing PNA lectin concentration [22].

Furthermore, measuring R_{eq}, the SPR binding response at steady state, as a function of analyte concentration, and fitting it to a Langmuir isotherm allows determining the equilibrium dissociation constant or the association constant (Fig. 2.3d):

$$R_{eq} = \frac{k_{ass} * R_{max} * c_A}{k_{ass} * c_A + k_{diss}} = \frac{R_{max} * c_A}{c_A + K_D} = \frac{R_{max} * c_A * K_A}{c_A * K_A + 1} \tag{2.7}$$

The range of kinetic constants that can be measured with precision with SPR are 10^3–10^7 M s^{-1} for the on-rate (k_{ass}) and 10^{-5} to 10^{-1} s^{-1} for the off-rate (k_{diss}).

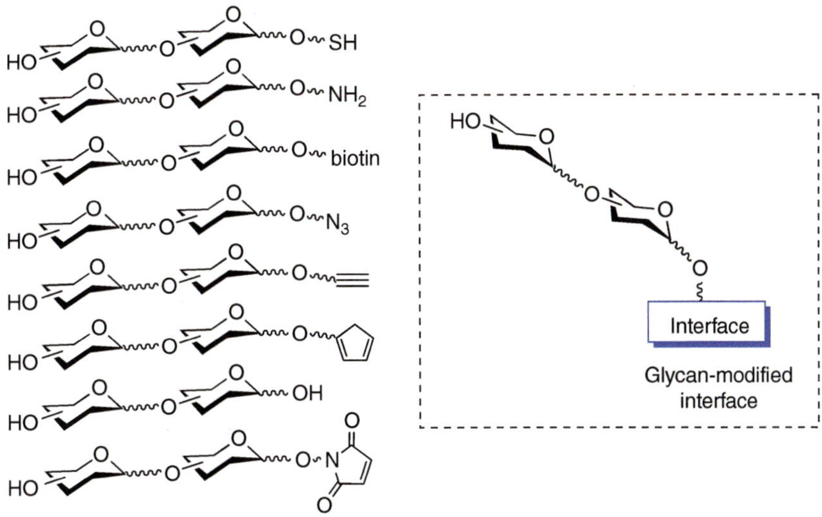

FIGURE 2.4 Formation of glycan interfaces and arrays through the use of differently functionalized and nonfunctionalized glycans.

Finally, SPR also allows determining the amount of immobilized ligand and/or bound analyte g, assuming that 1000 RU correspond to the binding of 1 ng mm^{-2} [13]:

$$g = \sqrt{\frac{10^{12} * 1000 * 10^{9} * \mathrm{MW}}{N_{\mathrm{A}} \Delta \mathrm{RU}}} \tag{2.8}$$

with MW being the molecular weight of the analyte, ΔRU the change of the SPR signal upon binding of analyte to surface, and N_A the Avogadro number ($6.023*10^{23}$ molecules mol^{-1}).

2.3 SENSOR SURFACE AND IMMOBILIZATION CHEMISTRY

One of the first steps in any SPR-based protocol concerns the way the receptor molecules are anchored onto the chip surface. In the case of glycan receptors, the sugars must be displayed at the reducing end for successful recognition. As the many hydroxyl groups may be difficult to differentiate, integration of linker systems to the anomeric center of glycans is of high importance but a challenging task. Over the years, a variety of different surface linking strategies have been developed; Figure 2.3 shows some of them. In the drive to pioneer the first-generation carbohydrate interfaces and microarrays, many contributors deserve mentioning. To date, no particular fabrication method has emerged as the clear winner, and new methods to link carbohydrates are actively being developed. In the case of SPR chips, the class of glycans used for immobilization is still largely based on thiolated glycans. "Click" chemistry approaches are one of the newer trends in the covalent linking of glycans [23–25]. The direct photochemical linking of unmodified carbohydrates has lately attracted much attention, as such a strategy allows the direct surface immobilization of any carbohydrate structure without the need of premodification [23,26–28]. The surface modification routes are outlined in the following.

2.3.1 Dextran-Modified SPR Interfaces for Linking Glycans: Biacore Chip

The development of a commercial SPR instrument by Biacore at the beginning of the 1990s made this method accessible for industry and research institutes. Introduced by Lofas and Johnsoon [29], Biacore proposed at the same time the use of an SPR interface modified with an activated carboxymethylated dextran (CM-dextran) matrix for the linking of ligands (Fig. 2.5a). Dextrans are hydrophilic, noncharged neutral polymeric carbohydrates, soluble in water in any proportion forming highly hydrated hydrogels. Due to these properties, dextrans display very low nonspecific interactions with bioreceptors. Owing to the high concentration of hydroxyl groups in the dextran matrix, chemical modification is possible without significantly affecting its hydrophilicity. The essentially nonbranched polymer chains are highly flexible, and ligands immobilized in dextran matrices are thus well accessible. The matrix is constructed by self-assembly of 1,ω-hydroxyalkythiol (16-mercaptohexadecane-1-ol) onto gold, followed by covalently linking of the dextran polymer by activation of the hydroxyl groups with epichlorohydrin under basic condition to yield epoxides, which by further reaction with bromoacetic acid results in the formation of surface-linked CM-dextran films (Fig. 2.5a) [6]. The thickness of the polymer matrix is about 100 nm, and the carboxylic acid groups can be further used to covalently link amine-terminated or thiol-terminated bioreceptors. The references that can be cited using this type of substrate, known as CM5 chip, are countless. It is also one of the first SPR interfaces used for the study of the interaction between surface-linked oligosaccharides and lectins [18,21,30]. Glycopeptides or lectins were linked to the carboxylic acid groups by first activation with *N*-hydroxysuccinimide, followed by carbodiimide-catalyzed amide formation between the NH_2 groups of the glycoprotein and the activated ester groups of the CM5 chip (Fig. 2.5b). One of the more recent examples on the use of the CM5 chip for glycobiology was for the study of monomeric gp120 and trimeric gp140 glycoproteins [31]. The presence of gp120/gp41 in the envelope of HIV-1 is of crucial importance for viral infectivity and transmission, as the gp120/gp41 envelope is densely glycosylated (up to 50% of the mass of gp120), and a variety of carbohydrate-binding agents (CBA) are used to block HIV entry into target cells. Despite the fact that the HIV envelope appears as a trimeric complex, binding studies were exclusively performed with the monomeric gp120 glycoproteins. Hoorelbeke *et al.* studied in more detail the impact of the oligomeric state of the gp120 envelope glycoprotein on ligand binding by SPR [31]. Surprisingly, it was found that CBA such as Griffithsin, mannose-specific lectins, sCD4, and others display comparable kinetic properties to monomeric gp120 and trimeric gp140, confirming that monomeric gp120 is a good surrogate molecule for native HIV-1 envelope trimer. The same group used an SPR-based competition assay to study the role of

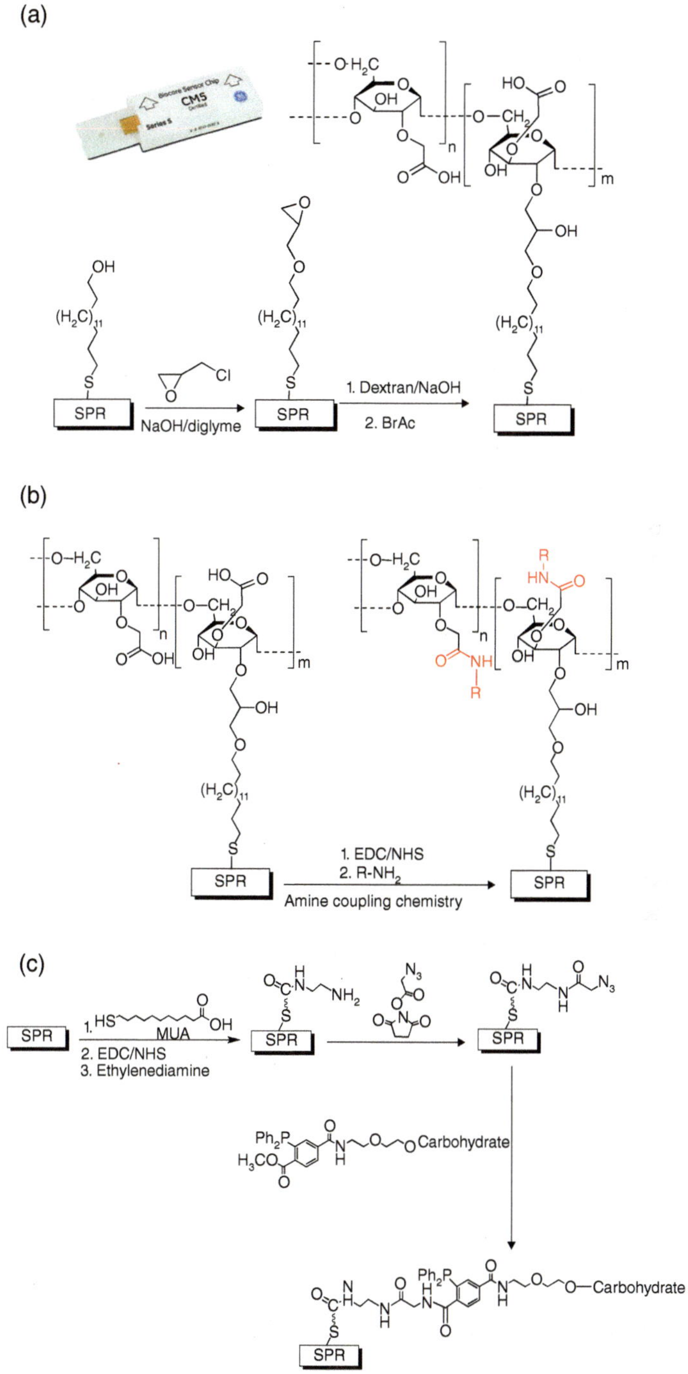

FIGURE 2.5 (a) Formation of surface-linked carboxymethylated dextran polymer films on gold (Biacore CM5 chip). (b) Amine-coupling chemistry. (c) Conjugation of carbohydrate epitopes to a CM5 interface using Staudinger ligation chemistry [33].

carbohydrate-binding sites of Griffithsin in the prevention of DC-SIGN-mediated capture and transmission of HIV-1 by immobilization of recombinant HIV-1(III)g gp120 and HIV-1(ADA) gp120 on a CM5 chip [32]. More recently, carbohydrate epitopes were immobilized on CM5 SPR interface using Staudinger ligation chemistry (Fig. 2.5c) [33]. This was accomplished by first introducing azide functionalities onto the CM5 interface, followed by reaction with phosphine-modified carbohydrate ligands. The advantage of this approach is that the chemistry employed is extremely mild and can be easily adapted to any biosystem.

2.3.2 Self-Assembled Monolayers Based on Thiolated Glycans

The other linking system widely used for the attachment of carbohydrates to a surface is based on the formation of alkanethiolate monolayers on a gold surface (Fig. 2.6a) [34,35]. Such gold–sulfur (Au–S) bonds are readily formed at nearly neutral pH and moderate temperature, enabling in an easy and fast manner the direct attachment of carbohydrates containing a thiol moiety either directly at the anomeric center or attached via a pendant "spacer" group [36]. The gold–thiol reaction may be regarded as an oxidative addition of the S–H bond to the surface, followed by a reductive elimination of hydrogen [35]. The S–Au bond is strong with a homolytic bond strength of ≈40 kcal mol^{-1} [35]. Two general strategies have been used for producing sugar SAMs on gold: (i) separate synthesis of the sugar derivative with a pendant alkanethiol group and (ii) direct chemical transformation of the functional self-assembled monolayer (SAM).

2.3.2.1 Separate Synthesis of Thiolated Sugars One of the first examples on the fabrication of glycan-modified interfaces is based on the direct immobilization of thiol-containing mannopyranoside and its interaction with *Con A* [37]. Exposure of the mannose-derived SAM to a carbohydrate-binding protein such as *Con A* led to a specific binding interaction as measured by SPR (Fig. 2.6b). In contrast, upon exposure of the mannose SAM to the L-fucose-specific lectin *Tetragonolobus purpureas*, no such binding was observed. These results suggested that highly ordered SAMs of specifically designed carbohydrate derivatives can be formulated to mimic natural cell surface structures and can be used to study selective molecular recognition interactions.

The possibility to form single-component as well as mixed SAMs allowing to control the surface-linked glycan density has made SAMs of particular interest for the preparation of carbohydrate interfaces. The effect of surface glycan density on carbohydrate–protein interactions was systematically studied by Dhayal and Ratner through the immobilization of thiolated trimannoside using oligoethylene glycol (OEG) of different chain lengths as diluting compound (Fig. 2.6a) [10]. Mannose-specific *Con A* was used to optimize the binding reaction with carbohydrate microarrays of different glycan densities. Figure 2.6f shows the SPR array of mixed glycan/OEG SAMs with varying concentrations of glycan and the corresponding SPR response after interaction with *Con A*. Proteins RNase A and B served as negative and positive controls for the *Con A* binding due to the presence of a large

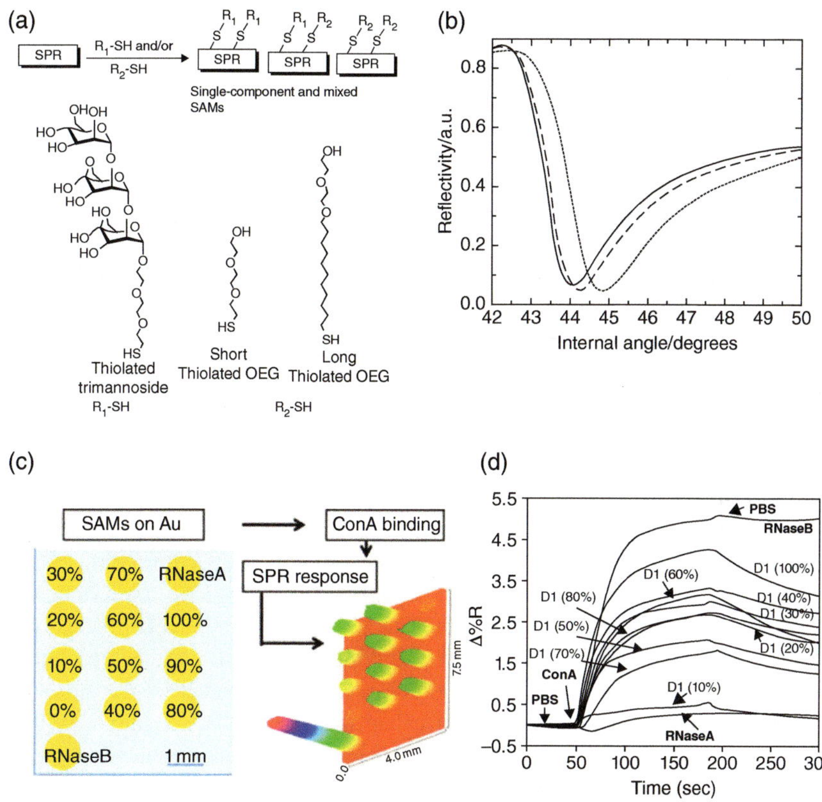

FIGURE 2.6 (a) Modification of gold-based SPR chips with thiolated glycans: formation of trimannoside-modified SPR chip using long and short OEG as diluting compounds. (b) Reflectivity curves for a gold SPR interface (a), after modification with thiolated mannopyranoside (b), followed by interaction with Con A (c). *Source*: Reproduced from Ref. 37 with permission from the American Chemical Society. (c) Image of glycan array prepared by spotting thiolated molecules onto gold and varying the percentage of trimannoside together with SPRi response after incubation with Con A. *Source*: Reproduced from Ref. 37 with permission from the American Chemical Society. (d) SPR sensorgram of Con A binding to a density profile of trimannoside in mixed SAMS with short OEG thiol. *Source*: Reproduced from Ref. 10 with permission from the American Chemical Society.

asparagine-linked glycan containing mannose residues in RNase B, which was completely absent in RNase A. Quantitative analysis of *Con A* binding to different surface densities of trimannoside-mixed SAMs with short OEG showed a markedly different profile from that of *Con A* binding to mixed sugar/OEG SAMs using the longer alkanethiol. In the case of long OEG, *Con A* binding is nearly nonexistent when the glycan concentration was less than 60%. In the case of short OEG, the relative binding of *Con A* rises significantly with increasing concentration of trimannoside from 0 to 40%. Beyond 40–60%, there is a plateau and slight decrease in overall *Con A* binding (Fig. 2.6d). At high surface densities, the immobilized glycan appears to become less readily bound by *Con A*. A number of explanations are possible, including steric hindrance at the surface by neighboring carbohydrates and carbohydrate–carbohydrate interactions between surface-bound glycans as well as multivalency effects.

2.3.2.2 Multistep Transformation of Functional SAMs The drawback of the separate synthesis of the thiolated sugar derivative arises from the tremendous synthetic effort of sugars anchored with a pendant alkanethiol, especially for more complex glycans. Furthermore, as the complexity of tethered bioreceptors increases, there is no guarantee that the molecules will pack to form a structurally well-defined monolayer. The chemical transformation of SAM of functional alkane thiols is thus a widely used approach. This method introduces sugar moieties onto the prefunctionalized SAMs using selected chemical reactions. Obviously, the identification of a good reaction is the critical determinant for the fabrication of sugar SAMs. Several interphase reactions can be identified for anchoring sugar units, including Diels–Alder reaction [24,17], disulfide exchange [38], "click chemistry," and photochemical linking of unmodified glycans [26,28] (Fig. 2.7).

The interest of a surface-anchored disulfide linkage for the immobilization of thiol-modified glycans was shown by Corn *et al.* (Fig. 2.7b) [38]. First, a densely packed SAMs of an amine-terminated alkanethiol (11-mercaptoundecylamine (MUAM)) was generated, which was treated with a heterobifunctional cross-linker (S-acetylthiopropionate) containing an amine-reactive succinimide ester and a protected thiol moiety. Reaction of the ester group with the primary amine on MUAM results in the formation of an amide bond to yield a monolayer with masked thiol groups. The thioester can be transformed to a free sulfhydryl surface with an effective nucleophile such as hydroxylamine. Activation of the surface sulfhydryl groups with 2,2'-dipyridyldisulfide followed by exchange reaction of the sulfhydroxyl-containing carbohydrate derivatives results in the glycan anchoring onto the SPR chip. Exposure of the surface to DTT results in the cleavage of the disulfide linkage and in the regeneration of the sulfhydryl surface. The mannose- and galactose-modified SPR arrays (Fig. 2.8a) were used to monitor the binding of two lectins (*Jacalin* and *Con A*) (Fig. 2.8b). This approach also allows the formation of mixed carbohydrate monolayers, composed of a mixture of mannose and galactose ligands. The binding of *Con A* or *Jacalin* to the mixed surfaces could be detected by SPR imaging (Fig. 2.8c). In both cases, a plateau in protein binding is seen, indicating that there is a density of immobilized carbohydrate ligands beyond which the addition of more binding epitopes does not increase the adsorption of protein to the surface. At high ligand densities, not all of the glycans residues are active (Fig. 2.8d). Only near 50% mannose substitution are the mannose residues are close enough on the surface such that on average an absorbed *Con A* can interact with two mannose residues.

The Cu(I)-catalyzed 1,3-dipolar Huisgen cycloaddition (CuAAC) of terminal azide to terminal alkyne to form triazole, termed "click" chemistry, has been established as an important tool for the formation of surface-immobilized glycans (Fig. 2.7c) [23–25]. The highly selective reaction between azide and alkyne chemical moieties that are both rare and generally inert in biological systems, forming a strong

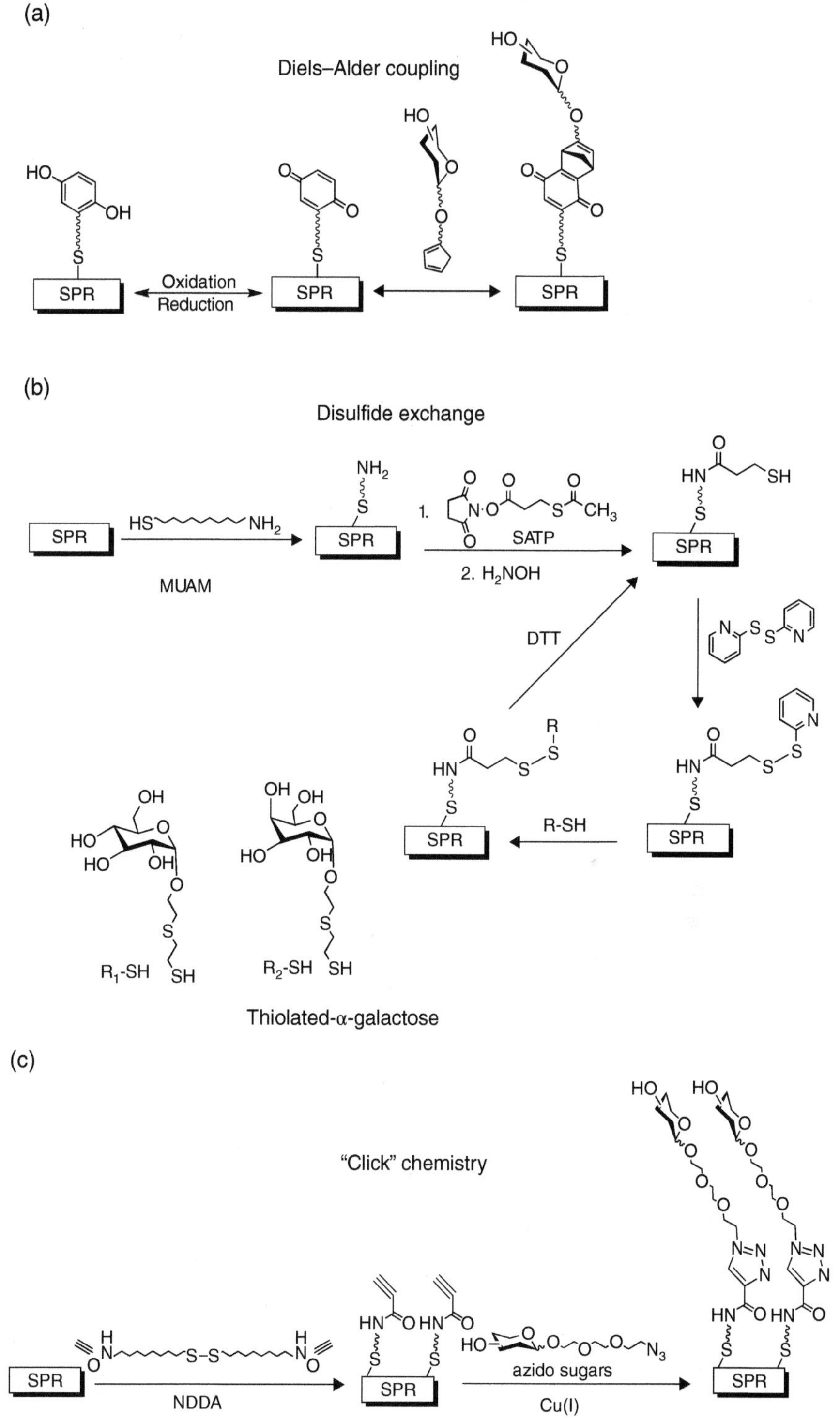

FIGURE 2.7 Different strategies for the preparation of prefunctionalized SPR interfaces and subsequent linking of glycans: (a) Diels–Alder coupling, (b) disulfide exchange, (c) "click" chemistry, and (d) photoactivation of perfluorophenylazide group.

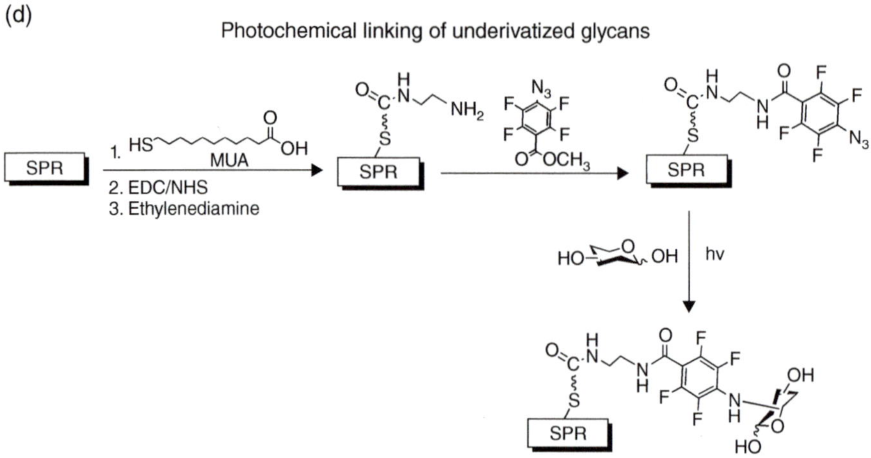

FIGURE 2.7 (*Continued*)

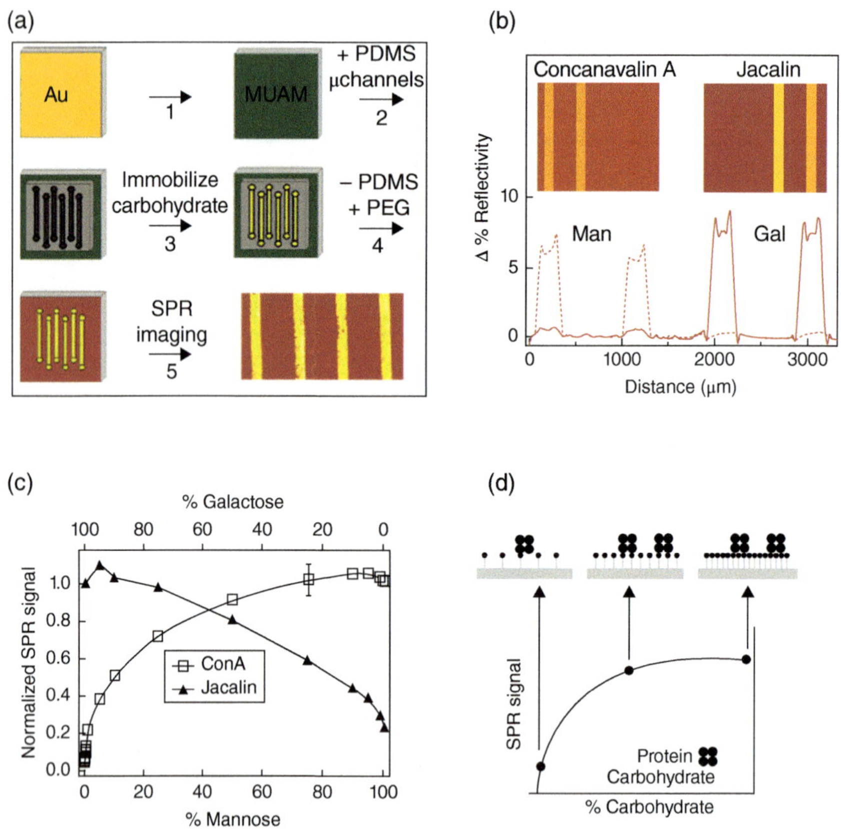

FIGURE 2.8 (a) Schematic representation of the 5-step process used to fabricate glycan arrays on gold using modification steps outlined in Figure 2.7b. (b) Line profile taken from SPR difference images of a two-component glycan array containing mannose (left) and galactose (right). (c) Normalized SPR signal obtained from *Con A* and *Jacalin* binding to mixed mannose/galactose surfaces. (d) Schematic model of the influence of ligand density on surface on binding amount of lectin. *Source*: Reproduced from Ref. 38 with permission from the American Chemical Society.

nonhydrolyzable triazole linkage between the two molecules, makes this method of particular interest. Azides are rather easy to incorporate synthetically in organic compounds as they are used as precursors to obtain amine-functionalized molecules, a feature that has resulted in large libraries of available carbohydrate azides. Among various options, Zhang *et al.* premodified gold with a long-chain alkyl disulfide linker carrying terminal alkyne groups as template for the fabrication of the carbohydrate SAM (Fig. 2.7c) [25]. The formation of mixed monolayers is possible using "click" chemistry of a mixture of azide-functionalized ligands. To eliminate nonspecific adsorption, azidotris(ethylene oxide)glycol can be, for example, integrated into the glycan monolayer.

All the discussed glycan interfaces require pretreatment of the carbohydrate. A direct immobilization strategy was recently proposed (Fig. 2.7d) [26,28]. This approach relies on the use of perfluorophenylazide (PFPA) as a photocoupling agent. PFPA are highly efficient photoaffinity agents with significant enhanced C–H and N–H insertion yields. PFPA derivatives can be readily prepared from commercially available starting materials and are stable when protected from direct or prolonged light exposure. This approach is also adaptable for the formation of carbohydrate arrays on SPR chips. Aqueous carbohydrate solutions can be spotted on PFPA-functionalized chips using a microprinter system forming spots of 250 μm [28]. After the spots were dried, the chip was exposed for 5 min at 365 nm with a power intensity of 8 mW cm^{-2} for photoimmobilization. The most striking advantage of this method is that ketoses and even alditols such as sucrose and mannitol can be readily conjugated to the PFPA-functionalized surfaces [26]. In addition, small molecules such as glucose, fructose, and lactose can be immobilized, which is not possible by the physical adsorption method. Because of the nonspecific reaction of PFPA, the possibility exists that the binding sites of the carbohydrate may be altered after the azide insertion and thus impact their interaction with the binding partners. It might be argued that such a coupling strategy might result in "uneffective" glycan interfaces. Indeed, the high reactivity of the perfluorophenyl nitrene intermediate, generated through UV irradiation of the azide, is known to give a wide range of potential reaction products, although it is the C–H insertion pathway that is reported to be preferred [39]. To investigate whether the immobilized ligands retain their binding affinity and selectivity, SPR chips spotted with *PNA*-binding lactose and galactose and *Con A*-binding carbohydrates (mannose, glucose) were incubated with both lectins. The immobilized galactose and lactose showed clear images after reaction with *PNA*, confirming that the photocoupling chemistry preserves the affinity difference of the immobilized carbohydrates.

2.3.3 Biotin–Streptavidin-Based Linker

The initial efforts to understand glycan–lectin binding used the biotin–streptavidin interaction to display oligosaccharides on surfaces as biotin–avidin chemistry allows for strong binding ($K_A = 10^{15}$ M^{-1}) (Fig. 2.9a). However, reports on the use of biotinylated glycans are not numerous [40,41] [42,43]. Linman *et al.* [40] synthesized biotinylated sialosides containing a flexible hexaethylene glycol spacer to closely mimic sialoside–lectin interactions (Fig. 2.9a). The SPR chip was first modified with biotin BSA and NeutrAvidin to allow the linking of the sugar. The specificity and sensitivity of lectin binding were characterized using *Sambucus nigra agglutinin* (SNA) and other lectins, including *Maackia amurensis* (MA), *Con A*, and *wheat germ agglutinin* (WGA). The results indicate that α–2,6-linked sialoside exhibits high binding affinity to SNA, while alteration in sialyl linkage and terminal sialic acid structure compromises the affinity by a varied degree. Multiple experiments could be performed on the same substrate using a glycine stripping buffer, which selectively regenerated the surface without damaging the sialoside.

Field and coworkers used SPRi of glycan arrays for the identification of carbohydrate-based ligands for potential ricin sensor development [43]. SPRi analysis of galactose-specific plant lectin *Ricinus communis* agglutinin 120 (RCA120) (a surrogate for the bioterrorism agent ricin) interaction with a 40 compound glycoarray highlighted the potential of 6-modified D-galactoside to serve as an efficient ligand.

2.3.4 Polypyrrole Linker

A convenient way to modify the surface of conducting materials such as gold is through electrodeposition of thin oligomeric or polymeric films such as polypyrrole from monomer solutions (Fig. 2.10a) [22,44]. The use and interest of electropolymerization for anchoring bioreceptors such as oligosaccharides onto biosensor interfaces are driven mainly by the following factors: (i) the polymer films are uniform, (ii) their thickness can be readily controlled, and (iii) the surface modification is limited to the

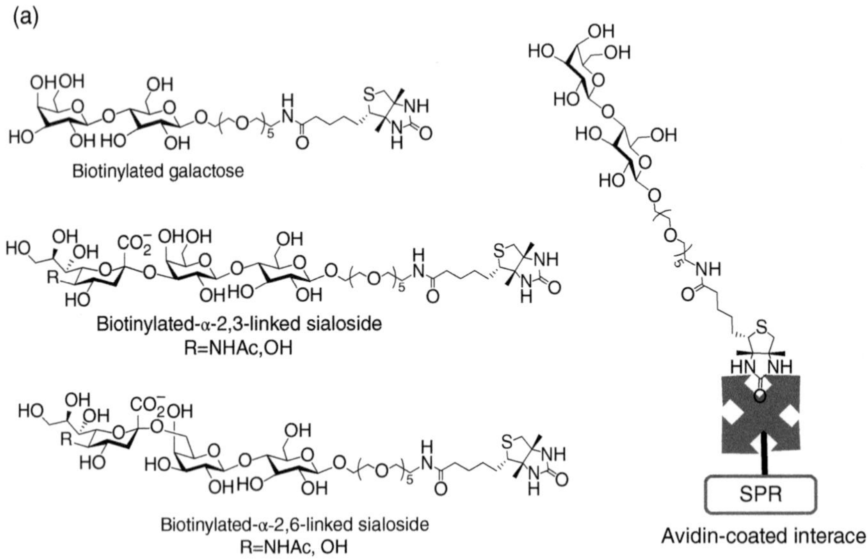

FIGURE 2.9 (a) Biotinylated glycans immobilized on streptavidin/NeutrAvidin-coated SPR surfaces. (b) SPR imaging responses for equilibrium binding of RCA120 to an array of 40 biotinylated natural glycans together with library of biotinylated glycans used. *Source*: Reproduced from Ref. 43 with permission from the Royal Society of Chemistry.

(b)

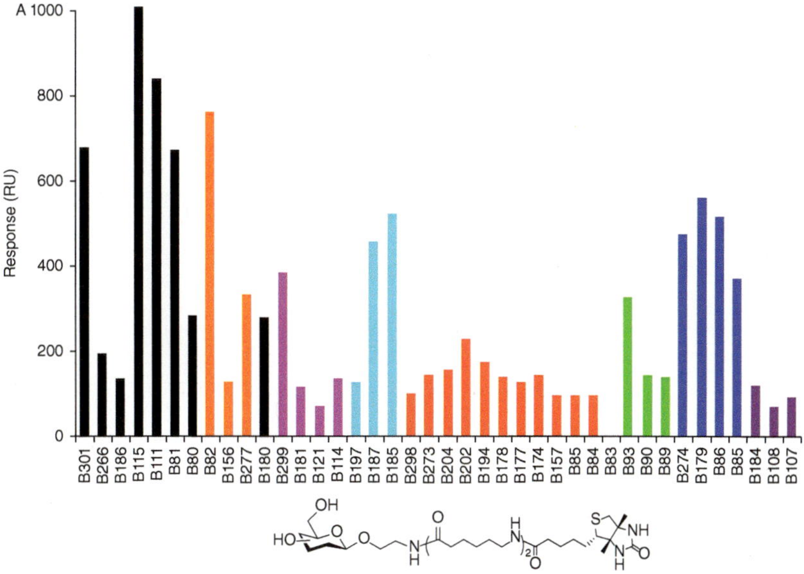

Group name	Glycotope structure	Common name
B301	Galβ1-4GlcNAcβ1-3Galβ1-3GlcNAc	L-NLe^{c}
B266	Galβ1-4[Fucα1-3]GlcNAc	Le^{x}
B186	Galβ1-4(6OSO3)Glc	6SuLac
B115	Galβ1-4GlcNAcβ1-3Galβ1-4Glc	LNnT
B111	(Galβ1-4GlcNAcβ)$_2$	Di-LN
B81	Galβ1-4GlcNAc	LN
B80	Galβ1-4Glc	Lac
B82	Galβ1-3GlcNAc	Le^{c}
B156	Galβ1-3[Fucα1-4]GlcNAc	Le^{a}
B277	Galβ1-3GlcNAcβ1-3Galβ1-4GlcNAc	Le^{c}-LN
B180	GalNAcβ1-4GlcNAc	LDN
B299	GlcNAcβ1-3Galβ1-3GlcNAc	GN-Le^{c}
B181	GlcNAcβ1-3Glaβ1-4GlcNAc	GN-LN
B121	GlcNAc	GN
B114	GlcNAcβ1-3Galβ1-4Glc	LNT$_2$
B197	[3OSO3]Galβ1-4Glc	3′SuLac
B187	[3OSO3]Galβ1-4(6OSO3)Glc	6,3′-di-SuLac
B185	(6OSO3)Galβ1-4Glc	6′SuLac
B298	Neu5Acα2-3[Neu5Acα2-3Galβ1-3GalNAcβ1-4]Galβ1-4Glc	GD$_4$
B273	Neu5Acα2-3GalNAcβ1-4GlcNAc	3′SLDN
B204	Neu5Acα2-3[Galβ1-3GalNAcβ1-4]Galβ1-4GlcNAc	GM$_2$(NAC)
B202	Neu5Acα2-3[Galβ1-3GlcNAcβ1-4]Galβ1-4Glc	GM$_1$
B194	Neu5Acα2-3[Galβ1-4GlcNAcβ1-3]$_3$	3′SLN-LN-LN
B178	Neu5Acα2-3[Galβ1-4GlcNAcβ]$_2$	3′-SLN-LN
B177	Neu5Acα2-3[GalNAcβ1-4]Galβ1-4Glc	GM$_2$
B174	Neu5Acα2-3Galβ1-3[Fucα1-4]GlcNAc	SLe^{2}
B157	Neu5Acα2-3Galβ1-4[Fucα1-3]GlcNAc	SLe^{a}
B85	Neu5Acα2-3Galβ1-3GlcNAc	3′Se^{c}
B84	Neu5Acα2-3Galβ1-4GlcNAc	3′SLN
B83	Neu5Acα2-3Galβ1-4Glc	3′SLac
B93	Neu5Acα2-6Galβ1-4GlcNAc	6′S(Gc)LN
B90	Neu5Acα2-3Galβ1-4GlcNAc	3′S(Gc)LN
B89	Neu5Acα2-3Galβ1-4Glc	3′S(Gc)Lac
B274	Neu5Acα2-6GalNAcβ1-4GlcNAc	6′SLDN
B179	Neu5Acα2-6[Galβ1-4GlcNAcβ]$_2$	6′-SLN-LN
B87	Neu5Acα2-6Galβ1-4GlcNAc	6′SLN
B86	Neu5Acα2-6Galβ1-4Glc	6′SLac
B184	Neu5Acα2-8Neu5Acα2-3[GalNAcβ1-4]Glaβ1-4Glc	GD$_2$
B108	Neu5Acα2-8Neu5Acα2-8Neu5Acα2-3Glaβ1-4Glc	GT$_3$
B107	Neu5Acα2-8Neu5Acα2-3Galβ1-4Glc	GD$_3$

FIGURE 2.9 (*Continued*)

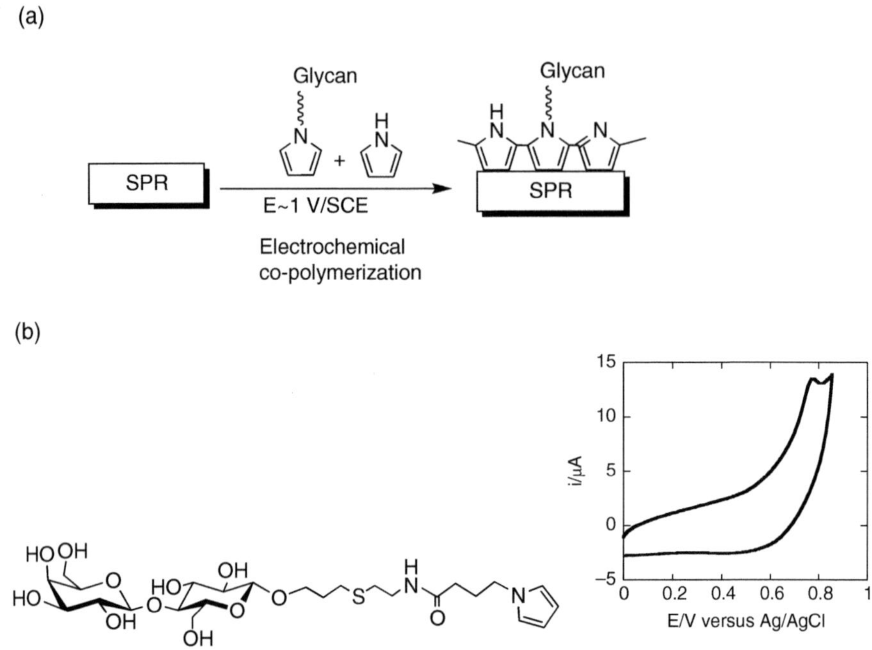

FIGURE 2.10 (a) Principle of the preparation of glycan chip through electropolymerization of pyrrole units. (b) Structure of pyrrole-lactosyl and cyclic voltammogram of pyrrole-lactosyl (3 mM) in $LiClO_4$ (0.1 M/water); $v = 100\,mV\,s^{-V}$, $A = 0.21\,cm^2$.

surface of the electrode. In addition, electrochemistry can be easily integrated with SPR measurements [22,45]. Both methods are compatible in the sense that they both rely on a conductive substrate: for electrochemistry, the gold film of the SPR interface is the working electrode, without disturbing its use for the generation of surface plasmons. Gondran *et al.* used lactose and 3′-sialyllactose-modified pyrrole for the generation of sensors specific to *Arachis hypogaea* (PNA) and MA agglutinins [22]. The electrochemical behavior of pyrrole-lactosyl is depicted in Figure 2.10b: upon oxidative scanning up to 0.85 V versus Ag/AgCl, the cyclic voltammogram of the pyrrole derivative displays an irreversible anodic peak at 0.77 V, which is attributed to the oxidation of the pyrrole unit leading to highly reactive radical cations.

2.3.5 Glass-Based SPR Interfaces: Lamellar Structures

While the surface chemistry developed on gold has been of great value [34,35], the limitations of working on gold are becoming more noticeable with increasingly complex fabrication requirements for biometric systems and arrays. Alternative routes and improvements were thus sought after. Silicon dioxide-based materials such as glass (silicate) are standard materials for biosensing being inexpensive and benefiting from a rich variety of well-developed attachment schemes based on silane-coupling chemistry. For many years, the use of such lamellar structured SPR chips (Fig. 2.11a)

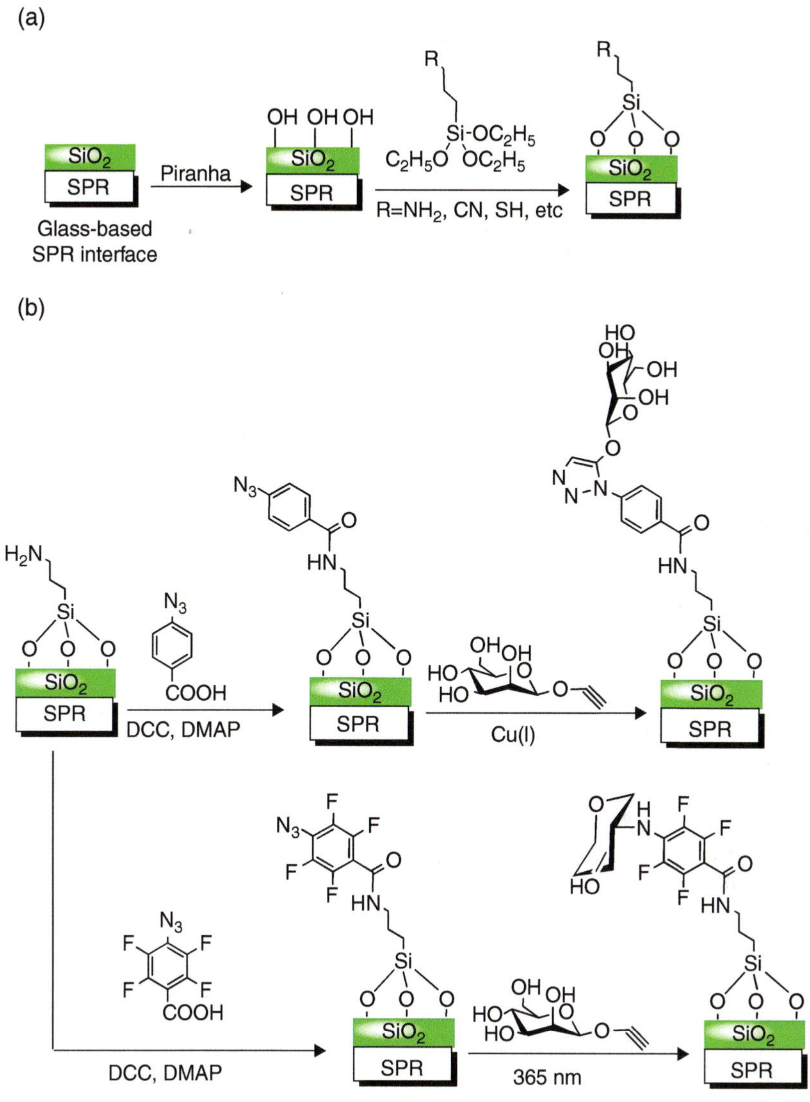

FIGURE 2.11 (a) Glass-based lamellar SPR interface and its modification using silane chemistry. (b) Formation of glycans interfaces using "click" chemistry or photocoupling. (c) SPR binding curves of *Lens culinaris* ($10\,\mu g\,ml^{-1}$) to mannose-modified SPR interfaces formed through UV irradiation of perfluorophenylazide-modified interface (upper curve) and linked to phenylazide via "click" chemistry (lower curve) [23].

for biosensing applications was hampered as the gold/SiO_x interface was in most cases not stable upon immersion in water and/or PBS buffer. These hurdles are overcome nowadays by using sol–gel-based approaches [46] and plasma-enhanced chemical vapor deposition (PECVD) based on the decomposition of a mixture of silane gas

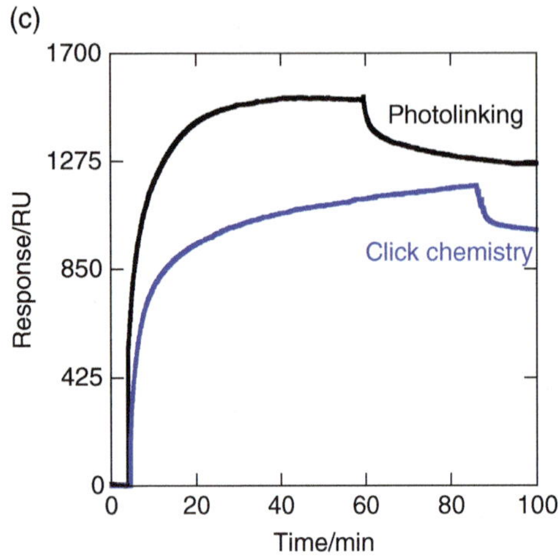

FIGURE 2.11 (*Continued*)

(SiH_4) and nitrous oxide (N_2O) near the substrate surface, enhanced by the use of a vapor containing electrically charged particles or plasma at 300°C [47,48]. The classical strategy for adding functionality to such interfaces is based on the reaction of the silicon oxide layer with functional silanes (Fig. 2.11a) [48].

These lamellar SPR interfaces were also used for the construction of glycan interfaces (Fig. 2.11b). Reaction of the silicon oxide layer with 3-aminopropyltrimethoxysilane results in the introduction of amine terminal groups; 4-azidobenzoic acid was linked through amide bond formation [23]. Mannose units were integrated onto this surface via "click" chemistry and photoinduced (irradiation at 365 nm) attachment of unmodified mannose. It was found that both modification methods give identical amounts of mannose on the respective interfaces, enabling the direct comparison of the SPR signal upon lectin binding. An immediate change in the SPR signal is observed upon injection of the mannopyranoside specific *Lens culinaris* (LC) lectin with both mannose-decorated surfaces (Fig. 2.11c). The random photoimmobilization seems to be thus less random than expected, having a strong tendency to take place preferentially at the reducing ends [23].

2.3.6 Graphene-Based SPR Interfaces

More recently, thin graphene layers are considered as attractive coatings for SPR interfaces [49]. Graphene has a very high surface-to-volume ratio, which is expected to be beneficial for efficient adsorption of biomolecules as compared to gold. The presence of graphene on the top of silver passivates the surface against oxidation, making possible the use of silver as plasmon active interface rather than gold. It is also hoped that graphene has a chemical effect, increasing the adsorption of molecules *via* π–π stacking and/or H-bonding interactions. While for a long time the formation of reproducible graphene-based SPR interfaces was a considerable hurdle to overcome, chemical and/or mechanical transfer of CVD-grown graphene sheets

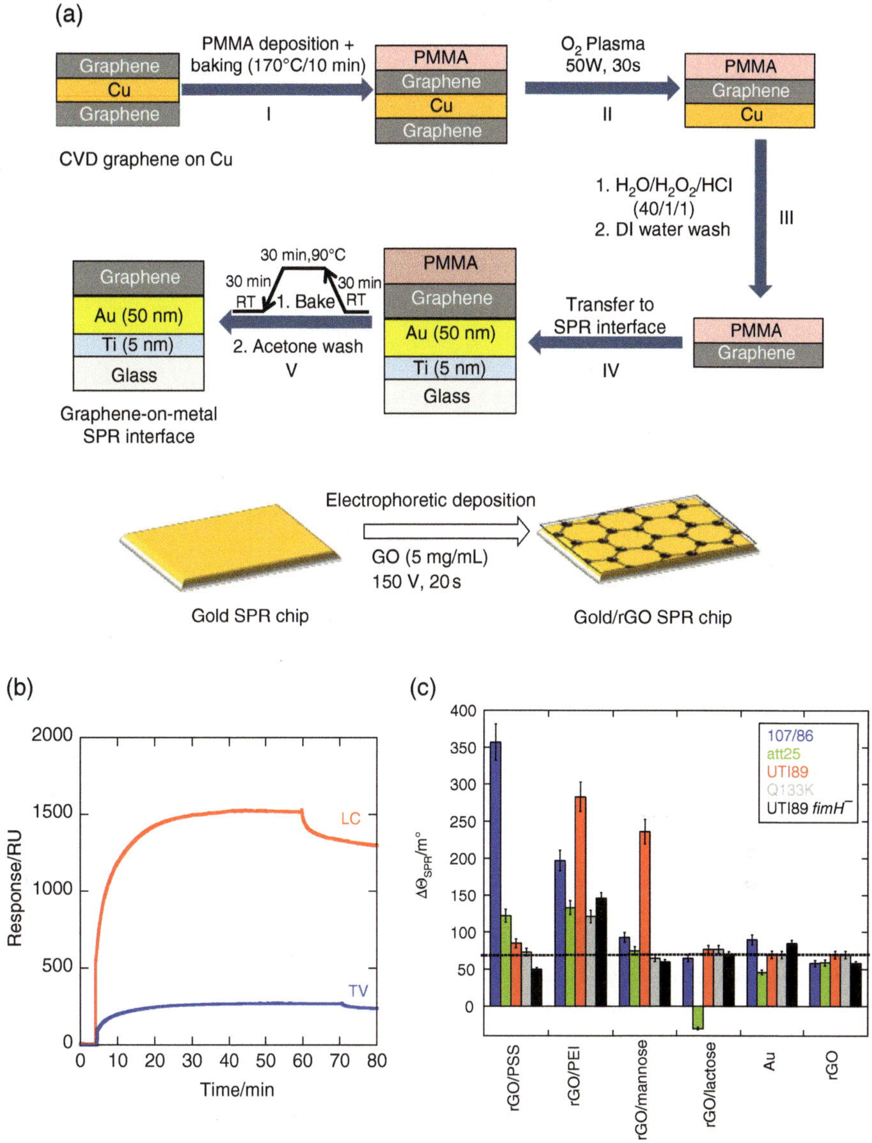

FIGURE 2.12 (a) Formation of graphene-based SPR interface: (top) transfer process of CVD-grown graphene on Cu onto gold SPR substrates and (bottom) electrophoretic deposition of reduced graphene oxide from a solution of graphene oxide (GO) [52]. (b) SPR binding curves of *Lens culinaris* ($500\,\mu g\,ml^{-1}$) and *Triticum vulgaris* ($500\,\mu g\,ml^{-1}$) on mannose-modified SPR interfaces. (c) SPR responses upon addition of different *E. coli* strains ($1.5\times10^{9}\,cfu\,ml^{-1}$) onto graphene-based SPR interfaces (107/86 (blue), att25 (green), UTI89 (red), UTI89 Q133K (gray), and UTI89 *fimH*⁻ (black) [52].

onto gold and silver [14,50] and electrophoretic deposition of reduced graphene oxide [51,52] have been successfully applied (Fig. 2.12a).

Graphene-based SPR interfaces were most recently investigated for their ability to bind lectins (Fig. 2.12b) as well as for the possibility to understand better the tropism of bacteria (Fig. 2.12c). Modification of graphene-based SPR interfaces with mannose by simple immersion of the SPR interface for 4 h into a 100 mM solution of mannose allowed the differentiation between mannose-specific *LC* (46 kDa) and *Triticum vulgaris* (*TV*, 36 kDa). The association constant K_A of the mannose-modified interfaces with *LC* lectin was estimated to be $(3.6 \pm 0.6) \times 10^6\ M^{-1}$. This indicates that the presence of noncovalently bound mannose allows tuning the surface selectivity in an easy manner.

Such interfaces allowed also in an easy manner to investigate the adhesion behavior of different *E. coli* strains [51]. The electrophoretically formed rGO SPR interfaces were modified with differently charged ligands such as positively charged polyethyleneimine (PEI) and negatively charged poly(sodium 4-styrenesulfonate) (PSS) as well as with glycans such as mannose or lactose. The SPR results suggested that local charge variations close to the binding pockets of the bacteria seem to determine bacteria affinity and that chemically based affinity effects have to be considered in addition (Fig. 2.12c). The strategy is adaptable to any other bacteria and more specifically if one is interested in examining bacterial affinities to molecules on a surface and quantitative detection of pathogens in solution. It should be possible to design novel experiments based on simplified model systems to answer many fundamental questions using this technique, combining a plasmonic readout and graphene-based surface functionalization schemes.

2.4 CONCLUSION

Through the use of unique surface functionalization schemes and advanced substrate design, the scope of SPR for the study of affinity binding events has changed considerably over time. The field of SPR for the characterization of low-affinity interactions as is the case between glycans and their associated binding partners has matured over the last 15 years tremendously. Probing carbohydrate-related interactions with sufficient specificity and sensitivity with SPR has become one of the routine analysis tools in glyconomics. This development was made possible through the development of surface chemistry strategies displaying the glycan moieties in a fashion that are readily accessible by the analyte. Technical support and interest from industry to investigate different aspects of SPR and to improve its performance have helped in addition to the success of SPR-based glycan sensing.

Although SPR-based glycan technology yields significant knowledge of glycan interactions, there are still obstacles to overcome in the production of universal carbohydrate arrays. In order for the carbohydrate interface to release its full potential as a tool for biomedical research and drug discovery, the glycan interface must be rigorously interrogated and optimized. The possibility to use "direct" approaches based on photochemical activation of PFPA function for the linking of unmodified carbohydrates and the use of graphene-based SPR has opened a new area for glycobiology. The development of other plasmonic-based glycan platforms such as those based on metallic nanostructures is currently emerging [53]. Enhanced response and visual detection of protein binding are possible with this approach and overcome the necessity of complicated optical setups. The way the glycans are presented on these nanostructures is again of high importance to achieve correct binding. The communal efforts of materials scientists, organic chemists, glycobiologists, as well as spectroscopists will be needed in the future to advance plasmonic-based glycan sensing and to optimize glycan-functionalized transducers to reach excellent selectivity.

ACKNOWLEDGMENTS

Financial support from the Centre National de la Recherche Scientifique (CNRS), the Université Lille 1, the Nord-Pas-de-Calais region, the Institut Universitaire de France (IUF), and the European Commission through the ITN project MATCON (contract no. 238201) is acknowledged.

REFERENCES

1. Wood, R., *Phil. Mag.* **1902**, *4*, 396–402.
2. Ritchie, R.; Arakawa, E.; Cowan, J.; Hamm, R., *Phys. Rev. Lett.* **1968**, *21*, 530–1532.
3. Otto, A., *Z. Physik* **1968**, *216*, 398–410.
4. Kretschmann, E.; Raether, H., *Z. Naturforsch.* **1968**, *23A*, 2135–2136.
5. Liedberg, B.; Nylander, C.; Lundstrom, I., *Sensors and Actuators B* **1983**, *4*, 229.
6. Lofas, S.; Johnsson, B.; Tegendal, K.; Ronnberg, I., *Colloids Surf. B* **1993**, *1*, 83–89.
7. Wink, T.; Van Zuilen, S. J.; Bult, A.; Van Bennekom, W. P., *Anal. Chem.* **1998**, *70*, 827.
8. Jordan, C. E.; Frutos, A. G.; Thiel, A. J.; Corn, R. M., *Anal. Chem.* **1997**, *69*, 4939–4947;Peterlinz, K. A.; Georgiadis, R. M.; Herne, T. M.; Tarlov, M. J., *J. Am. Chem. Soc.* **1997**, *119*, 3401–3402.
9. Love, K. R.; Seeberger, P. H., *Angew. Chem. Int. Ed.* **2002**, *41*, 3583; Taylor, M. E., Oxford Univ. Press, Oxford 2003.
10. Dhayal, M.; Ratner, D. M., *Langmuir* **2009**, *25*, 2181.
11. Shalabney, A.; Abdulhalim, I., *Laser Photonics Rev.* **2011**, *5*, 1–36.
12. Szunerits, S.; Boukerroub, R., *Introduction to Plasmonics: Advances and Applications.* CRC Press, Taylor and Francis Group, Boca Raton, **2015**.
13. Schasfoort, R. B. M.; Tudos, A. J., *Handbook of surface Plasmon resonance.* The Royal Society of Chemistry: **2008**.
14. Szunerits, S.; Maalouli, N.; Wijaya, E.; Vilcot, J.-P.; Boukherroub, R., *Anal. Bioanal. Chem.* **2013**, *405*, 1435–1443.
15. Szunerits, S.; Shalabney, A.; Boukherroub, R.; Abdulhalim, I., *Rev. Anal. Chem* **2012**, *31*, 15–28.
16. Mann, D. A.; Kanai, M.; Maly, D. J.; Kiessling, L. L., *J. Am. Chem. Soc.* **1998**, *120*, 10575.
17. Housemann, B. T.; Mrksich, M., *Chem. Biol.* **2002**, *9*, 443.

18. Haseley, S. R.; Talga, P.; Kaerling, J. P.; Vligenthart, J. F. G., *Anal. Biochem.* **1999**, *274*, 203.
19. MacKenzie, C. R.; Hirama, T.; Deng, S.-J.; Bundle, D. R.; Narang, S. A.; Young, N. M., *J. Biolog. Chem.* **1996**, *271*, 1527.
20. Duverger, E.; Frison, N.; Roche, A.-C.; Monsigny, M., *Biochimie* **2003**, *85*, 167;Myszka, D. G., *Curr. Opin. Biotechnol.* **1997**, *8*, 50.
21. Shinohara, Y.; F., K.; Shimizu, M.; Goto, M.; Tosu, M.; Hasegawa, Y., *Eur. J. Biochem.* **1994**, *223*, 189.
22. Gondran, C.; Dubois, M.-P.; Fort, S.; Cosnier, S.; Szunerits, S., *Analyst* **2008**, *133*, 206.
23. Maalouli, N.; Barras, A.; Siriwardena, A.; Bouazaoui, M.; Boukherroub, R.; Szunerits, S., *Analyst* **2013**, *138*, 805.
24. Sun, W.-L.; Stabler, C. L.; Cazalis, C. S.; Chaikof, E. L., *Bioconjugate Chem.* **2006**, *17*, 52.
25. Zhang, Y.; Luo, S.; Tang, Y.; Yu, L.; Hou, K.-Y.; Cheng, J.-P.; Zeng, X.; Wang, P. G., *Anal. Chem.* **2006**, *78*, 2001–2008.
26. Wang, H.-K.; Zhang, Y.; Yuan, X.; chen, Y.; Yan, M., *Bioconjugate Chem.* **2011**, *22*, 26.
27. Norberg, O.; Deng, L.; Yan, M.; Ramstrom, O., *Bioconjugate Chem.* **2009**, *20*, 2364–2370.
28. Tyagi, A.; Wang, X.; Deng, L.; Ramstrom, O.; Yan, M., *Biosens. Bioelectron.* **2010**, *26*, 344.
29. Lofas, S.; Johnsoon, B. J., *Chem. Commun.* **1990**, *21*, 1526.
30. Blanco, J. L. J.; Haseley, S. R.; Kamerling, J. P.; Vliegenthart, J. F. G., *Biochimie* **2001**, *83*, 653.
31. Hoorelbeke, B.; van Montfort, T.; LiWang, P. J.; Tanaka, H.; Igarashi, Y.; van Damme, E. J. M.; Sanders, R. W.; Balzarini, J., *FEBS Lett.* **2013**, *587*, 860.
32. Hoorelbeke, B.; Xue, J.; LiWang, P. J.; Balzarini, J., *PLOS one* **2013**, *8*, e64132.
33. Loka, R. S.; Cairo, C. W., *Carbohydrate Research* **2010**, *345*, 2641–2647, doi:10.1016/j.carrs.2010.09.020
34. Love, J. C.; Estroff, L. A.; Kriebel, J. K.; Nuzzo, R. G.; Whitesides, G. M., *Chem. Rev.* **2005**, *105*, 1103.
35. Ulman, A., *Chem. Rev.* **1996**, *96*, 1533.
36. Fritz, M. C.; Hahner, G.; Spencer, N. D.; Burli, R.; Vasella, A., *Langmuir* **1996**, *12*, 6074; Revell, D. J.; Knight, J. R.; Blyth, D. J.; Haines, A. H.; Russell, D. A., *Langmuir* **1998**, *14*, 4517.
37. Revell, D. J.; Knight, J. R.; Blyth, D. J.; Haines, A. H.; Russell, D. A., *Langmuir* **1998**, *14*, 4517.
38. Smith, E. A.; Thomas, W. D.; Kiessling, L. L.; Corn, R. M., *J. Am. Chem. Soc* **2003**, *125*, 6140.
39. Liu, L.-H.; Yan, M., *Acc. Chem. Res.* **2010**, *43*, 1434.
40. Linman, M. J.; Taylor, J. D.; Yu, H.; Chen, X.; Cheng, Q., *Anal. Chem.* **2008**, *80*, 4007.
41. Munoz, F. J.; Rumbero, A.; Sinisterra, J. V.; Santos, J. I.; André, S.; Gabius, H.-J.; Jiménez-Barbero, J.; Hérnaiz, M. J., *Glycoconjugate J.* **2008**, *25*, 633.
42. Suenaga, E.; Mizuno, H.; Penmetcha, K. K. R., *Biosens. Bioelectron.* **2012**, *32*, 195; Ilyas, R.; Wallis, R.; Soilleux, E. J.; Townsend, P.; Zehnder, D.; Tan, B. K.; Sim, R. B.; Lehnert, H.; Randeva, H. S.; Mitchell, D. A., *Immunobiology* **2011**, *216*, 126.
43. Fais, M.; Karamanska, R.; Allman, S.; Fairhurst, S. A.; Innocenti, P.; Fairbanks, A. J.; Donohoe, T. J.; Davis, B. G.; Russell, D. A.; Field, R. A., *Chem. Sci.* **2011**, *2*, 1952.

44. Mercey, E.; Sadir, R.; Maillart, E.; Roget, A.; Baleux, F.; Lortat-Jacob, H.; Livache, T., *Anal. Chem.* **2008**, *80*, 3476.

45. Zhang, N.; Schweiss, R.; Zong, Y.; Knoll, W., *Electrochim. Acta* **2007**, *52*, 2869; Szunerits, S.; Knorr, N.; Calemczuk, R.; Livache, T., *Langmuir* **2004**, *20*, 9236–9241.

46. Kambhampati, D. K. M.; Robertson, J. W.; Cai, M.; Pemberton, J. E.; Knoll, W., *Langmuir* **2001**, *17*, 1169–1175.

47. a) Szunerits, S.; Boukherroub, R., *Langmuir* **2006**, *22*, 1660; Szunerits, S.; Boukherroub, R., *Electrochem. Commun.* **2006**, *8*, 439–444; b) Szunerits, S.; Coffinier, Y.; Janel, S.; Boukherroub, R., *Langmuir* **2006**, *22*, 10716–10722; c) Szunerits, S.; Nunes-Kirchner, C.; Wittstock, G.; Boukherroub, R.; Chantal, G., *Electrochim. Acta* **2008**, *53*, 7805–7914; d) Zawica, I.; Wittstock, G.; Boukherroub, R.; S. Szunerits, S.; , *Langmuir* **2008**, *24*, 3922–3929; e) Zawica, I.; Wittstock, G.; Boukherroub, R.; Szunerits, S.; *Langmuir* **2007**, *23*, 9303–9309.

48. Manesse, M.; Stambouli, V.; Boukherroub, R.; Szunerits, S., *Analyst* **2008**, *133*, 1097–1103.

49. Choi, S. H.; Kim, Y. L.; Byun, K. M., *Opt. Expr.* **2010**, *19*, 458.

50. Penezic, A.; Deokar, G.; Vinaud, D.; Pichonat, E.; Subramanian, P.; Gašparovi, B.; Boukherroub, R.; Szunerits, S., *Plasmonics* **2014**, *9*, 677–683.

51. Subramanian, P.; Barka-Bouaifel, F.; Bouckaert, J.; Yamakawa, N.; Boukerroub, R.; Szunerits, S., *ACS Appl. Mater. and Inter.* **2014**, *6*, 5422–5431.

52. Subramanian, P.; Lesniewski, A.; Kaminska, I.; Vlandas, A.; Vasilescu, A.; Niedziolka-Jonsson, J.; Pichonat, E.; Happy, H.; Boukherroub, R.; Szunerits, S., *Biosens. Bioelectron.* **2013**, *50*, 239–243.

53. Bellapardrona, G.; Tesler, A. B.; Grunstein, D.; Hossain, L.; Kikkeri, R.; Seeberger, P. H.; Vaskevich, A.; Rubinstein, I., 2012, 84, 232–240; Liu, X.; Zhang, Q.; Tu, Y.; SZhao, W.; Gai, H., *Anal. Chem.* **2013**, *85*, 11851–11857.

3

CARBOHYDRATE-MODIFIED GOLD NANOPARTICLES

MIKKEL B. THYGESEN AND KNUD J. JENSEN

Department of Chemistry, Centre for Carbohydrate Recognition and Signalling, Faculty of Science, University of Copenhagen, Frederiksberg, Copenhagen, Denmark

3.1 INTRODUCTION

A central theme in functional analysis of carbohydrate recognition is the well-established requirement for clustering of the ligands. The low affinities characterizing most glycan–protein and glycan–glycan interactions, typically in the millimolar to high micromolar range, represent a challenge to structure–activity determinations based on traditional ligand-binding assays. In nature, the low affinities are increased by multivalent presentation of glycan ligands and their receptors to achieve high observed affinities (i.e., avidity) with high specificities [1]. Synthetic macromolecules and assemblies capable of mimicking the multivalent representation and interactions of glycan structures in nature are emerging as enabling tools for the general study of glycan interactions. Of particular importance in this field is the use of glycofunctionalized [2] dendrimers, liposomes, globular protein scaffolds, microarrays, and nanoparticles.

Carbohydrate-modified nanoparticles (glyconanoparticles) have been introduced as an attractive platform for studying and intervening in carbohydrate–protein and carbohydrate–carbohydrate interactions. A number of comprehensive reviews on glyconanoparticles have surveyed their synthesis and use [3]. By far, most glyconanoparticles consist of an inorganic core of nanoscale dimensions on top of which is a flexible organic layer, linker, or polymer, conjugated covalently or noncovalently

Carbohydrate Nanotechnology, First Edition. Edited by Keith J. Stine.

with carbohydrates. Different materials may be utilized for the core depending on the application of the glyconanoparticles, for example, noble metal nanoparticles, semiconductor quantum dots, or magnetic nanomaterials. The developing field of glyconanoparticle research constitutes perhaps one of the most intriguing research areas within biomolecular nanoscience due to the convergence of the inherent characteristics of nanoparticles and the aforementioned multivalency effects of carbohydrate interactions. Quantitative studies have revealed that optimized anchoring of glycans on nanoparticles may lead to binding enhancements of up to five to six orders of magnitude compared to the free glycans in solution [4], thus paralleling the best of synthetically tailored scaffolds [5]. The high degree of solubility and stability in biological and buffered aqueous media makes glyconanoparticles well suited for interactions with biological material [6]. In this chapter, we focus on carbohydrate-modified gold nanoparticles (AuNPs).

3.2 TUNABLE NANOSCALE PROPERTIES OF GOLD NANOPARTICLES

Between metal atoms and bulk metal, there are, roughly speaking, three size regimes of nanoscale metals, although the boundaries between them are not clearly defined: (i) gold clusters or nanoclusters, which contain an exact number of gold atoms in the core, ranging from a few to hundreds, and have diameters from subnanometer to approximately 1–2 nm. Small gold clusters may be regarded as "superatoms" [7] or metal molecules, however, the difficulties in purification of clusters have become a major obstacle to their practical use; (ii) gold nanoparticles or nanocrystals, which are characterized by having a relatively narrow size distribution, potentially in the 2–100 nm range but most typically in the 2–20 nm range; and (iii) colloidal gold or large gold nanoparticles, which typically are characterized by having broad size distributions with core diameters in the 20–100 nm range.

The vast majority of reported carbohydrate-modified gold nanoparticles have spherical geometry; however, alternative morphologies of nanostructures, such as nanorods or nanostars, may also be considered [8].

3.2.1 Quantum Confinement Effects of Small Gold Clusters

When the particle size approaches the Fermi wavelength of electrons of gold (~0.5 nm), the optical and electronic properties of the clusters display characteristic features due to the quantization of energy states of the entire particle. Of particular interest is the potential for tunable fluorescence emission in the visible range following optical excitation. Examples of such quantum dot behavior of gold clusters containing less than 100 gold atoms in the core have been reported [9]. Robust and highly emissive carbohydrate-modified gold nanoclusters of discrete molecular composition are prominent targets of current nanobiotechnological efforts.

3.2.2 Surface Plasmon Resonance Effects of Gold Nanoparticles

The deep red colors of colloidal gold (i.e., gold nanoparticles) have been cherished since antiquity and have found widespread use in coloring of glasses and ceramics. The colors originate from a broad absorption band centered in the visible spectrum of light. This band is attributed to surface plasmon (SP) resonance effects. For gold nanoparticles, the plasma oscillation of the free electrons (mainly 6s and 5d electrons) of the conduction band results in the formation of an oscillating dipole [10]. The two phases of the dipole consist of energy states immediately above and below the Fermi energy level, see Figure 3.1. The SPs of noble metal nanoparticles, such as gold and silver NPs, typically have resonance energies located in the visible spectrum of light, and hence, the absorption bands, termed SP bands, result in vibrant coloration of dispersions of these nanoparticles.

The location and shape of the SP band is strongly dependent on a range of factors [10,11], of which the most important are the following: (i) the mean nanoparticle size, which affects both location and shape of the SP band. For very small particles and

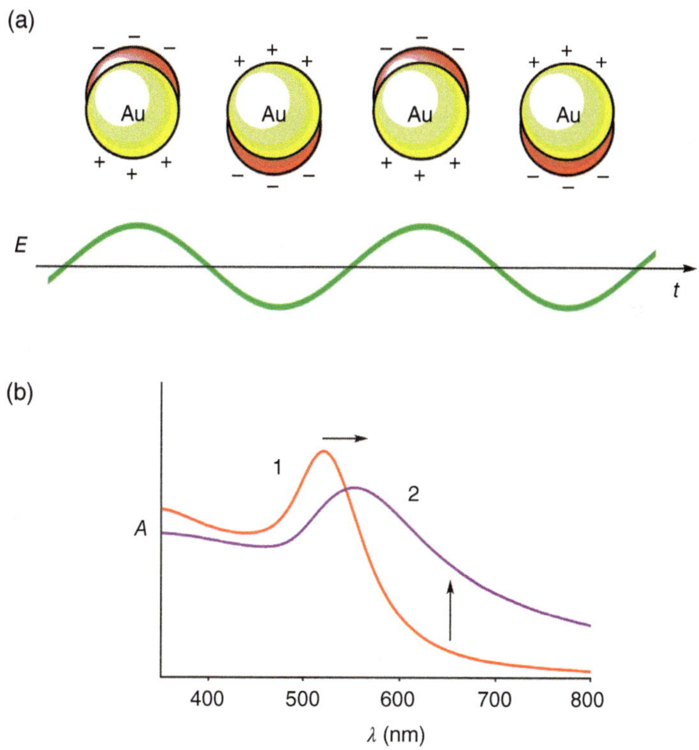

FIGURE 3.1 (a) Schematic representation of oscillating dipole resulting from optical excitation of gold nanoparticle surface plasmons. (b) Absorption spectrum of 12 nm citrate-stabilized gold nanoparticles (1) displaying SP band. The band shifts toward longer wavelength, and the peak is broadened upon aggregation (2). (a) Redrawn after Ref. 10.

clusters (<2 nm), the intensity of the SP band decreases due to the onset of quantum confinement effects (see in the following text). However, for gold nanoparticles in the 2–100 nm range, the wavelength position and extinction coefficient intensity of the SP band increase exponentially with increasing particle size, a fact which may be exploited for determinations of size and concentration [12]; (ii) the nanoparticle size distribution affects the shape of the SP band by providing sharp bands for preparations having a narrow distribution and broad bands for polydisperse preparations; (iii) the shape of the nanoparticles may affect both location and shape of the SP band; (iv) the mean interparticle distance has a dramatic effect on the SP band once the distance approaches the diameter of the particles. This effect has been explored in a number of application, exemplified in section 3.5; (v) the surface functionalization may affect the electronic environment, and especially in case of thiol anchoring, a distinct redshift of the SP band is normally observed [11]. Additionally, the dielectric properties and the temperature of the medium may affect the SP band. Collectively, these features of the SP band make it a highly sensitive means of quality control for gold nanoparticle bioconjugates.

3.2.3 Scattering Effects of Large Nanoparticles

Large gold nanoparticles (diameter typically >40 nm) display prominent Mie scattering [13] that can be exploited in bioimaging or as optical tags. For example, gold nanoparticles of 50 and 100 nm exhibit scattering at different wavelength maxima, yielding green or red scattering readouts, respectively [14]. Resonance light scattering probes based on metal nanoparticles display signal intensities many orders of magnitude more powerful than conventional fluorescent probes and with complete resistance to quenching [15]. Scattering effects have so far not been widely explored for glyconanoparticles.

3.3 METHODS FOR ASSEMBLY OF GLYCONANOPARTICLES

Covalent anchoring of carbohydrates to nanoparticles requires two key linkages: (i) the formation of a firm covalent bond between the carbohydrate and an aglycone spacer and (ii) formation of a bond to the metal surface, typically formation of Au—S bonds. The linker between the glycan and the surface should provide optimal presentation of glycans and prevent nonspecific interactions, for example, with proteins or surfaces.

Most methods for construction of glyconanoparticles have relied on neoglycoconjugates with thiol-terminated linkers, or disulfides thereof, either via (i) *in situ* reduction of metal salts in the presence of thiol-terminated glycoconjugates or (ii) ligand exchange of thiol-terminated glycoconjugates onto preformed nanoparticle cores (Fig. 3.2). Thus, the development of efficient and simple conjugation protocols to enable thiolation of biologically relevant carbohydrate ligands is central to the advancement of glyconanoparticle synthesis.

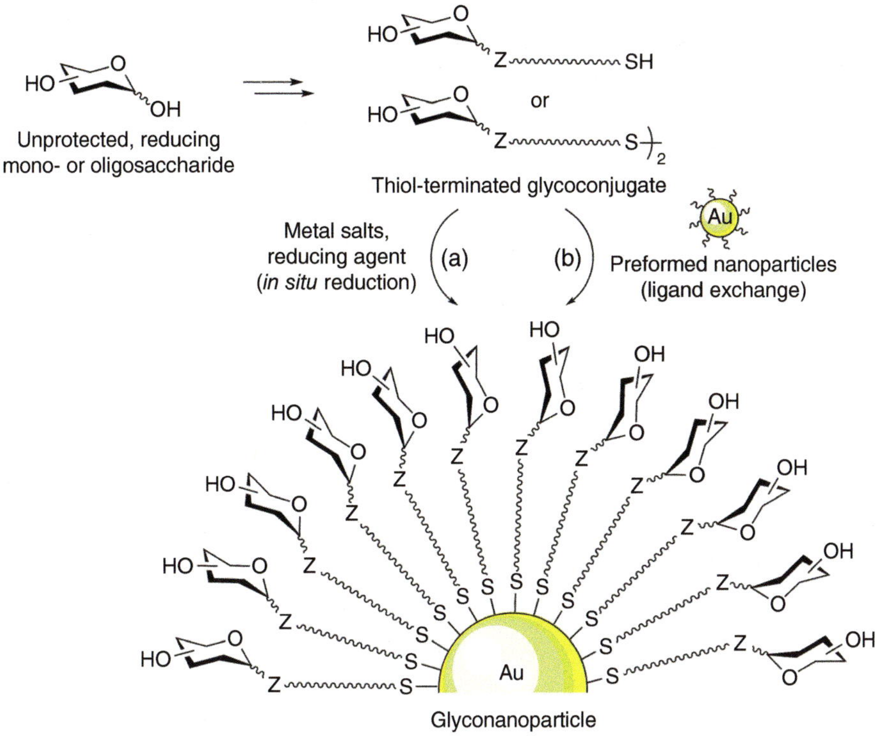

FIGURE 3.2 Synthesis of glyconanoparticles from thiol-terminated glycoconjugates by *in situ* reduction (a) or ligand exchange (b). Z = anomeric heteroatom. Redrawn after Ref. 3g.

3.3.1 *In Situ* Reduction of Metal Salts in the Presence of Thiol-Terminated Glycoconjugates

Gold glyconanoparticles were first described in 2001 by Penadés and coworkers [16]. The method of nanoparticle construction is based upon the landmark Brust–Schiffrin two-phase protocol for synthesis of thiol-derivatized gold nanoparticles [17]. Here, Au(III) ions from an aqueous phase are transferred, by way of a phase-transfer agent, to an organic phase, in which a thiol is dissolved. Initial complexation and potential partial reduction of Au(III) with the phase-transfer agent or thiol is followed by reduction to Au(0) after addition of sodium borohydride. Particle size control is mediated by complexes formed specifically in the organic phase [18].

The method adopted by the Penadés group entails complexation of tetrachloroauric(III) acid with the glycoconjugate disulfide or thiol in a methanol–water solution followed by reduction with sodium borohydride. As such, it may be considered a simplified version of the procedure of Brust and Schiffrin, operating in a homogeneous, polar phase due to the polarity of the glycoconjugates.

This procedure provides small glyconanoparticles with core diameters typically of ca. 1–3 nm. Discrete, monodisperse glyco-functionalized gold nanoclusters have not been reported to date for this size range. The obtained particles have only very weak SP resonance signatures and generally do not display quantum dot behavior in terms of fluorescence emission. They can normally be easily handled as powders and display high storage stability. Importantly, due to the small size and high carbohydrate–gold ratio, these particles are amenable to NMR spectroscopy as well as elemental analysis, which facilitate precise analysis of the presentation and ligand density. Larger particles may be obtained by varying the gold–thiol ratio, but this generally results in broad size distributions.

The main drawback of the procedure is that the method provides poor control over particle size because the nature of the thiol ligands influences nucleation and growth of the particles. This leaves it difficult to generate panels of glyconanoparticles having identical gold cores. In addition, the strongly reducing conditions may be deleterious to the thiolated ligands of choice. For simple oligosaccharides, however, this is normally not a major concern.

3.3.2 Self-assembly of Thiolated Carbohydrates onto Citrate-Stabilized Gold Nanoparticles

Citrate-stabilized gold nanoparticles are synthesized by the citrate reduction method, reported by Turkevich and coworkers [19] and later modified by Frens to allow particle size control [20]. Here, citrate functions as both reducing and capping agent. Formally, the reaction may be regarded as an oxidative decarboxylation of sodium citrate with concomitant reduction of tetrachloroauric(III) acid to Au(0). However, citrate ions additionally mediate the formation of gold nanoparticles by acting as a capping or stabilizing agent that controls particle growth from nucleation sites. A third role of citrate as a moderator of solution pH has recently been underscored to provide nanoparticle size control [21].

Molecular self-assembly of thiols on citrate-stabilized gold nanoparticles was initially reported by Whitesides and coworkers [22]. By using this strategy, gold glyconanoparticles, having core diameters of ca. 10–40 nm with relatively narrow size distributions, may be generated by self-assembly of thiol-terminated glycoconjugates onto preformed nanoparticles, as reported initially by Russell et al. [23]. The introduction of thiol-linked saccharides is performed by partial displacement of the citrate ions under formation of Au—S bonds. The low carbohydrate-to-gold ratio complicates direct NMR analyses. The obtained glyconanoparticles have strong SP resonance signatures that make them ideal for colorimetric bioassays.

3.3.3 The Au—S Linkage on Gold Nanoparticle Surfaces

The nature of the Au—S bonds has been the subject of debate. The oxidation state of the gold atoms of the nanoparticle cores has been shown by X-ray photoelectron spectroscopy to be exclusively in the Au(0) state thus excluding an ionic interaction [17]. The bond may be described as being dative to emphasize that both bonding electrons are provided by the thiolate anion. However, from the viewpoint of bond energy and electronegativity difference between Au and S, amounting 40–45 kcal/mol [24] and 0.04 units, respectively, the bond is rather covalent. The thiols on the surface are exchangeable, which may be used for the functionalization of AuNPs in so-called thiol place-exchange reactions [25]. X-ray crystallographic analyses of thiol-coated gold clusters have indicated that the Au—S bonds are organized in a coordination geometry, also referred to as "staple motifs," involving two sulfur atoms and three gold atoms, in which each sulfur atom coordinates two gold atoms [26]. This motif may account for the observed mobility of thiols on gold surfaces [27]. In addition, differences in gold crystal packing on faces, edges, and vertices may affect variations in the strength of the Au—S bonds.

3.3.4 Surface Presentation of Glycan Ligands

The optimal presentation of carbohydrate ligands on the surface of nanoparticles reflects a balance between ideal multivalent display of the glycan ligands for their interaction partners and suitable physicochemical properties of the nanoparticles that confers biocompatibility. Important factors for the multivalent display are (i) nanoparticle size, (ii) the number and density of carbohydrate ligands, and (iii) the length and flexibility of the linker connecting the carbohydrate to the nanoparticle surface.

Both surface area and surface curvature are directly dependent on nanoparticle size and this influences surface presentation. For very small nanoparticles and clusters, the surface-to-volume ratio is high, and a dominant part of the gold atoms is located on the surface. For example, the Au_{25} cluster has 24 gold atoms engaged in thiol binding [9a]. Nanoparticle size also determines the surface curvature, which affects the spatial arrangement of the glycoconjugates on the surface. Thus, the multivalent presentation of identical glycoconjugates on different gold cores yields different observed avidity of protein interaction [28]. The nature of the thiol linker may significantly influence the resulting number of glycoconjugates presented on the nanoparticle surface. This is particularly so for ligand displacement reactions, for example, of thiols on citrate-stabilized gold nanoparticles. Here, self-assembly of glycoconjugate alkanethiols results in incorporation of approximately 5–10 fold more sugar ligands in the resulting glyconanoparticles as compared to similar oligo(ethylene glycol) thiols (see, e.g., Refs. 29 and 30 for comparison). The density of the ligands may be further modulated by application of two-component thiol mixtures, as recently reviewed by Marradi et al. [3h].

3.3.5 Methods for Synthesis of Thiol-Functionalized Glycoconjugates

Procedures for glycoconjugate synthesis based on chemical O-glycosylation with fully protected glycosyl donors and thiol-terminated linkers can be efficiently applied for simple mono- and oligosaccharides. Different strategies for preparing the required thiol-terminated glycoconjugates have been examined in reviews on glyconanoparticles [3]. As the complexity of the carbohydrate ligand increases beyond simple oligosaccharides, installment of the thiol-terminated linker becomes challenging [31] due to the requirement for multistep protection and deprotection reactions and the low overall yields. In the following, we will focus on methods for installment of suitable thiol-terminated linkers by using simple, high yielding, and chemoselective conjugation conditions.

A few examples in the literature describe the application of glycosides carrying anomeric moieties that allow easy and efficient conjugation to thiol-terminated linkers for glyconanoparticle synthesis. For example, Kamerling and coworkers have developed a methodology based on the synthesis of allyl glycosides that are conjugated with 1,6-hexane-dithiol in a thiol–ene reaction at the end of oligosaccharide synthesis (Fig. 3.3a). The allyl glycoside is stable to conditions for oligosaccharide synthesis, and a range of thiol-terminated mono- and disaccharides could be prepared and used for synthesis of gold nanoparticles [32]. Likewise, Penadés and coworkers introduced a heterobifunctional linker based on thiourea coupling between aminoethyl glycosides and a terminal isothiocyanate group (Fig. 3.3b). Oligosaccharide synthesis could be conducted with benzyloxycarbonyl protection of the amino group [31].

Reducing carbohydrates include a structural feature that greatly facilitates conjugation. This is the presence of an aldehyde or ketone moiety at the "reducing end" that presents a single uniquely reactive functional group in carbohydrates and a group that is normally not present in other biopolymers. The reducing end aldehyde moiety, in the open-chain form, is in equilibrium with the corresponding hemiacetal in the ring-closed form of the carbohydrate. However, the electrophilic aldehyde moiety can still react with nucleophiles. The nucleophiles most commonly used for the reaction with the reducing end are aminooxy or hydrazide compounds, providing relatively stable oximes or hydrazides. Other nucleophiles can also react, for example, amines; however, the initially formed product is less stable and needs to be transformed, for example, by reductive amination or by the use of 1,2-amino thiols, which can form stable thiazolidines [33].

Reductive amination chemistry has been described by Kamerling and coworkers for the preparation of mercaptoethylamine-terminated oligosaccharides from unprotected, reducing mono- to hexasaccharides containing various degrees of branching. In this procedure, the reducing end monosaccharide unit is converted into an

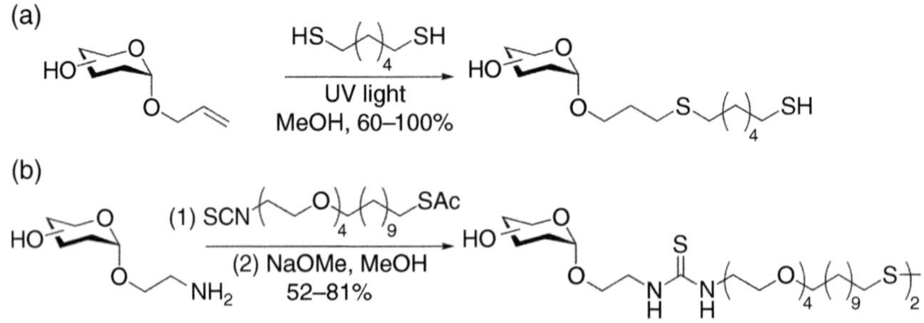

FIGURE 3.3 Chemoselective approaches to thiol-terminated neoglycoconjugates from *O*-allyl glycosides via thiol–ene coupling (a) [32] or from *O*-aminoethyl glycosides via thiourea coupling (b) [31].

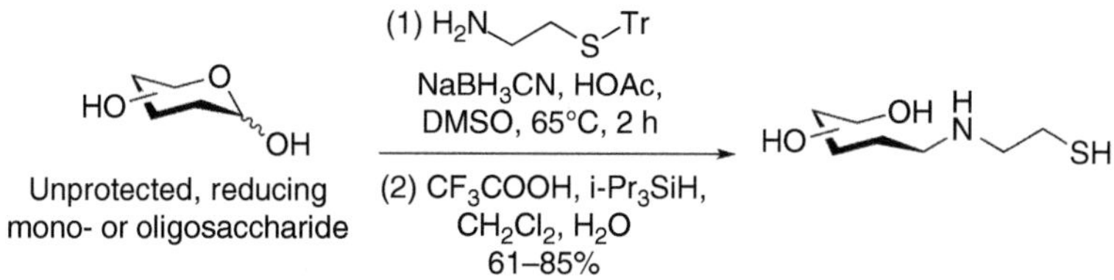

FIGURE 3.4 Chemoselective synthesis of mercaptoethylamine-terminated carbohydrates from reducing carbohydrates [35].

open-chain secondary amine (Fig. 3.4). These glycoconjugates were used in the preparation of gold glyconanoparticles via a modified Brust–Schiffrin protocol. The drawback of reductive amination protocols, however, is the potential lack of specific binding between proteins and short oligosaccharides, imposed by the reductively opened monosaccharide units [34].

The highly chemoselective reaction of glycans through their reducing ends with aminooxy groups to form oximes has become an important method in glycoconjugate synthesis. It is sometimes speculated that this leaves the monosaccharide at the reducing end solely in a nonnative open-chain form, as in the case of reductive amination. However, we and others have clearly demonstrated that carbohydrate oximes and oxime ethers exist as a dynamic mixture of several tautomeric forms involving both ring- and open-chain forms and that the closed ring forms can be recognized by proteins [30,36]. The open-chain (*E*)- and (*Z*)-forms generally predominate over the α- and β-pyranosyl forms, while α- and β-furanosyl forms may constitute only minor tautomers, the distribution being dependent on the monosaccharide unit (Fig. 3.5) [37]. Hydrolysis of oximes, that is, the reverse reaction of their formation, is known to occur, especially at pH values below 5–6. However, our group has shown [38] that at least some aminooxy reagents provide carbohydrate oximes with a remarkable stability.

The rate of formation of carbohydrate oximes can in certain cases be too low to be of practical use. It had previously been demonstrated that anilines catalyze the formation of oximes [39]. We demonstrated that aniline catalysis also can occur in the more complex formation of carbohydrate (aldose) oximes [40]. For carbohydrates, aniline catalysis is complicated because the aldehyde moiety is in equilibrium with the cyclic hemiacetal and because of the possibility of intermediate formation of unstable, cyclic phenyl glycosyl amines (Fig. 3.5). Aniline plays a dual role, as it provides an enhanced carbohydrate electrophile for the aminooxy nucleophile, in the form of a protonated iminium ion, and additionally shifts the equilibrium away from the hemiacetal. Aniline catalyzed the formation of carbohydrate oximes with up to 20-fold rate increases (at 100 mM aniline) and was especially useful for the formation of oximes of carbohydrates with slow reacting units, for example, GlcNAc, at the reducing end. This very useful methodology has enabled the formation of complex glycoconjugates [40].

In 2009, we introduced the application of chemoselective carbohydrate oxime formation for the anchoring of reducing glycans to gold nanoparticles [30].

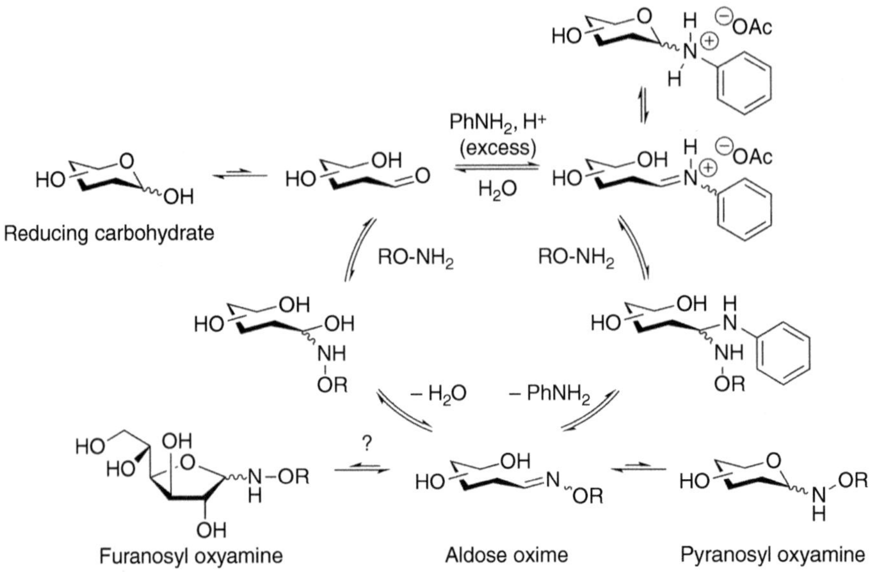

FIGURE 3.5 Nucleophilic catalysis of carbohydrate oxime formation by aniline in acetate buffer, pH 4.6. Oxime formation in the absence (left side) or presence of aniline (right side). The aldose oxime is in equilibrium with cyclic oxyamines. Redrawn after Ref. 40.

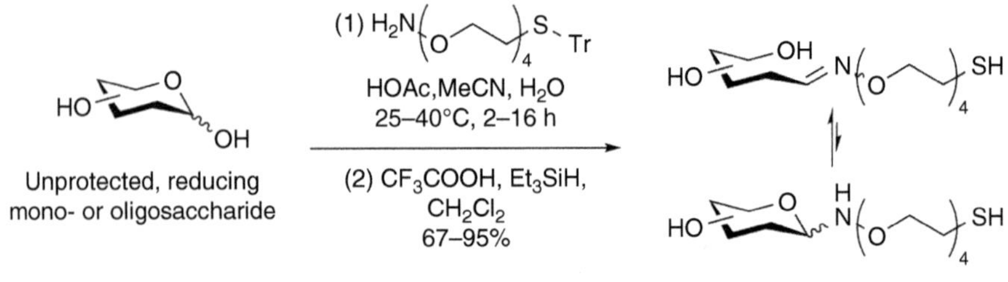

FIGURE 3.6 Chemoselective oxime formation with heterobifunctional linker [30].

The strategy is based on a heterobifunctional linker comprising an aminooxy group in one end, and a trityl-protected thiol in the other end of a tetra(ethylene glycol) spacer, which provides flexibility for the carbohydrate ligands and prevents unspecific binding with proteins (Fig. 3.6). Carbohydrate oxime formation with this linker has been conducted with a range of reducing mono- and oligosaccharides, in some cases under aniline catalysis to increase the reaction rate [30,40]. The trityl group functions as a lipophilic and UV active tag that facilitated easy analysis of reagent conversion and isolation of the oxime products. After removal of the trityl group, the thiol-terminated glycoconjugates were anchored on citrate-stabilized gold nanoparticles via ligand exchange. The obtained glyconanoparticles were soluble under a wide range of aqueous conditions. This approach provides very simple handling and high efficiency for the generation of functionalized glyconanoparticles for bioanalytical assays.

We studied the ability of proteins to interact via specific molecular recognition with carbohydrate oxime conjugates by comparing the functional binding of oxime ring-chain tautomers with analogues trapped in an open-chain configuration by reduction of the oximes to the corresponding hydroxylamines [30]. The results with the enzyme glucoamylase from *A. niger* showed that the equilibration of tautomers mediated productive enzyme binding at a stoichiometric level. With the lectin concanavalin A from *C. ensiformis*, the results showed that specific recognition of monosaccharide oximes could be obtained while these were attached to gold nanoparticle surfaces. Thus, carbohydrate oximes may provide efficient anchoring of glycans through the reducing end while maintaining a biologically relevant presentation of the glycan. These findings have general significance for construction of glycoconjugates.

3.4 DIRECT COUPLING OF CARBOHYDRATES ON REACTIVE NANOPARTICLE SURFACES

The previous part of this chapter has described methods that rely on preformed thiol-terminated glycoconjugate intermediates for glyconanoparticle synthesis. An alternative strategy that obviates this need is the application of chemoselective reactions between carbohydrates and reactive functional groups on linker or polymer-coated nanoparticles (Fig. 3.7). A key requirement for these reactions is that they should be chemoselective

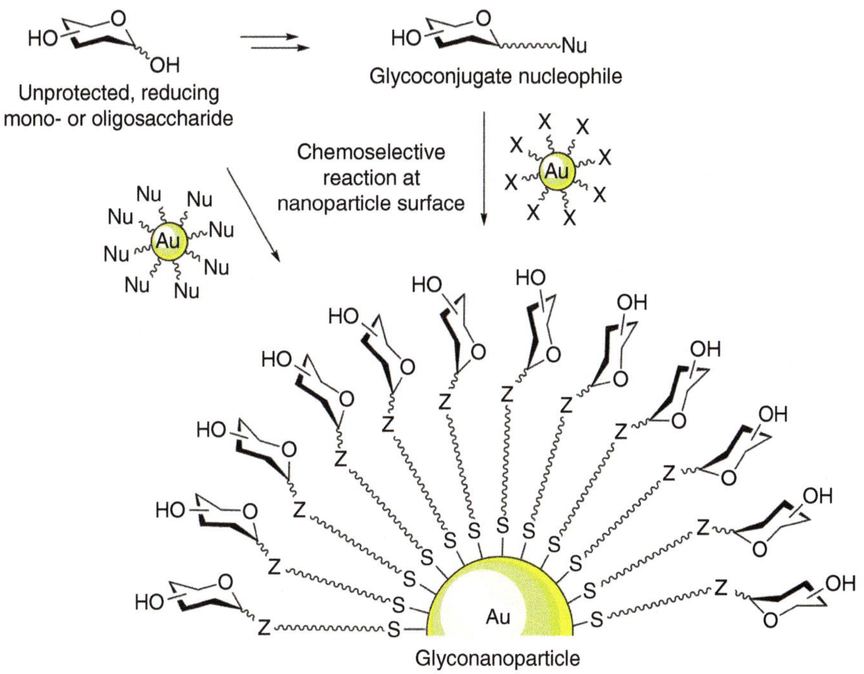

FIGURE 3.7 Synthesis of glyconanoparticles by chemoselective reaction at nanoparticle surface. Nu, nucleophile, X, electrophile, Z, covalent linkage. Redrawn after Ref. 3g.

and high yielding, that is, essentially by-product-free, due to the fact that glyconanoparticles carrying side products cannot be selectively removed. Ideally, the reaction conditions involving nanoparticles should be mild and not contain an excessive number of transformations in order to avoid degeneration or aggregation and to maintain a homogeneous population of nanoparticles.

In 2001, Kataoka et al. published a strategy, where aldehyde-functionalized gold nanoparticles were assembled in the presence of a heterobifunctional PEG polymer, carrying a thiol and an acetal-protected aldehyde moiety [41]. After liberation of the aldehyde groups on the nanoparticle surface under mildly acidic conditions, these particles were reductively aminated with unprotected 4-aminophenyl glycosides in the presence of borane-dimethylamine in aqueous phosphate buffer pH 7 (Fig. 3.8).

3.4.1 Chemoselective Reactions with Reducing Carbohydrates

Our group has developed a methodology for synthesis of gold glyconanoparticles where unmodified and unprotected glycans are anchored directly via a single chemoselective reaction on surface-bound aminooxy groups (Fig. 3.9). Furthermore, we have shown that oxime coupling on terminal aminooxy groups allows anchoring of various other aldehyde-containing biomolecules, for example, glycopeptide aldehydes. Anchoring is conducted under mild conditions in neutral aqueous solution [42].

Our methodology [42] is based on the anchoring of a heterobifunctional tetra(ethylene glycol) spacer, carrying a protected aminooxy group and a thiol group to provide water-soluble, citrate-stabilized gold nanoparticles. We applied a

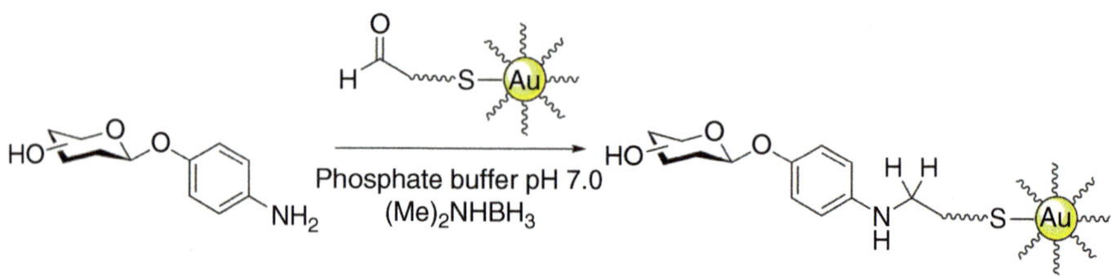

FIGURE 3.8 Chemoselective reactions on nanoparticle surface via reductive amination. Only one surface ligand is shown [41].

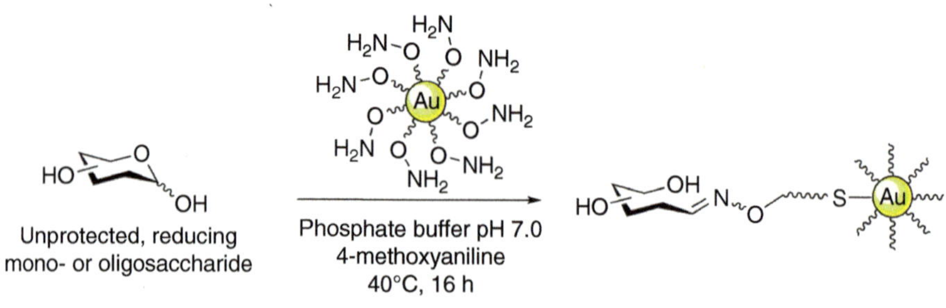

FIGURE 3.9 Chemoselective oxime formation with reducing carbohydrates on aminooxy-functionalized nanoparticles [42]. Only one surface ligand is shown.

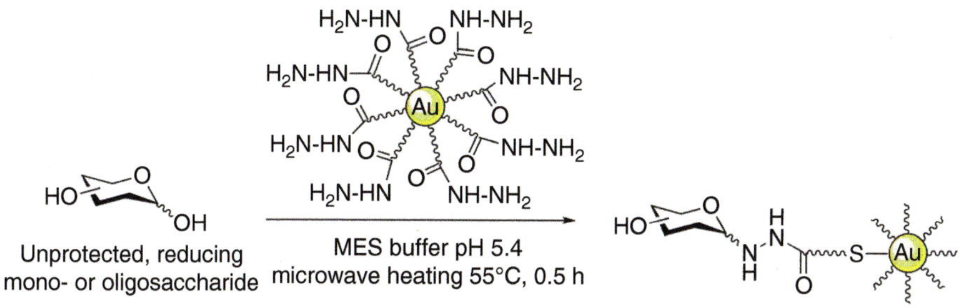

FIGURE 3.10 Chemoselective hydrazide formation on the surface of gold nanoparticles [43]. Only one surface ligand is shown.

4-carboxy-phthaloyl protecting group for the aminooxy moiety. The carboxylate group served to provide a high degree of aqueous solubility of the nanoparticles under neutral conditions. The protecting group could be removed by dilute (1 mM) aqueous hydrazine, and in addition, it allowed quantification of release of aminooxy groups on the nanoparticles. The mild conditions ensured that the nanoparticles were not compromised by aggregation. The aminooxy-terminated gold nanoparticles, termed reactive core–shell nanoparticles, were reacted with unprotected, reducing disaccharides. Carbohydrate oxime formation could be promoted by nucleophilic catalysis with 4-anisidine at 40°C.

Zhou at al. [43] have published a related strategy, however, without aniline catalysis, based on hydrazide-coated gold nanoparticles that may react chemoselectively with unprotected, reducing glycans under microwave heating conditions at pH 5.4 to obtain high reaction rates (Fig. 3.10).

3.4.2 Nonselective Methods

Ramström, Yan, and coworkers have developed a methodology based on photochemical coupling between unprotected carbohydrates and perfluorophenylazide groups on nanoparticle surfaces [44]. The azide moiety was activated by UV light, via conversion into a reactive nitrene intermediate that underwent insertion reactions into C—H bonds of the carbohydrate. The technique provides access to glyconanoparticles; however, the orientation of the anchored glycans cannot be controlled, and part of the structures may be presented in denatured forms in which they may not be able to interact specifically.

3.5 APPLICATION OF CARBOHYDRATE-MODIFIED GOLD NANOPARTICLES

A comprehensive account of all reported applications of gold glyconanoparticles is beyond the scope of this chapter. The reader is referred to Chapters 12, 13, and 16 in this book, as well as a number of excellent literature reviews [3]. In this section, we highlight a few examples of particular and recent interest.

3.5.1 Chemical Biology

Gold nanoparticle-based colorimetric assays have been applied for the detection of many glycan-binding proteins, that is, lectins and toxins. These proteins typically exist as homooligomers in solution and, as such, they may be applied in detection strategies involving only two components, that is, a glyconanoparticle containing the glycan of interest and a protein with affinity for this glycan (Fig. 3.11) [41]. The binding of glycans from two or more different nanoparticles to the same lectin molecule induces a reduction in the mean nanoparticle distance that results in a colorimetric change. Gold nanoparticle-based colorimetric assays have also been applied for the detection of glycan–glycan interactions [45].

Various types of nanoparticles, mainly gold and silver, have been used in conjunction with mass spectroscopic (MS) methods, primarily matrix-free laser desorption–ionization (LDI-MS) techniques, to improve the sensitivity of MS detection of biomolecules or to characterize biomolecule-functionalized nanoparticles. Examples of mass spectrometric applications of gold nanoparticles include monitoring of glycosyl transferase activity on oligo(ethylene glycol)-coated AuNPs by Nishimura et al. [47] and the direct analysis and epitope mapping of glyconanoparticle-bound proteins by Lin et al. [48].

Penadés and coworkers have also developed the use of nanoparticles consisting of a 4 nm diameter magnetic core with a 1 nm gold shell functionalized with carbohydrates (magnetic glyconanoparticles) [49]. Their aim was to identify *in vivo* endogenous neural precursor cells migrating toward a brain damage site. The monoclonal antibody Nilo2 recognizing cell surface antigens on neuroblasts was coupled to magnetic glyconanoparticles (mGNPs). The Nilo2–mGNP complexes allowed, by magnetic resonance imaging (MRI) in living animals, the *in vivo* identification of endogenous neural precursors at their niche, as well as their migration to a lesion site, an induced brain tumor.

Our group together with Martos-Maldonado and Berenguel has reported the convenient preparation of water-soluble gold nanoparticles coated with mannose–ferrocene conjugates for electrochemical sensing [29]. The preparation involved the

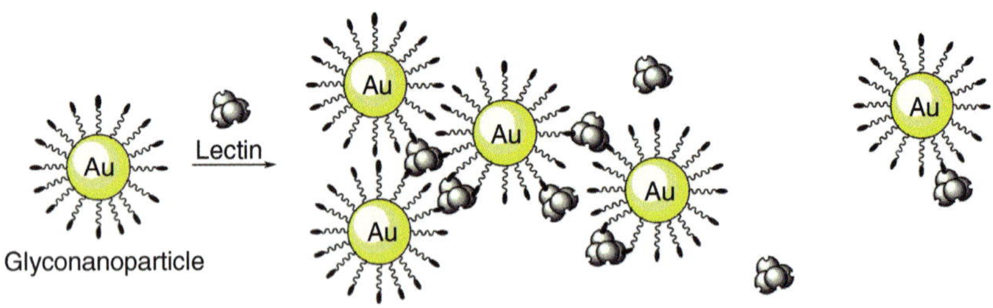

FIGURE 3.11 Schematic representation of lectin-induced colorimetric glyconanoparticle assay. The oligomeric (in this case tetrameric) lectin may bind to multiple glycan ligands (•), which are anchored to gold nanoparticles. This causes a reduction in the mean interparticle distance and a visible shift in the surface plasmon band. Various types of glyconanoparticle–lectin interactions are shown. Redrawn from Ref. 46.

use of unsymmetrical 1,1-bis-substituted ferrocene building blocks for sequential orthogonal coupling with propargyl mannopyranoside and disulfide-functionalized linkers to anchor the mannose–ferrocene conjugate to the metal surface. The mannose–ferrocene-coated gold nanoparticles exhibited sensing abilities toward mannose-binding lectin concanavalin A. This was obtained by both optical and electrochemical methods based on SPR shift properties of the metal and the electrochemical activity of the ferrocene moiety, respectively. Concanavalin A was detected with high sensitivity in the low nanomolar range by electrochemical methods. The redox-sensing abilities for mono- and bivalent mannose–ferrocene conjugates revealed in comparison that the multivalent presentation of both mannose and ferrocene units led to a remarkable enhancement of sensitivity, the limit of detection being at least three orders of magnitude lower than for the single conjugates.

Lin, Withers, and coworkers have recently shown that enzymatic glycosyltransferase activity can be assayed in high-throughput by label-free AuNP-based colorimetric detection [50]. By self-assembly of thiol-terminated sialyllactose (GM3) or disialyllactose (GD3) ganglioside analogues onto citrate-stabilized gold nanoparticles, they could generate glyconanoparticles that were substrates of polysialyltransferase (PST) from *Neisseria meningitidis*. Polysialylated nanoparticles were treated with an inactive, but still binding, mutant of the polysialic acid-specific trimeric phage tailspike endosialidase from *E. coli* phage K1F (EndoNF-DM), which served as a cross-linker of glyconanoparticles to generate the colorimetric response.

Davies and coworkers described the construction of gold nanoparticles covered by a polymer with Tn-antigen glycans attached as potential synthetic cancer vaccines that generate significant titers of antibodies selective for aberrant mucin glycans [51]. Immunological studies indicated that these nanomaterials generated strong and long-lasting production of antibodies that are selective to the Tn-antigen glycan and cross-reactive toward mucin proteins displaying Tn (see also Chapter 13).

3.5.2 Diagnostics

The Penadés group recently reported an *ex vivo* method to estimate the *in vivo* performance of potential MRI contrast agents [52]. They used 1.5–2.5 nm-sized AuNP cores covered by carbohydrate ligands. The 1,4,7,10-tetraazacyclodecane-1,4,7-triacetic acid (DO3A) ligand for gadolinium was anchored to the AuNP by ligand exchange. The final AuNPs contained 70–130 copies of carbohydrate and 14–34 copies of Gd chelators, while the Gd(III) content was 3.2–7.0% relative to the total mass of the nanoparticle, including gold.

Vargas-Berenguel and coworkers described glyconanoparticles as potential site-specific delivery systems for anticancer drugs [53]. They reported the preparation of novel gold nanoparticles containing multiple long, flexible linkers decorated with lactose and β-cyclodextrin. The interaction of these glyconanoparticles with β-D-galactose-recognizing lectins peanut agglutinin (PNA) and human galectin-3 (Gal-3) was demonstrated by UV–vis studies. Gal-3 is well known to be overexpressed in several human tumors.

Penadés and coworkers used gold-coated magnetite (glyco-ferrites) as multimodal fluorescent/magnetic glyconanoparticles, which were converted into specific probes by the covalent coupling of protein G and subsequent incubation with an IgG antibody [54]. The immunomagnetic-fluorescent nanoparticles were used for specific labeling of peripheral blood mononuclear cells in a complex biological medium, as human blood. They were able to label specifically peripheral blood mononuclear cells present in blood at a percentage as low as 0.10–0.17% in complex biological media.

Russell, Field, and coworkers have recently shown that gold glyconanoparticles, carrying a trivalent, α-2,6-thio-linked sialic acid–galactose derivative, were able to detect influenza virus (H3N2) within 30 min by acting as a plasmonic sensor [55]. Importantly, these nanoparticles were able to discriminate between human (α-2,6 binding) and avian (α-2,3 binding) influenza.

3.6 CONCLUSION

Carbohydrate-modified gold nanoparticles have emerged as highly promising and versatile molecular tools, as they can be prepared in different sizes, their optical properties depend on their size, the well-understood surface chemistry for anchoring of thiols though Au—S bonds, and the possibility for controlled multivalent display of carbohydrates. Carbohydrate-modified gold nanoparticles can be prepared from preformed, thiol-functionalized glycoconjugates or by direct coupling of unprotected, unmodified glycans to reactive shell nanoparticles. Highly chemoselective reactions allow the simple and convenient construction of glycoconjugates and carbohydrate-modified gold nanoparticles, even with complex carbohydrates. Applications of carbohydrate-modified gold nanoparticles have been demonstrated, especially for studying carbohydrate–protein interactions *ex vivo* and for *in vivo* or *in vitro* diagnostics.

REFERENCES

1. (a) Lee, Y. C.; Lee, R. T. Carbohydrate-protein interactions: Basis of glycobiology. *Acc. Chem. Res.* **1995**, *28*, 321–327; (b) Mammen, M.; Choi, S. K.; Whitesides, G. M. Polyvalent interactions in biological systems: Implications for design and use of multivalent ligands and inhibitors. *Angew. Chem. Int. Ed.* **1998**, *37*, 2755–2794; (c) Kitov, P. I.; Bundle, D. R. On the nature of the multivalency effect: A thermodynamic model. *J. Am. Chem. Soc.* **2003**, *125*, 16271–16284.
2. Bertozzi, C. R.; Kiessling, L. L. Chemical glycobiology. *Science* **2001**, *291*, 2357–2364.
3. (a) de la Fuente, J. M.; Penadés, S. Glyconanoparticles: Types, synthesis and applications in glycoscience, biomedicine and material science. *Biochim. Biophys. Acta Gen. Sub..* **2006**, *1760*, 636–651; (b) Larsen, K.; Thygesen, M. B.; Guillaumie, F.; Willats, W. G. T.; Jensen, K. J. Solid-phase chemical tools for glycobiology. *Carbohydr. Res.* **2006**, *341*, 1209–1234; (c) Sihelníková, L.; Tvaroska, I. Step by step towards understanding gold glyconanoparticles as elements of the nanoworld. *Chem. Pap.* **2007**, *61*, 237–255;

(d) García, I.; Marradi, M.; Penadés, S. Glyconanoparticles: Multifunctional nanomaterials for biomedical applications. *Nanomedicine* **2010**, *5*, 777–792; (e) Wang, X.; Ramström, O.; Yan, M. D. Glyconanomaterials: Synthesis, characterization, and ligand presentation. *Adv. Mater.* **2010**, *22*, 1946–1953; (f) Marradi, M.; Martín-Lomas, M.; Penadés, S. Glyconanoparticles: Polyvalent tools to study carbohydrate-based interactions. In *Advances in Carbohydrate Chemistry and Biochemistry*; Horton, D., Ed.; Elsevier Academic Press Inc: San Diego, CA, **2010**; Vol. *64*, pp 211–290; (g) Thygesen, M. B.; Jensen, K. J. Chemoselectivity and glyconanoparticles. In *Petite and Sweet: Glyco-Nanotechnology as a Bridge to New Medicines*; Huang, X.; Barchi Jr., J. J., Eds.; ACS Symposium Series; Oxford University Press: Oxford, **2011**; Vol. *1091*, pp 37–48; (h) Marradi, M.; Chiodo, F.; García, I.; Penadés, S. Glyconanoparticles as multifunctional and multimodal carbohydrate systems. *Chem. Soc. Rev.* **2013**, *42*, 4728–2745; (i) Adak, A. K.; Lin, H. J.; Lin, C. C. Multivalent glycosylated nanoparticles for studying carbohydrate-protein interactions. *Org. Biomol. Chem.* **2014**, *12*, 5563–5573.

4. Chien, Y. Y.; Jan, M. D.; Adak, A. K.; Tzeng, H. C.; Lin, Y. P.; Chen, Y. J.; Wang, K. T.; Chen, C. T.; Chen, C. C.; Lin, C. C. Globotriose-functionalized gold nanoparticles as multivalent probes for Shiga-like toxin. *Chembiochem* **2008**, *9*, 1100–1109.
5. Kitov, P. I.; Sadowska, J. M.; Mulvey, G.; Armstrong, G. D.; Ling, H.; Pannu, N. S.; Read, R. J.; Bundle, D. R. Shiga-like toxins are neutralized by tailored multivalent carbohydrate ligands. *Nature* **2000**, *403*, 669–672.
6. Barrientos, A. G.; de la Fuente, J. M.; Jiménez, M.; Solis, D.; Cañada, F. J.; Martín-Lomas, M.; Penadés, S. Modulating glycosidase degradation and lectin recognition of gold glyconanoparticles. *Carbohydr. Res.* **2009**, *344*, 1474–1478.
7. Walter, M.; Akola, J.; Lopez-Acevedo, O.; Jadzinsky, P. D.; Calero, G.; Ackerson, C. J.; Whetten, R. L.; Gronbeck, H.; Hakkinen, H. A unified view of ligand-protected gold clusters as superatom complexes. *Proc. Natl. Acad. Sci. U. S. A.* **2008**, *105*, 9157–9162.
8. Grzelczak, M.; Pérez-Juste, J.; Mulvaney, P.; Liz-Marzán, L. M. Shape control in gold nanoparticle synthesis. *Chem. Soc. Rev.* **2008**, *37*, 1783–1791.
9. (a) Parker, J. F.; Fields-Zinna, C. A.; Murray, R. W. The story of a monodisperse gold nanoparticle: $Au_{25}L_{18}$. *Acc. Chem. Res.* **2010**, *43*, 1289–1296; (b) Zheng, J.; Nicovich, P. R.; Dickson, R. M. Highly fluorescent noble-metal quantum dots. *Annu. Rev. Phys. Chem.* **2007**, *58*, 409–431.
10. Ghosh, S. K.; Pal, T. Interparticle coupling effect on the surface plasmon resonance of gold nanoparticles: From theory to applications. *Chem. Rev.* **2007**, *107*, 4797–4862.
11. Daniel, M. C.; Astruc, D. Gold nanoparticles: Assembly, supramolecular chemistry, quantum-size-related properties, and applications toward biology, catalysis, and nanotechnology. *Chem. Rev.* **2004**, *104*, 293–346.
12. Haiss, W.; Thanh, N. T. K.; Aveyard, J.; Fernig, D. G. Determination of size and concentration of gold nanoparticles from UV-Vis spectra. *Anal. Chem.* **2007**, *79*, 4215–4221.
13. Mie, G. Articles on the optical characteristics of turbid tubes, especially colloidal metal solutions. *Ann. Phys.* **1908**, *25*, 377–445.
14. Taton, T. A.; Lu, G.; Mirkin, C. A. Two-color labeling of oligonucleotide arrays via size-selective scattering of nanoparticle probes. *J. Am. Chem. Soc.* **2001**, *123*, 5164–5165.
15. Rosi, N. L.; Mirkin, C. A. Nanostructures in biodiagnostics. *Chem. Rev.* **2005**, *105*, 1547–1562.
16. de la Fuente, J. M.; Barrientos, A. G.; Rojas, T. C.; Rojo, J.; Cañada, J.; Fernández, A.; Penadés, S. Gold glyconanoparticles as water-soluble polyvalent models to study carbohydrate interactions. *Angew. Chem. Int. Ed.* **2001**, *40*, 2258–2261.

17. Brust, M.; Walker, M.; Bethell, D.; Schiffrin, D. J.; Whyman, R. Synthesis of thiol-derivatized gold nanoparticles in a 2-phase liquid-liquid system. *J. Chem. Soc. Chem. Commun.*, **1994**, 801–802.

18. Goulet, P. J.; Lennox, R. B. New insights into Brust-Schiffrin metal nanoparticle synthesis. *J. Am. Chem. Soc.* **2010**, *132*, 9582–9584.

19. Turkevich, J.; Stevenson, P. C.; Hillier, J. A study of the nucleation and growth processes in the synthesis of colloidal gold. *Discuss. Farad. Soc.* **1951**, *11*, 55–75.

20. Frens, G. Controlled nucleation for regulation of particle-size in monodisperse gold suspensions. *Nat. Phys. Sci.* **1973**, *241*, 20–22.

21. Ji, X. H.; Song, X. N.; Li, J.; Bai, Y. B.; Yang, W. S.; Peng, X. G. Size control of gold nanocrystals in citrate reduction: The third role of citrate. *J. Am. Chem. Soc.* **2007**, *129*, 13939–13948.

22. Weisbecker, C. S.; Merritt, M. V.; Whitesides, G. M. Molecular self-assembly of aliphatic thiols on gold colloids. *Langmuir* **1996**, *12*, 3763–3772.

23. Hone, D. C.; Haines, A. H.; Russell, D. A. Rapid, quantitative colorimetric detection of a lectin using mannose-stabilized gold nanoparticles. *Langmuir* **2003**, *19*, 7141–7144.

24. Whitesides, G. M.; Laibinis, P. E. Wet chemical approaches to the characterization of organic surfaces: Self-assembled monolayers, wetting, and the physical-organic chemistry of the solid-liquid interface. *Langmuir* **1990**, *6*, 87–96.

25. (a) Hostetler, M. J.; Green, S. J.; Stokes, J. J.; Murray, R. W. Monolayers in three dimensions: Synthesis and electrochemistry of omega-functionalized alkanethiolate-stabilized gold cluster compounds. *J. Am. Chem. Soc.* **1996**, *118*, 4212–4213; (b) Briñas, R. P.; Maetani, M.; Barchi, J., J. J. A survey of place-exchange reaction for the preparation of water-soluble gold nanoparticles. *J. Colloid Interface Sci.* **2013**, *392*, 415–421.

26. Jadzinsky, P. D.; Calero, G.; Ackerson, C. J.; Bushnell, D. A.; Kornberg, R. D. Structure of a thiol monolayer-protected gold nanoparticle at 1.1 angstrom resolution. *Science* **2007**, *318*, 430–433.

27. McCarley, R. L.; Dunaway, D. J.; Willicut, R. J. Mobility of the alkanethiol-gold (111) interface studied by scanning probe microscopy. *Langmuir* **1993**, *9*, 2775–2777.

28. Lin, C. C.; Yeh, Y. C.; Yang, C. Y.; Chen, G. F.; Chen, Y. C.; Wu, Y. C.; Chen, C. C. Quantitative analysis of multivalent interactions of carbohydrate-encapsulated gold nanoparticles with concanavalin A. *Chem. Commun.* **2003**, 2920–2921.

29. Martos-Maldonado, M. C.; Thygesen, M. B.; Jensen, K. J.; Vargas-Berenguel, A. Gold ferrocene glyconanoparticles for high-sensitivity electrochemical detection of carbohydrate-lectin interactions. *Eur. J. Org. Chem.* **2013**, 2793–2801.

30. Thygesen, M. B.; Sauer, J.; Jensen, K. J. Chemoselective capture of glycans for analysis on gold nanoparticles: Carbohydrate oxime tautomers provide functional recognition by proteins. *Chem. Eur. J.* **2009**, *15*, 1649–1660.

31. Martínez-Ávila, O.; Hijazi, K.; Marradi, M.; Clavel, C.; Campion, C.; Kelly, C.; Penadés, S. Gold manno-glyconanoparticles: Multivalent systems to block HIV-1 gp120 binding to the lectin DC-SIGN. *Chem. Eur. J.* **2009**, *15*, 9874–9888.

32. Carvalho de Souza, A.; Halkes, K. M.; Meeldijk, J. D.; Verkleij, A. J.; Vliegenthart, J. F. G.; Kamerling, J. P. Synthesis of gold glyconanoparticles: Possible probes for the exploration of carbohydrate-mediated self-recognition of marine sponge cells. *Eur. J. Org. Chem.* **2004**, 4323–4339.

33. Guillaumie, F.; Thomas, O. R. T.; Jensen, K. J. Immobilization of pectin fragments on solid supports: Novel coupling by thiazolidine formation. *Bioconj. Chem.* **2002**, *13*, 285–294.

34. Liu, Y.; Feizi, T.; Carnpanero-Rhodes, M. A.; Childs, R. A.; Zhang, Y. N.; Muiioy, B.; Evans, P. G.; Osborn, H. M. I.; Otto, D.; Crocker, P. R.; Chai, W. C. Neoglycolipid probes prepared via oxime ligation for microarray analysis of oligosaccharide-protein interactions. *Chem. Biol.* **2007**, *14*, 847–859.

35. Halkes, K. M.; de Souza, A. C.; Maljaars, C. E. P.; Gerwig, G. J.; Kamerling, J. P. A facile method for the preparation of gold glyconanoparticles from free oligosaccharides and their applicability in carbohydrate-protein interaction studies. *Eur. J. Org. Chem.* **2005**, 3650–3659.

36. Liu, Y.; Feizi, T.; Campanero-Rhodes, M. A.; Childs, R. A.; Zhang, Y.; Mulloy, B.; Evans, P. G.; Osborn, H. M. I.; Otto, D.; Crocker, P. R.; Chai, W. Neoglycolipid probes prepared via oxime ligation for microarray analysis of oligosaccharide-protein interactions. *Chem. Biol.* **2007**, *14*, 847–859.

37. Finch, P.; Merchant, Z. Structures of D-arabinose and D-glucose oximes. *J. Chem. Soc. Perkin Trans. 1* **1975**, 1682–1686.

38. Cló, E.; Blixt, O.; Jensen, K. J. Chemoselective reagents for covalent capture and display of glycans in microarrays. *Eur. J. Org. Chem.* **2010**, 540–554.

39. (a) Cordes, E. H.; Jencks, W. P. Nucleophilic catalysis of semicarbazone formation by anilines. *J. Am. Chem. Soc.* **1962**, *84*, 826–831; (b) Dirksen, A.; Hackeng, T. M.; Dawson, P. E. Nucleophilic catalysis of oxime ligation. *Angew. Chem. Int. Ed.* **2006**, *45*, 7581–7584.

40. Thygesen, M. B.; Munch, H.; Sauer, J.; Cló, E.; Jørgensen, M. R.; Hindsgaul, O.; Jensen, K. J. Nucleophilic catalysis of carbohydrate oxime formation by anilines. *J. Org. Chem.* **2010**, *75*, 1752–1755.

41. Otsuka, H.; Akiyama, Y.; Nagasaki, Y.; Kataoka, K. Quantitative and reversible lectin-induced association of gold nanoparticles modified with alpha-lactosyl-omega-mercapto-poly(ethylene glycol). *J. Am. Chem. Soc.* **2001**, *123*, 8226–8230.

42. Thygesen, M. B.; Sørensen, K. K.; Cló, E.; Jensen, K. J. Direct chemoselective synthesis of glyconanoparticles from unprotected reducing glycans and glycopeptide aldehydes. *Chem. Commun.* **2009**, 6367–6369.

43. Chuang, Y. J.; Zhou, X. C.; Pan, Z. W.; Turchi, C. A convenient method for synthesis of glyconanoparticles for colorimetric measuring carbohydrate-protein interactions. *Biochem. Biophys. Res. Commun.* **2009**, *389*, 22–27.

44. Wang, X.; Ramström, O.; Yan, M. D. A photochemically initiated chemistry for coupling underivatized carbohydrates to gold nanoparticles. *J. Mater. Chem.* **2009**, *19*, 8944–8949.

45. Reynolds, A. J.; Haines, A. H.; Russell, D. A. Gold glyconanoparticles for mimics and measurement of metal Ion-mediated carbohydrate-carbohydrate interactions. *Langmuir* **2006**, *22*, 1156–1163.

46. Schofield, C. L.; Mukhopadhyay, B.; Hardy, S. M.; McDonnell, M. B.; Field, R. A.; Russell, D. A. Colorimetric detection of Ricinus communis Agglutinin 120 using optimally presented carbohydrate-stabilised gold nanoparticles. *Analyst* **2008**, *133*, 626–634.

47. Nagahori, N.; Nishimura, S.-I. Direct and efficient monitoring of glycosyltransferase reactions on gold colloidal nanoparticles by using mass spectrometry. *Chem. Eur. J.* **2006**, *12*, 6478–6485.

48. Chen, Y. J.; Chen, S. H.; Chien, Y. Y.; Chang, Y. W.; Liao, H. K.; Chang, C. Y.; Jan, M. D.; Wang, K. T.; Lin, C. C. Carbohydrate-encapsulated gold nanoparticles for rapid target-protein identification and binding-epitope mapping. *Chembiochem* **2005**, *6*, 1169–1173.

49. Elvira, G.; Garcia, I.; Benito, M.; Gallo, J.; Desco, M.; Penadés, S.; Garcia-Sanz, J. A.; Silva, A. Live imaging of mouse endogenous neural progenitors migrating in response to an induced tumor. *PLoS One* **2012**, *7*, e44466.

50. Yu, C. C.; Huang, L. D.; Kwan, D. H.; Wakarchuk, W. W.; Withers, S. G.; Lin, C. C. A glyco-gold nanoparticle based assay for α-2,8-polysialyltransferase from *Neisseria meningitidis*. *Chem. Commun.* **2013**, *49*, 10166–10168.

51. Parry, A. L.; Clemson, N. A.; Ellis, J.; Bernhard, S. S. R.; Davis, B. G.; Cameron, N. R. 'Multicopy Multivalent' glycopolymer-stabilized gold nanoparticles as potential synthetic cancer vaccines. *J. Am. Chem. Soc.* **2014**, *135*, 9362–9365.

52. Candiota, A. P.; Acosta, M.; Simões, R. V.; Delgado-Goñi, T.; Lope-Piedrafita, S.; Irure, A.; Marradi, M.; Bomati-Miguel, O.; Miguel-Sancho, N.; Abasolo, I.; Schwartz Jr., S.; Santamaria, J.; Penadés, S.; Arús, C. A new *ex vivo* method to evaluate the performance of candidate MRI contrast agents: A proof-of-concept study. *J. Nanobiotechnol.* **2014**, *12*, 12.

53. Aykaç A; Martos-Maldonado, M. C.; Casas-Solvas, J. M.; Quesada-Soriano, I.; García-Maroto, F.; García-Fuentes, L.; Vargas-Berenguel, A. Cyclodextrin-bearing gold glyconanoparticles for the development of site specific drug delivery systems. *Langmuir* **2014**, *30*, 234–242.

54. Gallo, J.; García, I.; Genicio, N.; Padro, D.; Penadés, S. Specific labelling of cell populations in blood with targeted immuno-fluorescent/magnetic glyconanoparticles. *Biomater.* **2011**, *32*, 9818–9825.

55. Marín, M. J.; Rashid, A.; Rejzek, M.; Fairhurst, S. A.; Wharton, S. A.; Martin, S. R.; McCauley, J. W.; Wileman, T.; Field, R. A.; Russell, D. A. Glyconanoparticles for the plasmonic detection and discrimination between human and avian influenza virus. *Org. Biomol. Chem.* **2013**, *11*, 7101–7107.

4

QUANTUM DOT GLYCOCONJUGATES

NAN LI AND KAGAN KERMAN

Department of Physical and Environmental Sciences, University of Toronto, Scarborough, Toronto, Ontario, Canada

4.1 INTRODUCTION

Quantum dots (QDs), also known as semiconductor nanocrystals that are typically between 2 and 10 nm in diameter, are a significant advancement in nanomaterial technology [1–8]. The nanocrystals are constructed of atoms arranged in a periodic lattice, typically containing elements in groups II–VI or III–V on the periodic table, and possess dimensions that are smaller than the exciton Bohr radius [1–8]. The structural and material organizations of nanocrystals provide them with unique optical and electrical properties. With regard to bulk semiconductors, photon energy is used to generate electron excitation through the promotion of an electron across the band gap [9]. In the case of QDs, the band gap can be modified in a size-dependent manner. As the size of QDs approaches the Bohr exciton radius for its particular element, the band gap energy becomes inversely proportional to the size of the particle [10]. This, in turn, results in the ability to construct QDs that possess variation in their absorption and emission spectra for a variety of applications [11]. In addition, the electrostatic repulsion of the charges confined on a QD results in an inversely proportional relationship between particle size and the energy required to add excess charges to a QD [12–18].

As shown in Figure 4.1, the most commonly used types of QD in biological applications are the CdSe/ZnS nanocrystals [19–21]. They consist of a roughly spherical CdSe core coated with atomic layers of ZnS. ZnS layers enhance and protect the optical properties of the CdSe core. Namely, ZnS layers isolate the core from its microenvironment, protecting it from degradation, as well as raising the potential

Carbohydrate Nanotechnology, First Edition. Edited by Keith J. Stine.

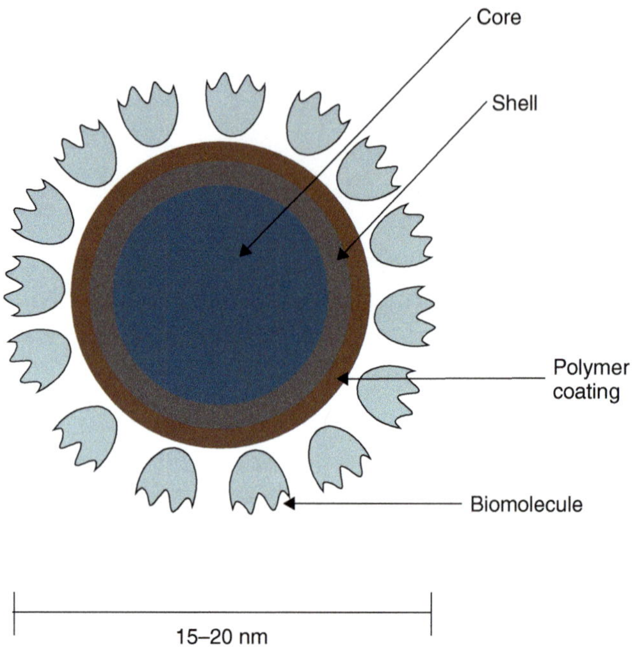

FIGURE 4.1 Common QD structure.

energy barrier for exciton confinement within the core. The popularity of CdSe/ZnS QDs owes largely to the already established and well-documented protocols making use of their visible spectrum photoluminescence that overlaps with numerous organic dyes, as well as their versatility in biomolecular applications [22–38]. Although CdSe/ZnS is the most common material used to construct QDs, other materials are currently being investigated. In addition, further enhancement of emission can be accomplished by structural variations such as core–shell–shell QDs and "Giant QDs" and by utilizing various band offsets and lattice parameters [21].

Part of the reason why many research efforts have focused on the biological applications of QDs is credited to their optical properties, which are heavily dependent on their size [19]. As the size of a QD is decreased, there is an increase in the electronic excitation energies and a concentration of oscillator strength into fewer transitions [19]. Using the relationship between energy and momentum in a bulk solid, it becomes easier to understand how a series of close transitions occurring at slightly variable energies in the bulk will be influenced by quantum confinement into a single concentrated transition in a QD. The onset shifts in absorption in colloidal prepared semiconductors such as CdS and CdSe (II–VI semiconductors) and InP (III–V semiconductors) allow for tuning across a major proportion of the visible spectrum. For example, Murray *et al.* [20] reported that the tuning of the band gap in CdSe can be modified from a deep red (1.7 eV) to green (2.4 eV) by reducing the cluster diameter from 200 to 20 Å. The luminescence emitted from these nanocrystals follows the absorption onset. At room temperature, one can observe, with efficiency at about 10%, a narrow band and a size-tunable luminescence, properties that are desirable for LED production. Research groups have constructed nanocrystals in thin layers, placing them into poly(*p*-phenylene vinylene) (PVP) [22]. PVP is known to be an electroluminescent polymer. As the nanocrystalline layer was subsequently treated with a coating of magnesium and silver, which served as the electron-injecting contacts, varying the applied voltage allowed for the variable colors of electroluminescence. Since nanocrystal emission is size dependent, these LEDs are beneficial for electronic and optoelectronic applications in that they can be tuned without significant changes in their production methods [23].

QDs exhibit narrow photoluminescent bands, which can be approximated by a Gaussian profile, displaying a full width at half maximum between 25 and 35 nm [23]. Compared to molecular fluorophores, QDs have been observed to exhibit much brighter luminescence that is resistant to photobleaching [23]. As a result, QDs allow for increased detection times, as well as enhanced reliability in the monitoring of complex samples. Although the blinking of fluorescence was thought to interfere with precise tracking, this obstacle could be overcome by creating aggregates of QDs [21,23].

An important property of semiconductor nanocrystals is their ability to vary the energy difference between valence and conduction bands by adjusting their particle size [12]. The strong correlation between the optical and electrochemical band gaps present in QDs has been elucidated using cyclic voltammetry (CV) [24]. In CV, irreversible anodic and cathodic waves were observed corresponding to the removal of an electron from the valence band and the injection of an electron into the conduction band [24]. As predicted by the quantum confinement of the charges, the difference between these peak potentials was inversely proportional to the size of the particle. In QDs, the electrochemical band gap was detected to be 0.4–0.5 eV smaller than the optical band gap, and this may be attributed to surface defects, which can likely act as trap states and holes for electrons [25]. Surface defects on QDs may play a key role in influencing the redox potentials of these particles. The potential roles of these surface defects may be most conspicuous in the outliers of both the smaller and larger particles in a batch of QDs. The moderately sized particles displayed much higher redox potentials, when compared to the others in any particular batch or family [25]. One method being investigated to partially remove these surface defects is to coat the core with a second semiconductor material, which would allow for much greater chemical stability. A study that used differential pulse voltammetry (DPV) to investigate the coated nanocrystals found a shift in both oxidative and reductive potentials in the positive direction, with oxidation potential being slightly decreased and reduction enhanced [26]. Anodic stripping DPV provided a cost-effective approach for the characterization of QDs [24–26].

4.2 SYNTHESIS OF QDs

This section will begin with an overview of the synthesis of the commonly used QDs. A safe experiment feasible in an undergraduate laboratory was discussed by Landry *et al.* [27]. Their experiment involved the synchronized injection of cadmium and selenium precursor solutions into a heated solution of octadecane. This process could take place at ambient atmosphere. The individual atoms of Cd and Se eventually formed clusters of CdSe, as the reaction proceeded in the forward direction. In a second procedure, oleylamine was included as an addition reagent to the solution. This reagent was reported to help reduce the trap state emission, in turn allowing for more exciton state emission [27].

In the synthesis reported by Wang *et al.* [28], a mechanism was suggested for the formation of CdTe QDs. $CdCl_2$ with 2.5 mol H_2O and TeO_2 were used as the precursors and cysteine as the capping ligand. The *in situ* cadmium–thiol complexes reacted with the reduced form of TeO_2 as Te_2^- to form the nuclear form of CdTe. The result of this synthesis was a tunable size-dependent emission from green to yellow [28].

Ceria (CeO_2) is a cubic fluorite-type oxide that has applications in catalysis, ion removal, and luminescent layers [29]. Using microwave-assisted solvothermal synthesis and a heteroleptic cerium alkoxide complex, CeO_2 QDs can be produced. This synthesis was reported to produce QDs with a desirable crystallinity and narrow size variability [29]. This refined technique significantly shortened the reaction time when compared to previously reported methods.

CdS is a material of special interest due to its combined properties of elevated photosensitivity and a band-gap energy (2.4 eV) that appears in the visible spectrum [30]. Entezari and Ghows [30] reported a conglomerate method of ultrasound and microemulsion to produce CdS QDs. The advantage of this technique was its efficiency in producing fine QDs of uniform shape. In addition, by increasing sonication time, the researchers were able to increase the surface area of QDs [30].

ZnO nanoparticles that were smaller than 7 nm exhibited quantum-size effect, in which the blueshift of light absorbance could be used to detect the enlargement of the band gap [31]. This property made ZnO nanoparticles desirable for UV protection films and chemical sensors [31]. ZnO/Si nanoparticles were synthesized through flame-sprayed pyrolysis, and through using this method, the size of nanocrystalline particles were closely monitored by the coprecipitation of silica in the flame [32].

The synthesis of SnS QDs has also been reported [33], with these semiconductor nanoparticles being especially useful for their applications in near-infrared detection, as well as possible medical applications in the treatment of hypothermia, in which substantial infrared absorption is required [33]. Colloidal SnS QDs have been reported to be synthesized with narrow size distributions using ethanolamine ligands and also exhibited size-dependent influences in absorbance and reflectance [33].

4.3 INTERFACIAL CHEMISTRY AND BIOCONJUGATION

Some of the most exciting applications of QDs lie within device fabrication and diagnostic applications [34–40]. This section will discuss these various applications while focusing on the interfacial and bioconjugate chemistry involved.

The use of Förster resonance energy transfer (FRET) is exceptional for the alteration of the bright luminescence presented by QDs, which have been immobilized at an interface. The immobilization of QDs via attachment to a suitable surface or

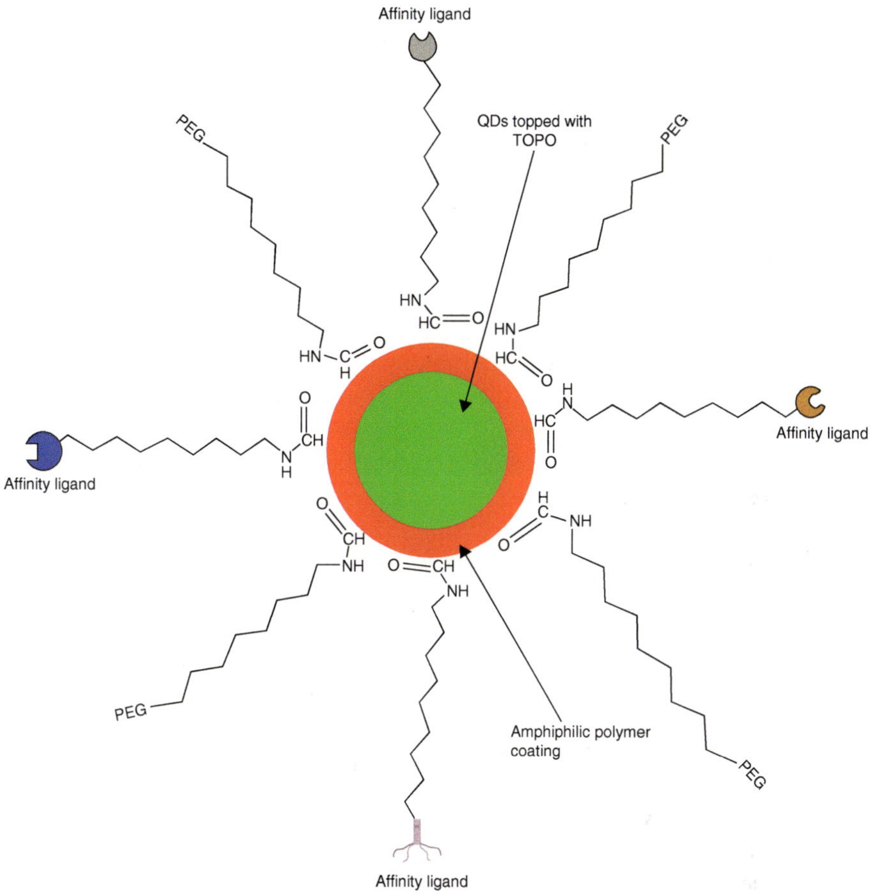

FIGURE 4.2 QD surface modifications for biological applications. Both trioctylphosphine and trioctylphosphine oxide (TOPO) are frequently used as the capping ligands. TOPO serves as solvent for the synthesis and solubilizes the growing nanocrystals. TOPO-coated QDs are typically soluble in chloroform, toluene, and (to a lesser extent) hexane. These QDs can also be used in biological applications, as the protective barrier of TOPO would prevent their degradation under physiological conditions.

molecular scaffold allows one to employ FRET techniques for the detection of specific nucleic acid sequences [34–39]. QD-FRET techniques have been reported to be useful for a wide variety of analytes, ranging in size and complexity from metabolites, nucleic acids, and small molecules to larger complexes such as enzymes and proteins [35–37]. The basic methodology used to produce an assay utilizing QDs in connection with FRET signaling begins with the passivation and protection of the core QD with a thin shell of a higher band-gap semiconductor. As shown in Figure 4.2, the surface of the QD is then altered with a bifunctional ligand or polymer. This addition provides aqueous stability as well as a suitable source of reactive groups for bioconjugation and the self-assembly of biomolecular complexes. Typically, a probe possessing selective interaction with the target of interest is bioconjugated with the QD. A fluorescent dye is then associated with the QD–target involvement via the bioconjugated reactive group; the interaction between the target molecule and the QD–dye complex results in a quenching effect, either turning the FRET on or off, allowing the determination or quantification of the target analyte [34–39].

The immobilization of QDs on surfaces is integral for successful development of bioassays. For example, the immobilized QDs compensate for the lack of colloidal stability inherent in QDs, providing increased resistance to degradation and an improvement in assay sensitivity. Two of the primary routes for immobilization are multidentate surface ligand exchange (MSLE) and biological tethers. MSLE relies on the direct coordination of interfacial functional groups with the QD of interest, while biological tethers involve high-affinity and selective interactions with biomolecules [34–39]. Past studies have shown that bidentate thiol ligands provided greater stability to the colloidal structure of CdSe/ZnS QDs, with the mechanism of this increased stability due to the reduction in the rate of desorption from ZnS shell [38]. An example of biological tethering as a means of immobilizing QDs is the well-described streptavidin–biotin system [39].

Self-assembly is a desirable property to possess, when attempting to produce hierarchal or complex molecular assemblies [39, 40]. Through the diazo reaction, it was reported that functional supramolecular assemblies of ZnO QDs formed micrometer- and millimeter-sized molecular structures. This process was completed through liquid–liquid interfaces, with the concerted effects of the diazo reaction and an arylamine treatment leading to the formation of self-assembled inorganic complexes [40].

A QD bioconjugate can be defined as a molecular assembly or scaffold, in which a QD has been attached to or complexed with a biomolecule (i.e., oligonucleotides, peptides, receptors), regardless of the type of bonding involved [36,37]. The applications for these bioconjugated QDs can be generally divided into two categories: passive and dynamic uses. The passive roles rely on the photoluminescence of QD for fluorescent signals to be used in imaging techniques, with an example being the use of QD–antibody conjugates in immunoassays [36,37]. The active incarnations of QD bioconjugates utilize QDs as both a means to interact with biomolecules and to monitor changes in the QD photoluminescence [37]. Typically, these active QD bioconjugates contain numerous dyes or electroactive biomolecules. The interaction of the target analyte with the biomolecules attached to the surface of QD would lead to conformational changes. These changes in the molecular shape lead to alterations in the mechanism or rate of energy transfer, ultimately changing the QD photoluminescence signal that is detected [36, 37].

Anchoring and hydrophilic groups are two integral components most surface ligands possess. The anchoring groups ensure a strong connection between the QD and the desired ligands, whereas hydrophilic groups are typically necessary in order to facilitate or improve the solubility of the bioconjugates in aqueous media. In many cases, a third functional unit on the ligand would be incorporated to allow for additional bioconjugations, providing additional versatility or the incorporation of ancillary labels. When deciding on what ligands to use for the creation of a water-soluble, biologically applicable QD, one must consider the following qualities: strong affinity to the QD, colloidal stability across a variety of solution conditions, minimized adsorption of nonspecific matter in the biological environment, minimized stimulation of the immune response in the biological environment, as well as additional economic considerations. It is also desirable to maintain a minimized final diameter and to possess the ability to incorporate additional units for further bioconjugations in order to provide versatile and multipurpose QD scaffolds [36–38].

It is especially important for a bioconjugated QD to be stable across a variety of pH and ionic strength conditions. This is due to the fact that endocytosis is a common method cells used to mediate the uptake of QDs into the cytosol [36–39]. It has also been reported that many of the biological coupling reactions, which take place in an aqueous environment, require an optimal pH, or specific ionic strength, for optimal binding. In other words, localized microenvironments within a cell may be slightly acidic or basic, reinforcing the need for the QD to be stable across a broad range of acidities [36–39].

N-Hydroxysuccinimide (NHS)- and ethylenediamine carbodiimide (EDC)-mediated condensation of carboxyl groups and amines forms a strong amide bond and is the most commonly employed method for the production of QD–antibody conjugations [37]. This preference is due to the fact that proteins globally present amine and carboxyl functional groups on their side chains as well as the fact that many QD preparations yield carboxyl and amine groups on their surfaces. Therefore, in many applications, the bioconjugation procedure included a buffered solution of the QD, protein, and EDC/NHS to produce the products as desired [37].

Another popular method employed to produce bioconjugated QDs involves the use of avidin–biotin chemistry. NeutrAvidin is commonly used due to the high association constant of ($K_a \sim 10^{15}$ M^{-1}) with biotin, wide pH and salt stability of the complex, ease of use, and the availability of many refined protocols [37]. Techniques such as the avidin–biotin chemistry and EDC/NHS have overthrown former techniques such as thiol binding due to its lack of stability and the ease in which the newer methods may be employed.

Studies have also investigated the need for functional group interconversion in QDs because of the limited functional groups on the surface of QDs that are available for direct display and the limitation of later bioconjugations. Therefore, there are numerous chemical reagents that can alter the existing functional groups present on a QD. For example, the amine, carboxyl, and thiol groups can be converted to display sulfhydryls, carboxylates, aldehydes, and numerous glycan groups [39–41].

4.4 CARBOHYDRATES

Carbohydrates are the most abundant organic molecules that comprise living systems [42]. Mainly described as polyhydroxy aldehydes/ketones or glycans, these molecules generally follow the empirical formula $C_nH_{2n}O_n$ where $n \geq 3$ [42]. The simplest carbohydrates are monosaccharides and when joined together form oligosaccharides or polysaccharides. An oligosaccharide is defined as containing 2–20 monosaccharide residues joined by glycosidic linkages, while a polysaccharide is defined as any linear or branched polymer of monosaccharides such as glycogen. Hence, the relationship between monosaccharides, polysaccharides, and oligosaccharides are analogous to amino acids and proteins, or nucleotides and nucleic acids [42, 43]. When macromolecules such as lipids or proteins contain covalently linked monosaccharides, they are known as glycoconjugates, which include glycolipids, glycoproteins, and proteoglycans [42]. Traditionally known for their role as energy suppliers through metabolism, glycan functions are thought to be split into two categories: the first is as structural and modular units, and the second is as specific molecular recognition [43]. These two functions involve glycans in a multitude of roles, such as cancer metastasis, cell adhesion, cell signaling, embryo development, protein function regulation, glycosylation, and cellular communications [44]. Glycans, unlike enzymes, don't have their information coded in the genome. Therefore, the recognition and binding of proteins to glycans are key to understanding their biological properties [43, 44]. The development of techniques such as X-ray crystallography, nuclear magnetic resonance, affinity chromatography, and isothermal titration calorimetry and the development of diagnostic agents like probes to recognize carbohydrate-based biomarkers have all been used to further the understanding of numerous carbohydrate functions and structures [43].

4.5 FLUORESCENCE SPECTROSCOPY OF QDs WITH CARBOHYDRATES

Fluorescence spectroscopy is one of most well-documented techniques for disease diagnosis, presenting a variety of advantages when compared to other biosensing methods. Presenting high sensitivity, fluorescence spectroscopy techniques also exhibit the ability to quickly quantify the target analyte through both digital analysis and semiquantitative visual inspection. These qualities present great promise for disease detection using real samples. For example, it is of utmost importance to monitor blood glucose levels in diseases like diabetes and certain types of cancers. Many methods have been proposed to detect glucose levels using near-infrared spectroscopy, colorimetry, and electrochemistry. A study reported by Li *et al.* [45] detailed a novel, fluorescence-based method for the detection of glucose under clinical conditions via solution of anionic carboxyl/hydroxyl-coated graphene-conjugated QDs (GQDs) and a cationic boronic acid-substituted bipyridinium salt. GQDs had high surface area, large diameter, and other characteristics unique to graphene and acted as the fluorescent labels, with *N,N′*-4,4′-bis(benzyl-2-boronic acid)-bipyridinium dibromide (BBV) serving as both a fluorescence quencher and glucose receptor. In solution, both the anionic GQDs and cationic BBV formed an electrostatic compound, effectively quenching the fluorescence signal. Experimental results suggested that the quenching effect was due to electron transfer, as opposed to FRET, due to the fast response time and lack of absorbance overlap between BBV and the GQD emission spectra. Results suggested that glucose inhibited the quenching effect of BBV by forming glucoboronate ester bonds with BBV, through

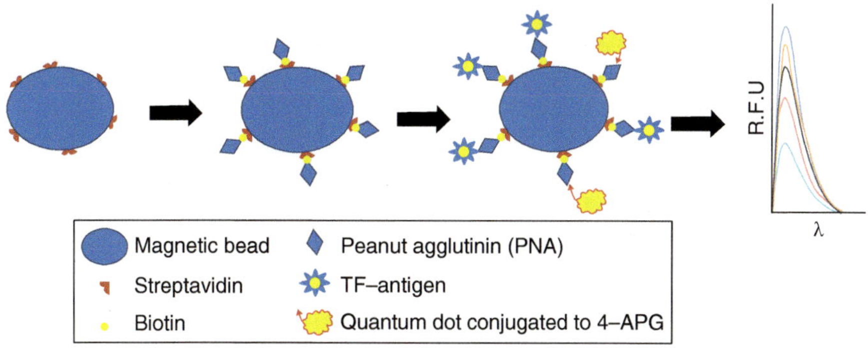

FIGURE 4.3 Conceptual illustration of the QD-based bioassay for the fluorometric detection of TF-antigen. The magnetic beads were coated with streptavidin and acted as the surface on which the PNA immobilized via biotin–streptavidin interaction. 4-Aminophenyl β-D-galactopyranoside (4-APG) was conjugated to QDs. The illustration further shows the competition between the TF-antigen and the 4-APG/QD complex for binding sites on PNAs that are immobilized on magnetic beads.

neutralization of the electrostatic interactions between BBV and GQD reporter. A linear relation in recovered fluorescence intensity was observed with BBV–GQD molar ratio of 20/1, producing a detection platform that was observable to the naked eye under UV light [45].

Current advances in biosensor-based research methods have opened up a path for the ultrasensitive detection of cancer using glycoprotein alterations, which are commonly found in malignant states. Thomsen–Friedenreich antigen (TF-antigen), a tumor-associated antigen, is commonly found on human carcinomas, and the detection of TF-antigen can be used as a new screening approach [46].

As shown in Figure 4.3, Li *et al.* [46] coupled QD-based bioassay with the magnetic beads for the ultrasensitive detection of TF-antigen. Initially, the biotinylated peanut agglutinin (PNA), which is a lectin from *Arachis hypogaea* (peanut), was immobilized onto streptavidin-coated magnetic beads. This was followed by the conjugation of 4-aminophenyl β-D-galactopyranoside (4-APG) to QDs to compete with the TF-antigen for binding sites on the PNA molecule. Transmission electron microscopy (TEM) images confirmed the successful conjugation of QDs with 4-APG, thus allowing the monitoring of the degree of PNA/TF-antigen interaction by measuring the fluorescence signal emission from the QDs. The procedure demonstrated the detection of TF-antigen in the 10^{-7} M range while enhancing the detection limit by eliminating excess QDs. Moreover, the magnetic bead/QD-based bioassay for fluorometric detection of TF-antigen is rapid, cost effective, and highly sensitive, thus addressing limitations of existing methods in this field.

While QD glycoconjugates present unique properties for the quantification and detection of various disease biomarkers, recent studies also present evidence for the use of QD glycoconjugates as drug delivery agents. Owing to the comparatively large surface area, graphene QDs serve as a favorable scaffold for the attachment of targeting molecules, with the possibility for application in mediating drug delivery. For example, chemotherapeutic techniques exhibit painful and debilitating side effects, which greatly hamper their therapeutic profile. A great deal of research has been put into both reducing the cytotoxic effects on normal cells and increasing the therapeutic efficacy of current chemotherapy treatments. Lu *et al.* [47] reported a study in which a novel breast cancer-targeting peptide SP90 was discovered via phage affinity selection for improved efficacy and discrimination in drug delivery. QDs conjugated with SP90 displayed tumor-targeting properties, suggesting their potential use in *in vivo* imaging of tumors. When compared to unlabeled QDs, SP90-conjugated QDs featured increased cellular uptake and postinjection radiance longevity. Another observed feature of interest was the biocompatibility of SP90-conjugated QDs, with low potential for prolonged bioaccumulation due to excretion. Additionally, the use of ligand-conjugated liposomal-encapsulated drugs was shown to be a very promising approach against cancer. The advantages accompanying this technique are as follows: the efficient delivery of 15,000 anticancer molecules into endosomes in low pH, increased therapeutic index, and enhanced drug accumulation in tumor tissues. Importantly, SP90-conjugated liposomes loaded with doxorubicin significantly reduced the toxicity of doxorubicin to white blood cells, while increasing tumor specificity and uptake. SP90-conjugated QDs produced liposomes in 90% of observed cells, compared to only 51% in cells incubated with control peptide liposomes. The phenomenon was accompanied by an increase in cell mortality rates and doxorubicin uptake and suggested the potential for improvements in drug delivery using target-specific peptides.

In addition to using QDs as a scaffold for the attachment of target-specific biomolecules in drug delivery, QDs possess great potential for use in fluorescence-based protein tracking and quantification techniques. Through the attachment of a reporter molecule to the QD scaffold, *in vivo* fluorescence-based bioassays can be developed for the detection of proteins. A bioassay for protein detection utilizing QDs was developed by Shao and Wang [48], which outlined GQDs as a novel platform for the detection of antithrombin (AT) via a heparin-mediated fluorescence anisotropy (FA) assay. In the experimental assay, polyethyleneimine-capped Mn-doped ZnS QDs (PEI-Mn-ZnS QDs) were used to target AT, a glycoprotein located in the blood plasma responsible for the inactivation of several enzymes required for the coagulation system. The determination of AT activity and content has traditionally been determined by chromogenic substrate methods, immunoassays, and more recently ELISA methods, although a fast and simple assay for AT is yet to be developed. The PEI-Mn-ZnS QDs featured very low background noise and persistent luminescence at 585 nm. The system featured a limit of detection (LOD) for AT as low as 0.9 nM and a dynamic range for AT concentrations between 8.6 and 21.5 nM, when utilizing a heparin concentration of 75 μM. The heparin-mediated FA assay was observed to produce varying LODs and linear range values based upon a temperature range of 4–37°C and heparin concentrations between 25 and 75 μM. At lower temperatures, sensitivity was found to be lowered but accompanied by a greatly increased linear range. By taking both sensitivity and detection range into account, the incubation temperature was set at 20°C. This observed property may present opportunities to further refine and individualize the AT detection method based upon required needs. The bioassay for AT detection was also found to be highly selective, producing very little response to the interfering proteins such as thrombin and albumin.

Huang *et al.* [49] reported two sensitive bioassays, which utilized FRET and a tyramide reaction to quantify glucose and hydrogen peroxide. Horseradish peroxidase (HRP) was first conjugated to QDs and could further detect tyramide-labeled carbocyanine dye (Cy5), which minimized the loss of resolution. Since deposition of many tyramide-labeled Cy5 was in very close proximity to the QD–HRP complex, QDs conjugated to HRP could be excited at 488 nm, and the energy was then donated to the accepting chromophore tyramide-labeled Cy5 at 605 nm. Finally, fluorescence intensity could be measured at 670 nm. It was found that the efficiency of FRET was the highest, when the concentration of HRP–QDs was 2.0×10^{-8} M, and that the highest peak of fluorescence intensity for tyramide was reached at 9.0×10^{-7} M, with a dynamic range between 10 and 100 nM, with an LOD of 10 nM. Therefore, an increase in intensity signified an increase in hydrogen peroxide concentration. In the second bioassay, the commercially available glucose (hexokinase method) detection kit was used to measure glucose levels in serum. The QD-based FRET with a glucose oxidase (GOx) enzyme catalyzed the oxidation of glucose. The bioassay used a reaction between glucose and O_2 in the presence of GOx enzyme, which produced hydrogen peroxide. Glucose concentration could be determined by measuring the increase in tyramide Cy5 fluorescence intensity, and the results indicated a greater sensitivity of at least one order of magnitude with respect to a commercially available hexokinase method.

By employing QDs to decrease excitation crosstalk, the FRET process becomes an ultrasensitive carbohydrate detection method. In a FRET-based detection method that utilized capillary electrophoresis, QDs acted as the electron donor and Cy5 dye as the electron acceptor [50]. Cy5 dye was first conjugated with biotinylated chondroitin sulfate-derived disaccharide using reductive amination reaction and then conjugated to streptavidin-labeled QDs. This allowed obtaining a cutoff point from QDs at 650 nm, distinguishing Cy5-labeled disaccharides at 662 nm. Thus, the spectral crosstalk interference was eliminated. By combining these methods, efficient FRET was achieved with an efficiency of 92%. In conclusion, high-resolution capillary electrophoresis proved to be effective in separating very small amounts of donor from the FRET complexes, thereby solving the problem of signal impurity that was associated with most FRET measurements.

Gao *et al.* [51] took advantage of the QD fluorescence quenching capabilities of doxorubicin. Using QDs, researchers were able to monitor doxorubicin release and selectivity of the drug, with the additional benefit of being able to image cancer cells. The drug delivery system was comprised of an electrostatic complex of positively charged $CuInS_2$ QDs and negatively charged poly(L-glutamic acid) (PGA) that acted as a structural scaffold to which doxorubicin was then conjugated. Following the synthesis of QD/PGA–doxorubicin nanoparticles, Fourier transform infrared (FTIR) spectra were used to confirm successful conjugation between doxorubicin and PGA. Stern–Volmer plots were utilized to quantify drug release efficiency, and photoluminescence was measured to quantify QD quenching. Results showed an increase in the concentration of doxorubicin by the addition of PGA–doxorubicin complex, which caused quenching of QDs and decreased the fluorescence intensity. Once the QD/PGA–doxorubicin complexes were taken up by the cell through endocytosis, PGA was degraded by hydrolase in lysosomes, which could break the covalent link between doxorubicin and PGA. Following drug administration, an MTT assay was performed to evaluate cytotoxicity; results showed no visible effect on cell viability at concentrations up to 9.05×10^{-7} mol L^{-1}, but after a 48-h incubation period, a decrease in viability was observed at a concentration of 1.81×10^{-7} mol L^{-1}. This phenomenon was attributed to the release of doxorubicin from the QD/PGA–doxorubicin complex. In addition, it was also found that increased hydrophilic properties of doxorubicin under acidic conditions resulted in the rapid release of the drug. To observe morphology, PC3M- and HepG2-treated cells were imaged using fluorescence microscopy. Results showed no change in cell structure, therefore demonstrating the good biocompatibility of QDs/PGA–doxorubicin nanoparticles.

4.6 ELECTROCHEMILUMINESCENCE OF QDs WITH CARBOHYDRATES

Electrochemiluminescence or electrogenerated chemiluminescence (ECL) is a means of harnessing a series of highly exergonic electrochemical reactions to produce excited electronic states that emit light upon relaxation. QDs provide an ideal medium for ECL, due to their unique optical and electronic properties, demonstrating great potential in the study of cell surface features. Han *et al.* [52] designed a novel strategy to analyze cell membrane carbohydrates by incorporating ECL behavior of QDs with the recognition ability of lectins to target analyte carbohydrates. A sensitive ECL cytosensor for detecting varying concentrations of cell membrane carbohydrates and monitoring carbohydrate expression patterns was developed using CdSe QDs. The cytosensor was developed using CdSe QDs covalently bonded with chitosan–gold nanoparticles that were attached to a glassy carbon electrode. Human leukemic K562 cells were used to test the specific recognition of the QD-bound lectins and showed to decrease the ECL emission, which corresponded to the inhibition of electron transfer reactions on electrode surface due to sensitive and specific binding between lectin and carbohydrates on cell membrane. The detection limits of lectin from *Triticum vulgaris* (wheat germ agglutinin (WGA)), concanavalin A from *Canavalia ensiformis* (Jack Bean, abbreviated as Con A), and PNA-modified electrodes were determined to be 1.1×10^3, 1.6×10^3, and 2.1×10^3 cells mL^{-1}, respectively, which were lower in comparison with an immunosensor chip method. Since carbohydrate expression on cell membrane can indicate the condition of the cells, ECL was utilized to monitor cell response to drugs and diseases. Hemin and 3′-azido-3-deoxythymidine (AZT) were exposed to cells to demonstrate the changes of ECL signals. ECL-based immunosensors could sensitively monitor dynamic cell changes while having the possibility of being expanded to utilize the functionalized QDs as sensitive and specific ECL signal amplifiers.

Glycoconjugates of QDs were sometimes incorporated into the detection system with other nanoparticles. Liu *et al.* [53] used ECL as a detection method for carbohydrates, when Au-coated magnetic nanoparticles (AuMNPs) served as the immobilizer for the primary antibody, Ab_1 · CdTe QDs were attached to a graphene sheet and then coupled with the secondary antibody, Ab_2. When graphene oxide that was mixed with positively charged poly(diallyldimethylammonium chloride) reacted with negatively charged QDs, they spontaneously self-assembled through electrostatic interactions. The antibody containing complexes were modified with bovine serum albumin (BSA) in order to minimize the nonspecific adsorption of interfering biomolecules. The final sandwich-type ECL immunosensor was completed when Au–MNP–Ab_1–BSA targeting the carbohydrates and incubated with QD–Ab_2–BSA. The range of detection was 0.005–50 U mL^{-1}, and the system also possessed an LOD of 0.0012 U mL^{-1}, making this immunosensor extremely sensitive. The ECL immunosensor method was simple and quick and had high sensitivity, but also featured high stability, primarily due to ECL signal amplification by QDs.

As shown in Figure 4.4, Zhang *et al.* [54] developed a novel ECL immunosensor consisting of CdTe QDs as the fluorescence label and Fe_3O_4 MNPs as the immobilizer, facilitating a detection system for pregnancy marker hormone, human chorionic gonadotropin (hCG). The developed system was found to be both sensitive and specific. MNPs served as a bioactive and selective surface for attachment of primary antibody, with glutaraldehyde as a linker. The complex was characterized by TEM. QD–PtRu was labeled with secondary antibody for hCG. BSA was used to block the nonspecific binding sites. Using a flow injection cell and an indium tin oxide (ITO) glass working electrode, the sandwich immunosensor was shown to be sensitive, stable, and reproducible. The range of detection was 0.005–50 ng mL^{-1} with the LOD at 0.8 pg mL^{-1} (signal-to-noise ratio of 3). When compared to ELISA method using various concentrations of hCG in human serum samples, there was no significant difference. Functionalized MNPs acted as the carrier for Ab_1, while QDs acted as the signal amplifier allowing a simple and sensitive method for detecting low levels of hCG. This was accomplished due to the application of magnetic properties to assemble and purify the immunosensor and the tunable emission range of the QDs with excellent signal amplification capabilities. This methodology suggested a novel way to combine bifunctional nanomaterials and exceptional magnetism to allow for the development of various bioassay applications.

CdTe QDs, graphene, and dextran–Fe_3O_4 MNPs were also used by Gan *et al.* [55] to develop an ultrasensitive immunosensor for carcinoembryonic antigen 19-9 (CA19-9). The primary antibody for CA19-9 was immobilized on dextran MNPs, which then acted as capture probes. Amino-graphene–CdTe QDs were connected with antibody through covalent bond and acted as the signal probe. The resultant CdTe QDs emitted an ECL signal when excited, and the graphene nanosheets used for the immobilization of QDs had excellent structure for this purpose, due to their biocompatibility, large surface area, good adsorption capacity, and good electron transfer capabilities. When CA19-9 was detected by antibody/dextran–Fe_3O_4 MNPs and then incubated with antibody/CdTe QDs, the sandwich-type immunosensor showed good reproducibility and stability. When tested on the screen-printed carbon

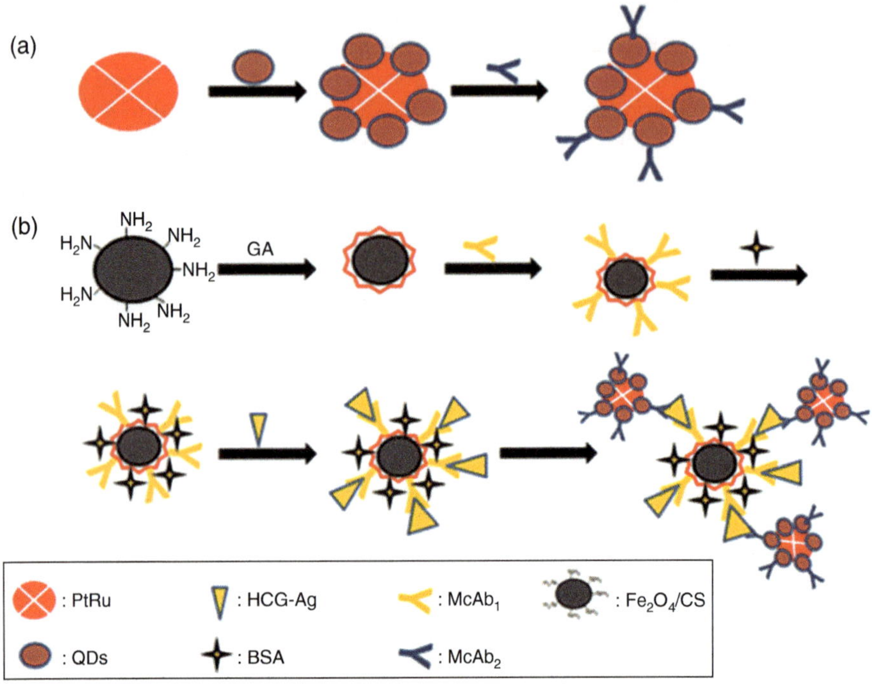

FIGURE 4.4 Conceptual illustration of the electrochemiluminescence (ECL) immunosensor for human chorionic gonadotropin antigen(hCG-Ag). (a) CdTe QD conjugated PtRu alloy was modified with $McAb_2$ to act as a label. (b) The chitosan (CS) coated magnetic nanoparticles (Fe_3O_4/CS MNPs) were modified with primary antibody for hCG-Ag. Once the hCg-Ag is captured, the PtRu based label could bind with hCG-Ag, and amplify the signal.

electrode, the immunosensor could be regenerated by simply adding or removing an external magnetic field. The immunosensor was tested by different concentrations of CA19-9 and found to have a detection range from 0.005 to 100 pg mL^{-1}, with a detection limit at 0.002 pg mL^{-1}. The immunosensor was tested for specificity by mixing CA19-9 sample solution with interfering substances, with the ECL intensity only reduced by 2.3%. When tested using various clinical samples, the novel method could be used with time-resolved fluorescence assay and showed consistent results.

Orndorff *et al.* [56] performed a study in which CdSe QDs targeting vascular cell adhesion molecule-1 (VCAM-1) were used to determine the source of reactive oxygen species, a primary cause of pulmonary inflammation. Lung inflammation is initiated by the activation and recruitment of neutrophils from the systemic circulation into the lung. In order to enter the lung, neutrophils must adhere to the lung endothelium and migrate to the tissue, where they subsequently release reactive oxygen species. One of the adhesion biomolecules is VCAM-1. It has been reported, however, that NADPH oxidase 2 (NOX2) has been a vital contributor to the formation of reactive oxygen species in pulmonary inflammation. In order to observe VCAM expression *in vivo* and *in vitro*, the CdSe QDs were coated with a PEG layer in order to conjugate with PAG adaptor protein, which can then couple with antibodies directed to VCAM or platelet endothelial cell adhesion molecule (PECAM). From *in vitro* tests on QD-VCAM and QD-PECAM, it was shown that QD-VCAM and QD-PECAM displayed high specificity. Further tests showed that in the presence of lipopolysaccharides, QD-VCAM had a significant association with NOX2 activity and VCAM expression. This showed that QD-VCAM was a better candidate to use for further investigations. Lastly, transgenic mice expressing endothelial NOX2 were used to determine the source of VCAM-inducing NOX2-derived reactive oxygen species production. Fluorescence signals were taken from QDs conjugated with VCAM. This procedure was carried out to determine where the greatest amount of VCAM was expressed. Results showed that QDs were less present on the null background as opposed to the transgenic samples, suggesting that NOX2 activation in pulmonary endothelium was sufficient to induce VCAM expression. Moreover, the results demonstrated that endothelial NOX2 induced VCAM expression in models of lipopolysaccharide administration. In summary, the findings of this study indicated that further investigation on the relationship between NOX2 and VCAM may help in the development of strategies necessary to prevent the progression of pulmonary inflammation.

Current chemotherapeutic cancer treatments are focused on targeting and disrupting vital cellular pathways present in both normal and cancerous cells. While this indiscriminate targeting method is effective in the treatment of a broad array of cancers, it also produces an equally large swath of undesirable, negative side effects. Adjusting or developing new delivery methods for these cytotoxic payloads may lead to greater efficacy in treatment via lowering the potential impact upon healthy cells or allowing greater concentrations of cytotoxins to reach their desired target. A novel method of delivering doxorubicin to the nucleus of cancerous cells was reported by Wang *et al.* [57] utilizing GQDs as fluorescent reporters and vehicles for doxorubicin to enhance its anticancer activity. GQDs maintain the intrinsic layered structural motif of graphene but with smaller lateral size and abundant periphery carboxylic groups and are more compatible with biological system and thus are promising nanomaterials for therapeutic applications. Doxorubicin–GQD conjugates were found to spontaneously form in solution, with a hypothesized mechanism of hydrophobic and π–π interaction between the planar GQD and doxorubicin molecules as observed using atomic force microscopy. The complex was stable over a broad pH range of 2.0–7.4, as a result of the hydrophobic environment required for the release of doxorubicin from the GQD conjugates. It was also determined that doxorubicin delivery into the nuclei was enhanced by increase in concentration of GQDs. The doxorubicin–GQD conjugates were found to greatly enhance the DNA cleavage of doxorubicin via increased DNA binding affinity, with time-dependent effects being observed at GQD concentrations between 60 and 150 $\mu g\ mL^{-1}$. Doxorubicin–GQD conjugates were also found to enhance the nuclear uptake of doxorubicin by allowing the use of endocytic internalization pathway(s), caveolae-mediated endocytosis being the primary pathway. Such nuclear uptake mechanisms were found to be unavailable to free doxorubicin, as evidenced by the greatly enhanced, localized nuclear fluorescence in MCF-7 cell lines. The over active drug efflux pumps were a generally accepted mechanism of drug resistance that were employed by many drug-resistant cells. This mechanism could be bypassed by the doxorubicin–GQD conjugates, avoiding drug resistance. Incubation with 150 μg mL^{-1} doxorubicin–GQD for 2 h leads to 98% of DNA to be found in nicked fragments, with an increase in cell-line mortality rates observed for both MCF-7 and MGC-803 cell lines. In conclusion, results suggested the potential for greatly enhanced therapeutic effects in the treatment of cancer when combining doxorubicin with GQDs, as both the delivery method and cytotoxic effectiveness of doxorubicin were greatly improved.

Hyaluronic acid (HA) is an ideal candidate for conjugation with QDs, as it is biodegradable and biocompatible and possesses a high affinity for the CD44 receptor (hyaluronan). Al-Nahain *et al.* [58] outlined a novel method for the attachment and formation of GQD–HA conjugates by utilizing catechol (L-DOPA, DN) as an adhesive. *In vivo* testing was accomplished via tail vein injections of GQDs and GQD–HA in female mice bearing A549 cell tumors. GQDs were found to accumulate in the liver, kidney, and tumor tissues, with the highest concentrations located in the liver tissues. GQD–HA was also found to accumulate in the liver, kidney, and tumor tissues but displayed greatly reduced accumulation in both liver and kidney when compared to GQDs. In addition to exploring the potential use of GQD–HA in targeted imaging, the efficacy in delivering doxorubicin was examined. It was found that HA increased the bright fluorescence from *in vivo* and *in vitro* investigations, since it acted as a targeting molecule and increased the accumulation of GQD in cancer cells. Doxorubicin uptake was quantitatively analyzed using UV spectroscopy at 480 nm, with an observed uptake of ~75% onto GQD surface. The release profile of doxorubicin from the surface of GQD–HA was observed to be both time dependent and pH sensitive. Full release of GQD–HA-loaded doxorubicin occurred within 48 h in slightly acidic conditions (pH 5.0) while remaining relatively stable at physiological conditions (pH 7.4). Overall, results suggested that the HA-conjugated GQDs were promising reagents for both the targeted delivery of cytotoxic drugs and *in vivo* fluorescence probing. QDs were also employed to detect carbohydrates, using test strips [59–63] and magnetic beads [64, 65].

4.7 ELECTROCHEMISTRY OF QDs WITH CARBOHYDRATES

Electrochemical techniques present unique opportunities in biosensors. In general terms, the target analyte undergoes a redox reaction on the electrode surface. This reaction generates a measureable current, depending upon how the reaction contributes or donates electrons to the system. Such systems offer relatively simple and fast diagnostic abilities while maintaining a high sensitivity. Examples of most commonly applied electrochemical techniques include potentiometric, amperometric, and impedimetric biosensors with a wide variety of detection capabilities.

Recently, Xiang *et al.* [66] synthesized reproducible QD barcodes to detect carcinoembryonic antigen (CEA). The encoding elements Zn^{2+}, Cd^{2+}, and Pb^{2+} were confined within a single QD using reverse micelle synthesis approach. In this

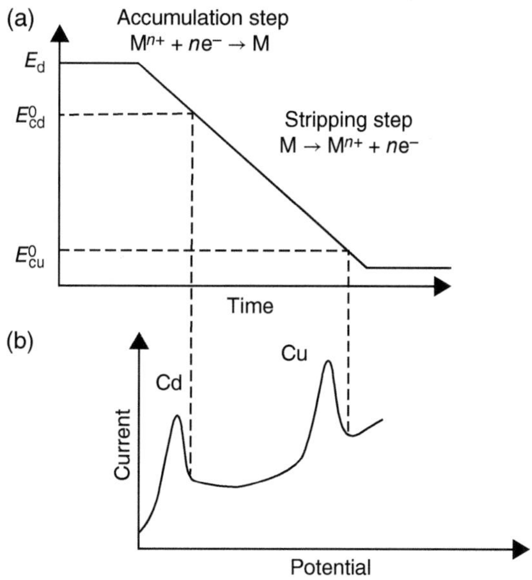

FIGURE 4.5 Anodic stripping voltammetry for cadmium and copper. The first step is accumulation step, where the metals are reduced at the electrode surface. In the stripping step, the metals are oxidized back into the solution.

proof-of-concept study, the CEA was introduced to prelabeled ELISA plates with anti-CEA film, and then the QD-conjugated CEA was introduced to form a "sandwich" complex. Voltammetric stripping could read the QD barcode based on the change in current intensity, and distinct peak potential Zn^{2+}, Cd^{2+}, and Pb^{2+} metal ions had well-resolved peaks at −1.0, −0.68, and −0.5 V, respectively. The increase in current intensity corresponded to a linear range from 0.01 to 80 ng mL^{-1} and a detection limit of 3.3 pg mL^{-1}.

As shown in Figure 4.5, the first step is an accumulation step where the metals such as Cd or Cu could form amalgams at the mercury electrode surface, and then the metals are stripped away and oxidized at distinct peak potential. Depending on the area and the height of the current peak, quantitative determinations can be performed at high sensitivity. There are many other studies that use stripping voltammetry to perform multiplexed detections such as DNA [67–69], and these methods could be also adapted to the detection of carbohydrates. QD-based barcode approach demonstrates the potential for a powerful multiplex detection tool for biomarkers.

A novel bioassay for fast, sensitive, and selective isolation and detection of the influenza vaccine hemagglutinin (IVHA) was developed by Krejcova *et al.* [70] using CdTe QDs that were complexed to IVHA. Magnetic beads were modified with glycans in order to selectively bind IVHA. Adsorptive transfer technique and differential pulse anodic stripping voltammetry were utilized to detect the protein from the influenza vaccine and also Cd(II) ions as part of the HA–CdTe complex. The proposed method of isolation through sonication was also tested by detecting the individual parts of the subsequently broken-up complex, again using the electrochemical techniques. The method is highly selective due to the selective binding of the glycan–IVHA complex, as the lowest detectable concentration of IVHA–CdTe was reported to be 62.5 pmol L^{-1}.

In various diseases, certain biomarkers are present in serum or urine samples, with changes in concentration revealing the severity or progress of the disease. Therefore, exploring an efficient method to detect multiple biomarkers present in one disease using a single test would be advantageous and cost effective. Tang *et al.* [71] proposed a method to simultaneously detect multiple biomarkers through a multiplexed stripping voltammetric immunoassay. First, trifunctionalized magnetic beads were used as an immunosensing probe for three different biomarkers at once. The biomarkers that were the targets of this study were sandwiched between magnetic beads and dendrimer-encapsulated QDs. CdS, ZnS, and PbS QDs were used to label analyte-specific antibodies, which were further used to distinguish each of the three biomarkers coimmobilized onto the magnetic beads. QDs present on the dendrimer were stripped from the dendrimer–QD complex through square-wave anodic stripping voltammetry in order to quantify the particular biomarker immunocomplexes. This study demonstrated a method for multiplexed electrochemical analysis using functionalized QDs for signal amplification, as opposed to using enzymes. Results displayed the simultaneous detection of three different biomarkers with a wide dynamic range of 0.01–50 U mL^{-1} along with a detection limit of 0.005 U mL^{-1}.

The photoelectrochemical detection methods use light to generate a current on the electrode through the excitation of active species. This current acts as an analytical signal that is based on the molecular architecture that exists between the electrode and the excited species. In this study conducted by Wang *et al.* [72], a sandwich assay was created on the surface of an ITO electrode using graphene-conjugated SnO_2 QDs. Graphene is an excellent conductor, which acted as a surface for the attachment of SnO_2 QDs. The resultant QD-containing nanostructures were excited using light in order to generate a current on the ITO electrode, which was measured for quantitative results. The amount of current was proportional to the amount of nanostructures adhering to the electrode surface through the antibody bridge and also proportional to the antibody concentration. The immunosensor had a wide linear range of 0.005–10 ng mL^{-1} with a detection limit of 0.036 pg mL^{-1} demonstrating potential for further applications using clinical samples.

Nanomaterials are recognized for their ability to transfer electrons directly between enzyme and electrodes. Du *et al.* [73] reported a study utilizing this property to compare the electrochemical and photoelectrochemical properties of three GOx-based biosensors consisting of different nanoparticles such as Au, CdS, and ZnS. The nanoparticles were electrodeposited onto an ITO electrode, and then the nanoparticle-modified surfaces were modified with GOx. The electrodeposited nanoparticles were characterized by scanning electron microscopy, UV–vis spectroscopy, and electrochemical impedance spectra. All three nanoparticles were shown to successfully promote electron transfer between GOx and ITO surfaces. However, the Au nanoparticles–GOx system suffered from the lowest sensitivity and CdS nanoparticles displayed toxicity, whereas ZnS appeared to be the most sensitive and the least toxic one.

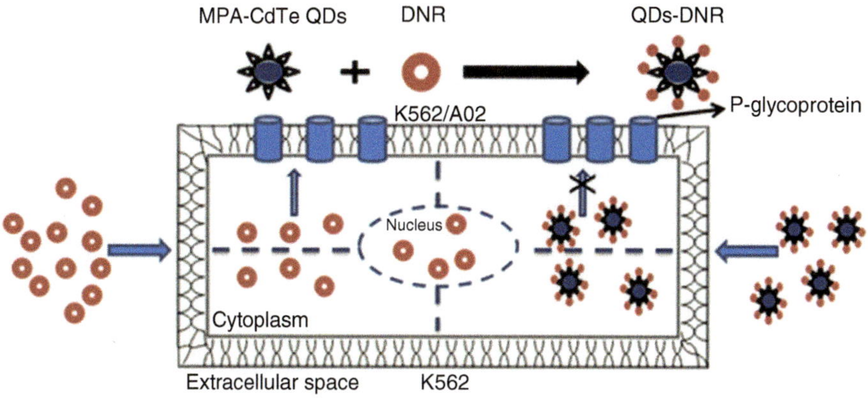

FIGURE 4.6 Conceptual illustration of CdTe QDs capped with negatively charged 3-mercaptopropionic acid (MPA) QDs to enhance the uptake of the drug daunorubicin (DNR) into the target cancer cells. The cytotoxic effect of MPA–CdTe QDs on sensitive K562 and drug-resistant K562/A02 cancer cell lines showed that the drug-resistant cell line was more sensitive to MPA–CdTe QDs, thus indicating the ability of the QD complex to inhibit the function of P-glycoprotein. Results from TIRFM showed an increase in the concentration of anticancer drug DNR upon incubation with MPA–CdTe QDs together, which indicated its efficiency as a carrier of DNR into leukemia cancer cells. These observations also suggest that the interaction between the negatively charged QD complexes had significantly suppressed the cytotoxicity of DNR on both drug-sensitive and drug-resistant leukemia cell lines.

As shown in Figure 4.6, the utilization of water-soluble CdTe QDs capped with negatively charged 3-mercaptopropionic acid (MPA) produced MPA–CdTe QDs, which were then used to enhance the drug uptake into the targeted cancer cells [74]. The efficiency of the biomarker and cancer treatments was assessed by evaluating the cytotoxicity of MPA–CdTe QDs using MTT assay, total internal reflection fluorescence microscopy (TIRFM), as well as electrochemical analysis using DPV and UV–vis spectroscopy. The cytotoxic effect of MPA–CdTe QDs on sensitive K562 and drug-resistant K562/A02 cancer cell lines was first evaluated by the MTT assay. The results suggested that the drug-resistant cell line was more sensitive to MPA–CdTe QDs, thus indicating that the QD complex could inhibit the function of P-glycoprotein. This inhibition could further facilitate the accumulation of anticancer drug daunorubicin in drug-resistant leukemia cells. The cellular interaction of MPA–CdTe QDs with drug-sensitive K562 and drug-resistant K562/A02 cell lines was then explored and visualized by the intracellular fluorescence microscopy in combination with TIRFM. Results from TIRFM showed an increase in the concentration of daunorubicin upon incubation with MPA–CdTe QDs together, which indicated it as an efficient facilitator for the delivery and accumulation of daunorubicin into leukemia cancer cells. To further support the findings from TIRFM, DPV experiments were performed to detect the electrochemical response of DNR and its interaction with the two leukemia cell lines in the presence and absence of MPA–CdTe QDs. Results suggested that the decrease in the DPV current was directly correlated to the amount of drug uptake in the cancer cells, which was consistent with the findings obtained from TIRFM. This, in addition, suggested an important synergistic effect of the MPA–CdTe QDs on facilitating the accumulation of daunorubicin on drug-sensitive and drug-resistant leukemia cells and confirmed the role of MPA–CdTe QDs as effective drug carriers for the inhibition of the leukemia cells under low drug concentrations, partially due to the negatively charged surface of the QDs. Observations obtained from the UV–vis spectroscopy of the daunorubicin residues after treating the two leukemia cell lines in the presence and absence of MPA–CdTe QDs also supported the previous findings. These observations indicated that the interaction between the negatively charged QD complexes had significantly enhanced the cytotoxicity suppression of daunorubicin on both drug-sensitive and drug-resistant leukemia cell lines.

While QDs possess many valuable intrinsic properties, their ability to act as a scaffold or foundation for the attachment of additional nanomaterials provides for an even greater repertoire of potential diagnostic and analytical techniques. QDs and carbon nanotubes (CNTs) are two popular nanomaterials in bioanalytical applications due to their unique chemical, physical, and electronic properties. Liu *et al.* [75] designed a novel electrochemical biosensing platform to detect glucose based on CdTe-modified CNTs. CNTs were used to compensate the poor conductivity of the CdTe QDs as CNTs could act as nanowiring of enzymes on the electrode surface. CdTe/CNT-modified electrode was first prepared by mixing CdTe QDs, CNTs, Nafion, and GOx in appropriate amounts and then modifying the mixture on the surface of a glass carbon electrode.TEM was applied to observe the dispersion of CdTe QDs on CNTs, and CV was applied to explore the electrochemical behavior of the CdTe/CNT-modified electrode. A pair of well-defined quasireversible redox peaks of GOx was obtained at the CdTe/CNT-modified enzyme electrodes with Nafion as a backbone to provide stability and to homogenize the thin films. In contrast, no redox peaks were observed at the Nafion–CdTe-modified electrodes, and some minor redox peaks at the Nafion–Gx-modified electrodes were observed. The observations indicated that the direct electron transfer of GOx was obtained through the incorporation of GOx into the Nafion–CdTe film, which may be due to the synergy effect of the Nafion–CdTe–CNTs composite matrix and the favorable orientation of GOx biomolecules in the composite film. As a result, the novel glucose biosensing platform constructed by using CdTe/CNT nanocomposites had a detection limit of 2.5×10^{-5} M, demonstrating potential application in diagnostics.

4.8 CONCLUDING REMARKS

Recent research has led to the production of a wide variety of QDs as a new class of fluorescent labels for numerous applications in biology and medicine. These highly bright fluorescent and stable QD probes are well suited for detecting protein biomarkers on cells and clinical tissue application. This chapter has reviewed the recent literature about different applications of QD glycoconjugates. While the intrinsic properties of QDs allow for many applications with electrochemical biosensors, the QDs also offer another unique advantage: innate ability to be used in fluorescence spectroscopy due to their broad absorbance and narrow emission spectra. This allows one to combine multiple imaging and detection techniques to provide a novel approach to studying cellular processes, a highly desirable trait toward *in vivo* applications. The extensive work on QD glycoconjugates has shown promising applications for diagnostics that may lead to the commercialization of these bioassays in the near future.

ACKNOWLEDGMENTS

Authors gratefully acknowledge financial support from NSERC Discovery Grants Program. Authors would also like to acknowledge Harrish Nithianandan, Nerosanth Selvarajah, and Suvethikah Nithianandan, who have generously given their time during the production of this chapter.

REFERENCES

1. Alivisatos, A. P. *Science* **1996**, *271*, 933.
2. Dabbousi, B. O.; Rodriguez-Viejo, J.; Mikulec, F. V.; Heine, J. R.; Mattoussi, H.; Ober, R.; Jensen, K. F.; Bawendi, M. G. *J. Phys. Chem. B* **1997**, *101*, 9463.
3. Delehanty, J. B.; Mattoussi, H.; Medintz, I. L. *Anal. Bioanal. Chem.* **2009**, *393*, 1091.
4. Fan, X.; Zhang, M.; Shafiq, I.; Zhang, W.; Lee, C.; Lee, S. *Crys. Growth Des.* **2009**, *9*, 1375.
5. Jennings, T. L.; Becker-Catania, S.; Triulzi, R. C.; Tao, G.; Scott, B.; Sapsford, K. E.; Spindel, S.; Oh, E.; Jain, V.; Delehanty, J. B.; Prasuhn, D. E.; Boeneman, K.; Algar, W. R.; Medintz, I. L. *ACS Nano* **2011**, *5*, 5579.
6. Koch, U.; Fojtik, A.; Weller, H.; Henglein, A. *Chem. Phys. Lett.* **1985**, *122*, 507.
7. Mädler, L.; Stark, W. J.; Pratsinis, S. E. *J. Appl. Phys.* **2002**, *92*, 6537.
8. Mattoussi, H.; Mauro, J. M.; Goldman, E. R.; Anderson, G. P.; Sundar, V. C.; Mikulec, F. V.; Bawendi, M. G. *J. Am. Chem. Soc.* **2000**, *122*, 12142.
9. Nirmal, M.; Brus, L. *Acc. Chem. Res.* **1999**, *32*, 407.
10. Noor, M. O.; Petryayeva, E.; Tavares, A. J.; Uddayasankar, U.; Algar, W. R.; Krull, U. *J. Coord. Chem. Rev.* **2014**, *263*, 25.
11. Reiss, P.; Protiere, M.; Li, L. *Small* **2009**, *5*, 154.
12. Algar, W. R.; Ancona, M. G.; Malanoski, A. P.; Susumu, K.; Medintz, I. L. *ACS Nano* **2012**, *6*, 11044.
13. Algar, W. R.; Prasuhn, D. E.; Stewart, M. H.; Jennings, T. L.; Blanco-Canosa, J. B.; Dawson, P. E.; Medintz, I. L. *Bioconjug. Chem.* **2011**, *22*, 825.
14. Algar, W. R.; Susumu, K.; Delehanty, J. B.; Medintz, I. L. *Anal. Chem.* **2011**, *83*, 8826.
15. Ali, M.; Pal, S. K.; Rahaman, H.; Ghosh, S. K. *Soft Matters* **2014**, *1*, 2767.
16. Amelia, M.; Lincheneau, C.; Silvi, S.; Credi, A. *Chem. Soc. Rev.* **2012**, *41*, 5728.
17. Blanco-Canosa, J.; Wu, M.; Susumu, K.; Petryayeva, E.; Jennings, T. L.; Dawson, P. E.; Algar, W. R.; Medintz, I. L. *Coord. Chem. Rev.* **2014**, *263*, 101.
18. Boeneman, K.; Mei, B. C.; Dennis, A. M.; Bao, G.; Deschamps, J. R.; Mattoussi, H.; Medintz, I. L. *J. Am. Chem. Soc.* **2009**, *131*, 3828.

19. Chan, W. C. W.; Maxwell, D. J.; Gao, X.; Bailey, R. E.; Han, M.; Nie, S. *Curr. Opin. Biotechnol.* **2002**, *13*, 40.
20. Murray, C. B.; Norris, D. J.; Bawendi, M. G. *J. Am. Chem. Soc.* **1993**, *115*, 8706.
21. Chen, L.; Algar, W. R.; Tavares, A.; Krull, U. *Anal. Bioanal. Chem.* **2011**, *399*, 133.
22. Colvin, V. L.; Schlamp, M. C.; Alivisatos, A. P. *Nature* **1994**, *370*, 354.
23. Talapin, D. V.; Lee, J.-S.; Kovalenko, M. V.; Shevchenko, E. V. *Chem. Rev.* **2010**, *110*, 389.
24. Inamdar, S. N.; Ingole, P. P.; Haram, S. K. *ChemPhysChem* **2008**, *9*, 2574.
25. Hynek, D.; Tmejova, K.; Milosavljevic, V.; Moulick, A.; Kopel, P.; Adam, V.; Kizek, R. *Int. J. Electrochem. Soc.* **2015**, *10*, 1117.
26. Sobrova, P.; Ryvolova, M.; Hubalek, J.; Adam, V.; Kizek, R. *Int. J. Mol. Sci.* **2013**, *14*, 13497.
27. Landry, M. L.; Morrell, T. E.; Karagounis, T. K.; Hsia, C.; Wang, C. *J. Chem. Educ.* **2014**, *91*, 274.
28. Wang, Y.; Liu, S.; Yang, K.; Zhou, L. *Colloid Polym. Sci.* **2012**, *291*, 1313.
29. Lehnen, T.; Schaefer, J.; Mathur, S. Z. *Anorg. Allg. Chem.* **2014**, *640*, 819.
30. Entezari, M. H.; Ghows, N. *Ultrason. Sonochem.* **2011**, *18*, 127.
31. Meulenkamp, E. A. *J. Phys. Chem. B* **1998**, *102*, 5566.
32. Wallace, R.; Brown, A. P.; Brydson, R.; Wegner, K.; Milne, S. J. *J. Mater. Sci.* **2013**, *48*, 6393.
33. Xu, Y.; Al-Salim, N.; Bumby, C. W.; Tilley, R. D. *J. Am. Chem. Soc.* **2009**, *131*, 15990.
34. Freeman, R.; Bahshi, L.; Finder, T.; Gill, R.; Willner, I. *Chem. Commun.* **2009**, 764.
35. Hermanson, G. T. *Bioconjugate Techniques*, 2nd edition, Academic Press: San Diego, **2008**.
36. Rosenthal, S. J.; Chang, J. C.; Kovtun, O.; McBride, J. R.; Tomlinson, I. D. *Chem. Biol.* **2011**, *18*, 10.
37. Sapsford, K. E.; Algar, W. R.; Berti, L.; Gemmill, K. B.; Casey, B. J.; Oh, E.; Stewart, M. H.; Medintz, I. L. *Chem. Rev.* **2013**, *113*, 1904.
38. Susumu, K.; Oh, E.; Delehanty, J. B.; Blanco-Canosa, J.; Johnson, B. J.; Jain, V.; Hervey, W. J.; Algar, W. R.; Boeneman, K.; Dawson, P. E.; Medintz, I. L. *J. Am. Chem. Soc.* **2011**, *133*, 9480.
39. Trindade, T.; O'Brien, P.; Pickett, N. L. *Chem. Mater.* **2001**, *13*, 3843.
40. Vriezema, D. M.; Comellas Aragon, M.; Elemans, J. A. A. W.; Cornelissen, J. J. L. M.; Rowan, A. E.; Nolte, R. J. M. *Chem. Rev.* **2005**, *105*, 1445.
41. Zhang, C.; Johnson, L. W. *Anal. Chem.* **2009**, *81*, 3051.
42. Varki, A.; Cummings R.; Esko, J. *Essentials of Glycobiology* 2nd edition, Cold Spring Harbor Laboratory Press: Cold Spring Harbor, **2009**.
43. Cheng, Y.; Li, M. Y.; Wang, S. R.; Peng, H. J.; Reid, S.; Ni, N. T.; Fang, H.; Xu, W. F.; Wang, B. H. *Sci. China Chem.* **2010**, *53*, 3.
44. Ghazarian, H.; Idoni, B.; Oppenheimer, S. B. *Acta Histochim* **2011**, *113*, 236.
45. Li, Y. H.; Zhang, L.; Huang, J.; Liang, R. P.; Qiu, J. D. *Chem. Commun.* **2013**, *49,* 5180.
46. Li, N.; Chow, A. M.; Ganesh, H. V. S.; Brown, I. R.; Kerman, K. *Anal. Chem.* **2013**, *85*, 9699.
47. Lu, R. M.; Chen, M. S.; Chang, D. K.; Chiu, C. Y.; Lin, W. C.; Yan, S. L.; Wang, Y. P.; Kuo, Y. S.; Yeh, C. Y.; Lo, A.; Wu, H. C. *PLoS One* **2013**, *8*, e66128.
48. Shao, M.; Wang, H. F. *Analyst* **2013**, *138*, 4618.
49. Huang, X.; Wang, J.; Liu, H.; Lan, T.; Ren, J. *Talanta* **2013**, *106*, 79.

50. Chang, Y.; Cai, C.; Li, L.; Miao, J.; Ucakturk, E.; Li, G.; Ly, M.; Linhardt, R. *J. Anal. Chem.* **2013**, *85*, 9356.
51. Gao, X.; Liu, Z.; Lin, Z.; Su, X. *Analyst* **2014**, *139*, 831.
52. Han, E.; Ding, L.; Lian, H.; Ju, H. *Chem. Commun.* **2010**, *46*, 5446.
53. Liu, W.; Zhang, Y.; Ge, S.; Song, X.; Huang, J.; Yan, M.; Yu, J. *Anal. Chim. Acta* **2013**, *770*, 132.
54. Zhang, Y.; Ge, S.; Wang, S.; Yan, M.; Yu, J.; Song, X.; Liu, W. *Analyst* **2012**, *137*, 2176.
55. Gan, N.; Zhou, J.; Xiong, P.; Li, T.; Jiang, S.; Cao, Y.; Jiang, Q. *Int. J. Mol. Sci.* **2013**, *14*, 10397.
56. Orndorff, R. L.; Hong, N.; Yu, K.; Feinstein, S. I.; Zern, B. J.; Fisher, A. B.; Muzykantov, V. R.; Chatterjee, S. *Am. J. Physiol. Lung Cell Mol. Physiol.* **2014**, *306*, L260.
57. Wang, C.; Wu, C.; Zhou, X.; Han, T.; Xin, X.; Wu, J.; Zhang, J.; Guo, S. *Sci. Rep.* **2013**, *3*, 2852.
58. Al-Nahain, A.; Lee, J.-E.; In, I.; Lee, H.; Lee, K. D.; Jeong, J. H.; Park, S. Y. *Mol. Pharm.* **2013**, *10*, 3736.
59. Zou, Z. X.; Wang, J.; Wang, H.; Li, Y. Q.; Lin, Y. *Talanta* **2012**, *94*, 58.
60. Li, Z.; Wang, Y.; Wang, J.; Tang, Z.; Pounds, J. G.; Lin, Y. *Anal. Chem.* **2010**, *82*, 7008.
61. Nian, H.; Wang, J.; Wu, H.; Lo, J. G.; Chiu, K. H.; Pounds, J. G.; Lin, Y. *Anal. Chim. Acta* **2012**, *713*, 50.
62. Liu, G.; Lin, Y.-Y.; Wang, J.; Wu, H.; Wai, C. M.; Lin, Y. *Anal. Chem.* **2007**, *79*, 7644.
63. Zhang, W.; Ge, X.; Tang, Y.; Du, D.; Liu, D.; Lin Y. *Analyst* **2013**, *138*, 5431.
64. Du, D.; Wang, J.; Wang, L.; Lu D.; Smith, J. N.; Timchalk, C.; Lin, Y. *Anal. Chem.* **2011**, *83*, 3770.
65. Wang, H.; Wang, J.; Timchalk, C.; Lin, Y. *Anal. Chem.* **2008**, *80*, 8477.
66. Xiang, Y.; Zhang, Y.; Chang, Y.; Chai, Y.; Wang, J.; Yuan, R. *Anal. Chem.* **2010**, *82*, 1138.
67. Kim, J.; Seo, K.; Wang, J. *Conf. Proc. IEEE Eng. Med. Biol. Soc.* **2004**, *1*, 137.
68. Hansen, J. A.; Wang, J.; Kawde, A. N.; Xiang, Y.; Gothelf, K. V.; Collins, G. *J. Am. Chem. Soc.* **2006**, *128*, 2228.
69. Wang, J.; Liu, G.; Merkoçi, A. *J. Am. Chem. Soc.* **2003**, *125*, 3214.
70. Krejcova, L.; Nejdl, L.; Hynek, D.; Krizkova, S.; Kopel, P.; Adam, V.; Kizek, R. *Molecules* **2013**, *18*, 15573.
71. Tang, D.; Hou, L.; Niessner, R.; Xu, M.; Gao, Z.; Knopp, D. *Biosens. Bioelectron.* **2013**, *46*, 37.
72. Wang, Y.; Li, M.; Zhu, Y.; Ge, S.; Yu, J.; Yan, M.; Song, X. *Analyst* **2013**, *138*, 7112.
73. Du, J.; Yu, X.; Di, J. *Biosens. Bioelectron.* **2012**, *37*, 88.
74. Zhou, Y.; Shi, L.; Li, Q.; Jiang, H.; Lv, G.; Zhao, J.; Wu, C.; Selke, M.; Wang, X. *Biomaterials* **2010**, *31*, 4958.
75. Liu, Q.; Lu, X.; Li, J.; Yao, X.; Li, J. *Biosens. Bioelectron.* **2007**, *22*, 3203.

5

CONJUGATION OF GLYCANS WITH CARBON NANOSTRUCTURES

ZACHARY P. MICHAEL[1], ALEXANDER STAR[1], AND SÉBASTIEN VIDAL[2]

[1] *Department of Chemistry, University of Pittsburgh, Pittsburgh, PA, USA*

[2] *Institut de Chimie et Biochimie Moléculaires et Supramoléculaires, Laboratoire de Chimie Organique 2—Glycochimie, UMR 5246, Université Lyon 1 and CNRS, Villeurbanne, France*

5.1 CARBON NANOMATERIALS

Carbon nanomaterials are a class of 0D, 1D, and 2D nanomaterials comprised of several allotropes of carbon. Buckminsterfullerenes (0D), carbon nanotubes (CNTs, 1D), and graphene (2D) each have unique and extraordinary physical and electrical properties, generating large interest in the past few decades. As a result, the number of research publications has risen drastically on a year-by-year basis, especially for CNTs and graphene (Fig. 5.1). As they consist of pure carbon, they are relatively stable in the environment but are capable of undergoing a plethora of chemistry to modify their properties for applications in photovoltaics, drug delivery, materials chemistry, biosensors, and more. In this context, we will examine the conjugation of carbon nanomaterials with glycans for a broad range of applications.

5.1.1 Buckminsterfullerenes

Buckminsterfullerenes, simply known as buckyballs or fullerenes, were first discovered in the form of C_{60} (Fig. 5.1, inset) by the late Richard Smalley and associates at Rice University in 1985 [2]. Typically produced through arc discharge or laser ablation of graphite, C_{60} is approximately 1 nm in diameter and forms a black, granular powder. Fullerenes absorb very well in the UV region due to the sp^2 character

Carbohydrate Nanotechnology, First Edition. Edited by Keith J. Stine.

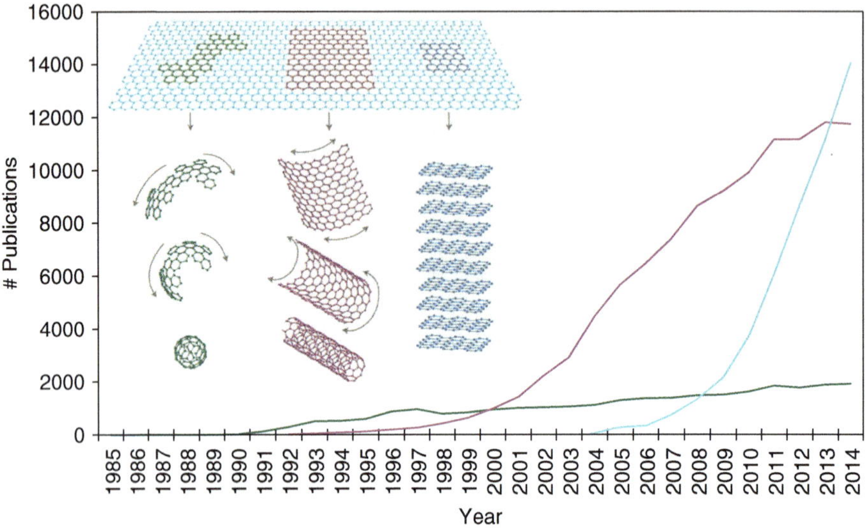

FIGURE 5.1 Publications by year for fullerenes, carbon nanotubes, and graphene from Scopus, accessed August 11, 2015. Inset structures of 0D C_{60} fullerenes (left), 1D single-walled carbon nanotubes (middle), and bulk graphite (right) as envisioned from a 2D sheet of graphene (top). *Source*: Reprinted by permission from Ref. 1, Copyright 2007, Macmillan Publishers Ltd.

and are nontoxic in pristine form [3], leading to many proposed applications in photovoltaics [4] and phototherapy techniques [5]. Furthermore, sugar-modified fullerenes have been shown to have applications in a range of therapeutic applications, which will be discussed further along.

5.1.2 CNTs

CNTs are cylindrical, 1D nanoparticles comprising of single, double, or multiple walls of sp^2-hybridized carbon. CNTs became an area of broad research interest after their structure was revealed by Sumio Iijima in 1991 [6]. CNTs are unique in that their physical and electronic properties are largely dependent on the number of walls and helical structure of the tube itself. For instance, single-walled carbon nanotubes (SWNTs) can be envisioned as a single piece of sp^2-hybridized carbon (graphene) rolled up into a cylinder (Fig. 5.1, inset). Depending on how the tube is rolled, an SWNT is formed of a specific diameter and chirality, which determines whether the tube is semiconducting or metallic as well as presenting characteristic absorption and emission spectra [7]. In addition to their unique electrical properties, CNTs are well known for their excellent mechanical properties demonstrating high Young's modulus and tensile strength. Due to these unique properties, CNTs have found myriad applications ranging from electronic and optical sensors [8], to drug delivery [9], to composite materials for antifouling coatings [10]. Unlike fullerenes, CNTs inherently exhibit some toxicological activity [11], but conjugation with polymers or glycans can greatly mitigate these effects for biological applications.

5.1.3 Graphene

Ideally, graphene is a single, atomically thin layer of sp^2-hybridized carbon (Fig. 5.1, inset). While graphene was long theorized as an interesting material, the first demonstrated synthesis was accomplished in 2004 via mechanical exfoliation, or the so-called "Scotch tape" method, of pyrolytic graphite by Novoselov et al. [12]. Pristine graphene, a zero band gap semiconductor, exhibits extraordinary physical and electronic properties, making it an ideal material for a broad range of applications [13] including sensors [14,15], photovoltaics [16], and composite materials [17]. These properties are highly tunable by adjusting the size and surface functionality of graphene flakes [15,18,19], thus the material is highly tunable to the desired application. Though graphene is the newest of the three materials discussed herein, the interest generated by it has been great.

5.2 CHEMISTRY

5.2.1 Toxicity

A major concern regarding carbon nanomaterials is their toxicity and persistence in the environment. General consensus on the toxicity of these materials is difficult to assert, as their persistence in the environment and danger to cells are largely dependent on the size, shape, and functionality of the material. Fullerenes are generally considered safe in their pristine form [3] due to their small size resulting in good uptake in cells. Gharbi et al. showed that C_{60} readily uptakes in rat liver cells (Fig. 5.2a) and found that fullerene-treated rats were less susceptible to necrosis from CCl_4 when compared to control rats [20]. Some covalently modified C_{60} compounds, on the other hand, have shown significant toxicity as compared to pristine fullerenes, highlighting the difficulty of assessing toxicity of these materials [3]. The toxicity of higher dimension materials, however, is much more difficult to assess.

CNTs have been widely studied for their potential toxicity to the environment. Pristine SWNTs have been shown to cause significant pulmonary toxicity when inhaled by mice [11], raising questions about their viability in industrial scale use and their impact on the environment. However, the mechanism leading to toxicity is complex, and factors including length dependence, diameter dependence, and surface functionality have been considered. The length of a nanotube has a great impact on the toxicity of the material. In the case of MWNTs, if the fiber is either short or tangled into a smaller bundle, they are able to be effectively cleared through the lymphatic system through phagocytosis [21] (Fig. 5.2b). Longer tubes are not able to be fully encapsulated in macrophages, leading to frustrated phagocytosis and accumulation of the material in the lungs. Diameter dependence is less clear; however, thicker MWNTs have been shown to be less toxic than their thinner counterparts [23].

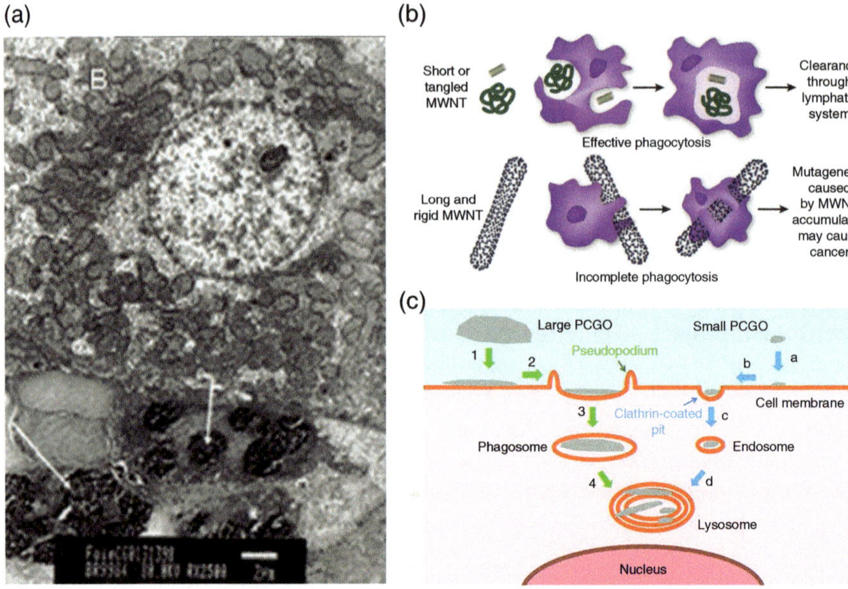

FIGURE 5.2 (a) TEM micrograph of C_{60} clusters in rat liver. *Source*: Reprinted with permission from Ref. 20, Copyright 2005, American Chemical Society. (b) Short or tangled MWNTs can be engulfed by macrophages and cleared through the lymphatic system, whereas long, rigid MWNTs cannot be fully engulfed, leading to frustrated phagocytosis. *Source*: Reprinted by permission from Ref. 21, Copyright 2008, Macmillan Publishers Ltd. (c) Cellular uptake of large and small protein-coated graphene oxide sheets. *Source*: Reprinted with permission from Ref. 22, Copyright 2012, The American Chemical Society.

Surface functionality also plays a large role when considering CNTs. Acid-oxidized S- and MWNTs have been shown to cause greater inflammatory responses than their pristine counterparts, possibly due to the higher rate of dispersion and therefore bioavailability, while noncovalently functionalized materials can significantly reduce toxicity. Of note, however, is that functionalized materials may also be more susceptible to natural degradation pathways outside of phagocytosis. Oxidized S- and MWNTs have the potential to be degraded by peroxidase enzymes naturally present in the environment. Horseradish peroxidase (HRP) and the human myeloperoxidase (MPO) have been shown to degrade oxidized CNTs over a period of several days [24], which is promising for the prospect of creating materials designed to be degraded by the body after a period of time.

Much like CNTs, the toxicity of graphene is largely dependent on its size and surface functionality. Interestingly, smaller graphene sheets have been found to be more cytotoxic than larger sheets, in contrast with studies on CNTs [25]. This may have to be due to the mechanism behind the toxic activity of graphene, but the question of what exactly is responsible for its effects is still under debate. Some proposed mechanisms include creation of reactive oxygen species, cutting of membranes by the sharp edges of graphene, and disruption of DNA [25]. By functionalizing graphene oxide (GO) with proteins, cells may uptake both large and small sheets and excrete them through lysozymes (Fig. 5.2c) [22]. In this context, functionalized graphene materials appear to be much safer than their pristine counterparts. Additionally, GO is also susceptible to degradation through peroxidase enzymes in the same manner as CNTs. In this case, however, the enzyme docks on the GO, forming holes in the sheets over a period of several days, until the sheet is completely destroyed [26].

5.2.2 Conjugation Techniques

The functionalization of carbon nanostructures implies their modification through either covalent bonds or noncovalent interactions in order to provide them further properties. The covalent approach requires the partial degradation (e.g., oxidation) of the conjugation pattern and therefore will affect the electronic properties of the carbon nanostructure. On the contrary, noncovalent functionalization does not perturb the intrinsic properties of the carbon nanostructures, but the strength of the noncovalent interaction used (mostly π–π stacking interactions) must be stable enough in order to provide a decent lifetime for the device designed.

5.2.2.1 Covalent Functionalization Covalent chemistry can be applied directly onto C_{60} mainly through cycloaddition reactions (e.g., [4+2] or [3+2]) and cyclopropanations [27,28]. The conjugated C_{60} derivatives display functional groups on their periphery and allow for the further derivatization toward the desired properties. CNTs can be functionalized through oxidation reactions generating defects in the 1D sp^2 carbon network with carboxylic acids appearing at the tip or in the frame of the nanotube. The conjugation of the newly formed acids with organic molecules is then the most convenient approach for the synthesis of covalently modified CNTs [29]. Similar approaches are used for the covalent modification of graphene through oxidations or cycloadditions, and again the overall structural features of the conjugated carbon atoms network are affected [30,31].

5.2.2.2 Noncovalent Functionalization The noncovalent functionalization of fullerenes has been approached through complexation with cyclodextrins [32,33] of cyclotriveratrylenes (CTV) [34] usually for water solubility applications. CNTs and graphene have attracted more attention and efforts for their noncovalent functionalization. Conjugation with biomolecules (e.g., enzymes, DNA) or carbohydrates allows for the interfacing of CNTs or graphene with biological systems for several applications. Such noncovalent functionalization can occur through hydrophobic contacts [35] but mainly using π–π interactions [36,37] or supramolecular wrapping [38].

5.3 GLYCOCONJUGATES

A detailed outline of the carbohydrate conjugation techniques to CNTs and graphene has been recently reviewed [39]. Recent applications of glycofullerenes have also been compiled [28]. The applications of such glycoconjugates are usually for water solubility of the carbon nanostructures but also for biomedical applications such as drug delivery or detection of bioanalytes.

5.3.1 Applications

5.3.1.1 ***Biosensors*** Due to their unique properties, CNTs and graphene have found a multitude of applications in biosensors [8,40,41]. With specific consideration for glycans, CNTs and graphene have been functionalized with glycoconjugates for the specific detection of biomedically relevant targets in both optical and electrochemical sensors [39]. As mentioned previously, these materials may be functionalized via covalent or noncovalent means; however, covalent modification involves disruption of the innate sp^2 structure, resulting in partial or total loss of electronic properties. Thus, for biosensor applications, noncovalent functionalization with pyrene/porphyrin conjugates or glycopolymers/lipids or by embedding in composites is highly preferred in order to preserve attractive properties.

Semiconducting CNTs have been widely exploited in field-effect transistor (FET) devices for the purpose of creating efficient and label-free biosensors [42]. Glycans are particularly attractive for creating functional sensors due to their highly specific interactions with bacterial cells. Through conjugation with porphyrins, an SWNT network was functionalized with specific glycoconjugates by Chen et al., creating specific sensors aimed at bacteria detection (Fig. 5.3a) [43]. Sensors show response to bacterial lectins that interact with the specific sugar, but not to control lectins. By changing out the type of porphyrin and sugar, sensors may be created for a broad range of targets. Similarly, Sudibya et al. functionalized SWNTs with pyrene and lipid-modified sugars with the purpose of directly interfacing with cells for the monitoring of cell apoptosis [45]. Optical methods are another broad area for biosensors using carbon nanomaterials.

Taking advantage of fluorescence resonance energy transfer (FRET), Lin and coworkers created a functionalized GO sensor [44]. In this particular sensor, GO is functionalized with a maltose-modified pyrene. Pyrene fluorescence is quenched when in close proximity to GO and recovers when concanavalin A (ConA) binds to maltose, thereby disrupting the π–π interactions and releasing the pyrene (Fig. 5.3b). SWNTs fluoresce in the near-infrared (NIR) region of the spectrum, and this fluorescence has been exploited for the purpose of creating optical sensors. Reuel et al. have embedded SWNTs in a chitosan matrix and utilized local fluorescence of SWNT bundles as individual sensors for biosensing purposes. This type of sensor array was applied to glycan–lectin-binding sensing [46] as well as detection of hypermannosylation and monitoring of IgG affinity distributions [47].

Glycosylated anthraquinones were also conjugated with GO through noncovalent interactions for the design of a screen-printed electrode [37]. The galactosyl ligands organized at the surface of the GO structure provided a dense array of ligands for recognition by the asialoglycoprotein receptors displayed by HepG2 cells. Conjugation of *N*-acetylglucosamine and *N*-acetylgalactosamine to GO with rhodamine [36] provided fluorescent glycoconjugates with an aromatic moiety capable of

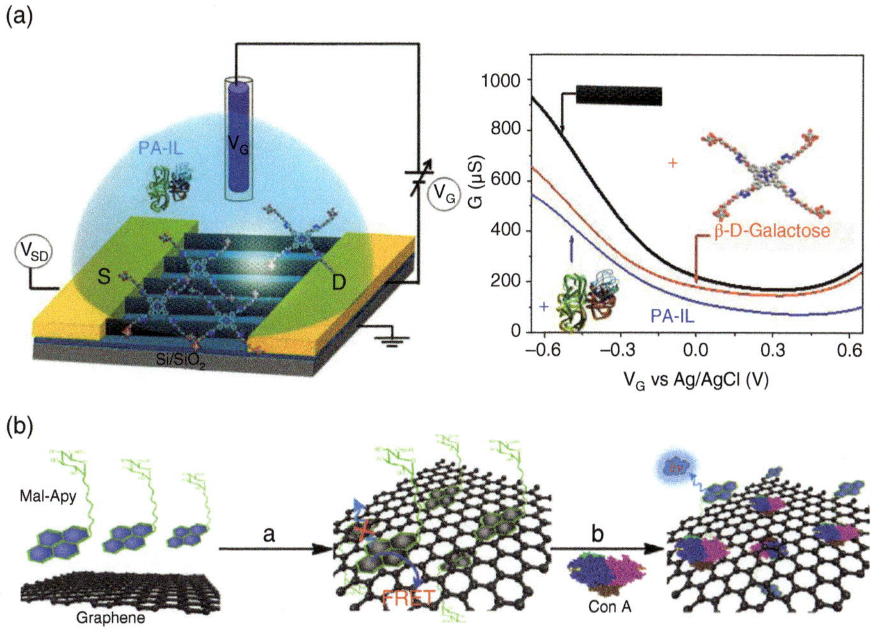

FIGURE 5.3 (a) Concept of glycoconjugate-functionalized CNT FET device for lectin detection (left). FET characteristics of a bare SWNT device, after noncovalent functionalization with a β-D-galactose glycoconjugate and after exposure to PA-IL. Inset structure of the glycoconjugate used (right). *Source*: Reprinted with permission from Ref. 43, Copyright 2011, The American Chemical Society. (b) Self-assembly of a maltose-grafted aminopyrene on graphene. Fluorescence of the pyrene is quenched when in contact with the graphene surface (a) and returns when concanavalin A competitively binds with maltose, removing the pyrene from the surface (b). *Source*: Reproduced with permission from Ref. 44, Copyright 2011, Elsevier.

noncovalent interactions with GO. The resulting fluorogenic conjugate could be used for the specific recognition of lectins from cancer cells.

Covalent conjugation of glycoclusters to CNTs and graphene was achieved by azide–alkyne cycloaddition onto preactivated surfaces with ethynyl-phenyl moieties [48]. The designed conjugates were assayed as biosensors toward concanavalin A through AFM, fluorescence, and UV–Vis studies and demonstrated the sensitivity and selectivity of the designed device.

5.3.1.2 Biofuel Cells Many carbohydrates are directly used by cells for the purpose of energy generation. Taking advantage of this and the electronic properties of CNTs and graphene, enzymatic biofuel cells have been targeted as a potential renewable energy source based on natural metabolism. As biofuel cells use specific enzymes and sugar targets, these types of fuel cells have the advantage of high selectivity, operation in mild conditions, and are generally safer than alternative hydrogen or methanol cells [49]. These types of biofuel cells are proposed for next-generation renewable battery sources for small electronics. As CNTs and graphene interact

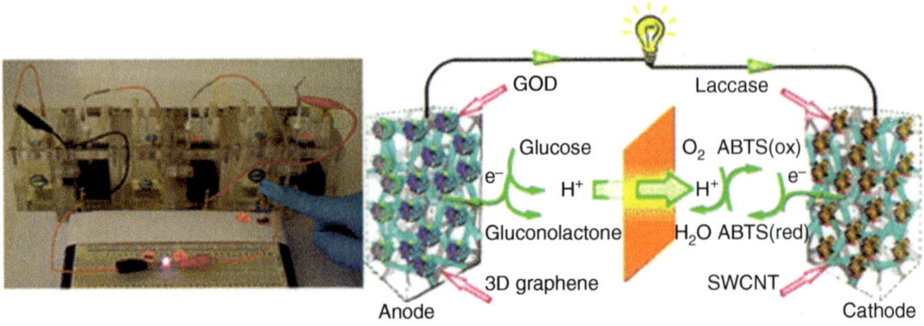

FIGURE 5.4 LED powered by three enzymatic biofuel cells (left) and a schematic of the electrode (right). *Source*: Reprinted with permission from Ref. 52, Copyright 2014, The American Chemical Society.

strongly with enzymes and are excellent for mediating electron transfer, they have been recently explored for applications in biofuel cells.

An early issue with usage of these materials related to poor dispersibility resulting in loss of available surface area for enzyme attachment. As such, 3D scaffolds of CNTs are popular for increasing the surface area and interacting with enzymes. Jose et al. manufactured a biofuel cell based on glucose oxidase, MWNTs, and electrospun gold fibers with high available surface area [50]. Another method involves embedding GO in a silica sol–gel to create a porous, membrane-free fuel cell [51]. More recently, Prasad et al. fabricated an enzymatic biofuel cell using laccase and glucose oxidase on a hybrid graphene–CNT scaffold [52]. Initially, graphene was grown via chemical vapor deposition (CVD) on nickel foam. After the nickel is etched away, oxidized SWNTs were introduced by soaking the graphene scaffold in a solution of nanotubes. The enzymes were then attached to their respective electrodes (Fig. 5.4) using coupling chemistry. By taking advantage of the close binding of SWNTs to active sites on the protein with the high conductivity of CVD grown graphene, they were able to produce a biofuel cell with excellent energy density capable of powering an LED (Fig. 5.4).

5.3.1.3 **In Vivo** ***Applications*** Graphene-based nanomaterials have been designed and applied to biomedical applications [35,53–55] such as cell imaging, photothermal and photodynamic therapies, and drug and gene delivery but with fewer *in vivo* applications, for instance drug delivery targeting tumors [9,56]. A selected series of examples will be discussed below.

In Vivo *Biosensing* Amylose was demonstrated to wrap around CNTs in an entropy-driven noncovalent supramolecular self-assembly process [57]. The similar observation was made for DNA helical oligomers composed of seven repeating AAAT tetrads [38]. The wrapping of the native DNA oligomer provided water solubility to SWNTs, but no fluorescent property could be exploited (Fig. 5.5). Conjugation with polyethylene glycol stabilized the NIR fluorescence properties of SWNTs for the *in vivo* biosensing of nitric oxide with a lower detection limit of 1 μM.

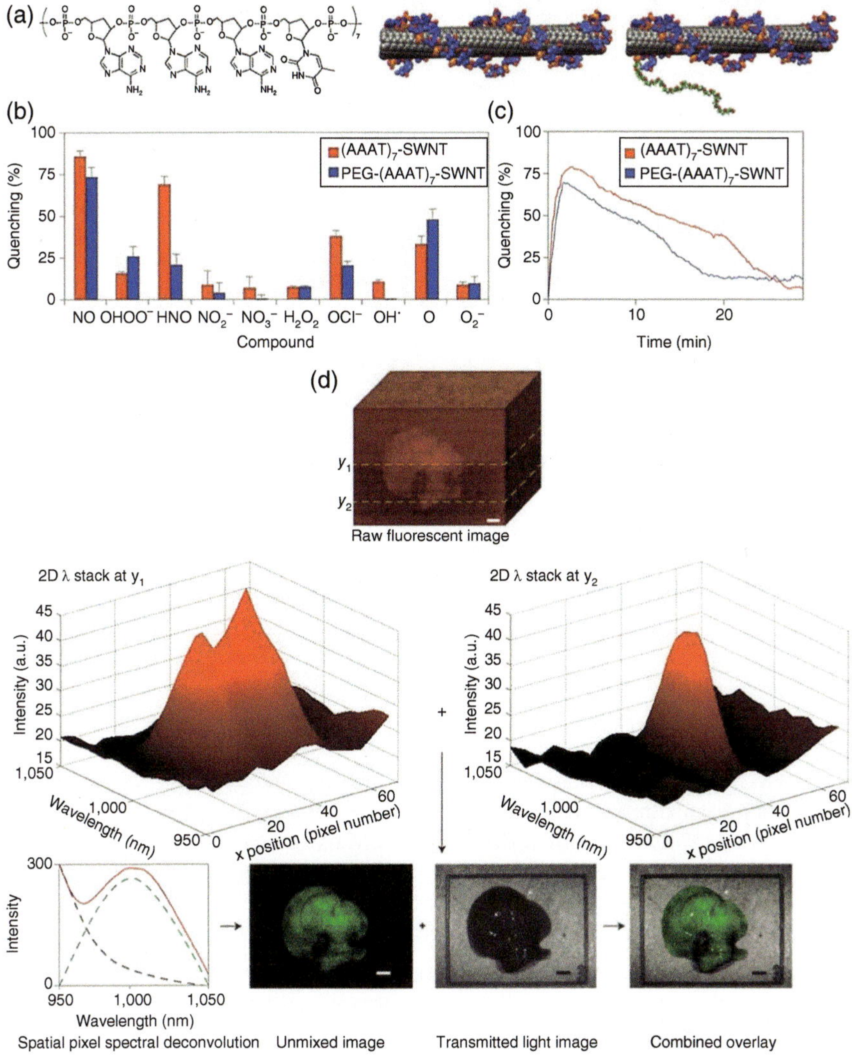

FIGURE 5.5 (a) Chemical composition of complex with d(AAAT)$_7$ (left), wrapped (AAAT)$_7$–SWNT (middle), and PEG–(AAAT)$_7$–SWNT (right). (b and c) Quenching activity of (AAAT)$_7$–SWNT and PEG–(AAAT)$_7$–SWNT sensors quantified by percent quenching of original fluorescence using a 785 nm photodiode following exposure to reactive nitrogen species and reactive oxygen species compounds (b) with error bars representing standard error and NO (c). (d) 2Dl imaging analysis for (AAAT)$_7$–SWNT in an excised mouse liver 30 min after a tail vein injection of PEG–(AAAT)$_7$–SWNT imaged with white-light excitation and an emission spectrum ranging from 950 to 1050 nm with a 10 nm step and 20 s accumulation time. The top image shows the raw data acquired followed by two cross sections from the y-axis. The bottom row depicts an example of the raw fluorescence (total curve) and its components (dashed peaked curve is SWNT fluorescence and dashed decaying curve is background fluorescence) followed by images of the unmixed data and transmitted light image and finally the combination of these two images to depict the location of the fluorescence in the tissue. Scale bars, 4 mm. *Source*: Reprinted by permission from Ref. 38, Copyright 2013, Macmillan Publishers Ltd on behalf of Cancer Research UK.

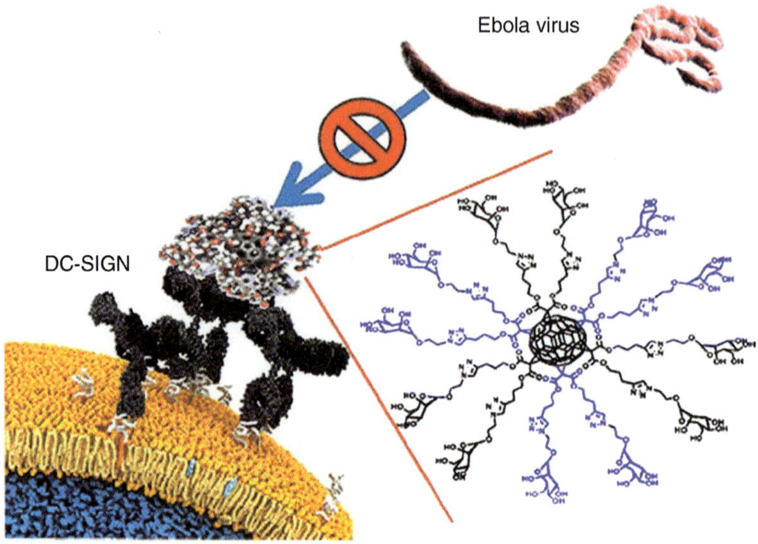

FIGURE 5.6 Schematic representation of the inhibition of Ebola infection with glycofullerenes. *Source*: Reprinted with permission from Ref. 58, Copyright 2013, The American Chemical Society.

Inhibition The design of multivalent glycofullerenes as the conjugation of multiple copies of a carbohydrate ligand for the efficient binding to lectins has been applied to a series of biological processes in the realm of glycobiology [28]. *Hexakis*-adducts of C_{60} displaying 12–6 mannoside residues (Fig. 5.6) have been reported as nanomolar inhibitors of viral infection mediated by DC-SIGN, a protein involved in HIV or Ebola viral infections [58].

Tissue Engineering Applications of CNTs in tissue engineering [59,60] have been recently reviewed in the context of neuroregeneration and repair [61] and for applications to the central nervous system [62]. For instance, chitosan was covalently conjugated to SWNTs through amide bonds between the amine groups of the glycopolymer and the carboxylic acid moieties generated on SWNTs after oxidation [63]. The designed conjugate provided a porous, biodegradable, and biocompatible chitosan. Association with hydroxyapatite afforded a composite matrix used for cell proliferation, which was twice as efficient as chitosan with suggested applications in bone tissue engineering [63].

5.4 CONCLUSION AND OUTLOOK

Nanotechnology has witnessed a dramatic increase of efforts and results not only in terms of fundamental knowledge in physics or chemistry but more importantly for applications in biology or medicine. Advances in synthetic methods have allowed for the design of complex architectures based on carbon nanostructures such as fullerenes, CNTs, or graphene. More specifically, conjugation with carbohydrates allowed for the interfacing of the resulting glycoconjugates with biological material such as proteins (e.g., enzymes, receptors) and microorganisms (e.g., viruses, bacteria) as well as cancer cells. The conjugation was accomplished through either covalent or noncovalent interactions, the latter being almost neutral in terms of conducting properties of the carbon nanostructures in order to protect the intrinsic properties of such conjugates. While fullerene-based glycoconjugates have found applications as enzyme inhibitors or receptor ligands in medicinal chemistry, CNTs and graphene were applied as drug delivery systems, biosensors, and biofuel cells. General aspects of carbon nanostructures and a review of selected applications in each field are provided herein.

REFERENCES

1. Geim, A. K.; Novoselov, K. S. *Nat. Mater.* **2007**, *6*, 183–191.
2. Kroto, H. W.; Heath, J. R.; O'Brien, S. C.; Curl, R. F.; Smalley, R. E. *Nature* **1985**, *318*, 162–163.
3. Kolosnjaj, J.; Szwarc, H.; Moussa, F. *Adv. Exp. Med. Biol.* **2007**, *620*, 168–180.
4. Li, C.-Z.; Yip, H.-L.; Jen, A. K.-Y. *J. Mater. Chem.* **2012**, *22*, 4161–4177.
5. Lee, D. J.; Ahn, Y. S.; Youn, Y. S.; Lee, E. S. *Polym. Adv. Technol.* **2013**, *24*, 220–227.
6. Iijima, S.. *Nature* **1991**, *354*, 56–58.
7. Odom, T. W.; Huang, J.-L.; Kim, P.; Lieber, C. M. *J. Phys. Chem. B* **2000**, *104*, 2794–2809.
8. Liu, S.; Guo, X. *NPG Asia Mater.* **2012**, *4*, e23.
9. Hong, H.; Yang, K.; Zhang, Y.; Engle, J. W.; Feng, L.; Yang, Y.; Nayak, T. R.; Goel, S.; Bean, J.; Theuer, C. P.; Barnhart, T. E.; Liu, Z.; Cai, W. *ACS Nano* **2012**, *6*, 2361–2370.
10. Zhang, J.; Xu, Z.; Mai, W.; Min, C.; Zhou, B.; Shan, M.; Li, Y.; Yang, C.; Wang, Z.; Qian, X. *J. Mater. Chem. A* **2013**, *1*, 3101–3111.
11. Lam, C. W.; James, J. T.; McCluskey, R.; Hunter, R. L. *Toxicol. Sci.* **2004**, *77*, 126–134.
12. Novoselov, K. S.; Geim, A. K.; Morozov, S. V.; Jiang, D.; Zhang, Y.; Dubonos, S. V.; Grigorieva, I. V.; Firsov, A. A. *Science* **2004**, *306*, 666–669.
13. Zhu, Y.; Murali, S.; Cai, W.; Li, X.; Suk, J. W.; Potts, J. R.; Ruoff, R. S. *Adv. Mater.* **2010**, *22*, 3906–3924.
14. Huang, Y.; Dong, X.; Liu, Y.; Li, L.-J.; Chen, P. *J. Mater. Chem.* **2011**, *21*, 12358–12362.
15. Chen, Y.; Michael, Z. P.; Kotchey, G. P.; Zhao, Y.; Star, A. *ACS Appl. Mater. Interfaces* **2014**, *6*, 3805–3810.
16. Gupta, V.; Chaudhary, N.; Srivastava, R.; Sharma, G. D.; Bhardwaj, R.; Chand, S. *J. Am. Chem. Soc.* **2011**, *133*, 9960–9963.
17. Feng, Y.; Zhang, X.; Shen, Y.; Yoshino, K.; Feng, W. *Carbohydr. Polym.* **2012**, *87*, 644–649.
18. Barone, V.; Hod, O.; Scuseria, G. E. *Nano Lett.* **2006**, *6*, 2748–2754.
19. Tan, X.; Chuang, H.-J.; Lin, M.-W.; Zhou, Z.; Cheng, M. M.-C. *J. Phys. Chem. C* **2013**, *117*, 27155–27160.

20. Gharbi, N.; Pressac, M.; Hadchouel, M.; Szwarc, H.; Wilson, S. R.; Moussa, F. *Nano Lett.* **2005**, *5*, 2578–2585.
21. Kostarelos, K. *Nat. Biotechnol.* **2008**, *26*, 774–776.
22. Mu, Q.; Su, G.; Li, L.; Gilbertson, B. O.; Yu, L. H.; Zhang, Q.; Sun, Y.-P.; Yan, B. *ACS Appl. Mater. Interfaces* **2012**, *4*, 2259–2266.
23. Lanone, S.; Andujar, P.; Kermanizadeh, A.; Boczkowski, J. *Adv. Drug Deliv. Rev.* **2013**, *65*, 2063–2069.
24. Kotchey, G. P.; Zhao, Y.; Kagan, V. E.; Star, A. *Adv. Drug Deliv. Rev.* **2013**, *65*, 1921–1932.
25. Seabra, A. B.; Paula, A. J.; de Lima, R.; Alves, O. L.; Durán, N. *Chem. Res. Toxicol.* **2014**, *27*, 159–168.
26. Kotchey, G. P.; Allen, B. L.; Vedala, H.; Yanamala, N.; Kapralov, A. A.; Tyurina, Y. Y.; Klein-Seetharaman, J.; Kagan, V. E.; Star, A. *ACS Nano* **2011**, *5*, 2098–2108.
27. Iehl, J.; Urbani, M.; Nierengarten, J.-F. Chemistry with fullerene building blocks in *Handbook of Carbon Nano Materials*, Francis D'Souza and Karl M. Kadish; World Scientific Publishing Co. Pte. Ltd., Singapore, **2011**; Vol. *1*, pp. 1–31.
28. Nierengarten, I.; Nierengarten, J.-F. *Chem.-Asian J.* **2014**, *9*, 1436–1444.
29. Tasis, D.; Tagmatarchis, N.; Bianco, A.; Prato, M. *Chem. Rev.* **2006**, *106*, 1105–1136.
30. Chua, C. K.; Pumera, M. *Chem. Soc. Rev.* **2013**, *42*, 3222–3233.
31. Eigler, S.; Hirsch, A. *Angew. Chem. Int. Ed.* **2014**, *53*, 7720–7738.
32. Andersson, T.; Nilsson, K.; Sundahl, M.; Westman, G.; Wennerstrom, O. *J. Chem. Soc., Chem. Commun.* **1992**, 604–606.
33. Yoshida, Z.-i.; Takekuma, H.; Takekuma, S.-i.; Matsubara, Y. *Angew. Chem. Int. Ed.* **1994**, *33*, 1597–1599.
34. Yang, F.; Chen, Q.; Cheng, Q.-Y.; Yan, C.-G.; Han, B.-H. *J. Org. Chem.* **2012**, *77*, 971–976.
35. Shi, S.; Chen, F.; Ehlerding, E. B.; Cai, W. *Bioconjugate Chem.* **2014**, *25*, 1609–1619.
36. Zhang, H.-L.; Wei, X.-L.; Zang, Y.; Cao, J.-Y.; Liu, S.; He, X.-P.; Chen, Q.; Long, Y.-T.; Li, J.; Chen, G.-R.; Chen, K. *Adv. Mater.* **2013**, *25*, 4097–4101.
37. Li, Z.; Deng, S.-S.; Zang, Y.; Gu, Z.; He, X.-P.; Chen, G.-R.; Chen, K.; James, T. D.; Li, J.; Long, Y.-T. *Sci. Rep.* **2013**, *3*, 2293.
38. Iverson, N. M.; Barone, P. W.; Shandell, M.; Trudel, L. J.; Sen, S.; Sen, F.; Ivanov, V.; Atolia, E.; Farias, E.; McNicholas, T. P.; Reuel, N.; Parry, N. M. A.; Wogan, G. N.; Strano, M. S. *Nat. Nanotechnol.* **2013**, *8*, 873–880.
39. Chen, Y.; Star, A.; Vidal, S. *Chem. Soc. Rev.* **2013**, *42*, 4532–4542.
40. Kruss, S.; Hilmer, A. J.; Zhang, J.; Reuel, N. F.; Mu, B.; Strano, M. S. *Adv. Drug Deliv. Rev.* **2013**, *65*, 1933–1950.
41. Kauffman, D. R.; Star, A. *Chem. Soc. Rev.* **2008**, *37*, 1197–1206.
42. Münzer, A. M.; Michael, Z. P.; Star, A. *ACS Nano* **2013**, *7*, 7448–7453.
43. Vedala, H.; Chen, Y.; Cecioni, S.; Imberty, A.; Vidal, S.; Star, A. *Nano Lett.* **2011**, *11*, 170–175.
44. Chen, Q.; Wei, W.; Lin, J.-M. *Biosens. Bioelectron.* **2011**, *26*, 4497–4502.
45. Sudibya, H. G.; Ma, J.; Dong, X.; Ng, S.; Li, L.-J.; Liu, X.-W.; Chen, P. *Angew. Chem. Int. Ed.* **2009**, *48*, 2723–2726.

46. Reuel, N. F.; Ahn, J.-H.; Kim, J.-H.; Zhang, J.; Boghossian, A. A.; Mahal, L. K.; Strano, M. S. *J. Am. Chem. Soc.* **2011**, *133*, 17923–17933.
47. Reuel, N. F.; Grassbaugh, B.; Kruss, S.; Mundy, J. Z.; Opel, C.; Ogunniyi, A. O.; Egodage, K.; Wahl, R.; Helk, B.; Zhang, J.; Kalcioglu, Z. I.; Tvrdy, K.; Bellisario, D. O.; Mu, B.; Blake, S. S.; Van Vliet, K. J.; Love, J. C.; Wittrup, K. D.; Strano, M. S. *ACS Nano* **2013**, *7*, 7472–7482.
48. Ragoussi, M.-E.; Casado, S.; Ribeiro-Viana, R.; de la Torre, G.; Rojo, J.; Torres, T. *Chem. Sci.* **2013**, *4*, 4035–4041.
49. Meredith, M. T.; Minteer, S. D. *Annu. Rev. Anal. Chem.* **2012**, *5*, 157–179.
50. Jose, M. V.; Marx, S.; Murata, H.; Koepsel, R. R.; Russell, A. J. *Carbon* **2012**, *50*, 4010–4020.
51. Liu, C.; Alwarappan, S.; Chen, Z.; Kong, X.; Li, C.-Z. *Biosens. Bioelectron.* **2010**, *25*, 1829–1833.
52. Prasad, K. P.; Chen, Y.; Chen, P. *ACS Appl. Mater. Interfaces* **2014**, *6*, 3387–3393.
53. Zhang, Y.; Nayak, T. R.; Hong, H.; Cai, W. *Nanoscale* **2012**, *4*, 3833–3842.
54. Yang, K.; Feng, L.; Shi, X.; Liu, Z. *Chem. Soc. Rev.* **2013**, *42*, 530–547.
55. Yang, K.; Feng, L.; Hong, H.; Cai, W.; Liu, Z. *Nat. Protoc.* **2013**, *8*, 2392–2403.
56. Hong, H.; Zhang, Y.; Engle, J. W.; Nayak, T. R.; Theuer, C. P.; Nickles, R. J.; Barnhart, T. E.; Cai, W. *Biomaterials* **2012**, *33*, 4147–4156.
57. Star, A.; Steuerman, D. W.; Heath, J. R.; Stoddart, J. F. *Angew. Chem. Int. Ed.* **2002**, *41*, 2508–2512.
58. Luczkowiak, J.; Muñoz, A.; Sánchez-Navarro, M.; Ribeiro-Viana, R.; Ginieis, A.; Illescas, B. M.; Martín, N.; Delgado, R.; Rojo, J. *Biomacromolecules* **2013**, *14*, 431–437.
59. Dvir, T.; Timko, B. P.; Kohane, D. S.; Langer, R. *Nat. Nanotechnol.* **2011**, *6*, 13–22.
60. Place, E. S.; Evans, N. D.; Stevens, M. M. *Nat. Mater.* **2009**, *8*, 457–470.
61. Fabbro, A.; Prato, M.; Ballerini, L. *Adv. Drug Deliv. Rev.* **2013**, *65*, 2034–2044.
62. Sharma, H. S.; Sharma, A. *Expert Rev. Neurother.* **2012**, *12*, 901–905.
63. Venkatesan, J.; Qian, Z.-J.; Ryu, B.; Ashok Kumar, N.; Kim, S.-K. *Carbohydr. Polym.* **2011**, *83*, 569–577.

6

SYNTHESIS OF GLYCOPOLYMERS AND RECENT DEVELOPMENTS

GOKHAN YILMAZ[1,2] AND C. REMZI BECER[3]

[1] *Department of Chemistry, University of Warwick, Coventry, UK*
[2] *Department of Basic Sciences, Turkish Military Academy, Ankara, Turkey*
[3] *School of Engineering and Materials Science, Queen Mary, University of London, London, UK*

6.1 INTRODUCTION

Glycopolymers, which are synthetic macromolecules with carbohydrate moieties, play a crucial role for many biological processes such as signal transmission, intercellular recognition, and fertilization [1–11]. Carbohydrates indeed exhibit a great coding capacity in terms of their monomeric units and their ability to be highly branched molecules [12–17]. Although the binding between carbohydrates and proteins is relatively weak, it could be greatly enhanced by the multivalent effect of densely packed carbohydrate molecules with unique functionalities, which is known as the "glycocluster effect" [17–25]. There is an increasing interest in synthetic glycopolymers that have a capability to interact with lectins as multivalent ligands in a similar manner to natural glycoproteins. Some recent studies have shown a sufficient control in chain length, architecture, monomer sequence, chain folding, and tertiary structures to achieve the synthesis of well-defined and multifunctional glycopolymers [26–31]. The first report on the synthesis of a glycopolymer was undertaken in 1944 [32]. The formation of a viscous polymer solution after heating tetra-allyl α-methyl glucosides was firstly introduced by Nichols and Yanovsky. Very soon after other reports on the polymerization of allyl ethers of carbohydrates emerged [33–35]. However, until the last decade, there had been limited attempts to react a functional polymeric backbone with a carbohydrate to obtain a glycopolymer because of the difficulty of

Carbohydrate Nanotechnology, First Edition. Edited by Keith J. Stine.

introducing sufficiently reactive pendant groups onto the polymer backbone to react with carbohydrates. This obstacle has been overcome by the combination of living polymerization and click chemistry during the previous years [36–39].

Generally, there are two advanced synthesis strategies to prepare glycopolymers, which are the polymerization of carbohydrate-bearing monomers without the necessity of deprotection steps and the synthesis of polymeric scaffolds bearing pendant reactive sites that are subsequently functionalized by carbohydrate moieties via controlled/living polymerization techniques in combination with click reactions such as copper-catalyzed azide–alkyne cycloaddition (CuAAC) [40,41]. According to these promising recent reports on the preparation of sequenced glycopolymers as well as their interactions with lectins, polymer scientists have been able to contribute to the design of the new generation sequence-controlled glycopolymers with multivalence and multifunctional and also complex glycoparticles with different architectures in nanometer dimension.

The main focus of this chapter is to provide a comprehensive review of all available routes that have been utilized for the preparation of glycopolymers. The most widely used controlled/living radical polymerization techniques for well-defined glycopolymers are nitroxide-mediated polymerization (NMP), atom transfer radical polymerization (ATRP), reversible addition–fragmentation chain transfer (RAFT), single-electron transfer-living radical polymerization (SET-LRP), ring-opening metathesis polymerization (ROMP), and the ionic ring-opening polymerization [42–53]. Besides, a large number of reports have emerged on the synthesis of glycopolymers via conventional free radical polymerization in both aqueous and nonaqueous media. These techniques are relatively more tolerant to prepare well-defined glycopolymers with different functionalities. Moreover, the most important advantage of these techniques is their applicability to use a very wide range of monomers under a large number of experimental conditions. During the last few years, the explosive number of publications devoted to prepare glycopolymers via these polymerization techniques that have a binding ability with different lectins prompted us to contribute this chapter.

6.2 SYNTHESIS OF GLYCOPOLYMERS

6.2.1 Controlled/Living Radical Polymerization

6.2.1.1 Ring-Opening Polymerization Glycomonomers based on 7-oxanorbornene or norbornene have been polymerized via ROMP technique successfully. The catalysts used in ROMP, such as ruthenium catalyst, are tolerant to mannose and glucose side groups. Therefore, the direct polymerization of unprotected glycomonomer could be undertaken via ROMP. Thus, it was possible to prepare the block glycopolymers. At the same time, isopropylidene-protected sugars have been polymerized in different catalyst systems and reaction conditions via ROMP. 3,6-Oxy -1,2,3,6-tetrahydrophthalic anhydride was reacted with a range of triethylsilyl-protected carbohydrates, which were deprotected prior to ROMP [54–56]. Depending on the final structure and further metal ligands, ROMP provides

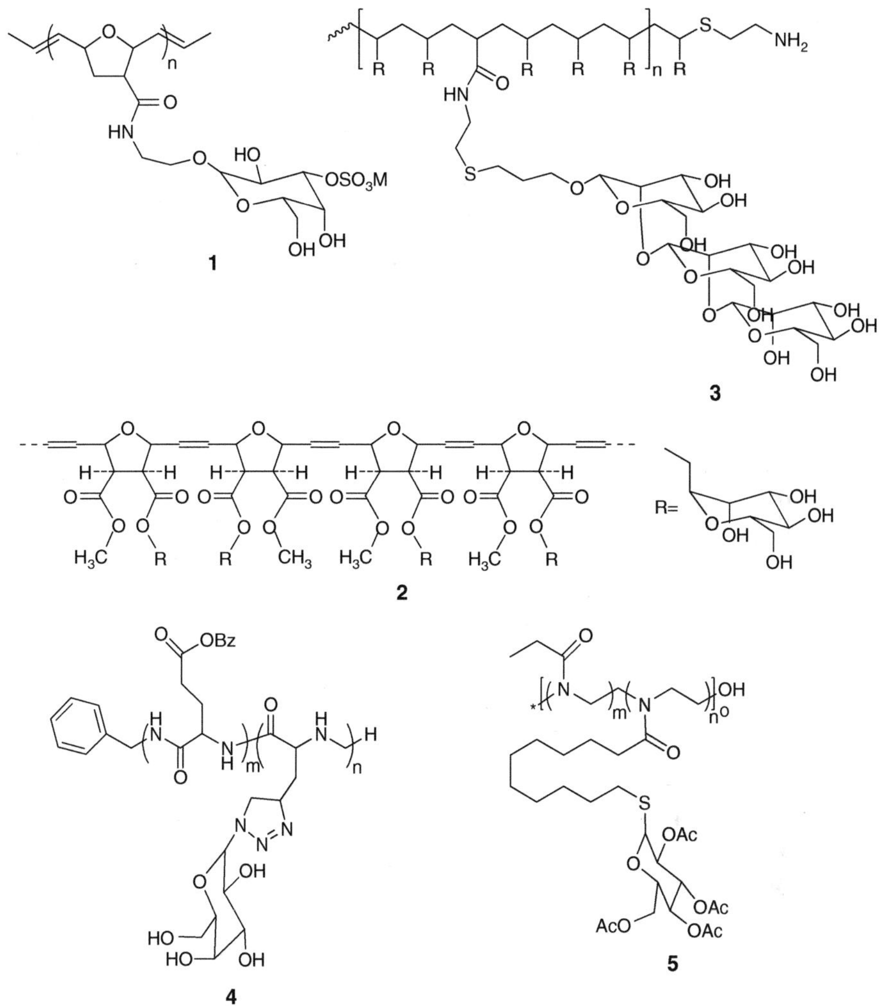

FIGURE 6.1 A schematic representation of glycopolymers synthesized via ring-opening polymerization.

polymerization to proceed in a living fashion and provides block coglycopolymers with narrow molecular weight distributions.

As illustrated in Figure 6.1, Kiessling *et al.* have succeeded to prepare mannose- and galactose-containing glycopolymers **1** via ROMP in water in the presence of $RuCl_3$ at 55°C [54]. These glycopolymers showed the control of multivalent interactions by altering the binding epitope density. Then, this work was extended to investigate the effect of the simultaneous existence of cis- and trans-isomers in the backbone. Moreover, they have discussed several mechanisms of multivalent ligand binding, that is, chelate effect, subsite binding, steric stabilization, statistical effect, and receptor clustering. The results showed that the glycopolymers synthesized via

ROMP acted as cell and lectin agglutination inhibitors. However, It should be rewritten like: The obtained molecular weights, typically around 10^6 g moL^{-1}, indicated that the chains were not living anymore. To prepare glycopolymers with different molecular weights, Grubbs ruthenium alkylidene catalyst was employed. Unfortunately, another obstacle came out by the contrasting solubility of catalyst and monomer for this time. More work has been carried out by the group of Kiessling on sulfated saccharides [57]. By using a similar strategy, ROMP was used to synthesize sulfated glycopolymers **2** showing efficient binding toward L- or P-selectin. Furthermore, they observed that glycopolymers synthesized in the presence of high content of ruthenium trichloride resulted in discoloration of polymers, which indicated the contamination of materials.

Another study reported by the Lipinski group has focused on the preparation of protein–glycopolymer conjugates **3** that are promising in mimicking structural and functional responsibilities of the polysaccharide–protein conjugate vaccines [58]. ROMP copolymerization of trisaccharide allyl glycoside monomer and acrylamide was undertaken in the presence of cysteamine and potassium persulfate. The ratio of monomers to cysteamine was chosen as 530:1 in order to obtain a polymer with molar mass of 100,000. End-group modification of the obtained polymers with azide moiety for bioconjugation was carried out in the presence of 6-azido-6-deoxy-D-galactose via reductive amination. The bioconjugation of propargylated chicken serum albumin (ChSA to the glycopolymer was performed at a ~1:1 ratio (protein to glycopolymer) via azide–alkyne Huisgen cycloaddition conjugation using bathophenanthroline/Cu^{+1} catalyst in 0.2 M Tris HCl at pH 8. In *in vitro* studies, these glycopolymer–ChSA bioconjugates bearing pendant oligosaccharide moieties showed the ability to enhance the immunogenicity of hapten and the immune response to antigens.

Heise and coworkers reported the synthesis of polypeptide block copolymers with different block length ratios using sequential ring-opening polymerization of benzyl-L-glutamate and propargylglycine (PG) N-carboxyanhydrides (NCA) **4** [59]. Recently, Schubert and coworkers reported the synthesis of glucose-substituted 2-oxazoline monomer and its polymerization via cationic ring-opening copolymerization (CROP) in the presence of 2-oxazoline-based monomers [60]. A series of well-defined block glycocopolymers **5** with different chain lengths was prepared in acetonitrile at 120°C for overnight.

Basically, even though ring-opening polymerization has a long history and been widely used for polymerization of different functional cyclic monomers, the use of it in the direct polymerization of carbohydrate-containing cyclic monomers is still limited. Another critical disadvantage of using ROMP is that the toxic heavy metals used in polymerization could potentially contaminate the final polymers; hence, the removal of these catalysts has to be scrutinized, especially for biological applications.

6.2.1.2 NMP Despite the fact that NMP is one of the earliest techniques in controlled radical polymerization, it is not often used for the synthesis of well-defined glycopolymers due to the relatively high polymerization temperatures that are required for the activation of alkoxyamine to initiate and propagate the

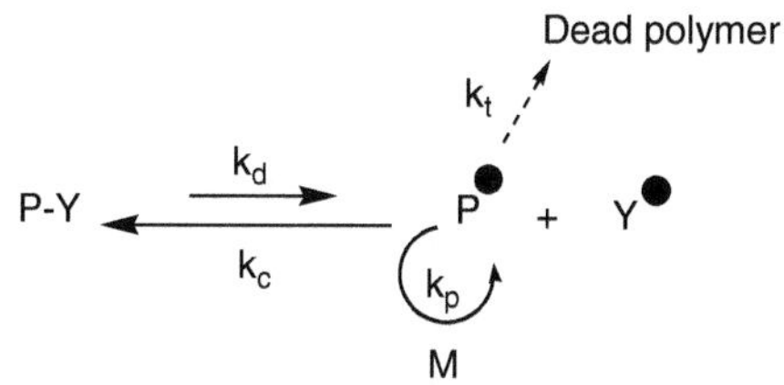

FIGURE 6.2 Activation–deactivation equilibrium in NMP.

polymerization. The reversible equilibrium between a growing radical, P˙., and a free nitroxide, Y˙., presents a thermal process: the (macro)alkoxyamine usually regenerate the propagating radical and the nitroxide by homolytic cleavage at temperatures higher than 90°C (Figure 6.2).

The most important advantage of NMP is that there is no requirement for a metal catalyst or chain transfer agent (CTAs); thus, it provides colorless and catalyst-free polymers without any need for a demanding purification step. Several different types of nitroxides, such as (2,2,6,6-tetramethylpiperidin-1-yl)oxy (TEMPO, N1), *N-tert*-butyl-1-phenyl-2-methylpropyl nitroxide (TIPNO, N2), *N*-tert-butyl-*N*-(1-diethylphosphono-2,2-dimethylpropyl)-*N*-oxyl (SG1, N3), and SG1-based alkoxyamine (BlocBuilder), have been designed to improve the range of polymerizable monomers under controlled/living conditions.

The first report on the synthesis of well-defined glycomonomer was published by Ohno and coworkers, where they have performed NMP of *N*-(*p*-vinylbenzyl)-[*O*-β-D-galactopyranosyl-(1-4)]-D-gluconamide (VLA), a styrene derivative with an oligosaccharide moiety, in *N,N*-dimethylformamide (DMF) at 90–105°C using di-tert-butyl nitroxide and dicumyl peroxide (DCP) radical initiators [61]. Even though they could not reach high monomer conversions, good control was achieved at various concentrations of di-tert-butyl nitroxide. Polydispersity indices (PDIs) increased with higher conversion, and no polymerization occurred when targeting higher molecular weights (target DP ≈ 100).

Chen and Wulff succeeded to polymerize four isopropylidene-protected sugar monomers in bulk at 130°C for 48 h using a TEMPO-derived alkoxyamine as the mediating agent and *N*-(a-methylbenzyloxy)-2,2,6,6-tetramethylpiperidine as an initiator [62]. In contrast to NMP conducted by Ohno and coworkers, no thermal decomposition of these monomers was observed. Various diblock copolymers with styrene were prepared using sugar-carrying homopolymers or polystyrene macroinitiators. PDIs of the obtained glycopolymers **6** were around 1.50. Thermogravimetric analysis showed that the monomers are thermally stable up to 150°C. This approach was taken one step further by Narumi and colleagues with the synthesis of triblock copolymers of 4-vinylbenzyl glucoside peracetate and styrene. The NMP of glycomonomer was carried out in chlorobenzene at 125°C for 5 h using (1S)-(+)-10-camphorsulfonic acid anhydride (CSA) as an accelerator and difunctional TEMPO-based mediator. These macroinitiators were used to chain extend with styrene for different segment lengths (M_n = 26 kDa, PDI = 1.17).

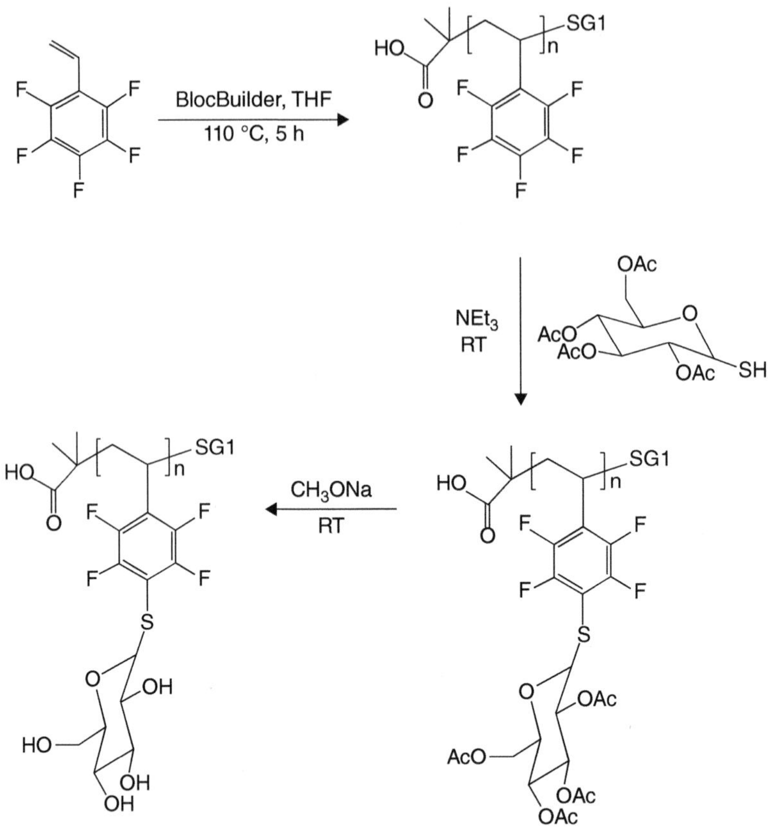

FIGURE 6.3 The synthesis of glycopolymers via the combination of NMP and thiol-para fluoro click chemistry. *Source*: Reproduced with permission from Ref. 63, Copyright 2009.

NMP can also be used with the advent of click chemistry to synthesize well-defined glycopolymers. The CuAAC reaction has a copper catalyst that requires a purification step, negating the benefits of NMP; therefore, metal-free click chemistry is a valuable tool for the synthesis of glycopolymers. One of the best examples of this concept was shown by Becer and coworkers [63]. They designed a new method that combines NMP and metal-free click chemistry where well-defined homopolymers of pentafluorostyrene (PFS) were functionalized by thiol-para fluoro click. PFS was polymerized in tetrahydrofuran (THF) using BlocBuilder alkoxyamine of desired chain length with low PDI (Figure 6.3). Subsequently, *p*-fluoro-thiol glucose (1.2 eq.) was clicked to poly(PFS) (1.0 eq. per PFS unit) in the presence of triethylamine (TEA) (3.0 eq.) under ambient conditions to yield well-defined glycopolymers **7**. Deacetylation of sugar units was performed in DMF and reacted with sodium methanolate at room temperature for 1 h. This reaction was confirmed by ^{13}C NMR and ^{1}H NMR measurements. Furthermore, the kinetics of the substitution reaction were

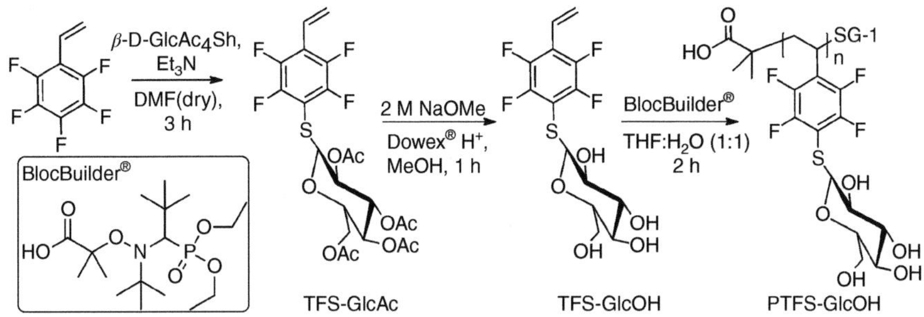

FIGURE 6.4 Chemical reaction pathways of glycomonomer synthesis and its polymerization using nitroxide-mediated polymerization. *Source*: Reproduced with permission from Ref. 64, Copyright 2011, American Chemical Society.

monitored by measuring ^{19}F NMR spectra at 40°C, and quantitative conversions were observed in less than 1 h. The main advantage of this method is that there are commercially available a wide range of primary amines and thiols that can be efficiently reacted with $-C_6F_5$ groups.

The same authors have extended this work, and thiol-para fluoro click reaction was performed before NMP to yield the glycomonomer [64]. The polymerization reaction was carried out in THF/water (H_2O) (1:1) mixture to solubilize the reactants due to the water-soluble deprotected glucose PFS monomer at 110°C (Figure 6.4). Fluorine NMR was used to characterize the *ortho*- and *meta*-fluorines of the tetrafluorostyryl units. Well-defined glycopolymers **8** were obtained with 24,000 g mol^{-1} molar mass and PDI of 1.16. The obtained glycopolymers were used as coating materials to stabilize magnetic iron oxide nanoparticles and cell–carbohydrate interactions.

Recently, the synthesis of precision glycopolymers was demonstrated by Lutz and coworkers using a combination of NMP polymerization of site-selective functionalization with reactive maleimides (MIs) and CuAAC [65]. Copolymerizations of styrene with the protected N-propargylmaleimides were performed to yield copolymers with a molar mass of 11.6–13.7 kDa. The alkoxyamine BlocBuilder was used as the initiator, and the time-controlled addition of the protected MIs (3 equivalents) during NMP in anisole at 120°C provided linear polystyrene chains presenting short but localized glycol functional regions (PDI = 1.23–1.24). At the beginning of the reaction, triisopropylsilyl-protected *N*-propargylmaleimide (TIPS-PMI) was introduced. Following the complete conversion of TIPS-PMI and further chain growth with styrene, the second functional monomer, triethylsilyl-protected N-propargylmaleimide (TES-PMI), was added. Finally, trimethylsilyl-protected N-propargylmaleimide (TMS-PMI) was added close to the end of the polymer chain due to their potential sensitivity to the reaction conditions.

After the synthesis of well-defined polymers via NMP, the next step was the stepwise orthogonal deprotection and functionalization of the three distinct alkyne sites. Firstly, the deprotection reaction for the removal of TMS group was carried out in a methanol/water/THF mixture with K_2CO_3 at 40°C for 7 h, and then the azidomannose was attached to the polymer via CuAAC click reaction. The TES group was then

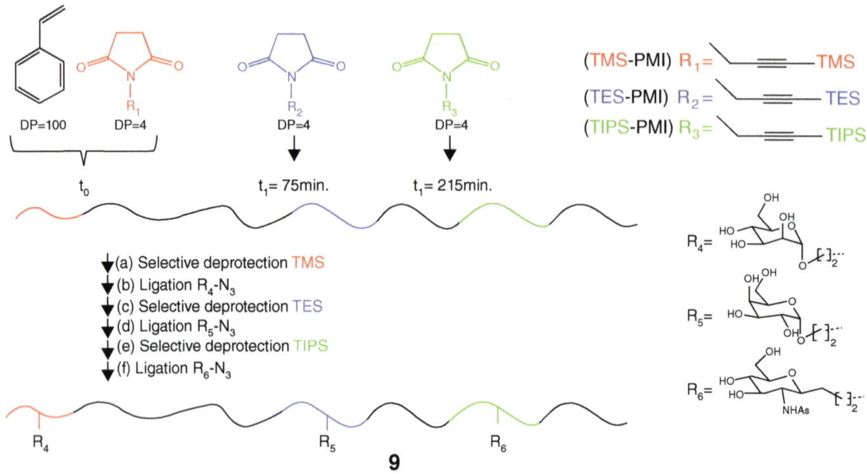

FIGURE 6.5 General strategy for the synthesis of single-chain sugar arrays. Experimental conditions: (a) K_2CO_3, MeOH/H_2O/THF, 40°C, 7 h; (b) CuBr, 4,4′-di-*n*-nonyl-2,2′-bipyridine, DMF, room temperature; (c) K_2CO_3, MeOH/H_2O/THF, 60°C, 89 h; (d) CuBr, 4,4′-di-*n*-nonyl-2,2’-bipyridine, azidomethyl phenyl sulphide, DMF, room temperature; (e) TBAF, THF, room temperature, overnight; (f) CuBr, 4,4’-di-*n*-nonyl-2,2’-bipyridine, 11-azido-3,6,9-trioxaundecan-1-amine, DMF, room temperature. DMF, *N,N*-dimethylformamide; TBAF, tetrabutylammonium fluoride.

removed by treatment with K_2CO_3 in a methanol/water mixture at 60°C for 89 h, and then the alkyne was reacted with the azidogalactose derivative. Lastly, the last protecting TIPS group was removed in THF with tetrabutylammonium fluoride at room temperature for overnight, and the clickable alkyne site reacted with the azido-N-acetylglucosamine derivative to yield precision glycopolymers **9** (Figure 6.5).

Analysis of the binding ability of the modified glycopolymers to hexose-specific lectins, namely, concanavalin A (Con A), peanut agglutinin (PNA), and wheat germ agglutinin (WGA), was done by using the quartz crystal microbalance with dissipation monitoring (QCM-D) technique. Interestingly, these lectins showed high binding affinity with mannose, galactose, and N-acetylglucosamine groups, respectively. This work is very important to understand the complexity of the glycocode and lead other research groups to the synthesis of unprecedented complex precision glycopolymers as glycan mimics.

Very recently, Stenzel *et al.* [66] reported a study based on the synthesis of galactose glycopolymers **10** for micelle and honeycomb structure formation. The copolymerization of acetylated galactose methacrylate monomer and styrene was undertaken in dioxane at 85°C using SG1 as a controlling agent. It was measured via ^{31}P NMR that 81% of the copolymer first block retains SG1 end functionality. Moreover, chain extension polymerization of galactose monomer with styrene was performed at 120°C to yield well-defined amphiphilic diblock copolymers. Acetylated galactose moieties on glycopolymers were deacetylated using sodium methoxide.

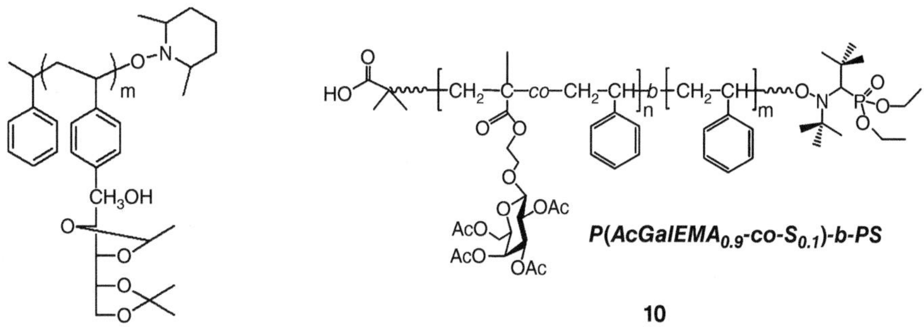

6

10

FIGURE 6.6 Glycopolymers prepared via NMP.

The obtained amphiphilic glycopolymers were used to form micelles and honeycomb structures, and these structures showed the binding affinity with PNA (Figure 6.6).

6.2.1.3 Cyanoxyl-Mediated Polymerization In the early 1990s, cyanoxyl (•OC≡N) radical was investigated and used as a control agent successfully for the polymerization of methacrylates, acrylates, and methacrylamides to form homopolymers as well as block polymers in relatively mild experimental conditions. The cyanoxyl radicals are generated by the one-electron reduction reaction of cyanate anions. Despite the fact that cyanoxyl-mediated polymerization has some advantages, such as tolerance to a broad range of functional groups and usage in aqueous solution and the wide lower range of reaction temperature (0–80°C), the use of the cyanoxyl-mediated polymerization was slightly restrained by the emergence of another type of oxygen-centered persistent radicals as mediators for controlled/living radical polymerization, namely, nitroxides.

Chaikof and coworkers succeeded in applying cyanoxyl-mediated polymerization for the first synthesis of glycopolymers from unprotected monomers [67]. The reduction of cyanate was generated by using *p*-chlorobenzenediazonium cations, and the formed *p*-chlorobenzene radicals initiated polymerization at the same time. In the following years, they have published several reports on the synthesis of well-defined glycopolymers **11** via cyanoxyl-mediated polymerization for different biological applications (Figure 6.7) [68]. More recently, the cyanoxyl-mediated free radical polymerization of 2-acrylaminoethyl lactoside and acrylamide was performed using 4-aminobenzyl-biotinamide initiators in the presence of BHF_4 and $NaNO_2$ in H_2O:THF (1:1) solution at 65°C for 16h [69]. In order to study biotinylated glycopolymer–streptavidin bioconjugation, the resulting glycopolymers were conjugated to biotin aryl free radicals with 75% conversion and a molecular weight distribution (M_W/M_N) of 1.3. The recognition ability between the biotinylated glycopolymer and streptavidin was analyzed. Furthermore, the formation of glycopolymer-coated surfaces for their lectin interactions was performed with the reaction of a glycopolymer solution with streptavidin-functionalized patterned PET membranes. According to these results, glycopolymer-based noncovalent neoglycoproteins can be used as a mimic of glycocalyx-mimetic surface.

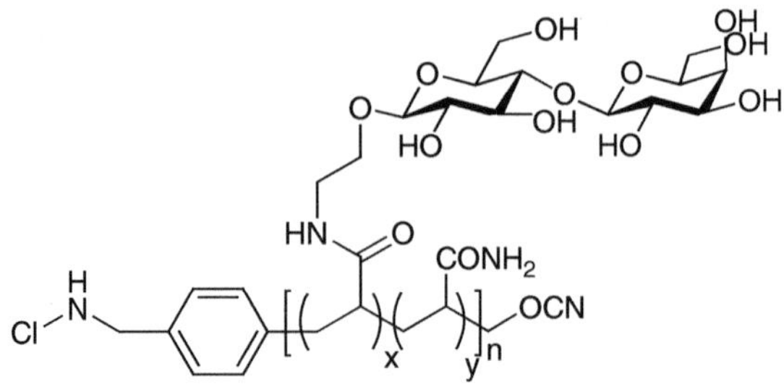

11

FIGURE 6.7 Chemical structure of *O*-cyanate chain end-functionalized lactose-based glycopolymer.

The same research group produced more biotinylated glycopolymers by using the same approach [70]. In this study, the cyanoxyl-mediated polymerization of lactose and sulfated lactose acrylamide monomers was initiated by biotin-functionalized arylamine initiator. The diazotization reaction of arylamine moiety was performed for the glycopolymer–protein bioconjugates. The high binding ability of biotinylated glycopolymers with streptavidin was obtained via HABA avidin assay. However, the measurements showed that there was a limitation for full occupancy of biotin-terminated glycopolymers in HABA avidin assay.

6.2.1.4 RAFT Polymerization RAFT polymerization using thiocarbonylthio compounds, including dithioesters and trithiocarbonates, was firstly reported by the CSIRO laboratory in 1998 [71]. After its discovery, it has become one of the most popular living polymerization processes. There are many articles reporting the synthesis of glycopolymers via RAFT. The main advantages of this technique are as follows: (i) It can be applied to a wide range of functional monomers, which allows polymers with precisely controlled structural parameters to be prepared such as random, block, gradient, and star copolymers as well. (ii) Higher molecular weights are easily achievable with the low PDIs. (iii) It is promising in bioapplications. RAFT polymerization is controlled by a reversible transfer reaction between a growing (macro)radical as an active species and a (macro)RAFT agent as a dormant species. The structure of the RAFT agent is crucial to its ability to introduce both α- and ω-functionality to polymers made by this technique. However, the introduction of terminal functionality to polymers is generally lower than 90% efficient at the α-terminus because of initiation from the thermal initiator. The possible issue is that reaction condition can decrease the bioactivity of synthesized glycopolymers. Most of RAFT polymerizations are carried out between 60 and 70°C, but it is also possible to conduct aqueous RAFT at ambient temperature. Therefore, RAFT polymerization has easily been employed for polymerization of glycomonomers in aqueous medium by many research groups.

Linear Glycopolymers The first report based on the RAFT polymerization of glycomonomer was published by Lowe and coworkers [72]. The glycomonomer (2-methacryloxyethyl glucoside) **12** was polymerized in aqueous solution using the

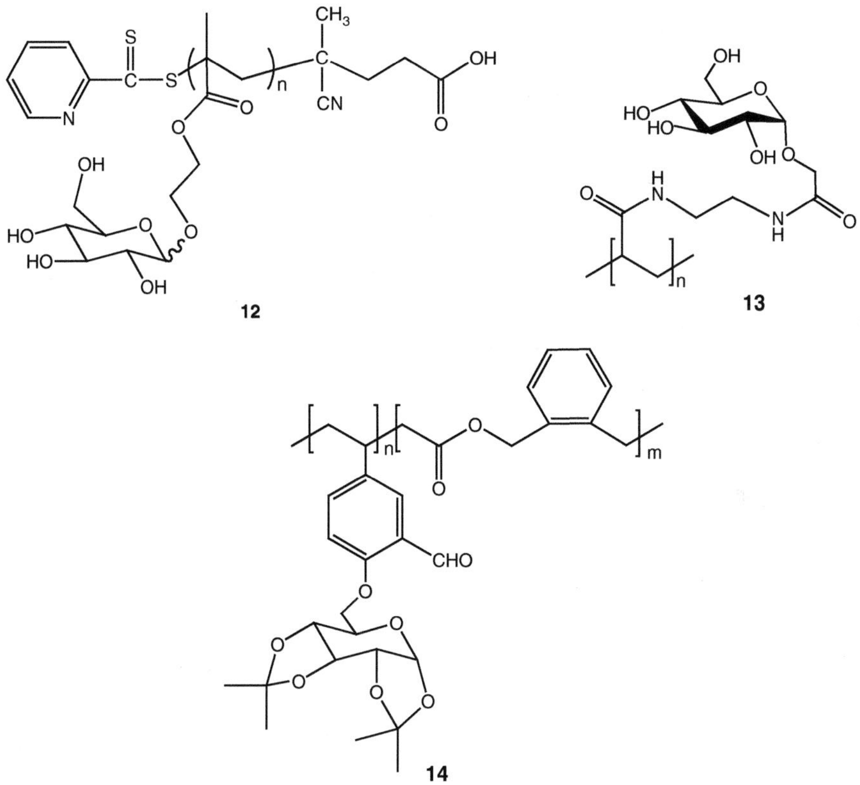

FIGURE 6.8 Linear glycopolymers prepared via RAFT.

most used RAFT agent (4-cyanopentanoic acid)-4-dithiobenzoate (CTP) and 4,4′-azobis(4-cyanopentanoic acid) (ACPA) as the water-soluble initiator. Basic polymerization condition was used to enhance the solubility of the RAFT agent and the initiator. Despite of the low conversion, the pseudo-first-order kinetics confirmed the good control on the polymerization. However, chain extension with 3-sulfopropyl methacrylate (SPMA) increased the PDI to 1.63 with a loss of active RAFT functionalities. In contrast to this, the chain extension of poly(SPMA) that used as a macro-RAFT agent with the glycomonomer gave a narrow distribution with a PDI of 1.18 (Figure 6.8).

Bernard *et al.* have reported to generate well-defined clickable glycopolymers in water/methanol mixtures by employing 3-benzylsulfanylthiocarbonyl sulfanylpropionic acid and trithiocarbonate RAFT agent at 70°C and 30°C [73]. Three glycomonomers—namely, *N*-(2-((α-D-glucopyranosyloxy)acetamido)ethyl) acrylamide (without azide group), (3,4,6-tri-*O*-acetyl-α-D-glucopyranosyloxy)-*N*-(tert-butylcarbonyloxy) acetamide (with azide group at second position), and *N*-(2-((6-deoxy-6-*N* azido-α-D-glucopyranosyloxy) acetamido)ethyl) acrylamide (with azide group at 6 position)—were polymerized via RAFT technique to yield well-defined homopolymers and block copolymers **13**. Although the RAFT polymerization of N-(2-((α-D-glucopyranosyloxy)acetamido)ethyl) acrylamide and *N*-(2-((6-deoxy-6-*N* azido-α-D-glucopyranosyloxy) acetamido)ethyl) acrylamide gave well-defined polymers with high molecular weight and narrow distribution at 70°C, glycomonomer (3,4,6-tri-*O*-acetyl-α-D-glucopyranosyloxy)-*N*-(tert-butyl-carbonyloxy) acetamide showed good control only at 30°C in the presence of 2,2′-azobis(4-methoxy-2,4-dimethylvaleronitrile) (V-70) as initiator. The reason of that is the 1,3-cycloaddition side reactions between the vinyl and the azide groups within the glycomonomers.

The synthesis of well-defined aldehyde-functionalized glycopolymers was reported by Lu and colleagues [74]. The RAFT polymerization of 1,2:3,4-di-*O*-isopropylidene-6-*O*-(2′-formyl-4′-vinylphenyl)-D-galactopyranose (IVDG) and 5,6-benzo-2-methylene-1,3-dioxepane (BMDO) was employed to yield statistical glycopolymers **14**. Polymerizations were carried out in anisole at 130°C using 1-phenylethyl phenyldithioacetate as RAFT agent and DCP as initiator.

Fluorescently Labeled Glycopolymers Mayhinney and coworkers developed RAFT-based one-step polymerizations to prepare tricomponent statistical fluorescent glycopolymers [75]. Three different monomers including *N*-(2-hydroxyethyl) acrylamide (HEAA), *N*-(2-aminoethyl) methacrylamide (AEMA), and *N*-(2-glyconamidoethyl) methacrylamides possessing varied pendant carbohydrates were used for the one-step tricomponent RAFT-based copolymerization. Totally, five different glycomonomers were synthesized by using the same procedure. RAFT copolymerization was carried out in DMF using CTP as a RAFT agent and ACPA as an initiator at 70°C for 24 h. Their subsequent modification with varied fluorescent labels due to functional primary amine groups on AEMA afforded a well-defined statistical tricomponent glycopolymers **15** with varied fluorescent labels demonstrating versatile reporters of specific lectins. The binding ability of these tricomponent glycopolymers was analyzed with *P. aeruginosa* lectin (PA-IL) bacteria and *Galanthus nivalis* plant lectin (GNL). While the α-mannose-containing polymer showed very strong binding with GNL, β-D-galactose-containing polymer showed enhanced binding ability with PA-IL. This new RAFT-based one-step polymerization was used to prepare a well-defined and wide range of tricomponent glycopolymers. Some galactose-based glycopolymers with fluorescent 4-(pyrenyl)butyl methacrylate (PyMA) were synthesized by Muller and coworkers [76]. Sequential RAFT polymerization was carried out using 2-cyanoprop-2-yl dithiobenzoate (CPDB) as a RAFT agent by initially polymerizing 6-*O*-methacryloyl-1,2:3,4-di-*O*-isopropylidene galactopyranose (MAIGal) in DMF for 22 h at 70°C. Subsequently, the introduction of MAIGal/PyMA (6:1) into the reaction was undertaken, and polymerization lasted for another 6 h to yield fluorescent glycopolymer PMAIGal-*b*-(PMAIGal-*co*-PPyMA) (Figure 6.9).

Another elegant strategy to synthesize a series of fluorescent mucin mimetics displaying a range of α-*N*-acetylgalactosamine (GalNAc) valencies was developed by Bertozzi and coworkers for the construction of a mucin-mimetic glycopolymer microarray [77]. To synthesize glycopolymer **16** as mucin mimicking, methyl vinyl ketone was polymerized via RAFT using biotin-containing trithiocarbonate CTA and 4,4′-azobis(4-cyanovaleric acid) (ACVA) as a radical initiator in 2-butanone at 65°C

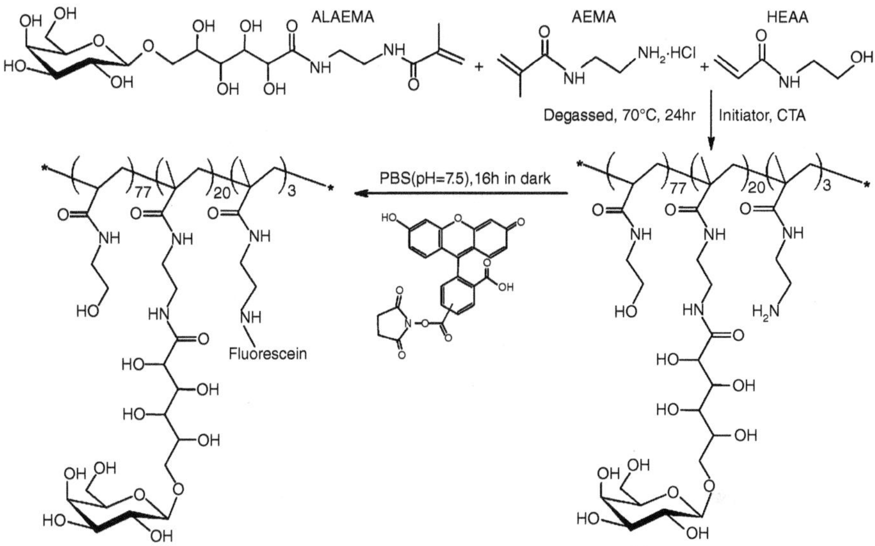

15

FIGURE 6.9 Illustration of the synthesis of fluorescent glycopolymers PMA-ALAEMA-fluorescein containing ß-galactoside as the pendant sugar. *Source*: Reproduced with permission from Ref. 75, Copyright 2014.

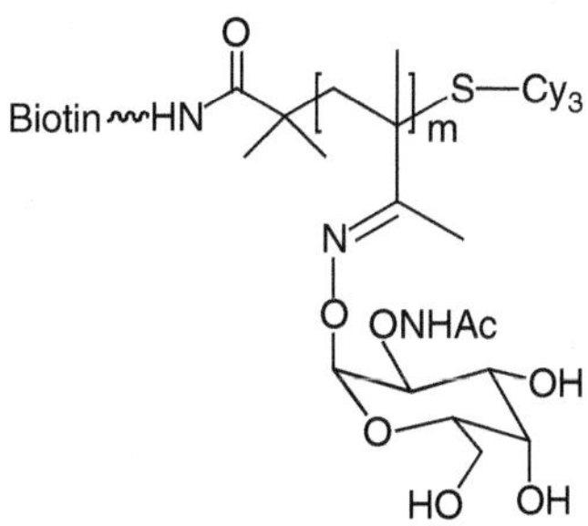

16

FIGURE 6.10 A mucin-mimetic glycopolymer synthesized via RAFT.

for 16.5 h. Streptavidin-coated microarray substrates were conjugated to biotin terminal functional groups of the obtained polymers. The opposite end of the polymer chains was functionalized with an MI-functionalized Cy3 dye because of their use as a fluorescent label. The final product was prepared in the presence of α-aminooxy-GalNAc under acidic conditions at 50°C for 20 h (Figure 6.10). After the attachment of the biotinylated mucin mimetics to the array surface with ~40–60% efficiency, their binding affinity was analyzed with four lectins (soybean agglutinin (SBA), Wisteria floribunda lectin (WFL), *Vicia villosa*-B-4 agglutinin (VVA), and *Helix pomatia* agglutinin (HPA)). Generally, while HPA showed stronger avidities than other lectins toward all the polymers irrespective of their GalNAc valency, SBA showed propensity to cross-link the high-valency mucin mimetics. Interestingly, increasing the surface density array did not show any significant enhancement for the binding affinity of all lectins.

Glycopolymer Bioconjugates Ahmed *et al.* prepared various cationic glycopolymers **17** with the low PDIs and the molecular weight from 3,000 to 30,000 $g\,mol^{-1}$ in the absence of protected group chemistry [78]. RAFT polymerizations were undertaken at 70°C with CTP as the RAFT agent and ACPA as the initiator. They have also developed anticancer agents from glycopolymers. The obtained glycopolymers **18** were conjugated with gold(I) phosphine to yield polymeric gold glycoconjugates bearing dithiocarbamate (DTC) compounds [79]. The same group copolymerized 3-gluconamidopropyl methacrylamide (GAPMA) with methacrylic acid (MA) to yield statistical glycopolymers **19** via RAFT polymerization in another work [80] (Figure 6.11).

The synthesis of pyridyl disulfide (PDS) functional well-defined glycopolymer **20** was reported by Shi's group [81]. The RAFT polymerization of 2-(2,3,4,6-tetra-o-acetyl-β-D-glucosyloxy)ethyl methacrylate (AcGlcEMA) glycomonomer was carried out in the presence of 4-cyanopentanoic acid dithiobenzoate (CPADB) and

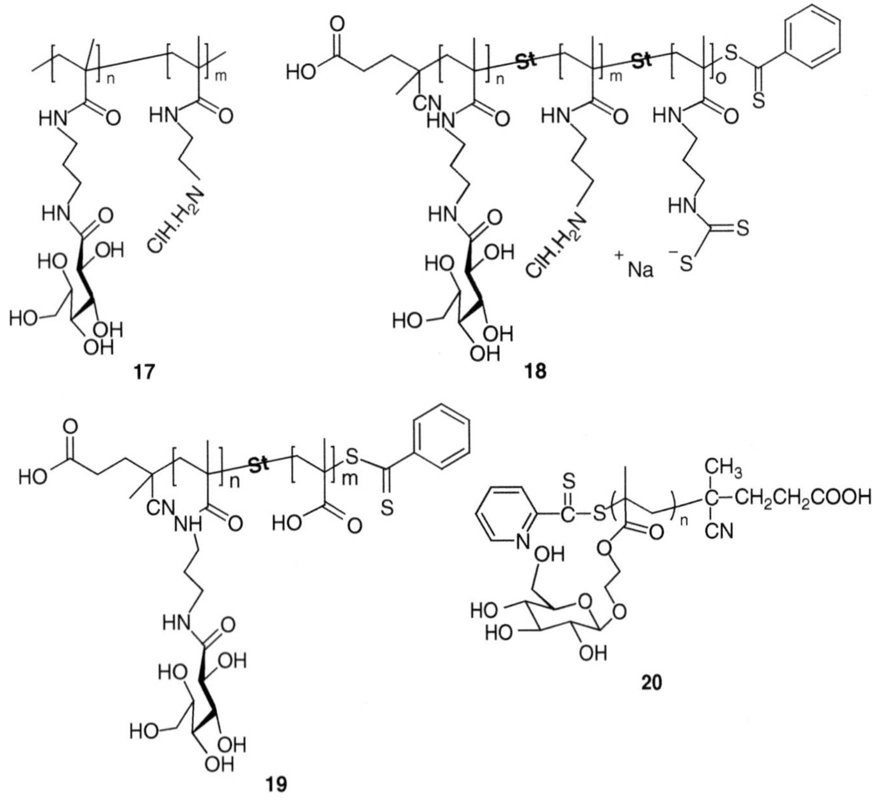

17 **18** **19** **20**

FIGURE 6.11 The glycopolymers polymerized via RAFT for bioconjugation.

ACPA in 1,4-dioxane at 70°C for 24 h. The molar ratio of CTA to initiator was chosen as 10:1, and the polymerization was stopped at low monomer conversion (50–60%) in order to obtain efficient dithioester end functionalities for the subsequent modification with PDS. The end group was modified with PDS in the presence of ethanolamine and subsequent reaction with an excess of 2,2′-dithiodipyridine (DTP). GSH was conjugated to PGlcEMA-PDS through the thiol–disulfide exchange reaction in 5 ml of phosphate buffer at pH 7.0 for 24 h under N_2 at room temperature. This reaction was confirmed by the disappearance of the significant signals of pyridyl ring protons at δ 7.4–8.5 ppm. The binding ability of PGlcEM–GSH bioconjugates with Con A was examined via UV–Vis spectroscopy, fluorescence quenching titration, and transmission electron microscopy (TEM). Moreover, *in vitro*, its antioxidant activity was investigated by using 1,1-diphenyl-2-picrylhydrazyl (DPPH) assay. Depending on the results, the peptide–glycopolymer bioconjugates are promising for the development of antioxidant delivery system, biomimetics, and biodetection.

Maynard *et al.* have been able to synthesize well-defined glycopolymers **21** that could be conjugated to a specific site on protein by inserting a specific reactive thiol group [82]. The glycomonomer 4,6-O-(4-vinylbenzylidene)-α,α-trehalose was synthesized by the attachment of trehalose to a styrene monomer via a 4,6-acetal linkage. Several thiol-reactive CTAs containing hydrophilic triethylene glycol spacer between the trithiocarbonate and pyridyl disulfide end group were prepared. The glycomonomer was polymerized via RAFT under [CTA]:[monomer]:[azodiisobutyronitrile (AIBN)] = 1:29:0.2 reaction ratio at 80°C in DMF for 6 h (Figure 6.12). The well-defined end-functionalized glycopolymers were obtained with molecular weights ranging from 4.2 to 49 kDa and low polydispersities. The conjugation reaction between thiolated lysozyme and trehalose polymers was carried out in Dulbecco's phosphate-buffered saline (DPBS) for 3 h at 4°C. Sodium dodecyl sulfate polyacrylamide gel electrophoresis (SDS-PAGE) analysis confirmed the coupling achievement and environmental stability of these bioconjugates. The results showed that this work is a crucial example to demonstrate that how the conjugation of well-defined glycopolymers has a big influence on the stability of proteins regarding the changing of environmental conditions.

Glycopolymers as Targeted Gene Delivery Vehicles Glycopolymers containing poly(2-deoxy-2-methacrylamido glucopyranose) (PMAG) and primary amine-containing poly(*N*-(2-aminoethyl) methacrylamide) (PAEMA) polymer blocks **22** using unprotected glucose-based glycomonomer were prepared by Reineke and coworkers [83]. The RAFT polymerization of these monomers was undertaken in aqueous solution using 4-cyano-4-(propylsulfanylthiocarbonyl) sulfanylpentanoic acid as the RAFT agent. Potential short interfering RNA (siRNA) delivery system was developed by using these glycopolymers. The same group has also reported the polymerization of 2-deoxy-2-methacrylamido glucopyranose (MAG) to achieve an adamantine-conjugated glycopolymer poly(2-deoxy-2-methacrylamido glucopyranose) (Ad-pMAG) as an adamantane RAFT agent [84]. A mixture of water, acetate buffer, and dimethyl sulfoxide (DMSO) was used for the polymerization with using ACPA initiator at 70°C. The obtained glycopolymers **23** showed the assembly

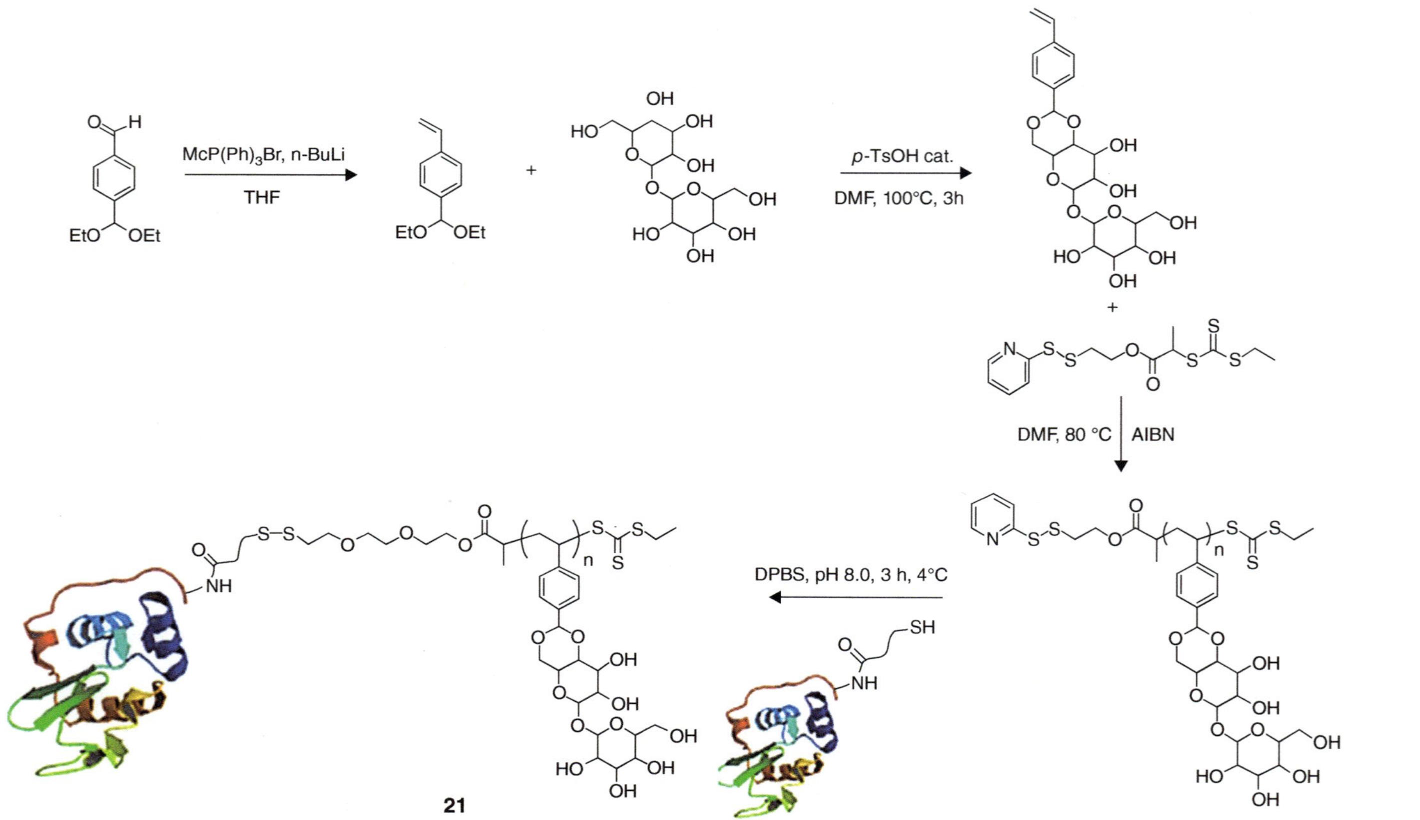

21

FIGURE 6.12 Disulfide end-functionalized glycopolymers from RAFT for selective conjugation to proteins. *Source*: Reproduced with permission from Ref. 82, Copyright 2012, American Chemical Society.

behavior through noncovalent interactions forming inclusion complex between hydrophobic adamantane and the β-cyclodextrin (βCD) core.

Another strategy for gene delivery system based on the synthesis of well-defined glycopolymers **24** grafted onto poly(L-lysine) (PLL) was proposed by Guo and coworkers [85]. Even though PLL has significant condensation capacity with plasmid DNA due to positively charged hydrophilic amino group in water, it is not usually chosen as a gene delivery vector because of high cytotoxicity and low transfection efficiency. However, here, it was modified by well-defined saccharide-containing polymers to reduce the cytotoxicity and increase the transfection efficiency of PLL significantly. Firstly, the glycomonomer 2-O-meth-acryloyloxyethoxyl-(2,3,4,6-tetra-O-acetyl-β-D-galactopyranosyl)-(1-4)-2,3,6-tri-O-acetyl-β-D-glucopyranoside (MAEL) was polymerized via RAFT in the presence of CPADB and AIBN in chloroform at 70°C for 24h. After the deprotection of the acetyl groups, the obtained glycopolymers **24** were grafted onto PLL through the amidation reaction between the carboxyl group of the dithioester residue on the glycopolymer and the amino group on PLL-mediated *N*-hydroxysuccinimide (NHS) and 1-ethyl-3-(3-dimethylaminopropyl)carbodiimide (EDC) to yield four different DPMAEL-g-PLL due to their different polyelectrolyte properties because of the N/P (charge ratio of amine to phosphate) ratios. The synthesized glycopolymers **24** modified PLL showed lower cytotoxicity than PLL according to the MTT assays with the mouse embryonic fibroblast cell line (NIH3T3) and human hepatoma cell line (HepG2). The complex formation ability of DPMAEL-g-PLL with pDNA was discovered by agarose gel electrophoresis. The results demonstrated that the modification of PLL with glycopolymers enhanced its biological properties for gene delivery application (Figure 6.13).

A delivery system of DNA and enzyme into cells was developed by using sugar-based nanogels [86]. Macro-RAFT agents of poly(3-gluconamidopropyl methacrylamide) (PGAPMA) and poly(2-lactobionamidoethyl methacrylamide) (PLAEMA) were initially synthesized using a dithioester RAFT agent. The homoglycopolymers **25** were chain extended with PEGMA, and an acid-degradable cross-linker was also added during the aqueous polymerization to yield glycopolymer nanogels.

Recently, Bulmus and collaborators developed a gene delivery system via the synthesis of spermine-like glycopolymer **26** via RAFT polymerization [87]. A new methacrylate monomer, namely, 2-((tert-butoxycarbonyl)(2-((tertbutoxycarbonyl) amino)ethyl)amino)ethyl methacrylate (BocAEAEMA), has been reported for the first time. After the RAFT polymerization of BocAEAEMA, P(BocAEAEMA) was used as a macro-RAFT agent for chain extension with glycomonomer mannose acrylate (ManAc). This chain extension reaction was undertaken in DMF at 70°C for 12h using AIBN. The amine groups were deprotected by removing BOC groups to yield P(AEAEMA)-b-P(ManAc). The ability of this well-defined glycopolymer **26** to form polyplex with DNA was analyzed. The glycopolymer block decreased this interaction between P(AEAEMA) and DNA regardless of their specific binding affinity with proteins. Another interesting result was that the size of P(AEAEMA)-*b*-P(ManAc) and DNA formed polyplex particles are smaller than the size of P(AEAEMA) and DNA formed particles according to dynamic light scattering (DLS) measurements most likely due to the hydrophilic property of the glycopolymer.

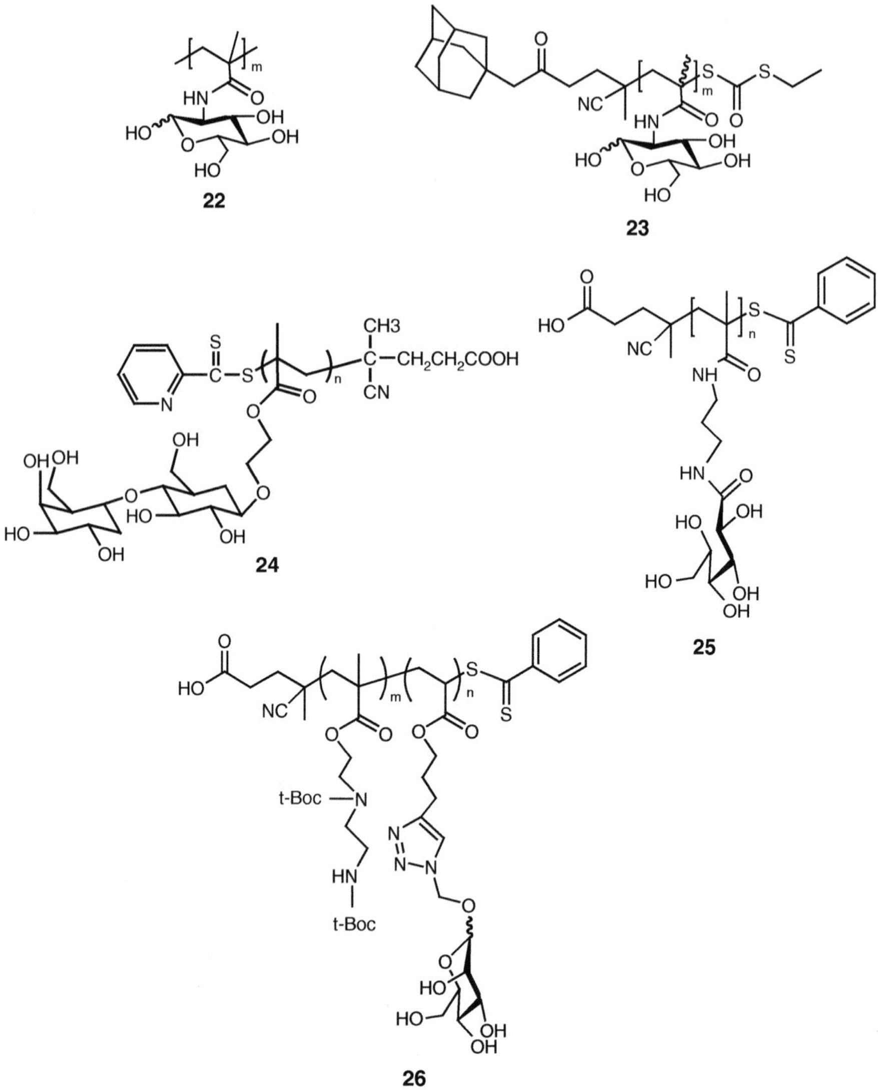

FIGURE 6.13 The glycopolymers as targeted gene delivery vehicles prepared via RAFT.

Glycopolymer-Grafted Nanoparticle Surface Chen and coworkers have recently been able to prepare glycopolymer-functionalized Ag nanoclusters (Gly-Ag NCs) that showed the fluorescent and cytotoxicity ability [88]. The synthesis of sugar- and acid-containing polymers **27** was carried out via RAFT polymerization in the presence of CPDN as the RAFT agent and AIBN as the initiator in DMF at 70°C (Figure 6.14). Then, these glycopolymers **27** were mixed with $AgNO_3$ and then placed into the CEM instruments. After microwave irradiation treatment, silver nanoclusters decorated with glycopolymers (PMAG-Ag NCs) were fabricated without using reducing agents.

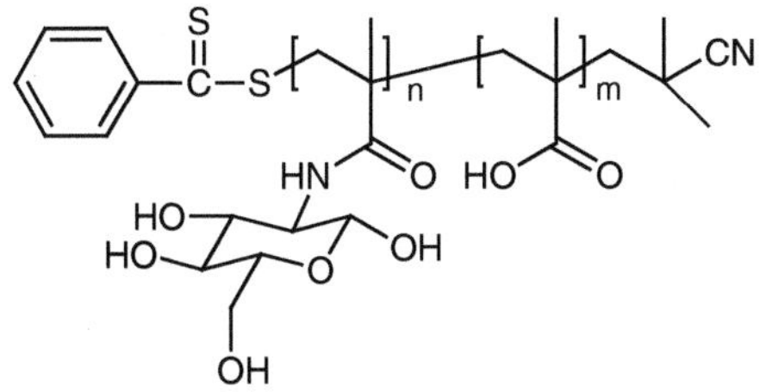

27

FIGURE 6.14 Well-defined glycopolymers synthesized via RAFT grafted onto Ag nanoclusters.

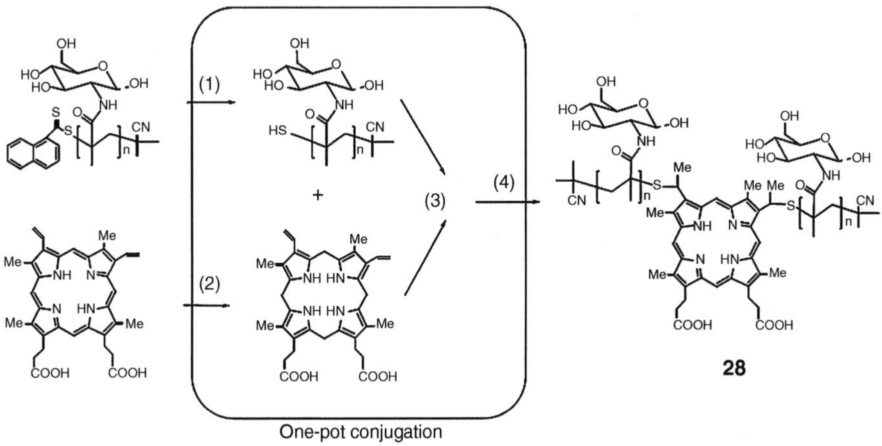

28

FIGURE 6.15 Schematic presentation of one-pot synthesis of glycopolymer–porphyrin conjugate: (1) reduction of RAFT end group to thiols, (2) reduction of PpIX to protoporphyrinogen, (3) thiol–ene reaction of protoporphyrinogen with thiol-terminal glycopolymer, and (4) the oxidation of protoporphyrinogen to afford the porphyrin–glycopolymer conjugate. *Source*: Reproduced with permission from Ref. 89, Copyright 2012.

Silver nanoclusters were chosen to decorate the glycopolymers due to their fluorescent and cytotoxic properties providing advantages in both cancer imaging and therapy. These nanoclusters showed significant binding ability and cytotoxicity against GLUT-overexpressing cancer cells K562. Furthermore, they inhibited their viability possibly through the enhancement of reactive oxygen species (ROS) production.

Promoted Self-Assembly of Glycopolymers The synthesis of self-assembled porphyrin–glycopolymer conjugates was developed by Chen *et al.* [89]. The combination of RAFT polymerization and one-pot conjugation reaction was used as a one-pot of multireactions. Firstly, RAFT polymerization of MAG was carried out using 2-cyanoprop-2-yl-1-dithionaphthalate (CPDN) as the RAFT agent and AIBN as the initiator. Then, the obtained glycopolymers **28** were conjugated with protoporphyrin (PpIX) via a one-pot reaction combining the multistep reactions including reduction of RAFT end group to thiols, converting PpIX to protoporphyrinogen; thiol–ene reaction of protoporphyrinogen with thiol-terminal glycopolymer; and the oxidation

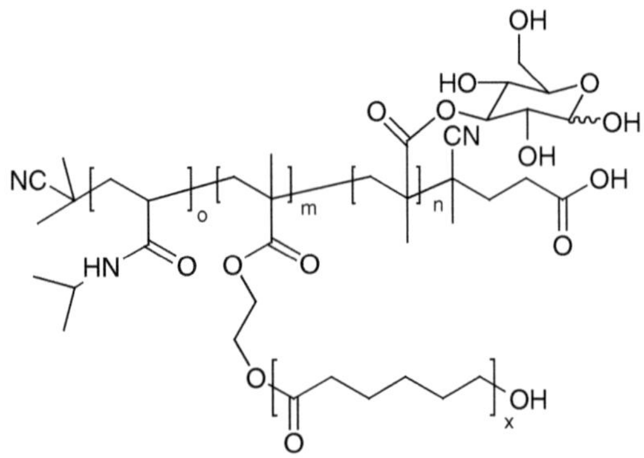

29

FIGURE 6.16 The self-assembly triblock glycopolymers synthesized via RAFT.

of protoporphyrinogen to yield the porphyrin–glycopolymer conjugate, respectively. Sodium mercury amalgam was used as a catalyst during the one-pot reaction (Figure 6.15).

These porphyrin–PMAG conjugates showed self-assembly behavior and formed micelles in water due to hydrophobic porphyrin in the middle and hydrophilic glycopolymer at both ends. Con A and anticancer effect for cancer cells (K562) were studied for the test of their binding ability. Glycoparticles showed high and specific binding ability with Con A. Moreover, the cytotoxicity test of the glycomicelles against K562 cells in different doses revealed that these self-assembled micelles killed these cancer cells under light irradiation and light treatment in a length-dependent manner. Therefore, this report is important and promising for the development of applications for cancer imaging and therapy.

Thermoresponsive micelles were prepared from "coil-comb-coil" triblock self-assembly glycopolymers [90]. These triblock glycopolymers **29** were synthesized via the combination of RAFT and ring-opening polymerization. Firstly, poly(3-O-methacryloyl-1,2:5,6-di-O-isopropylidene-D-glucofuranose) (PMAIpGlc) macro-RAFT agent was synthesized, and this was followed by two consecutive chain extension polymerizations of 2-hydroxyethyl methacrylate (HEMA) and *N*-isopropylacrylamide. The triblock glycopolymers **29** were grafted with ε-caprolactone at the middle block (HEMA section) to yield poly(3-O-methacryloyl-1,2:5,6-di-O-isopropylidene-D-glucofuranose)-*block*-poly(2-hydroxyethyl methacrylate-*graft*-poly(ε-caprolactone))-*block*-poly(N-isopropylacrylamide) (PMAIpGlc-*b*-P(HEMA-*g*-PCL)-*b*-PNIPAM) (Figure 6.16).

6.2.1.5 Copper-Mediated Controlled/Living Radical Polymerization Transition metal-mediated living radical polymerization (TMM-LRP) is another crucial route to synthesize glycopolymers, which are well-defined with different functionality. Functional groups can be introduced easily into polymers via TMM-LRP by introducing functionality to the initiator [91]. This process is based on a rapid exchange of a halide atom (especially Cl or Br) between a growing radical and a dormant species.

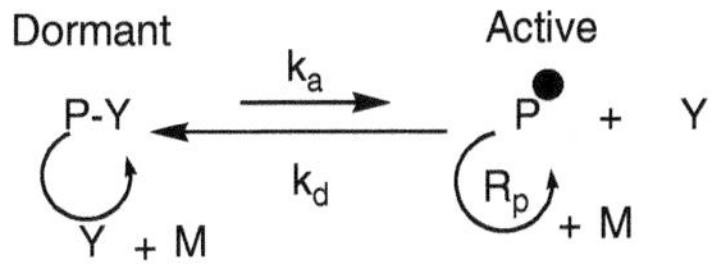

FIGURE 6.17 Activation–deactivation equilibrium in Cu-mediated living/radical polymerization.

To achieve a good control of the polymerization, the equilibrium is slightly shifted toward the dormant species [92] (Figure 6.17). In this way, the relatively stable dormant species could easily avoid side reactions. Various transition metals can be employed in TMM-LRP such as Cu, Ru, Fe, and Ni. However, copper has been the most commonly used of the transition metals because of its high efficiency and cheapness. Another important factor on the degree of control over the polymerization is the type of the ligand. Nitrogen-containing multidentate ligands have been commonly used for copper-mediated living/radical polymerizations. Highly polar solvents such as alcohols, DMSO, DMF, N-methyl-2-pyrrolidone (NMP), and mixtures with water are usually employed for the synthesis of glycopolymer in this technique because of the solubility of glycomonomer and the copper–ligand complex.

Cu-catalyzed living/radical polymerizations such as ATRP and SET-LRP are relatively more tolerant to different reaction solvents and functional in controlling the kinetic chain length and macromolecular. The most well-known technique is probably ATRP that utilizes the lower oxidation state copper(I) halide and (usually) nitrogen-based ligand complexes as the catalyst. Recently, new ATRP processes have been developed, namely, activators generated by electron transfer (AGET), activators regenerated by electron transfer (ARGET), electrochemically mediated ATRP (eATRP), and initiators for continuous activator regeneration (ICAR) ATRP [93]. However, simultaneous reverse and normal initiation (SR&NI) ATRP takes advantage as Cu(II); an alkyl halide and a radical initiator are initially present in the reaction medium. Therefore, it is possible to reduce the amount of copper complex and prepare different types of polymer architectures. One of the most important issues with Cu-mediated living/radical polymerizations is the purification of copper completely for biomedical applications.

The first paper that was reported in 1998 by Ohno *et al.* was based on the polymerization of a protected glycomonomer via ATRP [94]. The glycomonomer 3-*O*-methacryloyl -1,2:5,6-di-*O*-isopropylidene-D-glucofuranose (MAIpGLc) was polymerized in the presence of CuBr/4,4′-di-*n*-heptyl-2,2′-bipyridine catalyst in veratrole at 80°C to yield the glycopolymers **30**. The ATRP polymerizations of MAIpGLc were approximately linear due to the first-order kinetic plots. Liang and colleagues were another early research group to synthesize glycopolymers **31** via ATRP [95]. The polymerization of 2-(2′,3′,4′,6′-tetra-*O*-acetyl-β-D-glucopyranosyl) ethyl acrylate (AcGEA) was carried out in chlorobenzene using CuBr/2,2′-bipyridine catalyst and 1-phenylethyl bromide initiator at 80°C. The conversion reached 70% with low PDIs (Figure 6.18).

Precision Glycopolymers The synthesis of different types of precision multiblock glycopolymers **32** was performed by Haddleton *et al.* via SET-LRP [96]. Click chemistry of 3-azidopropyl acrylate (APA) to alkylated mannose, glucose, and fucose

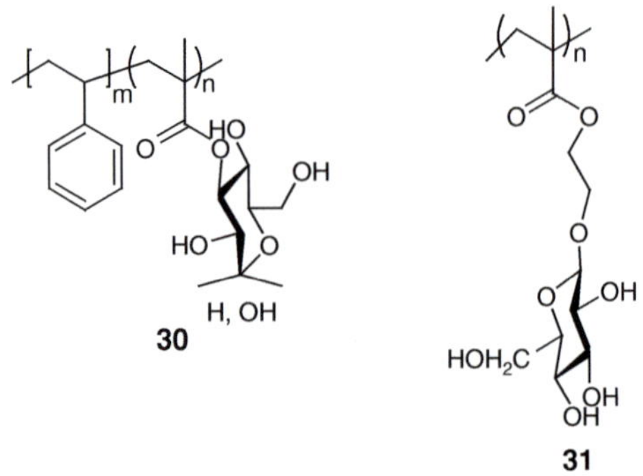

FIGURE 6.18 The first synthesized glycopolymers via ATRP.

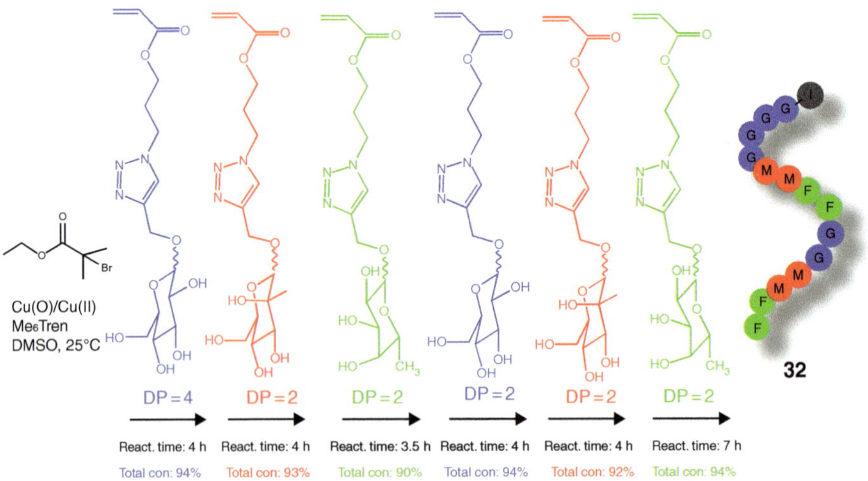

FIGURE 6.19 Schematic representation of the synthesis of multiblock glycopolymers by iterative addition of glycomonomers at defined time period (top) and SEC traces of each addition of glycomonomer (bottom). *Source*: Reproduced with permission from Ref. 96, Copyright, 2013.

was employed to prepare several types of glycomonomers, respectively. CuAAC click reaction was performed with $CuSO_4$ and sodium ascorbate in methanol/water mixture. The polymerization of glucose monomer (GluA) was undertaken using Cu(0)/Cu(II)/*tris*[2-(dimethylamino)ethyl]amine (Me_6TREN) as the catalyst system in DMF at 20°C. The obtained glycopolymers showed high chain end fidelity to allow for polymerization of more of the same or different glycomonomers in one pot. As depicted in Figure 6.19, chain extension was carried out by addition of mannose monomer (ManA) in 1 mL DMSO into the polymerization reaction mixture via cannula under nitrogen. Fucose monomer (FucA) was added in the same way after reaching full monomer conversion. This sequential addition of subsequent

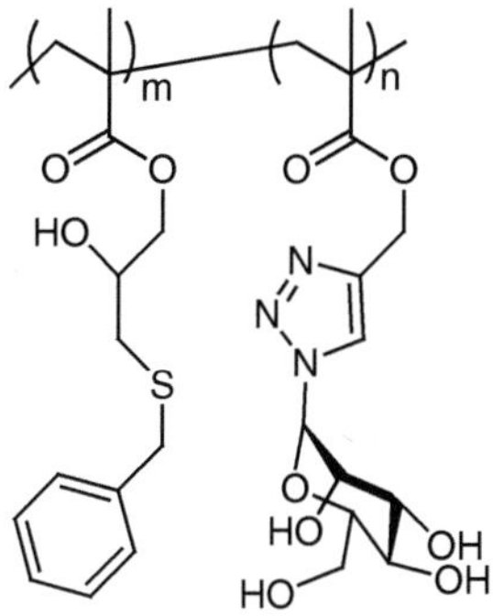

33

FIGURE 6.20 The precision multiblock glycopolymers prepared via the combination of Cu(0)-LRP with thiol–halogen, thiol–epoxy, and CuAAC click chemistry.

monomers were added in order one more time and continued until the glycopolymer reached to six short blocks (DP = 2 for each block; Mn = 7.6 kDa; PDI = 1.13).

The binding ability of these precision glycopolymers **32** were analyzed with the human lectin dendritic cell-specific intercellular adhesion molecule-3-grabbing nonintegrin (DC-SIGN) for the first time by using surface plasmon resonance (SPR). Even though the binding ability of the synthesized multiblock glycopolymers exhibited a clear dependence with the number-average degree of polymerization of each glycomonomer, the glycopolymers with different sugar units did not present enough binding affinity as well.

Recently, another strategy was reported on the synthesis of precision multiblock glycopolymers **33** by using the combination of copper(0)-mediated living radical polymerization (Cu(0)-LRP) with thiol–halogen, thiol–epoxy, and copper-catalyzed alkyne–azide coupling (CuAAC) click chemistry [97]. Well-defined alkyne and epoxide side-chain polymeric scaffolds were obtained by polymerization of glycidyl acrylate (GA) and acrylic acid 3-trimethylsilanyl-prop-2-ynyl ester (TMSPA) from Cu(0)-LRP using Cu(0) wire as the activator and additional $CuBr_2$ as the deactivator and Me_6TREN as the ligand in DMSO at ambient temperature, which allowed for further sequential thiol–halogen, thiol–epoxy, and CuAAC reactions to build functional glycopolymers **33** with a defined sequence and spatial orientation (Figure 6.20). These elegant works opened the door for the synthesis of complex precision glycopolymers to take it one more step forward. The only disadvantage of these works is that the scale was not sufficient to determine any effect of sequence on binding in the system due to the low degree of polymerization of each monomer.

Glycopolymer Bioconjugates Shi and coworkers proposed a very efficient approach to synthesize biotinylated copolymers containing glucose (Glu) and near-infrared fluorescent (NIFR) dye moieties via ATRP [98]. Firstly, azide-terminated poly(*tert*-butyl acrylate) was prepared in the presence of CuBr, *N,N,N′,N′,N′*-pentamethyldiethylenetriamine (PMDTA), and 2-bromo-2-methylpropionic acid

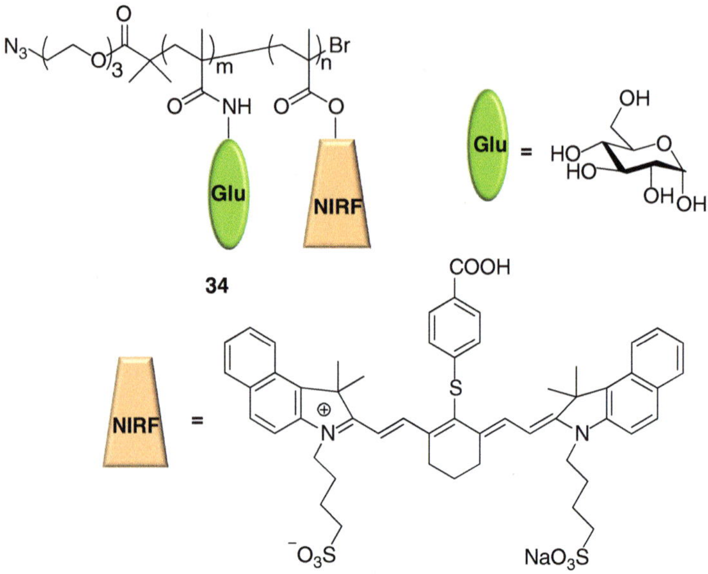

FIGURE 6.21 The poly((glucosamine)–(NIRF dye)) copolymers synthesized via ATRP.

2-[2-(2-azidoethoxy)ethoxy] ethyl ester for 2h at 60°C. The synthesis of poly(glucosamine)–poly(NIRF dye) copolymers **34** was performed with one-pot amidation with NIRF-NH_2 and Glu-NH_2. Finally, biotin alkyne was attached to azide-terminated copolymers via click chemistry. These biotin-terminated copolymers were employed to generate bioconjugates by using a biotin–avidin interaction. These fluorescent glycopolymer bioconjugates showed the ability for utilization for *in vitro* and *in vivo* imaging regarding different biological applications due to enhanced optical properties (Figure 6.21).

Maynard *et al.* have reported the synthesis of PDS end-functionalized glycopolymer **35** via ATRP [99]. PDS ATRP initiator was used with Cu(I)/Cu(II)/2,2′-bipyridine system as a catalyst in methanol/water mixture at 30°C to polymerize *N*-acetyl-D-glucosamine. The polymerization of *N*-acetyl-D-glucosamine proceeded with good control (M_n = 10.2kDa; PDI = 1.12). The bioconjugation of 5′-thiol-terminated double-stranded siRNA and α-PDS-terminated poly(N-acetyl-D-glucosamine) was conducted in 100mM sodium bicarbonate buffer at pH 8.5 and left for 24h at 21°C (Figure 6.22). This study is a good example for highly efficient synthesis and one-step conjugation system.

Two different strategies to prepare thiol-reactive well-defined neoglycopolymers **36** for bioconjugation were investigated by Haddleton *et al.* via a combination of TMM-LRP and Cu(I)-catalyzed click chemistry [100]. The first strategy included the polymerization of the glycomonomer bearing mannose units. The direct polymerization of glycomonomer was initiated with the MI-functionalized initiator in the presence of rhodamine B methacrylate using *N*-(ethyl)-2-pyridylmethanimide and Cu(I)Br as the catalytic system at ambient temperature. In the second route, the

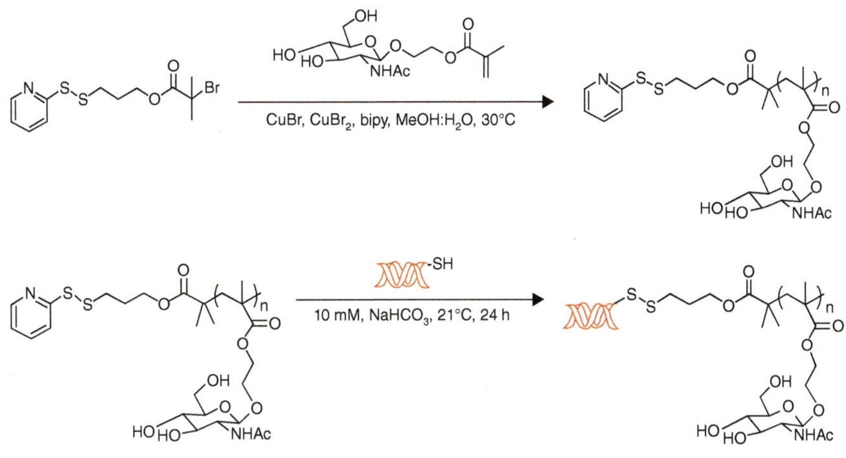

FIGURE 6.22 ATRP polymerization of N-acetyl-D-glucosamine with pyridyl disulfide initiator and conjugation of siRNA to thiol-reactive glycopolymer. *Source*: Reproduced with permission from Ref. [99], Copyright 2009.

postfunctionalization was used to synthesize well-defined neoglycopolymer **36** protein biohybrid materials. Firstly, trimethylsilyl-protected alkyne side-chain polymeric scaffolds were prepared via ATRP using again *N*-(ethyl)-2-pyridylmeth animide/Cu(I)Br as the catalytic system in toluene at 70°C (M_n = 8–30 kDa, PDI = 1.20–1.28). After deprotection of trimethylsilyl groups, alkyne groups were clicked with azidosugar derivatives to yield end-functionalized fluorescent neoglycopolymers. Bovine serum albumin (BSA) was chosen as a single thiol-containing model protein and conjugated to the obtained α-functional glycopolymers with a different density of D-mannose- and D-galactose-binding epitopes. The binding ability of these biohybrid materials with Con A was measured via SPR and enzyme-linked immunosorbent assay (ELISA) techniques regarding mannose-binding lectin (MBL). According to the results, the neoglycopolymer protein biohybrid materials exhibited more significant binding ability than native unmodified BSA (Figure 6.23).

Another versatile method was developed to synthesize well-defined and site-specific protein–glycopolymer **37** bioconjugates [101]. The ATRP homopolymerization of lactobionamidoethyl methacrylate (LAMA) was carried out in *N*-methyl-2-pyrrolidinone at 20°C by using a biotinylated poly(ethylene glycol) (biotin–PEG ATRP) initiator and CuBr/2,2′-bipyridine as a catalyst (Figure 6.24). The interaction of these biotin-terminated glycopolymers **37** was analyzed with streptavidin. The most important benefit of this method is that direct preparation of glycopolymers in the presence of free hydroxyl groups without a number of protection and deprotection steps is possible.

Maynard and colleagues have proposed the synthesis of tailor-made biotin-functionalized glycopolymers **38** by using ATRP initiators bearing a biotin moiety [102].

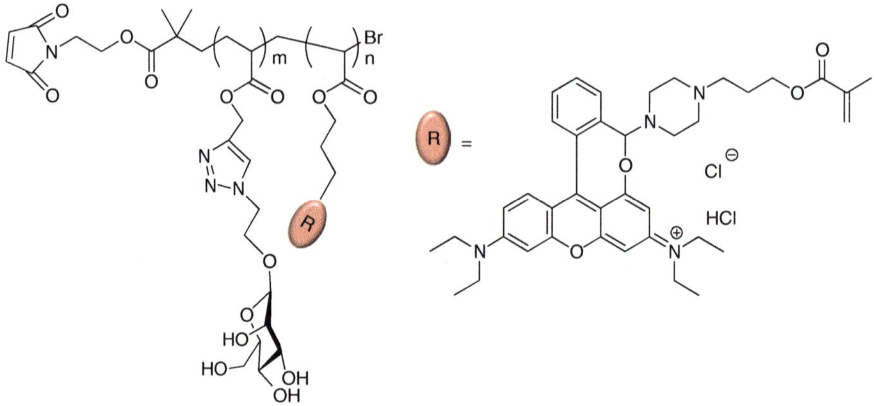

36

FIGURE 6.23 The well-defined glycopolymers prepared via the combination of TMM-LRP and Cu(I)-catalyzed click chemistry for bioconjugation.

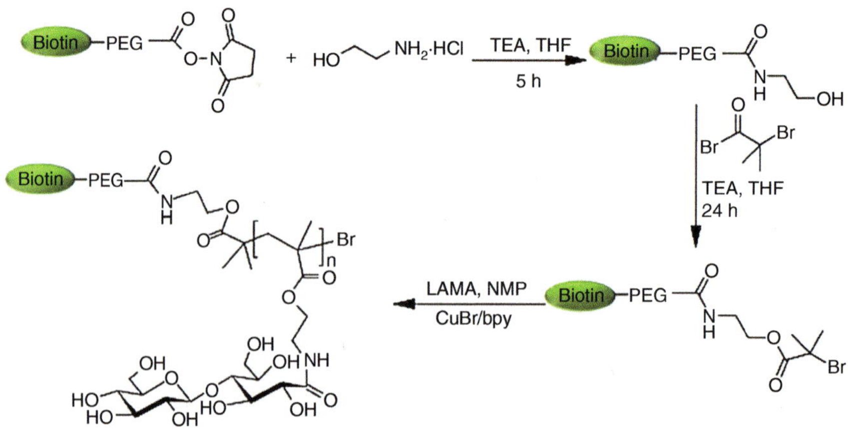

37

FIGURE 6.24 Homopolymerization of LAMA using biotin–PEG ATRP initiator. *Source*: Reproduced with permission from Ref. [101], Copyright 2006.

ATRP polymerization of the protected and unprotected methacrylate with a pendant *N*-acetyl-D-glucosamine (GlcNAc) unit was carried out at room temperature in DMSO using CuBr and Me_6TREN as a catalyst. After the deprotection reaction, the resulting well-defined glycopolymers (M_n = 11.8–68.8 kDa, PDI = 1.07–1.23) exhibiting GlcNAc side chains and biotin end groups were further used for bioconjugation with streptavidin. These glycopolymers showed high binding affinity with streptavidin. Therefore, this study is promising for the development of streptavidin-coated surfaces and nanoparticles to enhance biological interactions (Figure 6.25).

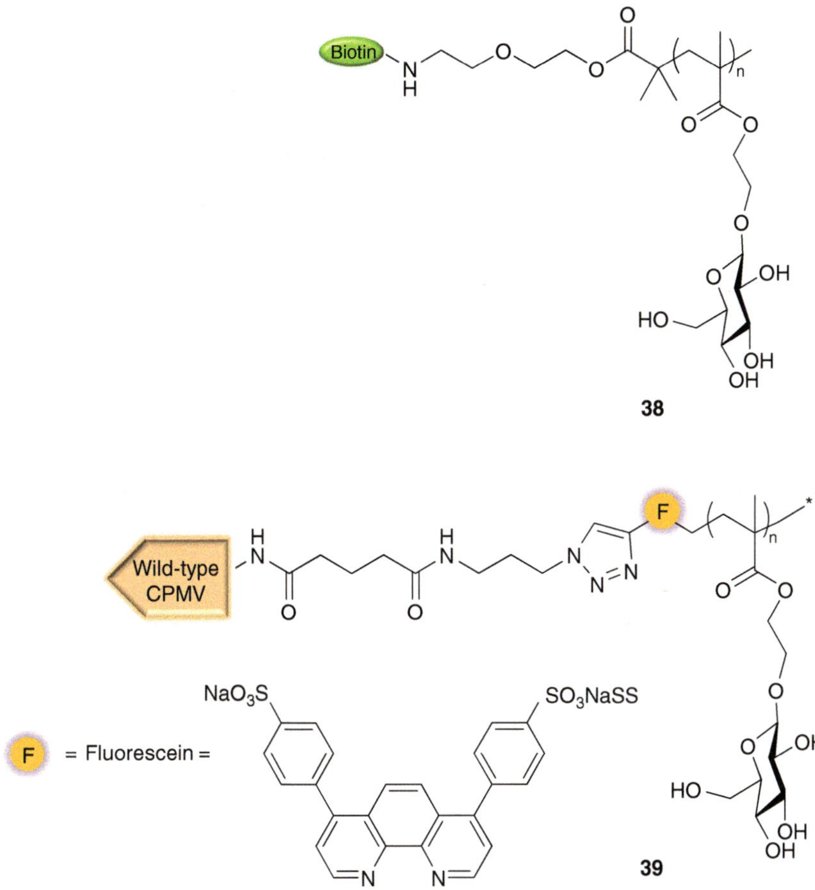

FIGURE 6.25 Glycopolymer bioconjugates with different proteins prepared via ATRP.

Viruses have a potential for a modification with one or multiple polymeric chains that could enhance the properties of virus therapeutics. From this perspective, Finn *et al.* have reported the synthesis of glycopolymer–virus bioconjugations **39** via the Cu^{I}-catalyzed ATRP and azide–alkyne cycloaddition reactions [103]. ATRP polymerization of methacryloxyethyl glucoside was performed with CuBr/bipy as a catalyst and azide-functionalized initiator in $MeOH/H_2O$ (3:2) at 25°C. The alkyne-terminated dye (fluorescein dialkyne) was attached to the azide-functionalized glycopolymer through CuAAC reaction using $CuSO_4$/sodium ascorbate catalytic system for further bioconjugation to virus. Cowpea mosaic virus (CPMV), a plant virus, having stability in a variety of conditions and compatibility with both hydrophobic and hydrophilic molecules was used to bioconjugate. The modification of CPMV with azide was derivatized with NHS with lysine side chains of capsid. After the coupling reaction between the obtained azide-functionalized virus and alkyne-terminated glycopolymer, the binding affinity of the obtained glycopolymer–virus bioconjugate **39** was analyzed with Con A.

Glycoparticles The new synthetic methodology to synthesize glycopolymers **40** with high specificity and selectivity as mimicking glycan architecture was reported by Gibson's group [104]. Firstly, glycidyl methacrylate (GMA) was polymerized via Cu(I)-mediated polymerization and introduced with sodium azide in DMF at 50°C to yield the azide functional groups. The reaction produced a secondary alcohol and reacted with acyl chlorides to install secondary motifs due to the fact that aromatic groups can bind the sialic acid site. Subsequently, the azide groups of the polymers were clicked with b-D-propargyl galactose by Cu(I)-catalyzed cycloaddition. In this way, a small library of glycopolymers **40** was prepared and used to study their binding affinity to the bacterial toxin (CTx) and PNA. The results showed that PNA showed higher binding ability than CTx. This report succeeded to use glycan-mimetic branching to introduce specificity and selectivity as well as affinity into synthetic glycopolymers.

Jiang *et al.* have prepared three self-assembled nanoparticles from triblock copolymers **41, 42,** and **43** with the same polymeric backbone but different sugar regioisomers as pendant groups [105]. Galactose was used as a well-defined constitutional isomer due to its anomeric position (1-Gal) or 6-position (6-Gal) and a mannopyranoside (1-Man) as a control. Firstly, rod block of poly(9,9-dioctylfluorene) macroinitiator was conjugated with the bromine-functionalized polyfluorene (PF) initiator from the two ends of it for further occurring the middle block of the glycopolymers. GMA was polymerized via ATRP in the presence of the synthesized macroinitiator to yield triblock copolymer (PGMA-b-PF-b-PGMA). The pendant epoxide groups of the PGMA block were reacted with NaN_3 for the sugars modified with alkyne click targeting. 1-(2′-Propargyl)-D-galactose (α:β = 10:3) and 1-(2′-propargyl)-D-mannose were prepared and used for the click chemistry that gave PF-1-Gal, PF-6-Gal, and PF-1-Man (Figure 6.26).

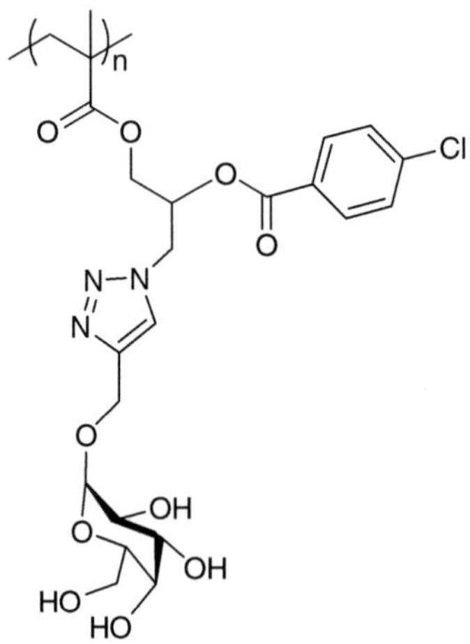

40

FIGURE 6.26 Glycopolymers as mimicking glycan synthesized via Cu(I)-mediated polymerization.

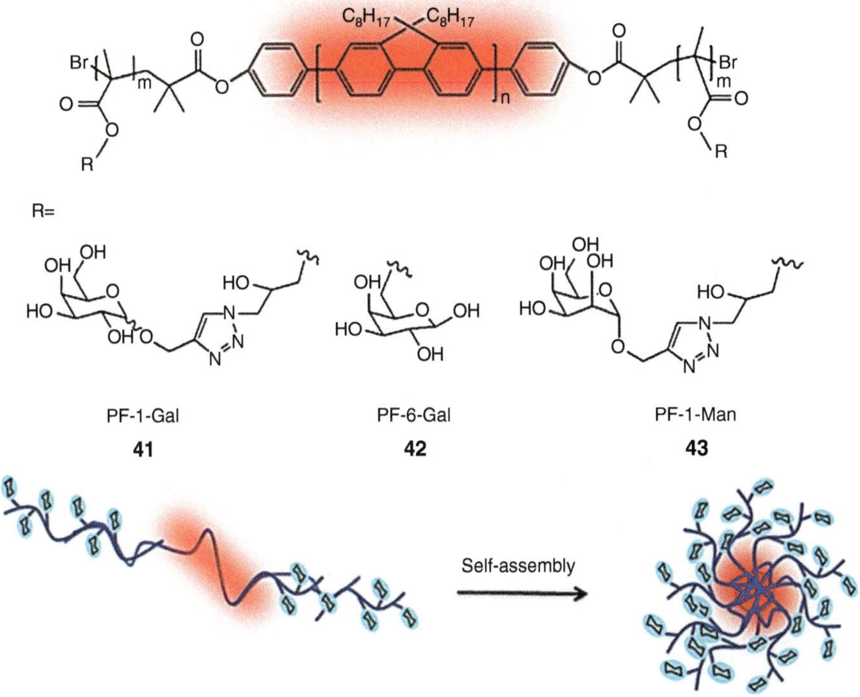

FIGURE 6.27 Chemical structures of PF-1-Gal, PF-6-Gal, and PF-1-Man and cartoon representation of the self-assembled nanoparticles.

The obtained glycopolymers (Figure 6.27) **41**, **42**, and **43** containing the same middle rod block and side coil blocks with similar DP showed self-assembly behavior in the water. Well-dispersed spheres took shape from the polymer self-assembled into nanoparticles with the glyco block as the shell and the rod block as the core. The binding ability of these formed nano-objects was examined using PNA and *Erythrina crista-galli* agglutinin (ECA). The asialoglycoprotein receptor (ASGPR) was used as a model of human lectin, and its binding affinity with nanoparticles was examined using a quartz crystal microbalance (QCM). As expected, all nanoparticles showed similar binding ability with ASGPR as found in previous investigations.

Another efficient approach was developed by Fernández-García and colleagues to prepare various amphiphilic block glycopolymers based on 2-{[(D-glucosamin-2-*N*-yl)carbonyl]oxy}ethylacrylate (HEAGI) and *n*-butyl acrylate (BA) and methyl methacrylate (MMA) via ATRP [106]. The monofunctional and difunctional macroinitiators of poly(BA) and poly(MMA) were used to make a block with the homopolymerization of the unprotected glycomonomer (HEAGI). The polymerization was carried out in DMF at 90 °C by using PMDTA and CuCl as a catalyst system. In this way, a small library of the well-defined amphiphilic block glycopolymers having di- and triblock glycopolymers with different hydrophobic blocks and varying hydrophilic block lengths was demonstrated. The self-assembly ability of the obtained

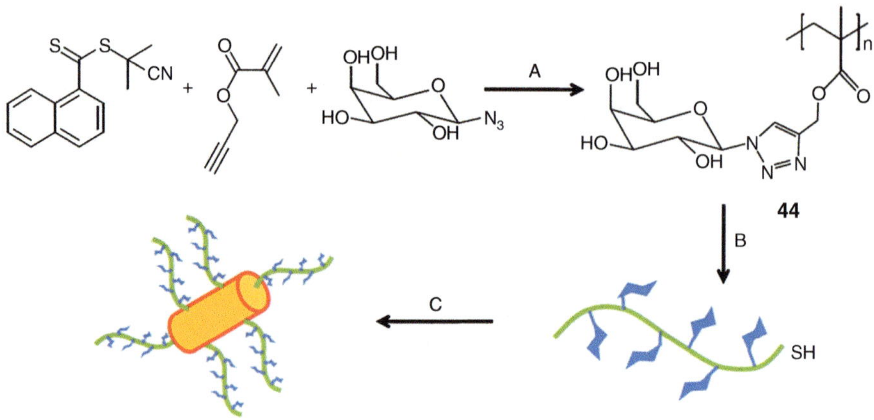

FIGURE 6.28 Synthesis of polymer-coated gold nanoparticle. Conditions: (A) copper powder, methyl 2-bromopropionate, DMSO, 25°C; (B) 2 equiv of R-NH_2 /NEt_3, DMF, 50°C, 24h; (C) gold nanorods, H_2O, 12h. *Source*: Reproduced with permission from Ref. 107, Copyright 2014.

amphiphilic block glycopolymers allowed to form micelles in aqueous solution. The DLS measurements revealed that the average diameter values are between 150 and 160nm due to strong hydrogen-bond interactions between hydroxyl groups and these polymers. The architecture of the block, diblock, or triblock glycopolymers did not show any significant influence on their interaction with Con A.

Glycopolymer-Grafted Nanoparticle Surface The Cu(0)-catalyzed one-pot reaction combining SET-RAFT for the synthesis of glycopolymers **44** was developed by Chen *et al.* for the first time [107]. Firstly, the polymerization and click reaction were carried out in the same time using CPDN as the RAFT agent and EBBr as the initiator in DMSO at 25°C. Subsequently, PMDTA was added and the reaction mixture was kept for 4h. The polymerization demonstrated good control with narrow polydispersity. (PDI=1.1–1.3) The relationship between the molecular weight and the monomer conversion was also linear. Therefore, this approach provided to design of polymers with special side-chain functionality. Another interesting result was that the click rate was significantly faster than the polymerization rate. The end-group reduction was performed to make the glycopolymers **44** being grafted to gold nanorods. Lastly, these thiol-terminated glycopolymers covered the surface of gold nanorods to form a self-assembled monolayer on the GNP surface due to the interaction of Au–S bond. The obtained glyconanorods were examined via TEM and DLS. These glycopolymer substituted GNRs showed a great binding affinity with PNA due to the sufficient numbers of galactose groups on the surface of the nanoparticles (Figure 6.28).

6.2.2 Conventional Free Radical Polymerization

Conventional free radical polymerization has been one of the most widely used polymerization techniques since the 1960s. Even though there are many elegant controlled/living radical polymerization techniques, free radical polymerization is still one of the favorite techniques to generate glycopolymers. The polymerization is initiated by free radical initiators and used to polymerize vinyl saccharides. The first paper based on the free radical homo- and copolymerization of glycomonomers was published by Kimura *et al.* in 1961 [108]. In the following years, the interest of the free radical polymerization of glycomonomers increased significantly due to some advantages of it. The main advantages are the tolerance to impurities and the flexible range of reaction conditions in terms of the solvents and temperatures. At the same time, high polydispersities of the obtained polymers and insufficient controlling of terminal functionalities are the most important disadvantages. The use of unprotected glycomonomers was widely carried out using free radical polymerization. Here, the recent few elegant studies were highlighted because of their biological investigations.

A small library of statistical copolymers, poly[2-(α-D-mannopyranosyloxy) ethyl-*co*-2-dimethylaminoethyl methacrylate]s (poly(ManEMA-*co*-DMAEMA)), **45** were prepared by free radical polymerization [109]. The resulting glycopolymers demonstrated the binding affinity with plasmid DNA while displaying a cluster glycoside effect. Further research was done to better understand the cellular uptake of these polymers; the same monomer was polymerized with HEMA [110]. The glycopolymers **46** were then quickly taken up by HeLa cells, a cell line that has a high affinity to mannose.

The thermoresponsive glycopolymers poly(*N*-isopropylacrylamide-co-6-O-vinyladipoyl-D-glucose) (poly-NIPAM-co-OVDG; PND) and poly(*N*-isopropylacrylamide-co-6-O-vinylazelaicoyl-D-glucose) (poly-NIPAM-co-OVZG; PNZ) were synthesized via free radical polymerization process [111]. The polymerization reaction was performed in anhydrous ethanol at 60°C for 8 h with the presence of AIBN as an initiator. Totally, six different copolymers were prepared with different ratios of NIPAM. Subsequently, for the preparation of nanofibers from these glycopolymers, electrospinning process was used with comprising blends of poly-NIPAM-co-OVDG/poly-NIPAM-co-OVZG with poly-L-lactide-co-ε-caprolactone (PLCL). The obtained nanofibers showed significant biocompatibility with HeLa cells and minimum cytotoxicity as well. Moreover, these glycopolymer nanofibers could be loaded with Con A, and the adsorbed Con A could easily be desorbed with a glucose solution and used to induce death in the cell population.

Fernández-García *et al.* have proposed to synthesize different amphiphilic glycopolymers **47** via free radical polymerization [112]. Poly(ethylene glycol) methacrylate (PEGMA) was prepared (Figure 6.29) and copolymerized with methyl acrylate (MA). Firstly, PEGMA was activated with *p*-nitrophenyl chloroformate in the presence of TEA in THF at 0°C for 24 h. After purification, the modified monomer with *p*-nitrophenyl carbonate groups was reacted with the glucosamine using TEA and hydroquinone in THF at ambient temperature to provide the glycomonomer. The free radical copolymerization of the glycomonomer and MA was undertaken using AIBN as an initiator in DMSO at 70°C. The different amphiphilic glycopolymers **47** with the different ratio of the monomers were prepared by using this approach. Moreover, the emulsion polymerization of MA and these glycopolymers allowed forming colloidal stability, spherical, monodisperse, and small latex particles with low efficiency. The binding ability of these polymer-coated particles with Con A was analyzed via fluorescence microscopy. It was found that the number of glyco units in the glycopolymer stabilizers affected the binding ability directly and significantly (Figure 6.30).

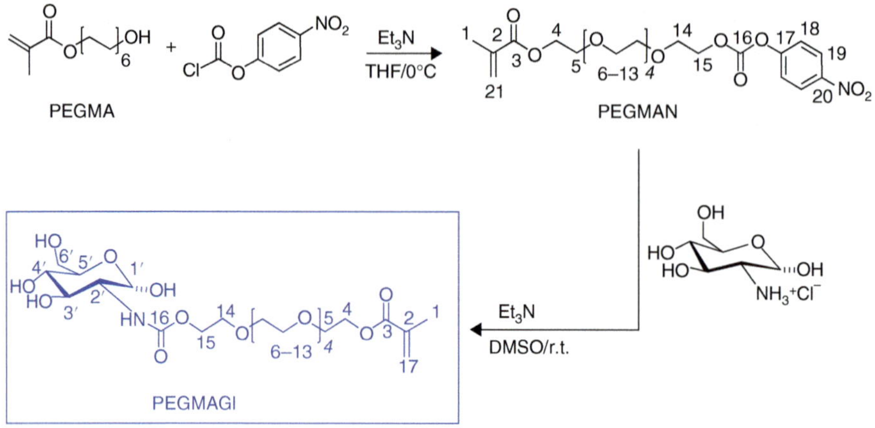

47

FIGURE 6.29 Outline of the synthesis procedure of the glycomonomer, PEGMAGI. *Source*: Reproduced with permission from Ref. 112, Copyright 2014.

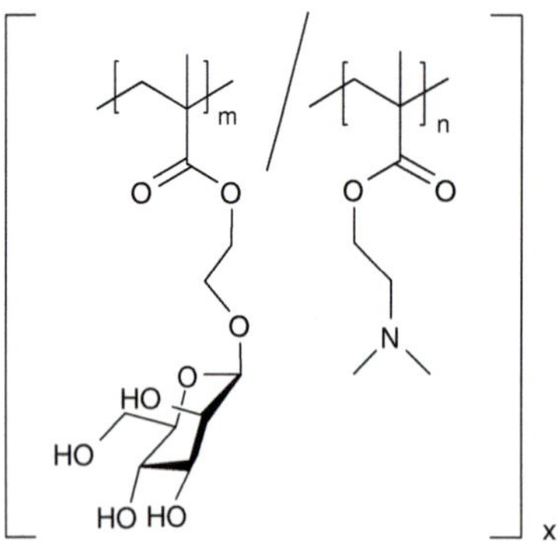

45

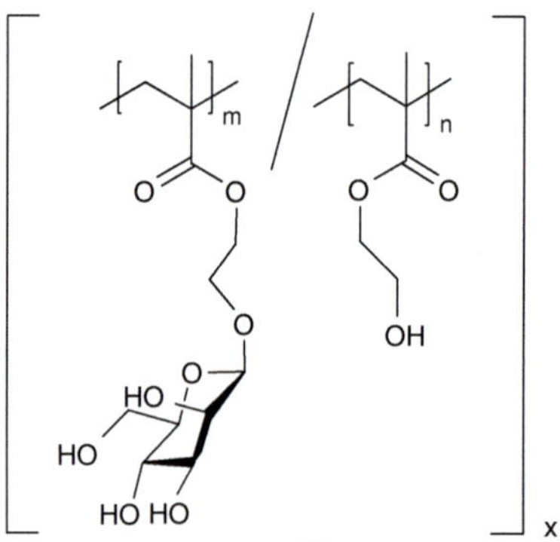

46

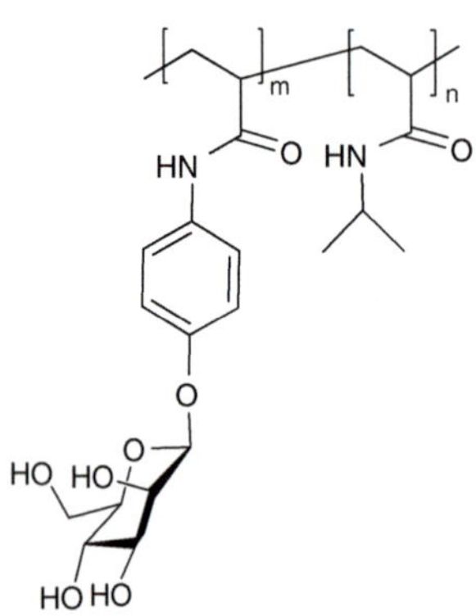

48

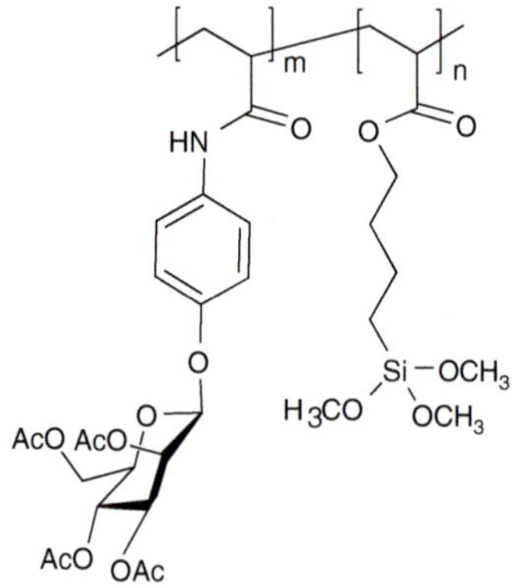

49

FIGURE 6.30 The glycopolymers synthesized via free radical polymerization.

The glycomonomer p-acrylamidophenyl-α-D-mannopyranoside (Man) was copolymerized with PNIPAm **48** via emulsion polymerization in the presence of a cross-linker [113]. Even though the resulting nanoparticles showed the specific interaction with Con A, the binding kinetic could be enhanced by the swelling of the nanoparticles. A similar monomer was also copolymerized with 3-(trimethoxysilyl)propyl methacrylate in another work [114]. In this study, the acetyl-protected monomer was employed, most likely to accommodate the solubility of the hydrophobic 3-(trimethoxysilyl)propyl methacrylate. The immobilization of these glycopolymers **49** on a silica surface could allow to develop a sensor that could detect Con A.

6.3 CONCLUSION

In this chapter, different types of polymerization techniques for the synthesis of various well-defined glycopolymer/particle architectures were highlighted to provide readers a general overview on the creation of different bioactive glycopolymer structures for different bio-related applications such as drug delivery purposes, biomaterials, gene therapy, and much more in the future. In particular, CLRP techniques not only provide glycopolymers with controlled degree of polymerization, but they also allow varying the architecture of glycopolymers. Very recent and elegant synthetic routes via ATRP, SET-LRP, and RAFT have allowed polymer chemists to prepare more complex glycopolymer architectures as a glycan mimic. In the last decade, functionalized self-assembled/decided nano-objects and scaffolds containing glycopolymers were designed to develop many biological and biomedical applications by using these synthetic routes. The combination of CLRP and click reaction techniques presents opportunities to prepare more complex and precision glycopolymers. Furthermore, glycopolymers could be functionalized with other compounds such as proteins or tags by postpolymerization modification reactions. However, there is still challenge to provide a precise control over the sequence of incorporated sugar units. To take the challenge one more step forward, controlled folding of the synthetic polymers and formation of the tertiary structures will be highly desirable in the near future.

ACKNOWLEDGMENT

We are grateful for the support from Turkish Land Forces.

REFERENCES

1. A. M. Granville, D. Quémener, T. P. Davis, C. Barner-Kowollik and M. H. Stenzel, *Macromol Symp*, **2007**, *255*, 81–89.
2. V. Ladmiral, E. Melia and D. M. Haddleton, *European Polymer Journal*, **2004**, *40*, 431–449.
3. M. Hetzer, G. Chen, C. Barner-Kowollik and M. H. Stenzel, *Macromolecular Bioscience*, **2010**, *10*, 119–126.

4. Z. Deng, S. Li, X. Jiang and R. Narain, *Macromolecules*, **2009**, *42*, 6393–6405.
5. L. M. Stefan, A. M. Pana, G. Bandur, P. Martin, M. Popa and L. M. Rusnac, *Journal of Thermal Analysis and Calorimetry*, **2013**, *111*, 789–797.
6. S. Slavin, J. Burns, D. M. Haddleton and C. R. Becer, *Eur. Polym. J.*, **2011**, *47*, 435–446.
7. K. Babiuch, R. Wyrwa, K. Wagner, T. Seemann, S. Hoeppener, C. R. Becer, R. Linke, M. Gottschaldt, J. R. Weisser, M. Schnabelrauch and U. S. Schubert, *Biomacromolecules*, **2011**, *12*, 681–691.
8. H. Shi, L. Liu, X. Wang and J. Li, *Polymer Chemistry*, **2012**, *3*, 1182–1188.
9. S. G. Spain, M. I. Gibson and N. R. Cameron, *Journal of Polymer Science Part A: Polymer Chemistry*, **2007**, *45*, 2059–2072.
10. L. L. Kiessling, J. E. Gestwicki and L. E. Strong, *Angew. Chem. Int. Ed.*, **2006**, *45*, 2348–2368.
11. M. Toyoshima and Y. Miura, *Journal of Polymer Science Part A: Polymer Chemistry*, **2009**, *47*, 1412–1421.
12. *The Sugar Code* (Ed: Gabius, H.-J.), *Wiley-Blackwell*, Weinheim, Germany, **2010**.
13. M. Ambrosi, N. R. Cameron and B. G.Davis, *Organic Biomolecular Chemistry*, **2005**, *3*, 1593.
14. H. J. Gabius, H. C. Siebert, S. André, J. Jiménez-Barbero and H. Rüdiger, *Chembiochem* **2004**, *5*, 740.
15. H. J. Gabius, S. André, H. Kaltner and H. C. Siebert, *Biochimica et Biophysica Acta*, **2002**, *1572*, 165.
16. D. P. Gamblin, E. M. Scanlan and B. G. Davis, *Chemical Reviwers*, **2008**,*109*, 131.
17. C. R. Bertozzi and L. L. Kiessling, *Science*, **2001**, *291*, 2357.
18. V. Ladmiral, G. Mantovani, G. J. Clarkson, S. Cauet, J. L. Irwin and D. M. Haddleton, *J. Am. Chem. Soc.*, **2006**, *128*, 4823–4830.
19. S. R. S. Ting, G. Chen and M. H. Stenzel, *Polymer Chemistry*, **2011**, *2*, 2917–2917.
20. C. Ruiz, M. Sánchez-Chaves, M. L. Cerrada and M. Fernández-García, *Journal of Polymer Science Part A: Polymer Chemistry*, **2008**, *46*, 7238–7248.
21. G. Chen, S. Amajjahe and M. H. Stenzel, *Chem. Commun.*, **2009**, *10*, 1198–1200.
22. A. Mateescu, J. Ye, R. Narain and M. Vamvakaki, *Soft Matter*, **2009**, *5*, 1621–1629.
23. Q. Chen, Y. Xu, Y. Du and B.-H. Han, *Polymer*, **2009**, *50*, 2830–2835.
24. S. G. Spain and N. R. Cameron, *Polym. Chem.*, **2011**, *2*, 60–68.
25. C. R. Becer, *Rapid Commun.*, **2012**, *33*, 742–752.
26. R. J. Pieters, *Org Biomol Chem*, **2009**, *7*, 2013–2025.
27. B. Voit and D. Appelhans, *Macromol. Chem. Phys.*, **2010**, *211*, 727–735.
28. M. Semsarilar, V. Ladmiral and S. b. Perrier, *Macromolecules*, **2010**, *43*, 1438–1443.
29. Y. Chen, G. Chen and M. H. Stenzel, *Macromolecules*, **2010**, *43*, 8109–8114.
30. J. Geng, F. Biedermann, J. M. Zayed, F. Tian and O. A. Scherman, *Macromolecules*, **2011**, *44*, 4276–4281.
31. Y. Miura, S. Onogi and K. Yamamoto, in Transactions of the Materials Research Society of Japan, Vol *33*, No 3, eds. S. Somiya and M. Doyama, **2008**, pp. 729–732.
32. P. L. Nichols and E. Yanovsky, *J Am Chem Soc*, **1944**, *66*, 1625–1627.
33. P. L. Nichols, A. N. Wrigley and E. Yanovsky, *J Am Chem Soc*, **1946**, *68*, 2020–2022.
34. E. A. Talley, M. D. Vale and E. Yanovsky, *J Am Chem Soc*, **1945**, *67*, 2037–2039.

35. P. L. Nichols and E. Yanovsky, *J Am Chem Soc*, **1945**, *67*, 46–49.
36. V. Ladmiral, G. Mantovani, G. J. Clarkson, S. Cauet, J. L. Irwin and D. M. Haddleton, *J Am Chem Soc*, **2006**, *128*, 4823–4830.
37. G. Chen, L. Tao, G. Mantovani, J. Geng, D. Nystroem and D. M. Haddleton, *Macromolecules*, **2007**, *40*, 7513–7520.
38. J. Geng, J. Lindqvist, G. Mantovani and D. M. Haddleton, *Angew. Chem. Int. Ed.*, **2008**, *47*, 4180–4183.
39. L. Nurmi, J. Lindqvist, R. Randev, J. Syrett and D. M. Haddleton, *Chem. Commun.*, **2009**, *19*, 2727–2729.
40. S. R. S. Ting, G. Chen and M. H. Stenzel, *Polym Chem*, **2010**, *1*, 1392.
41. V. Va ´zquez-Dorbatt, J. Lee, E. W. Lin and H. D. Maynard, *Chembiochem* **2012**, *13*, 2478.
42. S. Menon, R. M. Ongungal and S. Das, *Polym Chem*, **2013**, *4*, 623–628.
43. S. Menon and S. Das, *Polym Chem*, **2012**, *3*, 2619–2624.
44. Y. X. Wang, X. G. Zhang, P. E. Yu and C. X. Li, *Inter J Pharm*, **2013**, *441*, 170–180.
45. Q. Y. Meng, A. Haque, B. Hexig and T. Akaike, *Biomaterials*, **2012**, *33*, 1414–1427.
46. K. Nomura and R. R. Schrock, *Macromolecules*, **1996**, *29*, 540–545.
47. F. Zamora, K. Hakkou, A. Alla, M. Rivas, I. Roffé, M. Mancera, S. Muñoz-Guerra and J. A. Galbis, *J Polym Sci Part A-Polym Chem*, **2005**, *43*, 4570–4577.
48. S. N. Narla and X. L. Sun, *Org Biomol Chem*, **2011**, *9*, 845–850.
49. Z. Lu, L. Mei, X. Zhang, Y. Wang, Y. Zhao and C. Li, *Polym Chem*, **2013**, *4*, 5743–5750.
50. M. Obata, M. Shimizu, T. Ohta, A. Matsushige, K. Iwai, S. Hirohara and M. Tanihara, *Polym Chem*, **2011**, *2*, 651–658.
51. C. Zheng, Q. Q. Guo, Z. M. Wu, L. Sun, Z. P. Zhang, C. X. Li and X. G. Zhang, *Eur J Pharm Sci*, **2013**, *49*, 474–482.
52. S. F. Lou, H. Zhang, G. R. Williams, C. Branford-White, H. L. Nie, J. Quan and L. M. Zhu, *Coll Surf B: Biointerfaces*, **2013**, *105*, 180–186.
53. P. Escale, S. R. S. Ting, A. Khoukh, L. Rubatat, M. Save, M. H. Stenzel and L. Billon, *Macromolecules*, **2011**, *44*, 5911–5919.
54. K. H. Mortell, R. V. Weatherman and L. L. Kiessling, *J Am Chem Soc*, **1996**, *118*, 2297–2298.
55. M. Kanai, K. H. Mortell and L. L. Kiessling, *J Am Chem Soc*, **1997**, *119*, 9931–9932.
56. D. D. Manning, X. Hu, P. Beck and L. L. Kiessling, *J Am Chem Soc*, **1997**, *119*, 3161–3162.
57. M. C. Schuster, K. H. Mortell, A. D. Hegeman and L. L. Kiessling, *J Mol Cat A-Chemical*, **1997**, *116*, 209–216.
58. T. Lipinski, P. Kitov, A. Szpacenko, E. Paszkiewicz and D. R. Bundle, *Bioconjugate Chemistry* **2011**, *22*, 274.
59. J. Huang, C. Bonduelle, J. Thévenot, S. Lecommandoux and A. Heise, *J. Am. Chem. Soc.* **2011**, *134*, 119.
60. K. Kempe, T. Neuwirth J. Czaplewska, M. Gottschaldt,R. Hoogenboom and U. S. Schubert, *Polym. Chem.* **2011**, *2*, 1737.
61. K. Ohno, Y. Tsujii, T. Miyamoto, T. Fukuda, M. Goto, K. Kobayashi and T. Akaike, *Macromolecules*, **1998**, *31*, 1064 .
62. Y. M. Chen and G. Wulff., *Macromol Chem Phys*, **2001**, *202*, 3426.

63. C. R. Becer, K. Babiuch, D. Pilz, S. Hornig, T. Heinze, M. Gottschaldt and U. S. Schubert, *Macromolecules*, **2009**, *42*, 2387 .
64. S. R. S. Ting, E. H. Min, P. Escale, M. Save, L. Billon and M. H. Stenzel, *Macromolecules*, **2009**, *42*, 9422–9434.
65. N. Baradel, S. Fort, S. Halila, N. Badi and J.-F. Lutz, *Angew. Chem. Int. Ed.*, **2013**, *52*, 2335–2339.
66. K. Babiuch, R. Wyrwa, K. Wagner, T. Seemann, S. Hoeppener, C. R. Becer, R. Linke, M. Gottschaldt, J. Weisser, M. Schnabelrauch and U. S. Schubert, *Biomacromolecules*, **2011**, *12*, 681–691.
67. S. Hou, X.-L. Sun, C.-M. Dong and E. L. Chaikof, *Bioconjugate Chem.*, **2004**, *15*, 954–959.
68. S. N. Narla and X. L. Sun, *Org Biomol Chem*, **2011**, *9*, 845–850.
69. X.-L. Sun, K. M. Faucher, M. Houston, D. Grande and E. L. Chaikof, *Journal of the American Chemical Society*, **2002**, *124*, 7258.
70. S. Hou, X.-L. Sun, C.-M. Dong and E. L. Chaikof, *Bioconjugate Chemistry*, **2004**, *15*, 954.
71. J. Chiefari, Y. K. Chong, F. Ercole, J. Jeffery, T. P. T. Le, R. T. A. Mayadunne, G. F. Meijs, C. L. Moad, G. Moad, E. Rizzardo and S. H. Thang, *Macromolecules*, **1998**, *31*, 16, 5559.
72. A. B. Lowe, B. S. Sumerlin and C. L. McCormick, *Polymer*, **2003**, *44*, 22, 6761.
73. O. Abdelkader, S. Moebs-Sanchez, Y. Queneau, J. Bernard and E. Fleury, *J Polym Sci Part A-Polym Chem*, **2011**, *49*, 1309–1318.
74. N. Y. Xiao, H. Liang and J. Lu, *Soft Matter*, **2011**, *7*, 10834–10840.
75. W. Wang, D. Chance, V. Mossine and T. Mawhinney, *Glycoconj J*, **2014**, *31*, 133.
76. A. Kobayashi, T. Akaike, K. Kobayashi and H. Sumitomo, *Makromol Chem-Rapid Commun*, **1986**, *7*, 645–650.
77. K. Godula and C. R. Bertozzi, *Journal of the American Chemical Society* **2012**, *134*, 15732.
78. M. Ahmed and R. Narain, *Biomaterials*, **2011**, *32*, 5279–5290.
79. M. Ahmed, S. Mamba, X. H. Yang, J. Darkwa, P. Kumar and R. Narain, *Bioconjugate Chem.*, **2013**, *24*, 979–986.
80. M. Ahmed and R. Narain, *Biomaterials*, **2013**, *34*, 4368–4376.
81. H. Shi, L. Liu, X. Wang and J. Li, *Polymer Chemistry*, **2012**, *3*, 1182.
82. R. J. Mancini, J. Lee and H. D. Maynard, *Journal of the American Chemical Society*, **2012**, *134*, 8474.
83. A. E. Smith, A. Sizovs, G. Grandinetti, L. Xue and T. M. Reineke, *Biomacromolecules*, **2011**, *12*, 3015–3022.
84. D. J. Buckwalter, A. Sizovs, N. P. Ingle and T. M. Reineke, *Acs Macro Lett*, **2012**, *1*, 609–613.
85. D. Zhou, C. Li, Y. Hu, H. Zhou, J. Chen, Z. Zhang and T. Guo, *International Journal of Biological Macromolecules*, **2012**, *50*, 965.
86. M. Ahmed and R. Narain, *Mol Pharmaceutics*, **2012**, *9*, 3160–3170.
87. I. Kurtulus, G. Yilmaz, M. Ucuncu, M. Emrullahoglu, C. R. Becer and V. Bulmus, *Polymer Chemistry*, **2014**, *5*, 1593.
88. W. Lu, W. Ma, J. Lu, X. Li, Y. Zhao and G. Chen, *Macromolecular Rapid Communications*, **2014**, *35*, 827.

89. J. Lu, W. Zhang, L. Yuan, W. Ma, X. Li, W. Lu, Y. Zhao and G. Chen, *Macromolecular Bioscience*, **2014**, *14*, 340.
90. Y. Luo, L. Liu, X. B. Wang, H. T. Shi, W. H. Lv and J. Y. Li, *Soft Matter*, **2012**, *8*, 1634–1642.
91. M. Kamigaito, T. Ando and M. Sawamoto, *Chem Rev*, **2001**, *101*, 3689.
92. M. Ouchi, T. Terashima and M. Sawamoto, *Chem Rev*, **2009**, *109*, 4963.
93. K. Matyjaszewski, *Macromolecules*, *45*, 4015.
94. K. Ohno, Y. Tsuji and T. Fukuda, *Polymer Chemistry*, **1998**, *36*, 14, 2473.
95. Y. Z. Liang, Z. C. Li, G. Q. Chen and F. M. Li, *Polymer Int.*, **1999**, *48*, 9, 739.
96. Q. Zhang, J. Collins, A. Anastasaki, R. Wallis, D. A. Mitchell, C. R. Becer and D. M. Haddleton, *Angew. Chem. Int. Ed.*, **2013**, DOI:10.1002/anie.201300068.
97. Q. Zhang, A. Anastasaki, G. Z. Li, A. J. Haddleton, P. Wilson and D. M. Haddleton, *Polymer Chemistry*, **2014**.
98. W. Shi, S. Dolai, S. Averick, S. S. Fernando, J. A. Saltos, W. L'Amoreaux, P. Banerjee and K. Raja, *Bioconjugate Chemistry*, **2009**, *20*, 1595.
99. V. Vázquez-Dorbatt, Z. P. Tolstyka, C.-W. Chang and H. D. Maynard, *Biomacromolecules*, **2009**, *10*, 2207.
100. J. Geng, G. Mantovani, L. Tao, J. Nicolas, G. Chen, R. Wallis, D. A. Mitchell, B. R. G. Johnson, S. D. Evans and D. M. Haddleton, *Journal of the American Chemical Society*, **2007**, *129*, 15156.
101. R. Narain, *Reactive and Functional Polymers*, **2006**, *66*, 1589.
102. V. Vázquez-Dorbatt, H. D. Maynard, *Biomacromolecules*, **2006**, *7*, 2297.
103. S. S. Gupta, K. S. Raja, E. Kaltgrad, E. Strable and M. G. Finn *Chemical Communications*, **2005**, 4315.
104. M. W. Jones, L. Otten, S. J. Richards, R. Lowery, D. J. Phillips, D. M. Haddleton and M. I. Gibson, *Chemical Science*, **2014**, *5*, 1611.
105. P. Sun, Y. He, M. Lin, Y. Zhao, Y. Ding, G. Chen and M. Jiang, *ACS Macro Lett.* **2013**, *3*, 96.
106. A. Muñoz-Bonilla, O. León, V. Bordegé, M. Sánchez-Chaves and M. Fernández-García, *Journal of Polymer Science Part A: Polymer Chemistry*, **2013**, *51*, 1337.
107. J. Lu, W. Zhang, S.-J. Richards, M. I. Gibson and G. Chen, *Polymer Chemistry*, **2014**, *5*, 2326.
108. S. Kimura, M. Imoto, *Macromol. Chem.*, **1961**, *50*, 1, 155.
109. M. Obata, T. Kobori, S. Hirohara and M. Tanihara, *Polymer*, **2012**, *53*, 4672–4677.
110. M. Obata, M. Shimizu, T. Ohta, A. Matsushige, K. Iwai, S. Hirohara and M. Tanihara, *Polym Chem*, **2011**, *2*, 651–658.
111. L. Wang, G. R. Williams, H.-l. Nie, J. Quan and L.-m. Zhu, *Polymer Chemistry*, **2014**, *5*, 3009.
112. M. Alvárez-Paino, R. Juan-Rodríguez, R. Cuervo-Rodríguez, A. Muñoz-Bonilla and M. Fernández-García, *Journal of Colloid and Interface Science*, **2014**, *417*, 336.
113. Y. Hoshino, M. Nakamoto and Y. Miura, *J Am Chem Soc*, **2012**, *134*, 15209–15212.
114. H. Seto, Y. Ogata, T. Murakami, Y. Hoshino and Y. Miura, *ACS Appl Mater Interfaces*, **2012**, *4*, 411–417.

7

GLYCOCLUSTERS AND THEIR APPLICATIONS AS ANTI-INFECTIVE AGENTS, VACCINES, AND TARGETED DRUG DELIVERY SYSTEMS

JUAN MANUEL CASAS-SOLVAS AND ANTONIO VARGAS-BERENGUEL
Department of Chemistry and Physics, University of Almería, Almería, Spain

7.1 INTRODUCTION

The seminal contributions reporting on the enhanced affinity observed in the interaction of clustered carbohydrates with their protein receptors, also having clusters of binding sites, referred to such phenomenon as glycoside cluster effect [1]. Since then, glycoside clusters or glycoclusters became the topic of a very active research in the fields of glycochemistry and glycobiology [2]. Carbohydrate–protein interactions mediate a wide diversity of crucial biological events, but paradoxically interactions between saccharides and proteins are typically very weak [3,4]. Nature usually overcomes such low affinity between glycans and their specific protein receptors (lectins) through glycoside cluster effects transforming such interactions into very high-affinity bindings [5,6]. This phenomenon actually transcends the carbohydrate chemistry, referred to as multivalent effect in its widespread version, and is ubiquitous in biological systems where multivalent ligands of carbohydrate or noncarbohydrate nature bind to their respective multivalent receptors in a stronger and more selective fashion than their monovalent analogues [7]. In order to study and obtain a deeper understanding of these multivalent glycoside interactions, a great deal of research

Carbohydrate Nanotechnology, First Edition. Edited by Keith J. Stine.

has been devoted to synthesize multivalent glycomimetics of various architectures. In this regard, an important challenge when designing multivalent architectures is the optimization of the several factors involved in the high affinity and selectivity of the multivalent systems [2]. Those factors include not only the ligand density but also the spatial arrangement of the ligands, the linker, the scaffold, and the homo- and heteromultivalency [8], among others. Numerous excellent reviews dealing with multivalent neoglycoconjugates have been published [8–18], including the very comprehensive one by Chabre and Roy [2]. A large number of different synthetic scaffolds have been used for building such multivalent structures, giving rise to a large variety of modes of carbohydrate display with different topologies and valencies. Examples of scaffolds used are carbohydrates, dendrimers, cyclodextrins, calixarenes, oligopeptides, benzene derivatives, tetraphenylethylene, DNA, pentacyclen, azobenzene, triazine, corannulene, glycophanes, and porphyrins, to name but a few. This chapter will focus on the most relevant and recent reports on glycoclusters and their applications in the context of the study of carbohydrate–protein interactions and their uses for therapeutic purposes. For this review, we consider glycoclusters as multivalent glycoconjugates, regardless of the number of peripheral moieties, built on a well-defined central core to which saccharides are linked through tethers that in most of the cases do not contain repetitive units. Essentially, the potential biomedical applications of glycoclusters is based on the rational that the multivalency will allow for an enhanced avidity for a particular biological target. Glycoclusters with very high affinity for cell surface receptors of pathogens can favorably compete with them resulting in promising infection inhibitors. Such high affinity for a cell surface has been also used for the efficient targeting of drug delivery systems. Moreover, vaccines containing multiple copies of antigen glycosides can elicit immune response. This chapter will concentrate specifically on the uses of glycoclusters as pathogen infection inhibitors, vaccines, and targeted drug delivery systems.

7.2 INHIBITORS OF BACTERIAL AND VIRAL ADHESION

High-affinity multivalent glycomimetics that inhibit the adhesion of pathogens to cell surface receptors can be potential therapeutics. This approach is based on (i) the identification of pathogen surface carbohydrates that are key for the cell pathogen recognition process and (ii) the synthesis of a cluster glycoside consisting of a convenient scaffold displaying such carbohydrates. The resulting glycomimetics are expected to prevent the adhesion of the pathogen by interacting with their cell surface lectin receptors.

The influenza virus infects mainly vertebrates by adhering to the cell surface through the multivalent binding interaction between virus surface hemagglutinin (HA) and cell surface sialic acid [19]. Molecules that could inhibit the virus adhesion are potential anti-infective drugs. Pioneering work by Roy *et al.* [20] described the powerful inhibitor activity of dendritic sialosides based on hyperbranched L-lysine cores in experiments with influenza A. The results showed that even at the divalent level of clustering, the glycomimetic is five times more potent than a monosialoside.

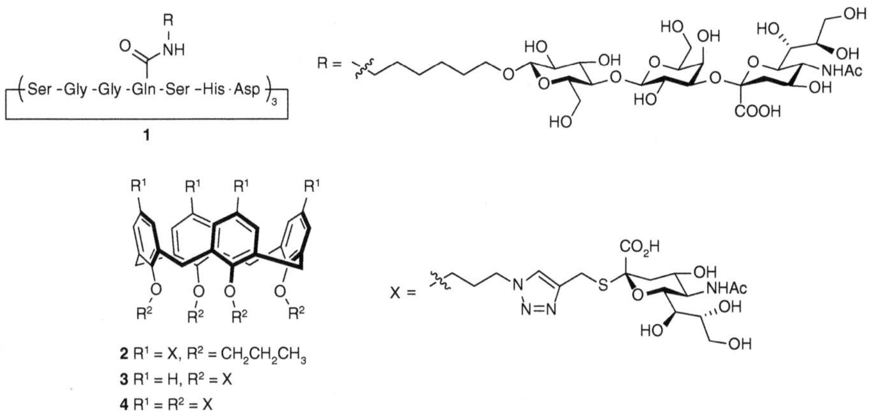

FIGURE 7.1 Influenza virus hemagglutinin (HA) inhibitors based on a cyclopeptide (**1**) and calix[4]arene (**2–4**) showing sialyl groups as vector moieties.

In the search for potential HA blockers, Nishimura and coworkers [21] designed glycoclusters by appending one, two, or three sialotrisaccharide units to glutamine moieties on cyclopeptides of different sequences and sizes (e.g., **1**, in Fig. 7.1). For all cases, distances between sialic acid residues were estimated to be between 50 and 70 Å. These distances are slightly larger than that for the separation between HA binding sites (40–50 Å). Inhibition assays for hemagglutination on chicken erythrocytes induced by influenza virus A/PR/8/34 (H1N1) and surface plasmon resonance (SPR) binding assays for the interaction with the HA protein indicated that only bi- and trivalent conjugates based on the cyclic $(\text{Ser-Gly-Gly-Gln-Ser-His-Asp})_3$ sequence were capable to bind HA protein.

Calixarenes are also very convenient macrocyclic scaffolds for building glycoclusters. Tetra- and octavalent sialoside clusters appended to a *cone*-blocked calix[4]arene through methyl- and propyl-based linkers were synthesized and evaluated as HA blockers [22]. Interestingly, when hemagglutination inhibition assays were performed against the influenza A virus, only those conjugates with propyl-based spacers (**2–4**) showed good inhibition properties, indicating that a minimum linker length is needed for the multivalent recognition. However, glycocluster effect operating in these compounds seemed to be moderate when relative inhibition properties per sialic acid moiety are considered, probably because of the unsuitable length of the spacers. Nevertheless, the three conjugates **2–4** proved to be efficient antiadhesive agents in MDCK cells.

Similarly to viral infection, bacterial adhesion occurs through specific interactions between proteins and glycoproteins or glycolipids at the surfaces of the bacteria and the target cell. Thus, a strategy to avoid the bacterial infection is the development of drugs that inhibit such interactions. In addition, many bacterial pathogens synthesize toxins that serve as primary virulence factor. Toxins can be single proteins or organized as oligomeric protein complexes with distinct AB structure–function properties [23]. The inhibition of the toxin attachment on the cell surface should be enough to avoid the infection.

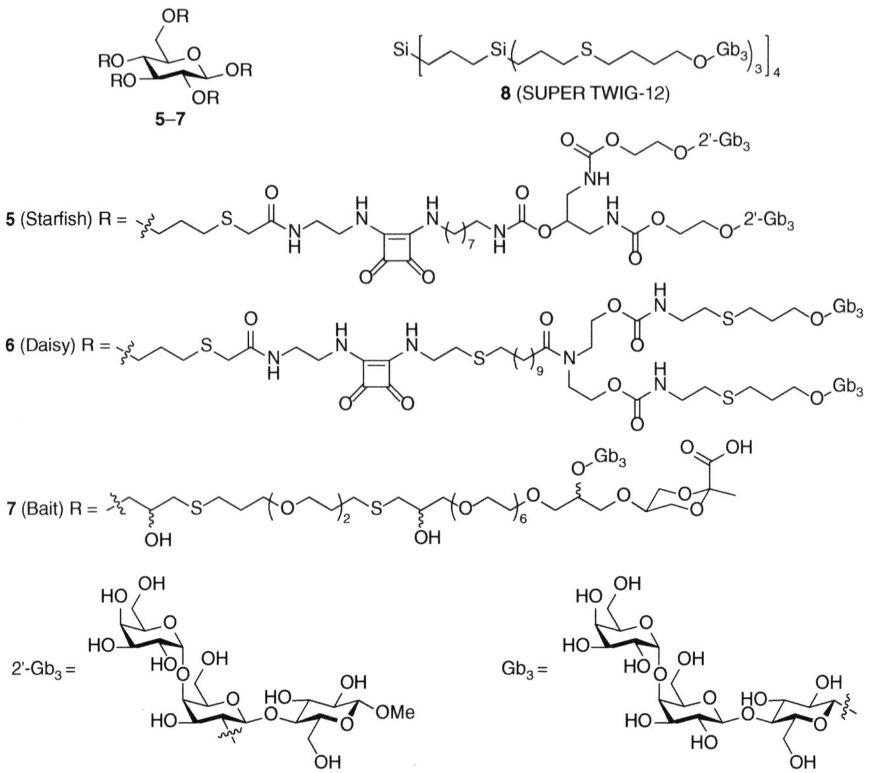

FIGURE 7.2 B_5 pentamer Shiga-like toxins (Stx-1/2) inhibitors centered on glucose (Starfish **5**, Daisy **6**, and Bait **7**) or carbosilane (SUPER TWIG-12 **8**) cores using globotriaosyl (Gb_3) as binding residues.

In particular, Shiga-like toxins (SLT-1/2 or Stx-1/2), also called verotoxins (VTs), are produced by some *Escherichia coli* strains called *shigatoxigenic*. They are named after Shiga toxin (Stx), which is produced by *Shigella dysenteriae* type 1, due to their high homology to the latter. These toxins present AB_5 structures and display selectivity toward the glycolipid globotriaosylceramide Gb_3 [α-D-Gal(1→4)β-D-Gal(1→4)β-D-Glc(1→*O*-ceramide]. As an inhibitor for the B_5 pentamer, Bundle's group designed a decavalent glycocluster called Starfish **5** (Fig. 7.2) [24], composed of a glucose unit as central core to which 10 Gb_3 moieties are linked at position 2′ through divalent linkers. This design was intended to bind simultaneously recognition sites 1 and 2 present in each B monomer. However, crystallographic studies drew a different picture where each arm binds two B-subunit monomers from separate toxin molecules through their recognition sites 2, resulting in 2:1 AB_5:Starfish **5** sandwich-like structures. Inhibition abilities of Starfish were about 10^6 times more potent than monovalent Gb_3 itself toward Stx-1. Thus, Starfish **5** was able to protect Vero cells against lethal doses of Stx-1 and Stx-2 over 2-day incubation periods. A modified version of Starfish, nicknamed Daisy (**6**), was later described by these authors [25].

Daisy displayed slightly longer arms, and Gb_3 moieties were linked through the reducing terminus instead of position 2′. Interestingly, these changes induced somewhat weaker *in vitro* inhibitory abilities than Starfish (**5**), but *in vivo* administration protected mice against both Stx-1 and Stx-2 in a significant higher extension. A heterodifunctional analogue to Starfish (**5**) and Daisy (**6**), displaying five Gb_3 moieties and five cyclic pyruvate ketal of glycerol residues, was also synthesized and called Bait (**7**) [26]. This derivative was able to bring together AB_5 Shiga-like toxins and similarly B_5 structured serum amyloid P component (SAP), one of the most abundant proteins in human serum, which would then act as a "trap" for the toxins in the so-called specific aggregation inhibition strategy. ELISA indicated that Bait (**7**) (IC_{50} = 140 μM) increased its inhibitory potency by 35 times in the presence of SAP (IC_{50} = 4 μM).

The use of diverse carbosilane scaffolds allowed for the synthesis of multivalent glycosides presenting different number (3, 6 and 12) and spatial arrangements (fan, dumbbell, and ball shape) of trisaccharide globotriaose Gb_3 residues (SUPER TWIGs) [27]. Hexa- and dodecavalent conjugates (Fig. 7.2), where distances from trisaccharide residues to central silicon carbon (~30 Å) matched predicted gap between Gb_3 binding sites on the Shiga toxin B-subunit pentamer, showed 30-fold higher affinity than trivalent SUPER TWIGs. These two conjugates markedly inhibited the cytotoxic activity of Stx1 and Stx2 to Vero cells *in vitro* without affecting their viability. In mice, hexa- and dodecavalent SUPER TWIGs completely suppressed the lethal effect of Stx2. Interestingly, the dodecavalent conjugate showed higher activity *in vitro*, but the hexavalent analogue performed better *in vivo*. In a further study [28], a SUPER TWIG presenting 18 trisaccharide residues also demonstrated to be another potent Stx neutralizer *in vivo*.

Cholera toxin is also an AB_5 protein that contains five galactoside binding sites and is able to recognize ganglioside GM1. The toxin is secreted by *Vibrio cholera* causing the disease of cholera [29]. The heat-labile enterotoxin (LT-B_5), which is found in *E. coli*, is structurally similar to cholera toxin (CTB) and binds also to GM1. To prepare multivalent inhibitors of LT-B_5 and CTB, Fan and coworkers [30–33] have followed a modular approach employing aza crowns as scaffolds (Fig. 7.3). LT-B_5 and CTB are symmetrical proteins, which contain five identical subunits presenting one binding site each. Distances between two nonadjacent binding sites are 45 Å. Using 4,7,10-trioxa-1,13-tridecanediamine as flexible modules [30], these authors synthesized a series of pentavalent galactoclusters (**9–12**) scaffolded on the fivefold symmetric 15-azacrown-5 (pentacyclen), which present very well-defined effective length between nonadjacent galactoside units, varying from ~30 to ~45 Å. ELISA results of the LT-B_5-GM1 binding inhibition of these compounds showed that IC_{50} values decreased with the increase of the spacer length, indicating a positive correlation between linker distance and affinity. The highest affinity (10^5-fold versus galactose) was achieved when the linker length was similar to the estimated distance in LT-B_5. Fan and coworkers also prepared an analogue containing *m*-nitrophenyl-α-D-galactoside (MNPG), a ligand 100-fold better than galactose [31]. Such glycocluster **13** proved to be 18-fold more active than the equivalent compound **10**. In a further step, decavalent analogues of these pentavalent glycoclusters were designed to

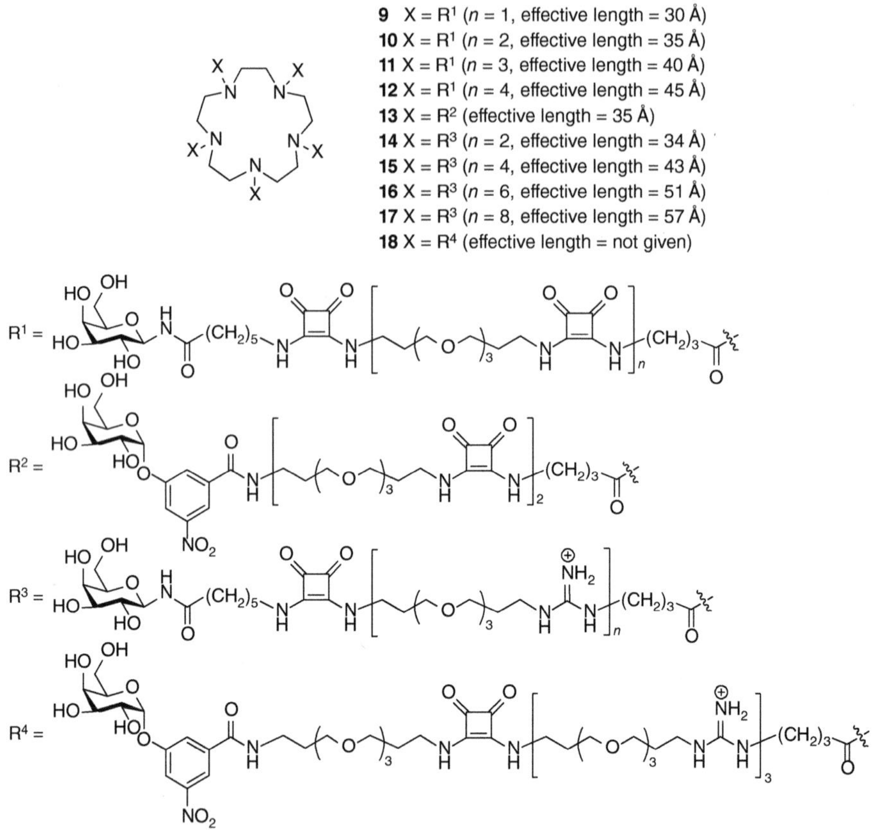

FIGURE 7.3 Cholera toxin B pentamer (CTB) inhibitors **9–18** designed on a pentacyclen core displaying five galactose residues.

target two cholera toxin B pentamers simultaneously [32]. Their IC_{50} values were one order of magnitude lower than those of their counterparts.

Subsequent attempts to improve the inhibitory properties of pentacyclen-based galactoclusters following a rational design involved the synthesis of pentavalent guanidinium-bridged derivatives **14–18** (Fig. 7.3) [33]. MNPG-based ligand **18** was the first ligand displaying an IC_{50} lower than the natural ligand GM1.

Fan's group also grafted *N*-galactoside residues to Lys moieties of three different cyclic decapeptides through a series of four lengths of the previously mentioned linkers. These macrocycles contained five Lys moieties alternated with glycine (Gly), γ-aminobutyric acid (Abu), or ε-aminohexanoic acid (Ahx) in order to tune ring size and force the rings to adopt expanded conformations in solution [34]. The abilities to block the cholera toxin CTB pentamer from *E. coli* binding to GM1 were tested by ELISA and compared to those performed by analogous pentacyclen-based pentavalent ligands **14–17**. As the cyclic decapeptide core size increased, the number of modules needed in the linker for an optimal inhibition decreased, so the effective

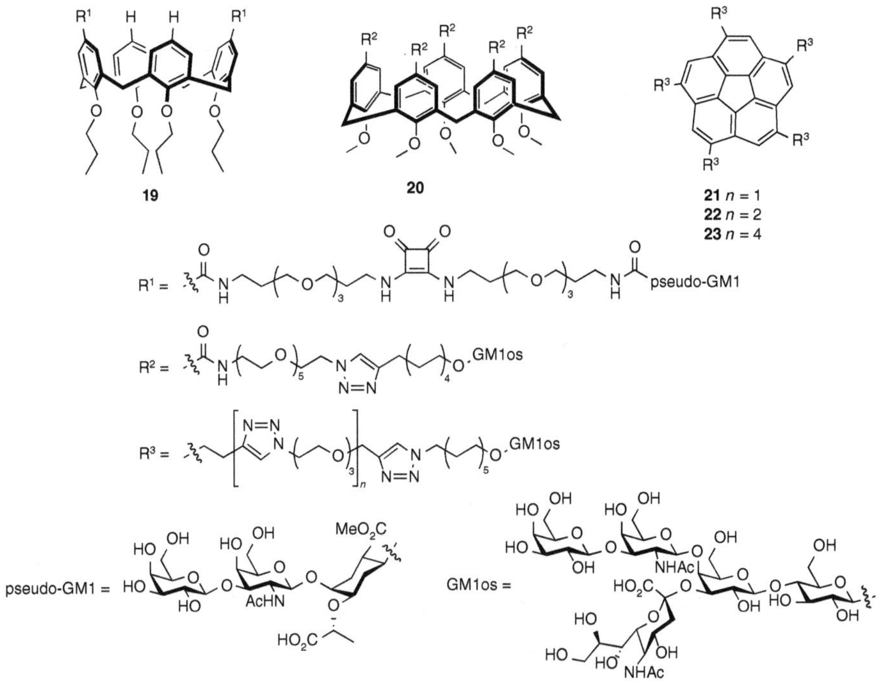

FIGURE 7.4 Cholera toxin B pentamer (CTB) inhibitors **19–23** presenting natural ganglioside GM1 oligosaccharide (GM1os) and unnatural pseudo-GM1 mimic residues grafted on calix[4]arene, calix[5]arene, and corannulene scaffolds.

distance between nonadjacent galactosamine moieties remained around 45 Å. Optimal distance yielded in all cases inhibitory powers around 100.000 times higher than that for monovalent galactose.

Using a similar linker, Ariosio *et al.* [35] have conjugated two units of pseudo-GM1 at the upper rim of a *cone*-locked calix[4]arene (Fig. 7.4). Such pseudo-trisaccharide is an unnatural mimic of the pentasaccharide GM1os of the natural ganglioside GM1 that is recognized by the cholera toxin CTB pentamer. The binding ability of this compound to CTB was studied by fluorescence spectroscopy titration and ELISA assays, and results indicated that the glycocluster displays only a slightly higher affinity for CTB (K_d = 48 nM) than GM1os (K_d = 219 nM). However, when compared to the monovalent unnatural pseudo-GM1 (K_d = 190 μM), relative binding potency per branch enhanced in a factor of ~4000, probably due to the appropriate orientation and length of the sugar moieties provided by the calix[4]arene core and the spacer, respectively. In order to match the CTB valency, a newer glycocluster (**20**) based on a *cone*-blocked calix[5]arene was subsequently prepared [36]. Five GM1os pentasaccharide units were grafted on the upper rim of the macrocycle but in this case through triazole-based linkers of an appropriate length (31 atoms). Such spatial arrangement allows the simultaneous interaction of the five GM1os units with the five B-subunits of a single toxin. The inhibition potency of this conjugate toward the interaction of

CTB with the natural ligand GM1 was measured by ELISA. Compound **20** showed a relative inhibitory potency 10^5 times (20,000 times per residue) higher than that for a monovalent GM1os analogue. This result indicates that GM1os-calix[5]arene is one of the most potent known CTB inhibitors ever reported.

Pentasaccharide GM1os were also grafted through similar triazole–ethylene glycol linkers of different lengths to *sym*-corannulene yielding pentavalent glycoclusters **21–23** (Fig. 7.4) [37]. Although the interaction of these systems to the cholera toxin CTB pentamer was hampered by aggregation processes, high, nanomolar inhibitory potencies were obtained, which were clearly affected by the linker length. In the case of GM1os family, conjugates **21–23** showed relative inhibitory potencies per sugar residue of 770, 3700, and 2600, respectively, when referred to monovalent GM1os.

As previously pointed out, a precise matching of the valency and spacing between the carbohydrate ligands and CTB is crucial in the design of an effective inhibitor for this toxin. Turnbull and coworkers postulated that the best scaffold for this aim would be the protein itself [38]. Thus, they modified an inactive mutant CTB protein unable to bind GM1os with five units of such oligosaccharide attached to its five N-terminal threonine residues located on the protein surface toward the ligand-binding face and equally spaced around the protein between the GM1 binding sites. The synthesis of this conjugate was easy and relatively short, ensuring a potential large-scale application. ELLA experiments toward CTB binding to GM1 reported the lowest IC_{50} value so far (104 pM), validating the proposed strategy. The formation of 1:1 CTB:neoglycoprotein heterodimers with an apparent K_d value of 30 nM was estimated by ITC, DLS, and analytical ultracentrifugation (AUC).

The surface of many bacteria is covered by the so-called type 1 fimbriae, a sort of adhesive pili presenting the mannose-binding lectin FimH, which mediate the bacterial adhesion through the interaction with mannose residues of the host cell glycocalyx. Lindhorst and coworkers [39] prepared a series of penta- to undecavalent mannosylated glycoclusters centered on mono-, di-, and trisaccharides (trehalose, melibiose, raffinose, and glucose). However, such glycoconjugates displayed similar inhibitory properties for the adhesion of *E. coli* bacteria to their host cells, regardless of the number of mannose residue or the spatial arrangement.

Also on saccharide cores such as di-, tri-, and heptavalent linear β-(1→4)-glucosides, Almant *et al.* synthesized clusters of heptyl α-D-mannoside (HM) (one of the most potent FimH inhibitors reported to date) [40]. Inhibition of hemagglutination (HIA) assay on guinea pig erythrocytes reported an inhibition potency enhancement with valency, the highest increase (64-fold, 9-fold per HM residue) corresponding to the heptavalent conjugate **24** (Fig. 7.5).

Gouin and coworkers prepared further tri- and tetra-(heptyl mannoside) clusters **25** and **26** based on a pentaerythritol scaffold [41]. These compounds were evaluated as antiadhesives toward type 1-piliated *E. coli* (covered by mannose-recognizing FimH lectins) by HIA and bladder binding assay (BBA). Results from HIA indicated that tri- and tetravalent conjugates were around twofold more potent inhibitors than heptyl mannoside (HM) for the hemagglutination of guinea pig erythrocytes promoted by *E. coli*. However, higher values were obtained when human bladder cells were used, indicating increases of 31- and 64-folds for the tri- and tetravalent conjugates, respectively.

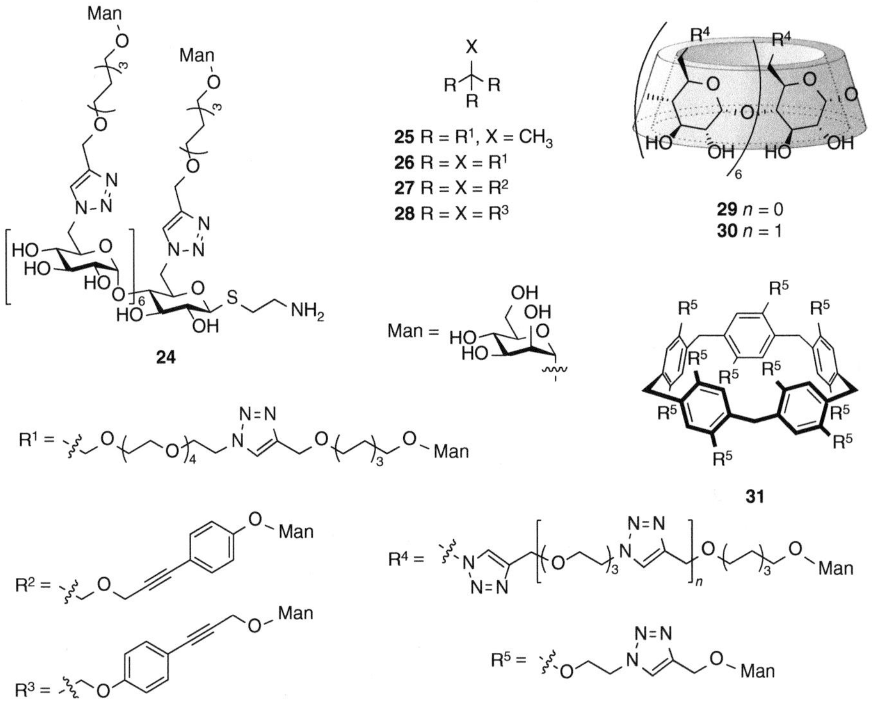

FIGURE 7.5 Antiadhesive cluster mannosides based on linear β-(1→4)-glucosides, pentaerythritol, β-cyclodextrin, and pillar[5]arene cores **24–31** against FimH lectin, present in type 1-piliated bacteria such as *Escherichia coli*.

Similarly, Touaibia *et al.* used pentaerythritol as a core for the preparation of a series of tetra- and hexavalent mannosylated glycoconjugates containing 1,2,3-triazolyl- or phenylethynyl-based linkers [42]. SPR experiments using soluble FimH lectin revealed that conjugate **27** was the best ligand, with a K_d = 0.45 nM (1277-fold better than that for mannose). Inversion of the spacer orientation (compound **28**) yielded K_d = 273 nM. Finally, inhibitory potencies against the hemagglutination of guinea pig or rabbit erythrocytes by type 1-piliated clinical isolate of *E. coli* UTI89 were evaluated by HIA. Again, compound **27** showed the highest potency, with an inhibition titer of 3 μM (1000-fold better than mannose).

More recently, heptavalent HM clusters **29** and **30** based on β-cyclodextrin (β-CD) with two linkers of different length and flexibility were studied [43]. The study on the *in vitro* interaction with FimH revealed that the glycocluster with the longest spacer showed a higher effective valency, close to the structural valency, although both compounds were able to promote bacterial clustering through cross-linking interactions with cells in solution in a similar fashion. *In vivo* experiments for prevention of infection in mice bladder showed that both compounds caused a 10-fold reduction in the level of infection at 10 μM concentrations.

The inhibitory capability of a pillar[5]arene glycoconjugate **31** containing 10 mannose residues (Fig. 7.5) against the adhesion of uropathogenic *E. coli* to guinea pig red blood cells has been tested by hemagglutination inhibition assay [44]. A relative potency per sugar moiety of 6.7 was found when compared to a monovalent analogue.

Another bacteria of interest is *Pseudomonas aeruginosa*, an opportunistic multidrug resistant (MDR) human pathogen and one of the leading causes of nosocomial infections worldwide [45]. Cell recognition is mediated by galactose-binding PA-IL (or LecA) and fucose-binding PA-IIL (or LecB) lectins, such lectins becoming biological targets of interest for the design of inhibitors and antiadhesives. For example, Vidal and coworkers prepared a series of seven lower-rim galactosyl calix[4]arenes using a triethylene glycol–triazolyl–methylene linker as ligands for PA-IL lectin [46]. Such series of glycoconjugates included several valencies but also several topological arrangements of the galactoside residues. *Cone*-blocked calix[4]arene was grafted with one, two (1,2 and 1,3 positions), three, and four saccharides. In addition, tetravalent *1,3-alternate-* and *partial cone*-calix[4]arene conjugates were also prepared. Considering only the tetravalent conjugates (Fig. 7.6), which display higher affinity than the rest, the 1,3-*alternate* conformation **33** showed the highest binding ability (K_d = 176 nM), followed by the *partial cone* **34** (K_d = 200 nM) and the *cone* **32** conjugates (K_d = 420 nM), yielding a 213-, 188-, and 89-fold in affinity enhancement per carbohydrate residue, respectively. This trend is in agreement with the higher affinity found by other authors for the case of 1,3-*alternate* calix[4]arene derivatives (see in the following text). These conjugates are thus more potent ligands than the best natural ligand of PA-IL (α-linked digalactoside).

These authors also carried out the synthesis of calix[4]resorcarene derivatives having four galactosyl and lactosyl units on the lower rim of the macrocycle through the benzylic positions [47]. Calix[4]resorcarene conformation was fixed in two different topologies: the *rccc* boat (**35** and **36**) and the *rctt* chair (**37** and **38**) stereoisomers. Studies on the binding abilities of both galactosyl chair and boat conjugates toward PA-IL lectin showed a 314-fold and 244-fold improvement, respectively, when compared to a monovalent analogue. However, the influence of the macrocyclic topology on the binding properties was barely observed.

Gening *et al.* prepared a family of cyclic oligo-(1→6)-β-D-glucosamines consisting of four monosaccharide units containing four attached galactoside ligands through spacer arms [48]. HIA, ELLA, and ITC were used to study their binding abilities toward PA-IL lectin from *P. aeruginosa*. Cluster **42**, having a rigid, short aromatic linker, in contrast with flexible and hydrophilic ethylene glycol tethers of derivatives **39–41**, showed the lowest IC_{50} value (57 nM), with a relative potency per residue 1210-fold higher than that for the corresponding monovalent analogue. This glycocluster showed the highest reported affinity for PA-IL (K_d = 80 nM).

In a similar fashion to that described previously, fucosylated glycoclusters have been designed as potential biofilm inhibitors toward *P. aeruginosa* PA-IIL lectin. For example, a tetravalent α-L-fucosylated calix[4]arene **43** was prepared by Consoli *et al.* (Fig. 7.6) [49]. The ability of this compound to inhibit the formation of the bacteria biofilm was evaluated by minimum biofilm inhibitory concentration (MBIC) assay. The activity was dose dependent, and the biofilm formation was inhibited in 73% at 200 μM.

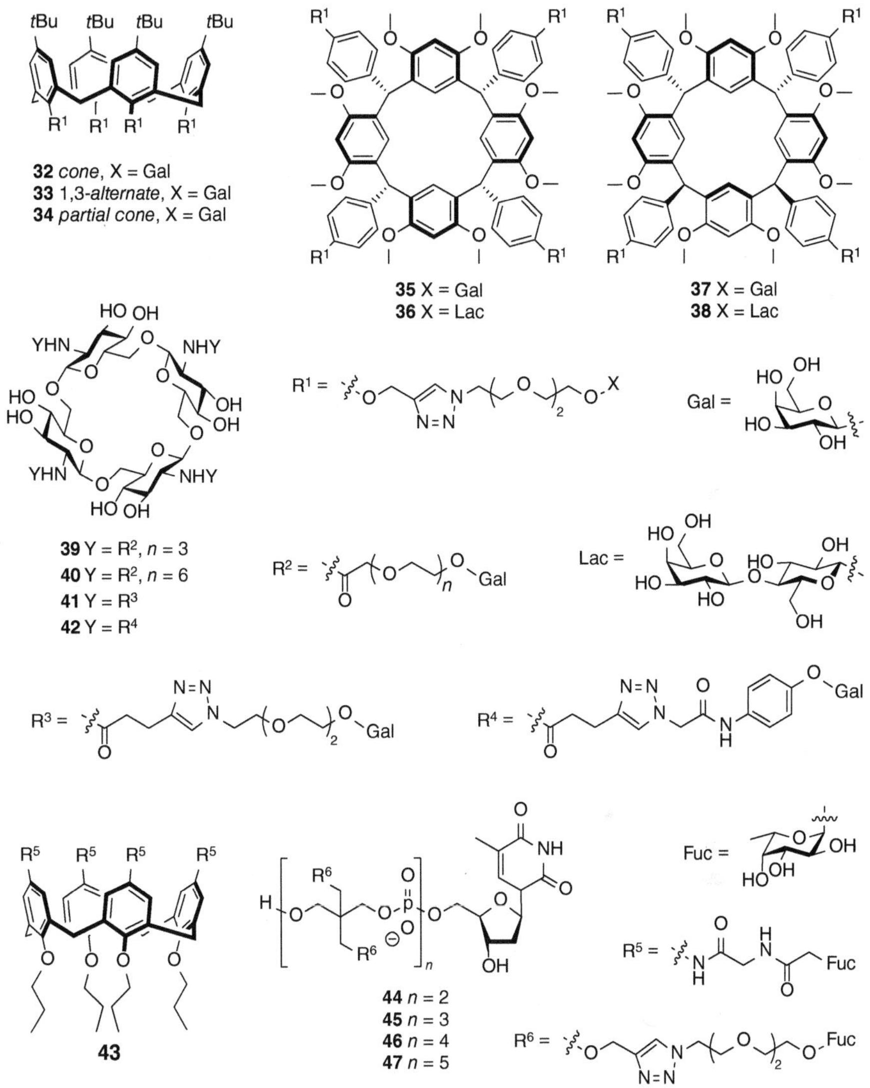

FIGURE 7.6 High-affinity ligands for galactose-binding PA-IL (LecA) and fucose-binding PA-IIL (LecB) lectins from *P. aeruginosa* (**32–42** and **43–47**, respectively) based on calix[4]arene, calix[4]resorcarene, and pentaerythritol phosphodiester oligomer (PePOs) cores.

Tetra- to decavalent α-L-fucosyl clusters **44–47** based on pentaerythrityl phosphodiester oligomers (PePOs) cores were prepared by Morvan *et al.* [50]. The structures also presented a thymidine moiety at the pseudo-3′-end in order to facilitate the determination of the conjugates concentration by UV spectroscopy. These compounds proved to inhibit the interaction of PA-IIL lectin with polymeric α-L-fucose. However, the increased binding found for these multivalent

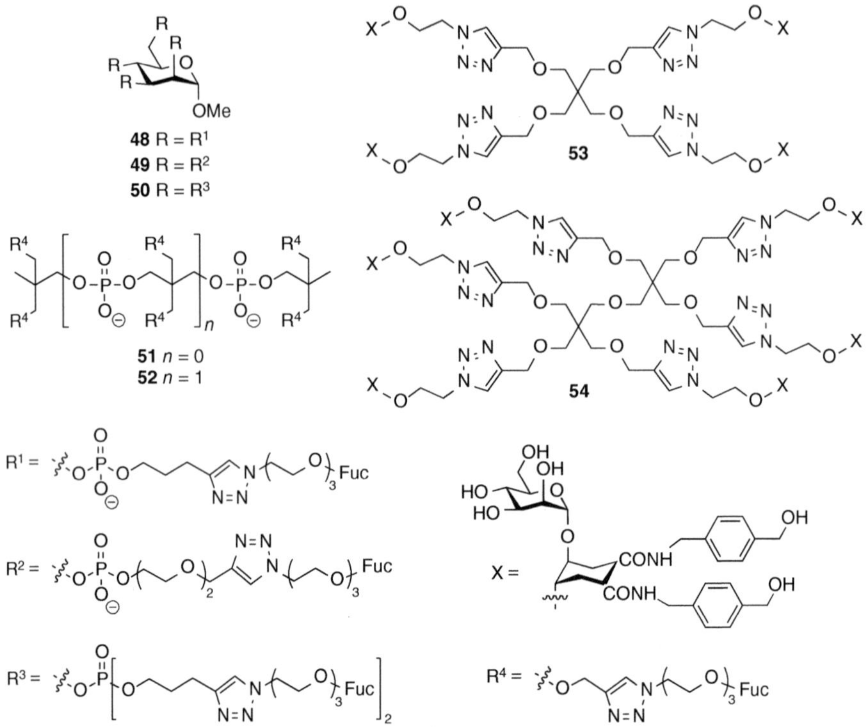

FIGURE 7.7 Fucosylated antiadhesive agents against BambL lectin from *Burkholderia ambifaria* bacteria (**48–52**) centered on mannose and pentaerythrityl phosphodiester oligomers (PePOs), as well as HIV-1 and dengue viruses inhibitors (**53** and **54**) bearing a nonnatural pseudo-disaccharide DC-SIGN antagonist.

carbohydrate ligands was attributed to the *macromolecular* effect [39] rather than the cluster or multivalent effect.

Fucosylated glycoclusters have also been reported as antiadhesives toward *Burkholderia ambifaria* bacteria, both using mannose (**48–50**) and a branched phosphodiester (**51** and **52**) as scaffold and bearing four, six, or eight sugar residues (Fig. 7.7) [51]. The binding abilities of these conjugates toward this bacterial fucose recognition lectin (BambL) were thus evaluated by means of HIA, SPR, and ITC. As previously observed by the authors in similar studies [52], these techniques yielded nonconsistent values, although all conjugates demonstrated higher binding (between 9 and 22 folds) and inhibitory (between 16 and 256 folds) potencies than the monovalent methyl α-L-fucoside.

Apart from bacteria, viruses are also desirable targets for inhibitors. An interesting approach for the development of HIV-1 and dengue virus (DV) inhibitors has been report by Varga *et al.* [53]. A powerful synthetic pseudo-disaccharide DC-SIGN antagonist that mimics the natural trimannoside ligand was used to prepare a series of multivalent derivatives having from three to eighteen valencies based on

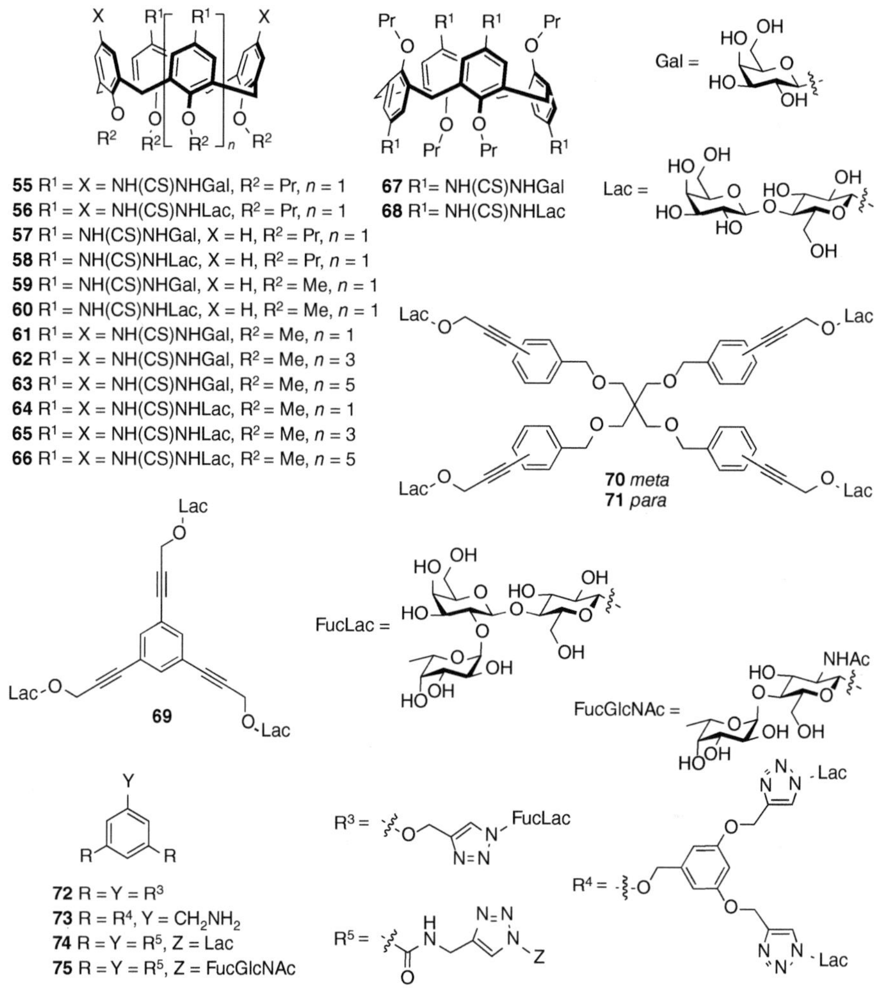

FIGURE 7.8 High-affinity ligands for galactose-binding AB toxin *Viscum album* agglutinin (VVA) and tumor-overexpressed lectins Gal-1, Gal-3, Gal-7, Gal-8, and Gal-9, consisting on the attachment of galactose, lactose, 2′-fucosyllactose, and 4-fucosylglucosamine residues on calix[*n*]arene (*n*=4, 6, or 8), pentaerythritol, or 1,3,5-benzene cores.

pentaerythritol cores (Fig. 7.7). Among the compounds tested as *in vitro* inhibitors, tetra- and hexavalent derivatives **53** and **54**, respectively, turned out to be the most active. While tetravalent **53** completely inhibited HIV *trans*-infection at 100 μM concentration, hexavalent **54** derivative only required 10 μM to perform the same result. The latter was also able to reduce DV serotype-2 infections by 85%. This result is remarkable since no clinical treatments are available for DV infection.

The inhibitory efficiency of glycocalixarenes **55–68** (Fig. 7.8) toward galactoside-binding AB toxin *Viscum album* agglutinin (VAA), but also lectins present in colon cancer such as galectins (Gal) 1, 3, and 4, was found to depend on the orientation of the sugars, the size of the ring, and the flexibility of the scaffold, apart from the valency [54]. This extensive study was developed from divalent to octavalent galactosyl- and lactosylthioureido calix[*n*]arenes ($n=4$, 6, and 8). In addition to finding decreases of IC_{50} values up to 300-fold relative to free sugar, selectivity depending on the lectin employed was also observed. In most cases, the highest inhibition potencies were found toward Gal-4, followed by VAA and Gal-3, whereas Gal-1 gave the poorest results. Interestingly, calix[6]- and calix[8] arenes being more flexible and dynamic gave the best inhibitory effects toward all the lectins tested, a result that contrasts with those reported by Consoli *et al.* on similar conjugates but using WGA as lectin [55]. In addition, tetravalent *cone*- and *alternated*-blocked calix[4]arene derivatives also showed sharp differences as inhibitors, despite of having the same valency. In general, *cone* conformation resulted in better inhibitory abilities except for the case of Gal-1, for which the *alternated* conjugate turned out to be the best ligand, whereas the *cone* analogue was even worse than the monovalent carbohydrate. This behavior for Gal-1 is similar to those results collected by Vidal and coworkers for PA-IL lectin using longer, more flexible linkers [46]. These trends are more or less maintained when the conjugates were tested in inhibition assays involving human tumor cell lines. Thus, this series of glycocalixarenes allows the possibility of targeting galectins differently: glycoconjugate **56** as blocker of Gal-3 binding to colon cancer cells, flexible **65** and **66** for Gal-4 binding to pancreatic carcinoma cells, and **68** as inhibitor of Gal-1 and -4. Selection between these three galectins could be further enhanced by replacing lactose residues with 3′-*O*-(*p*-methoxybenzyl)-LacNAc on *cone*-blocked calix[4]arene and calix[6]arene scaffolds. While the former showed a marked selectivity toward Gal-3, the latter displayed an improved efficiency toward Gal-4 [56].

Roy and coworkers synthesized tri- and tetravalent lactosyl clusters based on 1,3,5-benzene and pentaerythritol cores (**69–71**), respectively, by Sonogashira reaction (Fig. 7.8) [57]. Their inhibitory abilities were tested toward dimeric Gal-1 and Gal-7, monomeric Gal-3, the β-trefoil mistletoe lectin VAA, and a natural β-galactoside-specific immunoglobulin G fraction from human serum. Whereas Gal-1 and Gal-7 were not affected by these glycoclusters, Gal-3 was effectively inhibited by the trivalent derivative **69** ($IC_{50}=30.8\,\mu M$, relative potency = 7.6 per residue; IC_{50}(lactose) = 700 μM). Weaker effect was shown by the two tetravalent analogues **70** and **71** ($IC_{50}=125\,\mu M$ for *meta* and 62.3 μM for *para*). In contrast, these tetravalent clusters exhibited the strongest inhibition in the case of VAA ($IC_{50}=40\,\mu M$, relative potency = 12.5 per residue for both compounds; IC_{50}(lactose) = 2000 μM). *meta*-Derivative **70** was also a good inhibitor for the natural immunoglobulin G fraction ($IC_{50}=199\,\mu M$, relative potency = 3.8 per residue; IC_{50}(lactose) = 3000 μM).

Murphy and coworkers reported a series of tri- and tetravalent lactosyl clusters based on 1,3,5-benzene and tris(3-aminopropyl)amine, as well as a trivalent 2′-fucosyllactosyl cluster based on 1,3,5-benzene [58]. Their binding and inhibition abilities were tested toward a series of different galactoside-binding proteins. Best results were obtained by trivalent fucosylated and tetravalent lactosylated compounds (**72** and **73**) toward chicken galectin CG-3 (63- and 42-fold enhancement per sugar residue when compared to lactose). Both derivatives also bound chicken galectin CG-8 with values between 10 and 19 times higher than lactose.

Trivalent lactosyl conjugates **74** and **75** demonstrated enhanced inhibitory ability toward human Gal-1 by hemagglutination assay on type O red blood cells, with a relative potency per lactose moiety around 13. However, no cluster effect was observed when monomeric Gal-3 was used, with a relative potency per lactose residue of 1.1 [59,60]. Inhibition of adhesion of these and other lectins (VAA, Gal-8, and Gal-9) to human tumor cells (T-cell leukemia and a colon adenocarcinoma) assays, however, showed a remarkable activity of the lactosyl cluster **74** toward Gal-3, Gal-8, and Gal-9, but not to Gal-1 [61].

7.3 CARBOHYDRATE CLUSTERS AS SYNTHETIC VACCINES

An approach to develop synthetic vaccines involved the preparation of multivalent antigenic carbohydrates. Such multivalent display is crucial to cross-link with immunoglobulin receptors to generate anticarbohydrate antibody responses [62].

A group of tumor-associated antigens of carbohydrate nature have been identified and characterized by virtue of their reactivity with antibodies and lectins. Some of these so-called tumor-associated carbohydrate antigens (TACAs) are showed in Figure 7.9 [63]. It should be pointed that Tn, STn, and TF antigens are in fact glycosides of serine (Ser) or threonine (Thr), although only their saccharide fractions are

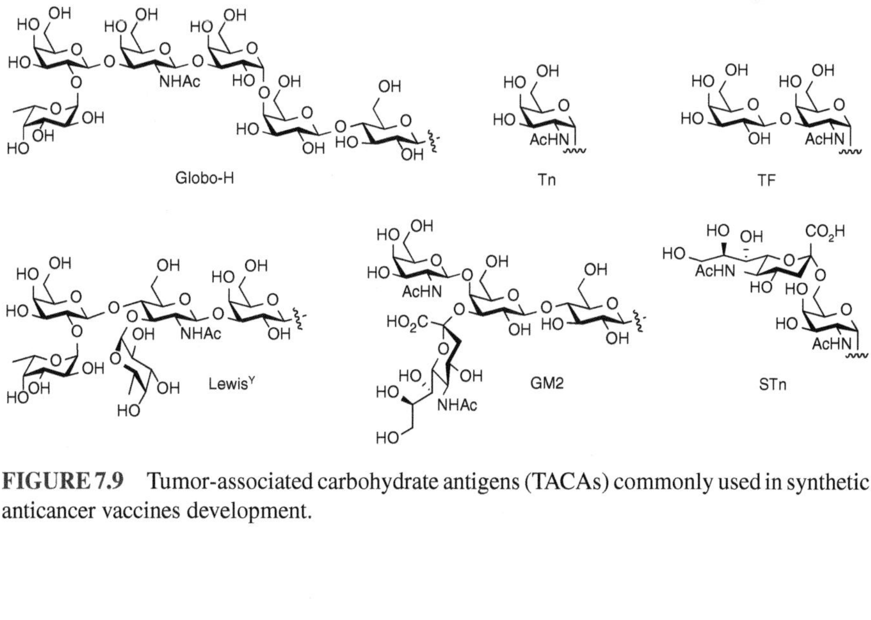

FIGURE 7.9 Tumor-associated carbohydrate antigens (TACAs) commonly used in synthetic anticancer vaccines development.

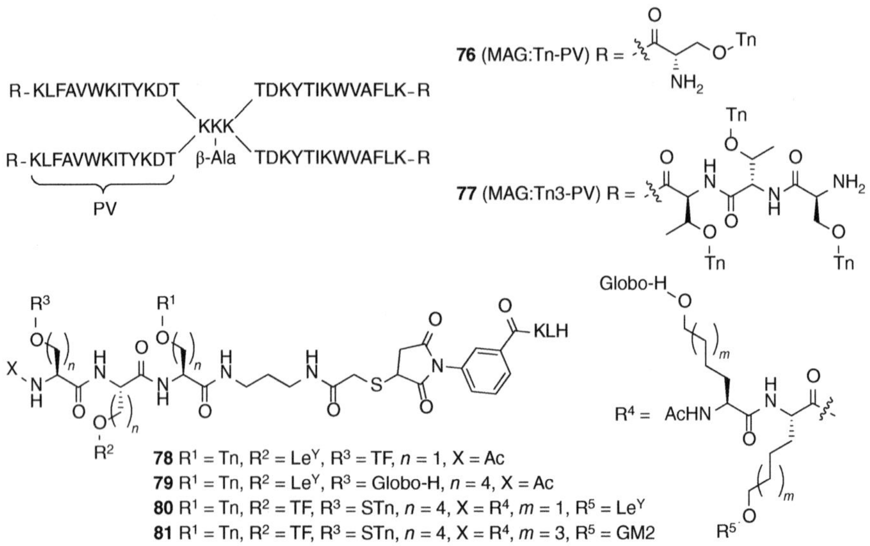

FIGURE 7.10 Anticancer vaccines based on the appendage of one or several TACAs on branched (**76** and **77**) or lineal (**78–81**) oligopeptides, using the PV oligopeptide from the poliovirus type 1 or keyhole limpet hemocyanin (KLH) carrier protein, respectively.

depicted for practical purposes. In a pioneer work, Leclerc and coworkers designed tetrameric multiple antigenic glycopeptide (MAG) **76** as a potential anticancer vaccine, consisting of a lysine tripeptide core with four arms linked to the PV oligopeptide from the poliovirus type 1 (Fig. 7.10) [64]. Such MAG is bearing four tumor Tn epitope units (α-GalNAc-*O*-Ser/Thr) appended at the amino terminus (MAG:Tn-PV). ELISA experiments led to the conclusion that the Tn saccharidic epitope increased T-cell stimulation by 10,000-fold when compared to the Tn non-bearing analogue. Immunization of mice with MAG:Tn-PV in the presence of the mild and harmless adjuvant aluminum hydroxide induced the production of T-cell dependent IgG anti-Tn antibodies capable of recognizing native Tn antigen present in human Jurkat T-lymphoma and LS180 adenocarcinoma cell lines. Dodecavalent analogue (MAG:Tn3-PV) **77** elicited a much more potent and persistent response than that described for MAG:Tn-PV and increased mice survival against TA3/Ha adenocarcinoma cells to 70–90% depending on the dose [65].

Danishefsky and coworkers developed a general strategy for the preparation of fully synthetic multiantigenic carbohydrate-based cancer vaccines consisting of oligopeptides presenting three (**78** and **79**) [66] or five (**80** and **81**) [67,68] different representative members of the four main classes of TACAs conjugated to the carrier protein KLH (Fig. 7.10). Using QS-21 as adjuvant, vaccines **78**, **79**, and **81** induced the production of titers corresponding to all of the carbohydrate antigens showed in their structure, while derivative **80** was not successful in generating an immunoresponse against one of the antigens (Lewis^{Y}) [69]. Moreover, fluorescence-activated cell sorter (FACS) showed the induced antibodies to react strongly with cancer cell lines expressing those antigens. All four constructions **78–81** revealed notably superior responses than those given by the pooled combination of the corresponding monovalent vaccines, except for Tn mucin. Vaccine **81** is currently under phase I clinical trial in ovarian cancer patients.

Leclerc's and Bay's groups [70] have developed an enzymatic method for the *in vitro* glycosylation of human mucin MUC6 from the MCF7 breast cancer cell line with different densities of the tumor epitope Tn. MUC6 protein is naturally expressed in gastric tissues, but it has been detected aberrantly glycosylated in several types of tumors such as intestinal, pulmonary, colorectal, and breast carcinomas. Although this method afforded different densities of glycosylation varying from 14 to 54 α-GalNAc units per protein molecule depending on the enzyme source and the protein acceptor (a major drawback of this vaccine preparation strategy), all protein glycoconjugates were recognized by two anti-Tn monoclonal antibodies specific to human cancer cells. Sera from mice immunized with these glycoclusters showed IgG antibodies that recognized the Jurkat human tumor cell line.

Dumy and coworkers have developed an anticancer vaccine designed as a Tn antigen cluster regioselectively appended on two different 10-residues addressable functionalized templates (RAFTs), backbone-cyclized decapeptides that present a stable β-sheet conformation in solution [71]. These templates served as scaffolds for the attachment of four Tn epitopes, as well as one (**82**) or two (**83**) poliovirus (PV) peptides (Fig. 7.11). Resulting glycoclusters were recognizable by two anti-Tn monoclonal antibodies. Sera from mice immunized with these two glycoclusters showed similar levels of Tn-specific IgG antibodies, indicating that the second PV epitope does not induce any immune response enhancement. Flow cytometry studies confirmed that elicited IgG antibodies were able to bind to the human Jurkat tumor cell line expressing the Tn antigen. As a further extension of this strategy, a four Tn epitopes containing RAFT was appended with a linear CD8+ T-cell peptide from ovalbumin ($OVA_{257-264}$), a CD4+ T-cell peptide (PADRE), and a palmitic acid moiety (PAM) as built-in lipid adjuvant that induces DC maturation. Resulting ovalbumin glyco-lipopeptide **84** (OVA-GLP) elicited significant levels of RAFT-specific antibodies in sera from mice even in the absence of any external adjuvant. Furthermore, 90% OVA-GLP-immunized mice rejected mouse melanoma cells (B16) expressing ovalbumin, and 90% nonimmunized, B16-inoculated mice evidenced tumor growth delay when treated with OVA-GLP [72,73]. Very recently, two saccharide epitopes, Tn and TF, were simultaneously combined on the RAFT platform in five different ratios (from 4:0 to 0:4); however, the immunological ability of these glycoclusters has not been evaluated yet [74].

Given the sensitivity of TACAs to the action of glycosidases, the use of modified TACAs (such as C- or S-glycosides) or glycosidase-resistant mimics was suggested. In this context, BenMohammed and coworkers prepared the new RAFT-based tetravalent glycocluster **85** presenting four Tn antigen mimetic MIM_Tn residues and the immunostimulant peptide epitope OvaPADRE [75]. Sera from immunized mice with this conjugate using CpG_{1826} as adjuvant showed significant

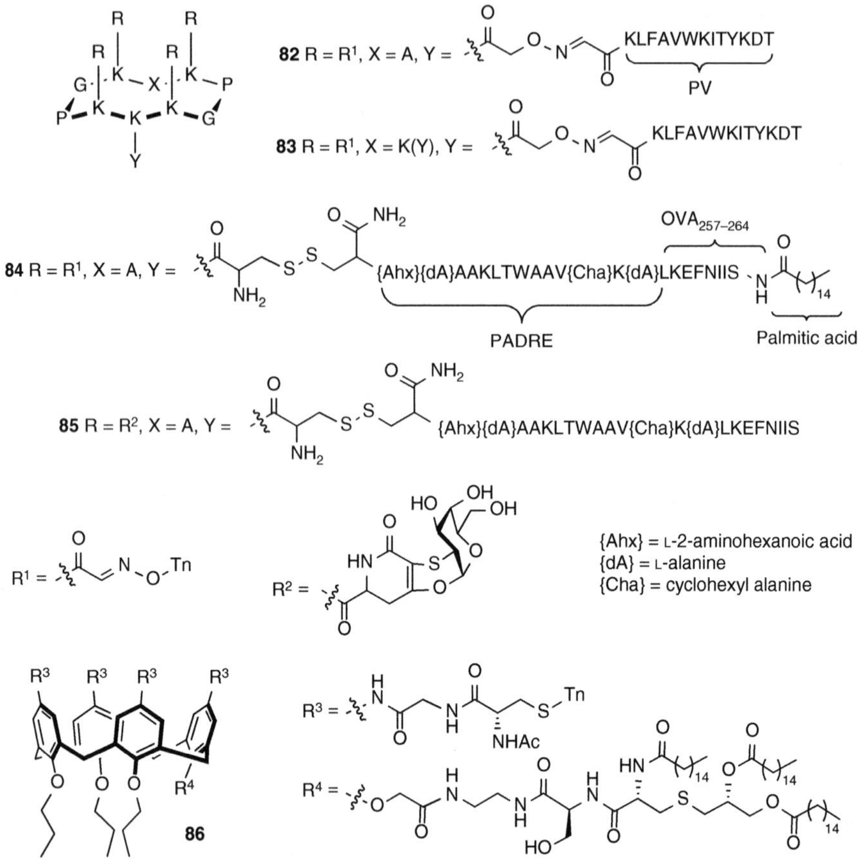

FIGURE 7.11 Anticancer vaccines containing Tn and Tn antigen mimetic MIM_Tn clusters grafted on 10-residue addressable functionalized template (RAFT) (**82–85**) and calix[4]arene (**86**) as scaffolds.

levels of mucin-specific IgG/IgM antibodies that bound Tn-expressing cancer cell lines both from mouse (NT2 and TA3HA) and human (MCF7). Furthermore, immunized mice showed a survival of 70% upon tumor implantation after 2 months through a B cells-mediated mechanism.

Spadaro and coworkers [76] synthesized *cone*-blocked calix[4]arene glycoconjugate **86** bearing four S-Tn (α-GalNAc-S-Ser) antigen residues and an immunoadjuvant moiety, tripalmitoyl-S-glycerylcysteinylserine (P_3CS) at the opposite rim of the macrocycle (Fig. 7.11). Sera from mice immunized with this structure showed the presence of an increased concentration of Tn-specific IgG antibodies, 4 times higher than nonimmunized mice.

The use of carbohydrate clusters as vaccines against HIV has also been investigated. As the major carbohydrate present on the surface of the HIV envelope spike protein gp120, glycan $Man_9GlcNAc_2$ is expected to stimulate the production of

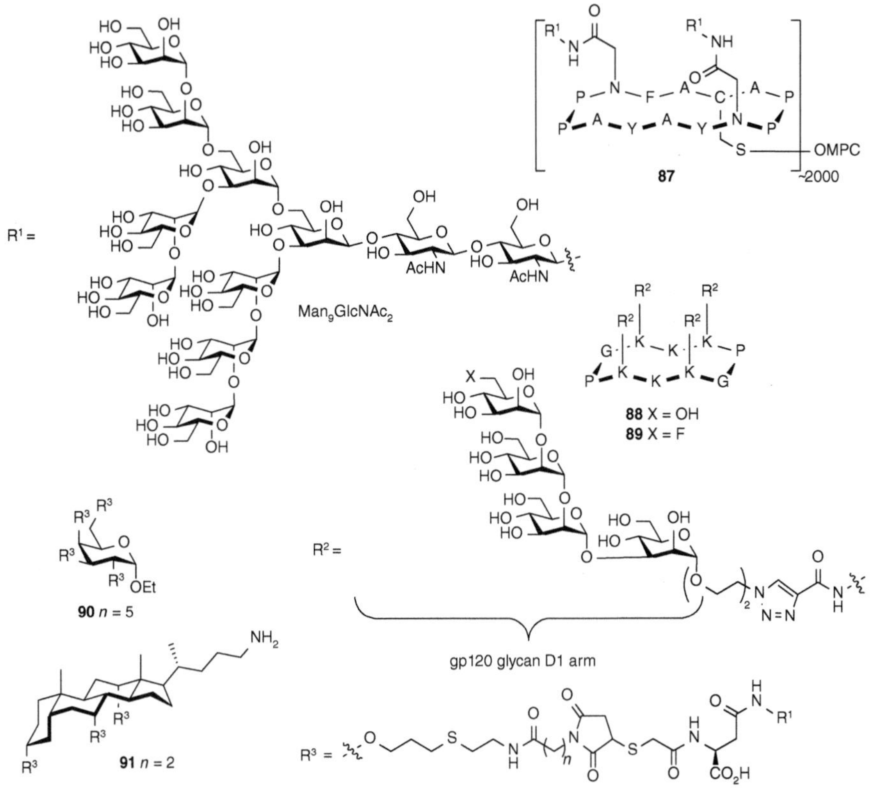

FIGURE 7.12 Carbohydrate-based vaccines against HIV designed as clusters of gp120 glycan ($Man_9GlcNAc_2$) or its structurally simpler D1 tetramannoside arm grafted on different scaffolds such as 14- (**87**) and 10-RAFTs (**88** and **89**), galactose (**90**), and cholic acid (**91**).

2G12 antibody, rarely excreted in human HIV immunological response and capable of neutralizing a broad range of HIV strains. Danishefsky and coworkers synthesized a fully synthetic HIV antigen **87** consisting of two copies of high-mannose glycan $Man_9GlcNAc_2$ assembled on a β-sheet-shaped 14-residues cyclopeptide (Fig. 7.12) [77]. Glycan–cyclopeptide conjugate was covalently coupled onto the outer membrane protein complex (OMPC) from *Neisseria meningitides*, a highly effective immunostimulatory carrier protein. This conjugation was estimated to give ~2000 copies of the glycopeptide per OMPC unit. The mimetic-OMPC conjugate showed a clear response, although it was not as robust as that of the positive control, gp160. Further studies presented evidence, from both direct and competitive binding assays, that there was no significant recognition of the glycopeptides, although certain sera did contain antibodies that could compete with 2G12 for binding to recombinant gp120 [78]. Similar derivatives were prepared by Wang and coworkers but with substantial differences in the structure and valency of the oligosaccharide [79] that included the fluorination at 6-position of the terminal mannose residue of gp120 glycan D1 tetramannoside arm in order to enhance the immunogenicity of the saccharide **89**. But the estimation of the binding abilities of this cluster to human 2G12 human antibody led to higher values for a nonfluorinated analogue **88** used as control.

Wang's group had already explored the use of galactose-centered and cholic acid-scaffolded analogues containing the whole $Man_9GlcNAc_2$ glycan (Fig. 7.12) [80,81]. They prepared a compound with four $Man_9GlcNAc_2$ moieties appended on positions 2, 3, 4, and 6 of the galactose core (**90**), as well as a cholate derivative having three glycan moieties at the cholate positions 3α, 7α, and 12α (**91**). The inhibition studies for these tetravalent and trivalent glycoclusters of 2G12 binding to gp120 showed IC_{50} values 73-fold and 46-fold higher than the monomeric $Man_9GlcNAc_2Asn$, respectively, but their relative affinities, in the μM level, were much lower than the affinity of 2G12 for gp120 in the nM level. When the galactose-centered glycocluster **90** was grafted with a T-helper epitope from tetanus toxoid or the carrier protein KLH, it was found that sera from immunized rabbits with such glycoclusters showed higher IgG type antibody titers than those immunized with a monovalent analogue. However, the majority of antibody responses were directed to the linkers [82].

As a further step in developing an anti-HIV vaccine, a number of multivalent 33-mer glycopeptides containing 3–5 copies of Man_9 epitope ($Man_9GlcNAc$) for 2G12 anti-HIV antibody were prepared by a directed evolution-based strategy (SELMA), in which a DNA backbone evolves to optimally cluster the epitope glycans [83]. Multiple mRNA oligopeptides directly linked to their encoding mRNA strands and showing a random distribution of 1–6 homopropargylglycine moieties were prepared by ribosomal translation from a DNA library. Propargylated glycine units were then grafted with the 2G12 Man_9 epitope and tested as 2G12 ligands. Those showing highest binding abilities were reverse transcribed to cDNA, from which a second-generation DNA library was generated. Iterative repetition of this process led to the identification of the 10 highest-affinity conjugates for the target 2G12 antibody, which were finally prepared without the mRNA tags. These glycopeptides showed tight binding abilities toward 2G12 antibody (K_D = 0.5–5 nM) and inhibitory properties toward the interaction of 2G12 with gp120 protein. These results represent an enhancement of up to 360,000-fold as compared to monovalent Man_9 glycan.

Dendritic cells (DCs), well-known T-lymphocyte stimulators, accomplish antigen uptake through a mechanism, among others, that involves the association with a surface mannose-binding C-lectin named DC-SIGN. As a consequence, mannosylation of antigens has been proposed as a strategy to improve the efficacy of vaccines as it would promote their internalization and recognition by the immune system. Following this concept, Grandjean *et al.* decorated the $HA_{307–319}$ peptidic epitope from influenza hemagglutinin with an octavalent mannose cluster based on a branched Lys-Ala oligopeptide as scaffold (Fig. 7.13) [84]. Human DCs from peripheral blood proved to specifically uptake fluorescent-labeled analogues of this mannocluster. The influence of the glycocluster valence was also studied by varying the number of epitopes from 2 to 8. Interestingly, highest uptake was achieved by

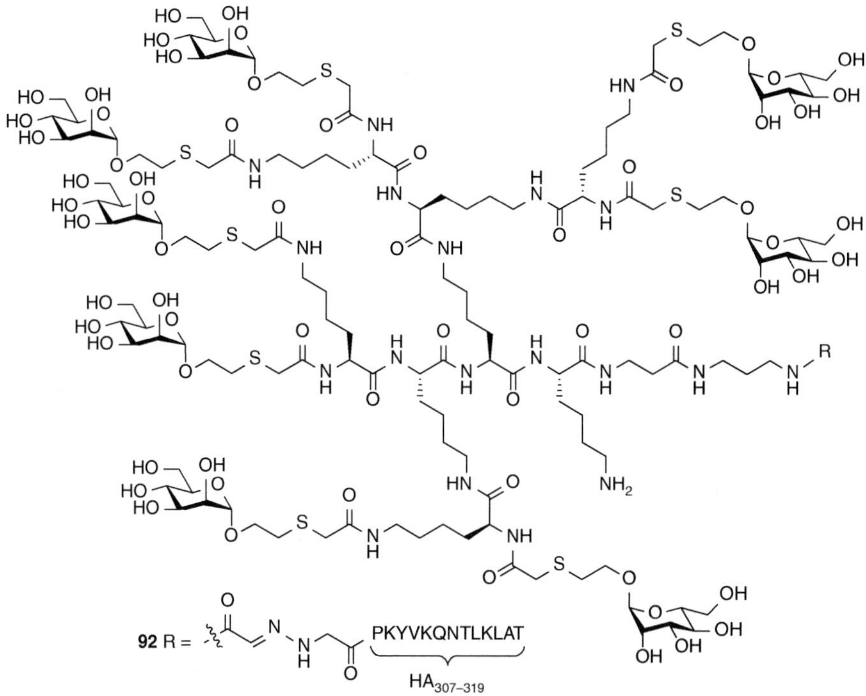

FIGURE 7.13 Mannosylated $HA_{307\text{-}319}$ peptidic epitope from influenza virus hemagglutinin.

the tetravalent glycomimetics (quinic and shikimic acids, as mannose mimetics) and the octavalent mannocluster **92** [85].

Berti and coworkers have recently reported multivalent synthetic vaccines against *Candida albicans* fungi based on a carrier protein from diphtheria toxin (CRM_{197}). Such conjugate presents a controlled number of β-(1→3) glucan hexasaccharide moieties linked through some of its tyrosine or lysine residues (Fig. 7.14) [86]. Modification of tyrosine residues was achieved in an average of 4 out of the 18 available linking points when a monovalent linker was used (**93**), whereas 6 of them were modified when a double spacer was employed (**94**). Attachment to lysine residues yielded a hexavalent glycoprotein (**95**). These glycoconjugates were expected to induce the production of antilaminarin antibodies. Laminarin is a natural β-(1→3)-β-(1→6) glucan present on the *C. albicans* surface, which mediates fungal adhesion to human epithelial cells. Sera from immunized CD1 mice using MF59 as coadjuvant showed high levels of IgG antilaminarin antibodies by ELISA. Neoglycoproteins **93** and **94** having carbohydrate derivatives linked to the tyrosine residues induced higher IgG levels, indicating that the attachment site of the glycan moiety was important. Sera were positively tested as inhibitors for the adhesion of *C. albicans* to human epithelial (HEP-2) cells. Interestingly, lysine-modified conjugate **95** (62% inhibition) proved to be as effective as the highly protective, anti β-glucan murine monoclonal antibody mAb 2G831 used as positive control.

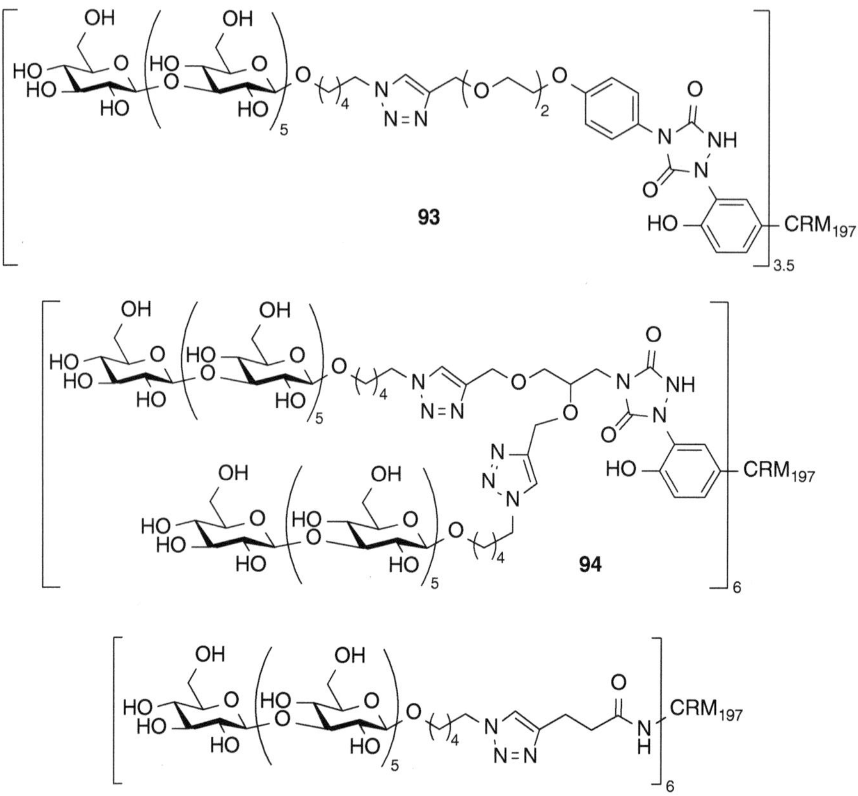

95

FIGURE 7.14 Synthetic vaccines **93–95** against *Candida albicans* based on the attachment of a controlled number of β-(1 → 3) glucan hexasaccharide moieties to tyrosine or lysine residues of CRM_{197} carrier protein from diphtheria toxin.

7.4 CARBOHYDRATE CLUSTERS AS DRUG DELIVERY SYSTEMS

Targeted drug delivery systems are engineered technologies for delivering therapeutic agents in a manner that increases their relative concentration in specific biological sites while reducing it in the remaining tissues. Synthetic glycoclusters that are able to carry therapeutics in a covalent and/or noncovalent fashion can be used as delivery systems directed to the specific carbohydrate moieties receptors. Thus, endogenous lectins are attractive targets for drug delivery, and multivalency can favorably increase the avidity of their interaction with the targeting ligand.

Several types of glycoclusters have been explored for targeted delivery of genes to the specific cells and/or for augmenting gene transfer efficiency. Aoyama and coworkers [87] proposed the use of octavalent lactose, maltose, and cellobiose calix[4] resorcarene-scaffolded glycoclusters **96–98** (Fig. 7.15) to create "artificial viruses" called "glycoviruses." These glycoconjugates self-aggregate to form spherical and

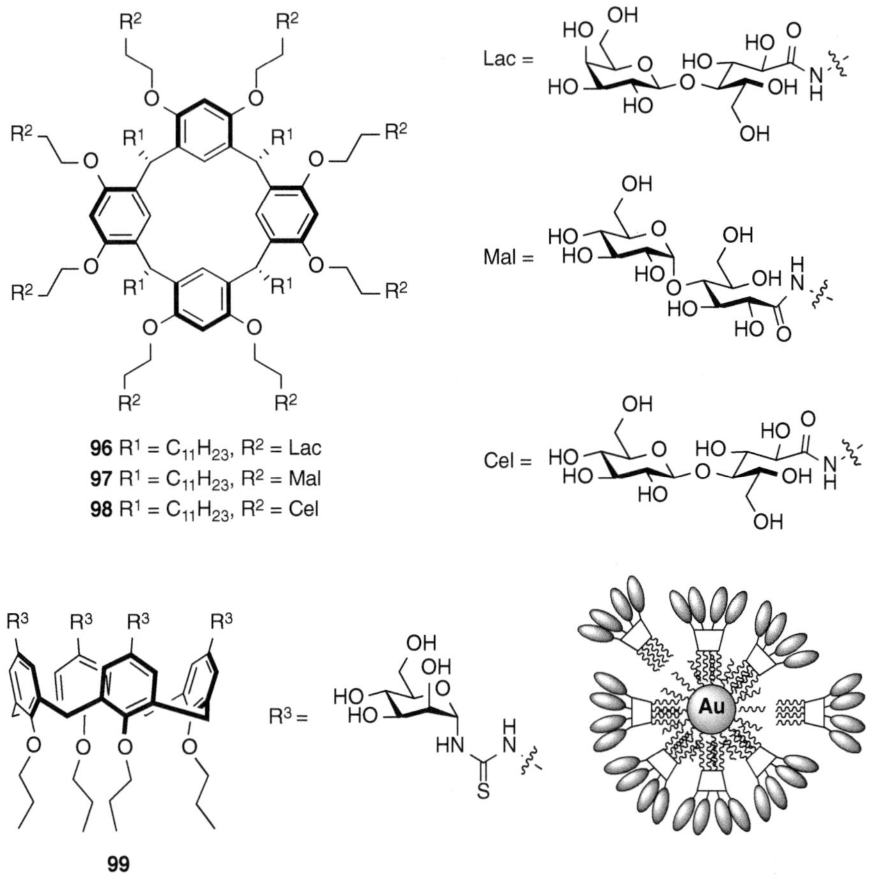

FIGURE 7.15 Artificial glycoviruses based on the supramolecular aggregation of calix[4] resorcarene-scaffolded glycoclusters **96–98** as DNA delivery systems. Amphiphilic glycocalix[4]arene **99** and its noncovalently self-assembly on *n*-dodecanyl AuNPs as a potential drug delivery system.

neutral nanoparticles (NPs), which are strongly adhesive on DNA aligning 2 glycoNPs along the major groove in every successive pitch, located at north and south or east and west pitch by pitch, resulting in a neutral glycoviral particle (GP). Such particle contains a single molecule of DNA efficiently packed in a viral size ($d \approx 50$ nm). GPs derived from **98** are nonadhesive with each other and keep monomolecularity, whereas those derived from **96** and **97** undergo reversible saccharide-dependent self-aggregation resulting in GPs aggregates of big size ($d \approx 200$ nm for **96** and ≈300 nm for **97**). Only monomeric GPs **98** possess substantial transfection ability in HeLa cells, where the activities are size regulated. However, when hepatic HepG2 cells, which possess specific receptors for β-galactose residues, were used, the activity of the GPs from **96** increased to reach a similar value than that for **98** due to size/receptor compensation, while activities of GPs **97** and **98** kept unchanged.

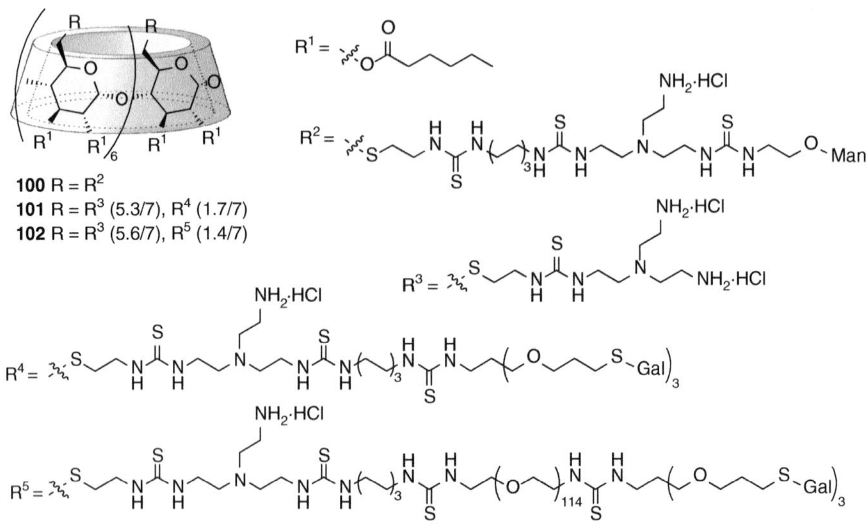

FIGURE 7.16 Polycationic amphiphilic β-CD-based mannosyl (**100**) and galactosyl (**101** and **102**) clusters as pDNA and mRNA nanocarriers targeting macrophage mannose (MMR) and asialoglycoprotein (ASGPr) receptors, respectively.

The ability of amphiphilic glycoclusters to aggregate in water has also been exploited for the preparation of supramolecular gold glyconanoparticles (glycoAuNPs). Avvakumova *et al.* have recently reported the noncovalent self-assembly of the amphiphilic *cone*-glycocalixarene **99** bearing four mannose residues on the surface of *n*-dodecanyl-decorated AuNPs [88]. The resulting glycocalixarene-AuNP was three-fold more efficiently uptaken by HeLa cells than the glycoAuNP that resulted from the assembly of a mannose–benzene conjugate on the *n*-dodecanyl-decorated AuNPs. This result indicated the relevance of using clustered multivalent glycosides on the AuNPs.

A polycationic amphiphilic β-cyclodextrin (β-CD)-based mannosyl cluster **100** bearing seven protonatable amino groups on the primary face was tested as a site-specific gene delivery system (Fig. 7.16) [89]. This new glycocluster was expected to combine the enhancing, although nonspecific, effect on the cellular uptake and transfection efficiency promoted by the overall positive surface electrostatic potential with the specific and selective recognition by cell membrane lectin receptors provided by the attached carbohydrates. Such conjugate forms stable cationic snake-like glyco-β-CD complexes (glyco-β-CDplexes) in the presence of pDNA. These glyco-β-CDplexes were recognizable by ConA and human macrophage mannose receptor (MMR) lectins. Interaction studies on resident peritoneal macrophages from mice indicated that adhesion of the glyco-β-CDplexes can operate in a nonspecific, electrostatic-mediated fashion or in a specific, lectin-conducted mode. MMR-mediated transfection of mouse leukaemic monocyte macrophage cells using a luciferase gene was achieved more efficiently than the reference commercial compound JetPEI™-Hepatocyte. In a further step, triantennary galactose dendrons

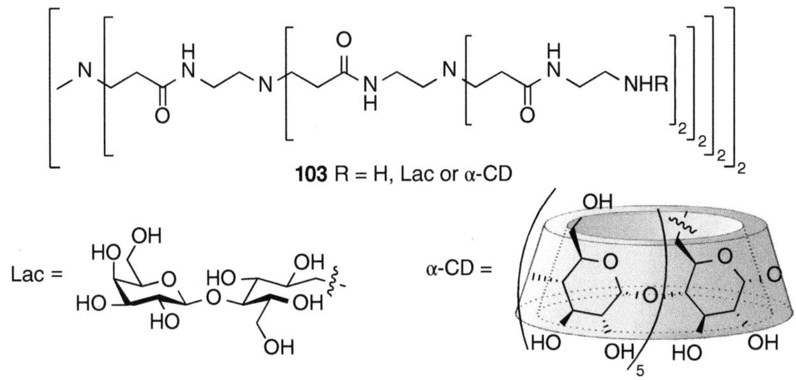

FIGURE 7.17 ASGPr-mediated gene delivery system in HepG2 cells using G2-PAMAM dendrimer as core.

were randomly attached on such amphiphilic β-CD (**101**) [90]. A PEG_{5000}-glycodendron was also used for the functionalization of the β-CD derivative leading to **102** in order to provide stabilization against serum proteins. Interestingly, these conjugates were able to form complexes with pDNA, but their transfecting abilities toward HepG2 cells were quite modest probably due to the difficulties in translocation of the pDNA into the nucleus. In contrast, very efficient asialoglycoprotein receptor (ASGPr)-mediated transfection was found when mRNA (which is translated in the cytoplasm) was used. Small changes in the arrangement of the cationic elements of such amphiphilic β-CDs have a strong influence on their gene transfer capabilities [91].

A second-generation (G2) PAMAM dendrimer functionalized with a random number of lactose and primary rim-grafted α-CD residues (Fig. 7.17) was tested as hepatocyte-selective nonviral vector for gene delivery in HepG2 cells and mice by measuring Renilla luciferase activity after transfection of pDNA [92]. Different degrees of substitution of the lactose moiety (1.2, 2.6, 4.6, 6.2, and 10.2) were tested. *In vitro* experiments pointed an optimal degree of substitution of 2.6 for the transfection, with low cytotoxicity and excellent hepatocyte selectivity through the interaction with ASGPr. In contrast, A549 and KB cells did not show any activity. The role played by the cyclooligosaccharide in the transfection process is assumed to enhance endosomal escape of the pDNA complex through the membrane disruption. In mice, luciferase activity mainly cumulated in the liver. Better results were found when a G3 analogue with a degree of substitution of the lactose moiety of 1.2 was used, although only *in vitro* experiments were performed with this multivalent system [93].

The use of a CD as scaffold of cluster glycosides offers the advantage of providing to the glycocluster the ability to form inclusion complexes with a large variety of drugs in aqueous solution. Thus, the combination of the ability to encapsulate drugs and being biorecognizable has converted such glycoclusters in the focus of attention of an extensive research [11,18]. The group of García-Fernández synthesized β-CD-based cluster mannosides **104–109** (Fig. 7.18) in which only one of the seven

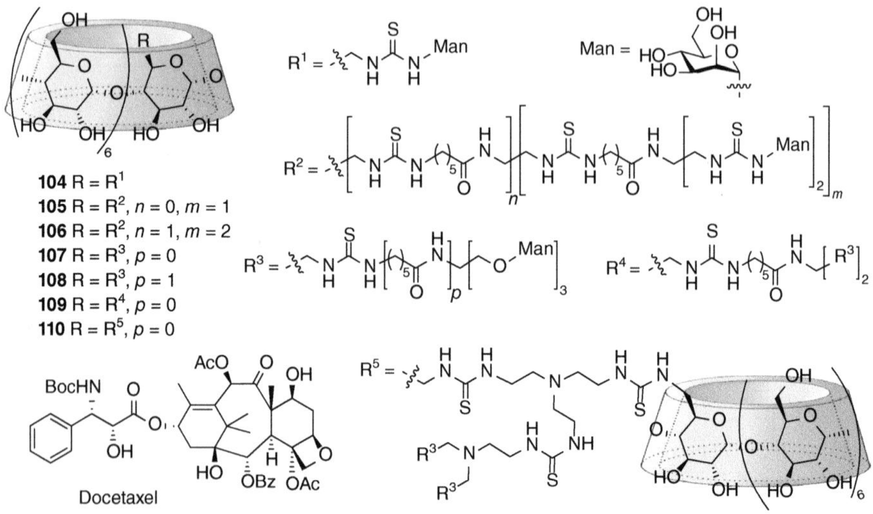

FIGURE 7.18 Primary face monosubstituted β-CD derivatives **104–109** and β-CD dimer **110** as mammalian mannose/fucose-specific cell surface receptor targeted delivery systems for water-insoluble anticancer drug docetaxel.

α-D-glucose moieties of the macrocycle was modified [94,95]. The solubilization experiments in water with anticancer drug docetaxel showed similar values to those obtained previously for monovalent β-CDs and higher than those for heptavalent per-6-substituted β-CD analogues. The IC_{50} values obtained by ELLA assays for Con A affinity showed the expected enhancement of lectin-binding strength for the higher-valent β-CD (up to 22-fold for hexavalent to monovalent derivatives) due to the cluster effect. In addition, results showed that the Con A affinity for **109**-docetaxel complex is twofold higher than that for the free host **109**. Such increase in Con A affinity was attributed to the interaction of a dodecavalent 2:1 **109**-docetaxel complex with the lectin. In a further step, the authors synthesized the *bis*-β-CD-glycocluster **110**. While the binding affinity of **109** and **110** toward Con A had similar IC_{50} values (ca. 10 μM for both), estimated association constant (K_c) values from solubility diagrams for glycocluster **109** and *bis*-β-CD-glycocluster **110** were 4000 and $1.5 \times 10^5\,M^{-1}$, respectively. The drug targeting potential of glycoclusters-β-CD **109** and **110** was then evaluated with a mammalian mannose/fucose-specific cell surface receptor from macrophages obtaining similar results to those obtained for the affinity of the docetaxel complexes to Con A.

Persubstitution of the β-CD at C-6 has been used as a very convenient way to synthesize glycoclusters (Fig. 7.19). Pioneer reports [96,97] described the synthesis of several per-6-glycosyl-thio-β-CDs having α/β Glc and Gal (**111–115**). The biorecognizable properties of such per-glyco-β-CDs were studied by testing their inhibitory power toward specific lectins (*Kluyveromyces bulgaricus*). Results showed that the heptavalent glycoclusters displayed better inhibitory properties than the monovalent analogues. Per-glucosyl-β-CD **114** formed stable complexes with *p*-nitrophenol and

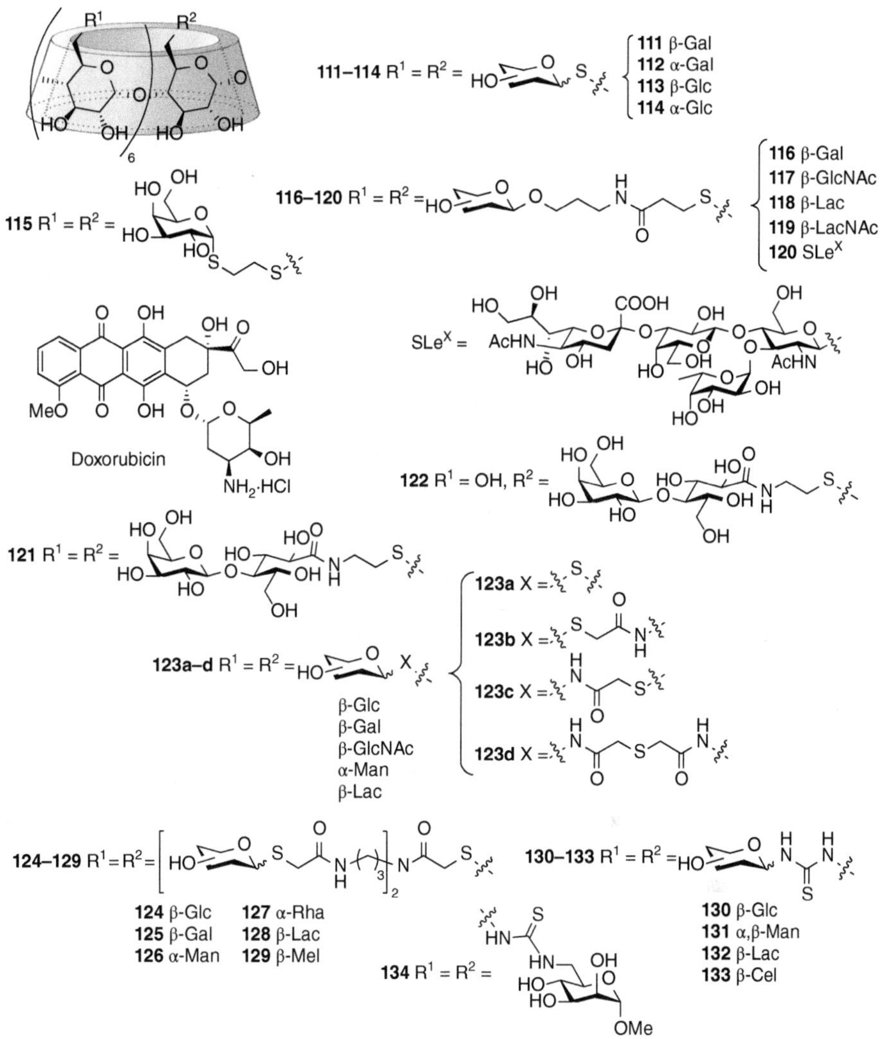

FIGURE 7.19 β-CD glycoclusters persubstituted on primary face (**111–121** and **123–134**).

prednisolone (K_a 2×10^3 and 0.2×10^3 M^{-1}, respectively) but less stable than those formed with the native β-CD.

A similar synthetic strategy was applied for the synthesis of heptavalent persubstituted β-CDs in which the carbohydrate ligands are bound to C-6 through spacer arms of different length and nature. Furuike *et al.* [98] prepared glycoclusters **116–119** with Gal, *N*-GlcNAc, lactose, and *N*-acetyllactosamine. Lectin binding studies with these glyco-β-CDs were performed by hemagglutination inhibition test of the wheat germ agglutinin (WGA) and *Erythrina corallodendron* lectin (ECorL), GlcNAc- and *N*-acetyllactosamine-specific lectins, respectively. The inhibitory ability for glycoCDs **119** and **117** was 4- and 40-fold higher, respectively, than the monovalent structures. Factors on the lectin binding abilities such as the carbohydrate epitope density and topology were also studied by measurements of the inhibition activities of glyco-β-CDs **116–119** against galactoside-specific agglutinin from *V. album* L lectin; Gal-1, Gal-3, and Gal-7; and lactoside-binding immunoglobulin G [99]. Authors found for heptavalent glyco-β-CDs notable inhibitory potencies per galactosyl residue relative to free lactose, as well as galectin binding selectivity.

A chemoenzymatic approach was carried out by Furuike *et al.* for an efficient synthesis of glyco-β-CDs having seven sialyl Lewis X units [100]. SPR inhibitory experiments performed using immobilized SLeX-containing neoglycoprotein (SLeXn-BSA) and E-selectin showed that glycocluster **120** exhibited a highly enhanced inhibitory ability (IC_{50} of 1.5 mM) relative to SLeX.

Yasuda *et al.* [101] have reported the synthesis of both per-6-galactosyl-glucono-amide-CD **121** and its monovalent analogue **122** (Fig. 7.19). The authors carried out SPR measurements that showed that both compounds form ternary complexes with PNA and the anticancer drug doxorubicin (DXR). The K_a value for the binding of **121** to PNA was two orders of magnitude higher than that for **122**. Interestingly, the persubstituted β-CD showed a K_a value for the binding to DXR of an order of magnitude higher than the monovalent β-CD, which was attributed to the formation of an inclusion complex in the cavity formed by the carbohydrate appendages.

A series of heptavalent and tetradecavalent β-CD-based *S*- and *N*-glycoclusters **123a–d** and **124–129**, having spacers of different length and nature, was synthesized (Fig. 7.19) [102–105]. ITC experiments gave the K_a values and the thermodynamic parameters for the inclusion complexation of such glycoclusters with sodium 8-anilino-1-naphthalensulfonate (ANS) and 2-naphthalenesulfonate (NS) as guest models. While persubstitution of the β-CD at C-6 with the appended carbohydrates led to a slight increase of the K_a values for ANS relative to β-CD, a higher steric congestion on the primary face resulted in lower K_a values for NS. These ITC studies also showed that the spacer arms that link the appended saccharide units to the β-CD core had an influence on the complex stability according to their chemical nature. In addition, an increase of the valency of the glyco-β-CDs from 7 to 14 ligands gave rise to low stable complexes. The binding properties of glyco-β-CDs **123a–d** and **124–129** toward several plants lectins were evaluated using inhibition experiments (ELLA) and ITC, among other techniques. *S*-Glyco-β-CDs, but not *N*-monosaccharide-β-CDs, showed good to excellent inhibition properties with the corresponding specific pea, PNA, and Con A lectins. PNA formed stable cross-linked complexes with Gal_7-β-CD **123b**, Lac_7-β-CDs **123a-d**, and Lac_{14}-β-CD **128** but not with the galactosylamine Gal_7-β-CDs **123c** and **123d**. The strength of the Gal_7-β-CDs·PNA complexes was attributed to the cluster effect. ITC results also showed that the use of 14-valent glyco-β-CDs did not contribute to the improvement of the binding complex stability relative to the heptavalent clusters.

García-Fernández, Defaye, and coworkers developed another successful strategy for the efficient attachment of carbohydrates on β-CD based on the formation of thiourea bridges (Fig. 7.19) [106,107]. Following this strategy, the synthesis of per-glycosyl-thioureido-β-CDs **130–134** has been performed in high yields.

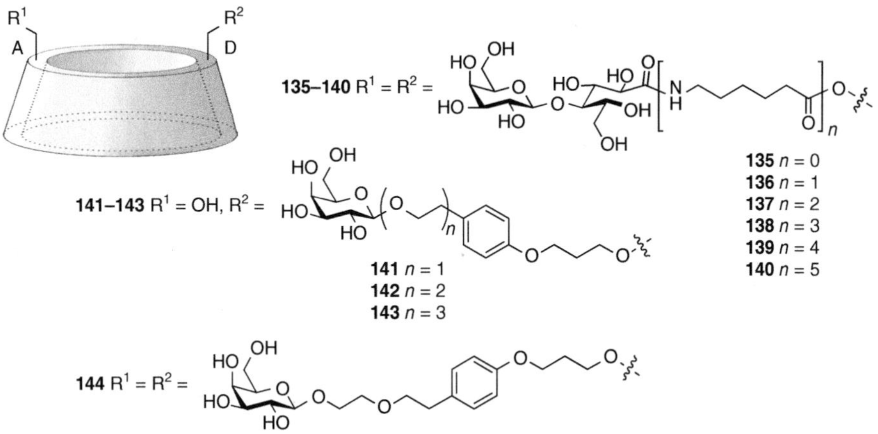

FIGURE 7.20 Mono- (**141–143**) and divalent (**135–140** and **144**) galactosylated β-CD derivatives as DXR delivery systems.

Docetaxel solubilization experiments in water showed solubility values higher than those obtained for the native β-CD but lower than those for the monovalent glyco-β-CD analogues. Unexpectedly, ELLA assays showed that heptavalent glyco-β-CDs **131** and **134** exhibited a much lower inhibitory ability than the monovalent β-CD analogues (IC_{50} = 3200 and 2900 μM, respectively) for the binding toward Con A.

As an alternative to the persubstituted glyco-β-CDs having seven or more carbohydrate units, divalent 6A,6D-galactose-β-CD derivatives having different spacer arm lengths were synthesized (Fig. 7.20) [108]. The formation of complexes with PNA lectin and the anticancer drug DXR was quantitatively evaluated by SPR and atomic force microscopy (AFM). Among these *bis*-Gal-β-CDs, **139** was found to have the optimum length for association with PNA lectin [K_a value for **139** (n = 4) $4.6 \times 10^6 M^{-1}$]. Their complex stability with the drug gradually increased along with the spacer lengths [K_a value for **140** (n = 5) $5.1 \times 10^4 M^{-1}$]. Similarly, Yamanoi and coworkers [109] prepared galactosylated *mono*- and *bis*-branched-β-CD conjugates (**141–143** and **144**, respectively) displaying *p*-phenyl ethylene glycol-based arms of different lengths between the carbohydrate moiety and the macrocycle. Both the drug encapsulation and the lectin-specific recognition abilities were evaluated by SPR experiments using immobilized DXR and PNA lectin, respectively. PNA binding K_a values were estimated to be $5.4-9.2 \times 10^4 M^{-1}$ for **141–143** and $1.3 \times 10^5 M^{-1}$ for **144**, evidencing no significant influence of the spacer length or the number of branches over the recognition of the lectin. The inclusion of the DXR into the cavity yielded K_i values between $1.1 \times 10^5 M^{-1}$ and $1.5 \times 10^5 M^{-1}$ for conjugates **141–143**, while derivative **144** showed 100-fold higher inclusion constant ($1.5 \times 10^7 M^{-1}$) due to the presence of two aromatic rings in the structure, which likely enhances π–π interactions with the drug.

Seeberger and coworkers [110] designed a galactosylated dendrimer attached to the primary face of a β-CD macrocycle (**145**) as potential targeted drug delivery system toward ASGPr, which is exclusively expressed on parenchymal hepatocytes

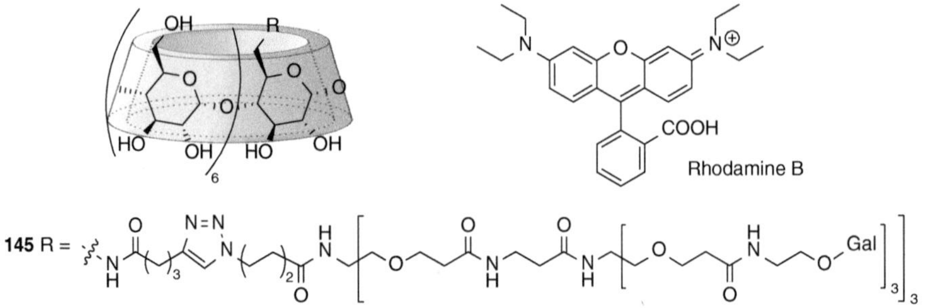

FIGURE 7.21 Nonavalent galactosylated β-CD conjugate **145** as a HepG2 hepatocyte-targeted delivery system for rhodamine B and anticancer drug DXR.

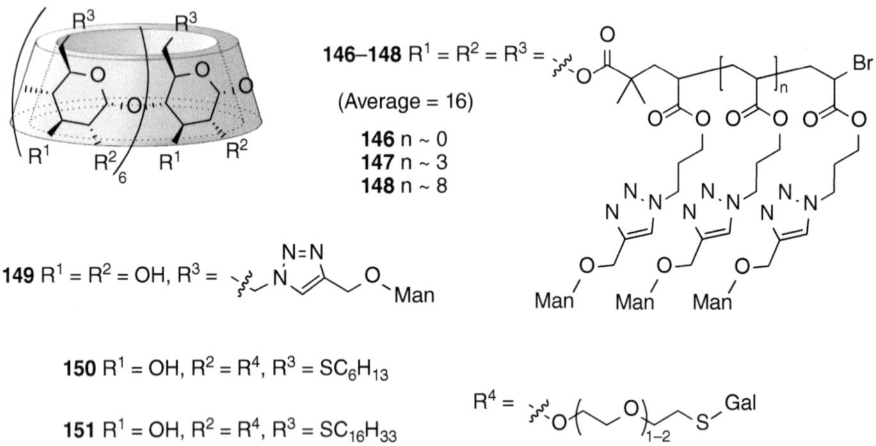

FIGURE 7.22 β-CD-based star mannosylated glycopolymers **146–148** as DC-SIGN-mediated delivery systems for anticancer agent 4-dihydroxyanthraquinone and anti-HIV drug Saquinavir, as well as amphiphilic β-CD derivatives **150** and **151** grafted in their secondary faces with galactose residues as nanoparticles- and vesicles-forming nanocarriers.

(Fig. 7.21). Evaluation of its delivery ability was performed by its uptake into hepatocellular carcinoma HepG2 cells using fluorescent dye rhodamine B as drug model. A mannosylated analogue was used as a negative control. Flow cytometry confirmed that galactosylated conjugate was efficiently uptaken by HepG2 cells indicating receptor-mediated endocytosis. The inclusion complex of this conjugate with anticancer drug DXR proved to induce apoptosis when incubated with HepG2, whereas the conjugate itself did not.

Haddleton, Bercer, and coworkers [111] prepared a series of β-CD-based star glycopolymers (Fig. 7.22) containing an average degree of substituted hydroxyl groups at the periphery of β-CD approximately equal to 16 and three different degrees of polymerization equal to 2, 5, and 10 (**146–148**). SPR technique was used to test binding abilities of these glycopolymers to DC-SIGN lectin. Affinity of these conjugates was much higher than the heptavalent β-CD analogue **149**. The conjugate

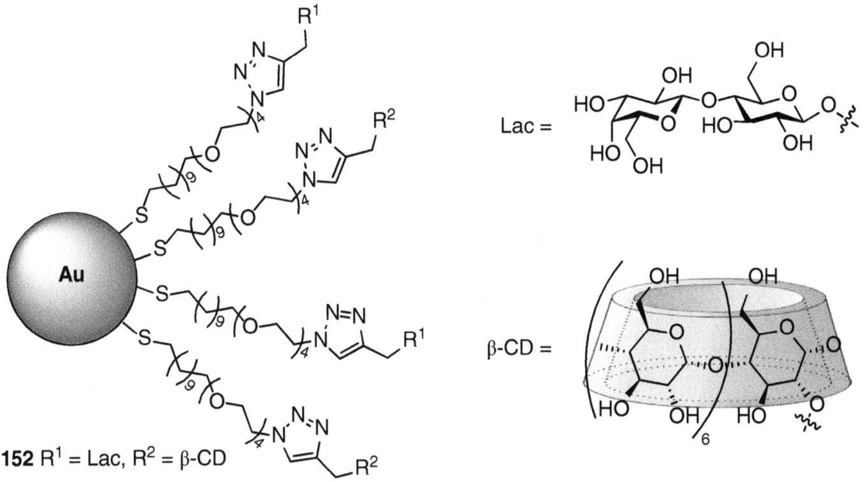

FIGURE 7.23 Dually functionalized gold nanoparticle **152** bearing simultaneously lactose residues as targeting ligands for Gal-3 lectin and β-CD macrocycles as encapsulating moieties for anticancer drug methotrexate (MTX).

showing a DP of 10 yielded a dissociation constant of 0.22 nM, only twofold higher than that described for the natural DC-SIGN ligand gp120. In addition, these glycopolymers were able to encapsulate 4-dihydroxyanthraquinone (DHA), a potential anticancer chemotherapeutic agent, as well as the anti-HIV drug Saquinavir as judged from UV–Vis and ^{1}H NMR spectroscopy, respectively.

Mazzaglia *et al.* [112] synthesized amphiphilic β-CDs **150** and **151** having alkylthio chains at the primary face and galactosylthio-oligo-(ethylene glycol) at the secondary rim (Fig. 7.22). These amphiphilic β-CDs formed NPs and vesicles, respectively, showing multivalent effect in their binding to a lectin. The binding of these amphiphilic β-CD colloids to galactose-specific lectins from *P. aeruginosa* and to PNA was investigated by MALDI-MS, fluorescence spectroscopy, and SPR, respectively. β-CD **151**-based vesicle showed better binding abilities than β-CD **150**-based NP. It was suggested that a binding optimization effect based on a dynamic and reversible self-assembly process might occur upon the binding interaction between the aggregates and the lectin. Further studies confirmed that balance between hydrophilic ethylene glycol groups and hydrophobic alkyl chains controls the self-organization that the conjugates adopt in water. Usually, a polydisperse mixture of small micelles (~4 nm) or nanospheres (20–30 nm) is obtain, which eventually aggregates into vesicles (100–200nm) or larger structures (500–600nm). Both vesicles and micellar aggregates present compartments (up to 20% v/v) able to encapsulate small molecules such as the fluorescent dye rhodamine 6G [113].

The concept based on the combination of the β-CD ability for drug complexation and the targeting properties of multivalent carbohydrates evolved to the preparation of gold nanoparticles (AuNP) containing multiple copies of lactose and β-CD (Fig. 7.23) [114]. Gal-3 demonstrated to interact with the lactose-containing AuNPs as it induced evident changes in their optical properties. The estimated loading capability of such glycoNPs for anticancer drug methotrexate via UV–Vis experiments indicated that the presence of β-CD moieties on the AuNPs surface enhanced their abilities to load such drug.

ACKNOWLEDGMENT

The authors acknowledge the Andalusian Government (Consejería de Economía, Innovación y Ciencia, Grant FQM6903), the Spanish Ministry of Economy and Competitiveness—ERD Fund (Grant CTQ2013-48380-R), and the Marie Curie ITN program (CYCLON Hit 608407) for financial support.

REFERENCES

1. Lee, Y. C.; Lee, R. T. *Acc. Chem. Res.* **1995**, *28*, 321–327.
2. Chabre, Y. M.; Roy, R. *Adv. Carbohydr. Chem. Biochem.* **2010**, *63*, 165–393.
3. Dwek, R. A. *Chem. Rev.* **1996**, *96*, 683–720.
4. Lis, H.; Sharon, N. *Chem. Rev.* **1998**, *98*, 637–674.
5. Lee, R. T.; Lee, Y. C. *Glycoconjugate J.* **2000**, *17*, 543–551.
6. Lundquist, J. J.; Toone, E. J. *Chem. Rev.* **2002**, *102*, 555–578.
7. Mammen, M.; Choi, S.-K.; Whitesides, G. M. *Angew. Chem. Int. Ed.* **1998**, *37*, 2754–2794.
8. Jiménez Blanco, J. L.; Ortiz Mellet, C.; García Fernández, J. M. *Chem. Soc. Rev.* **2013**, *42*, 4518–4531.
9. Ortiz Mellet, C.; Defaye, J.; García Fernández, J. M. *Chem. Eur. J.* **2002**, *8*, 1982–1990.
10. Lahmann, M. *Top. Curr. Chem.* **2009**, *288*, 17–65.
11. Vargas-Berenguel, A.; Ortega-Caballero, F.; Casas-Solvas, J. M. *Mini-Rev. Org. Chem.* **2007**, *4*, 1–14.
12. Peri, F. *Chem. Soc. Rev.* **2013**, *42*, 4543–4556.
13. Spinelli, N.; Defrancq, E.; Morvan, F. *Chem. Soc. Rev.* **2013**, *42*, 4557–4573.
14. Hatano, K.; Matsuoka, K.; Terunuma, D. *Chem. Soc. Rev.* **2013**, *42*, 4574–4598.
15. Galan, M. C.; Dumy, P.; Renaudet, O. *Chem. Soc. Rev.* **2013**, *42*, 4599–4612.
16. Sansone, F.; Casnati, A. *Chem. Soc. Rev.* **2013**, *42*, 4623–4639.
17. Chabre, Y. M.; Roy, R. *Chem. Soc. Rev.* **2013**, *42*, 4657–4708.
18. Martínez, A.; Ortiz Mellet, C.; García Fernández, J. M. *Chem. Soc. Rev.* **2013**, *42*, 4746–4773.
19. Weis, W.; Brown, J. H.; Cusack, S.; Paulson, J. C.; Skehel, J. J.; Wiley, D. C. *Nature* **1988**, *333*, 426–431.
20. Roy, R.; Zanini, D.; Meunier, S. J.; Romanowska, A. J. Chem. Soc., *Chem. Commun.* **1993**, 1869–1872.
21. Ohta, T.; Miura, N.; Fujitani, N.; Nakajima, F.; Niikura, K.; Sadamoto, R.; Guo, C.-T.; Suzuki, T.; Suzuki, Y.; Monde, K.; Nishimura, S.-I. *Angew. Chem. Int. Ed.* **2003**, *42*, 5186–5189.
22. Marra, A.; Moni, L.; Pazzi, D.; Corallini, A.; Bridi, D.; Dondoni, A. *Org. Biomol. Chem.* **2008**, *6*, 1396–1409.

23. Proft, T., ed. *Microbial Toxins: Current Research and Future Trends*. Caister Academic Press, Norfolk, **2009**.
24. Kitov, P. I.; Sadowska, J. M.; Mulvey, G.; Armstrong, G. D.; Ling, H.; Pannu, N. S.; Read, R. J.; Bundle, D. R. *Nature* **2000**, *403*, 669–672.
25. Mulvey, G. L.; Marcato, P.; Kitov, P. I.; Sadowska, J.; Bundle, D. R.; Armstrong, G. D. *J. Infect. Dis.* **2003**, *187*, 640–649.
26. Solomon, D.; Kitov, P. I.; Paszkiewicz, E.; Grant, G. A.; Sadowska, J. M.; Bundle, D. R. *Org. Lett.* **2005**, *7*, 4369–4372.
27. Nishikawa, K.; Matsuoka, K.; Kita, E.; Okabe, N.; Mizuguchi, M.; Hino, K.; Miyazawa, S.; Yamasaki, C.; Aoki, J.; Takashima, S.; Yamakawa, Y.; Nishijima, M.; Terunuma, D.; Kuzuhara, H.; Natori, Y. *Proc. Natl. Acad. Sci. U.S.A.* **2002**, *99*, 7669–7674.
28. Nishikawa, K.; Matsuoka, K.; Watanabe, M.; Igai, K.; Hino, K.; Hatano, K.; Yamada, A.; Abe, N.; Terunuma, D.; Kuzuhara, H.; Natori, Y. *J. Infect. Dis.* **2005**, *191*, 2097–2105.
29. Cuatrecasas, P. *Biochemistry* **1973**, *12*, 3547–3558.
30. Fan, E.; Zhang, Z.; Minke, W. E.; Hou, Z.; Verlinde, C. L. M. J.; Hol, W. G. J. *J. Am. Chem. Soc.* **2000**, *122*, 2663–2664.
31. Merritt, E. A.; Zhang, Z.; Pickens, J. C.; Ahn, M.; Hol, W. G. J.; Fan, E. *J. Am. Chem. Soc.* **2002**, *124*, 8818–8824.
32. Zhang, Z.; Merritt, E. A.; Ahn, M.; Roach, C.; Hou, Z.; Verlinde, C. L. M. J.; Hol, W. G. J.; Fan, E. *J. Am. Chem. Soc.* **2002**, *124*, 12991–12998.
33. Zhang, Z.; Pickens, J. C.; Hol, W. G. J.; Fan, E. *Org. Lett.* **2004**, *6*, 1377–1380.
34. Zhang, A.; Liu, J.; Verlinde, C. L. M. J.; Hol, W. G. J.; Fan, E. *J. Org. Chem.* **2004**, *69*, 7737–7740.
35. Arosio, D.; Fontanella, M.; Baldini, L.; Mauri, L.; Bernardi, A.; Casnati, A.; Sansone, F.; Ungaro, R. *J. Am. Chem. Soc.* **2005**, *127*, 3660–3661.
36. Garcia-Hartjes, J.; Bernardi, S.; Weijers, C. A. G. M.; Wennekes, T.; Gilbert, M.; Sansone, F.; Casnati, A.; Zuilhof, H. *Org. Biomol. Chem.* **2013**, *11*, 4340–4349.
37. Mattarella, M.; Garcia-Hartjes, J.; Wennekes, T.; Zuilhof, H.; Siegel, J. S. *Org. Biomol. Chem.* **2013**, *11*, 4333–4339.
38. Branson, T. R.; McAllister, T. E.; Garcia-Hartjes, J.; Fascione, M. A.; Ross, J. F.; Warriner, S. L.; Wennekes, T.; Zuilhof, H.; Turnbull, W. B. *Angew. Chem. Int. Ed.* **2014**, *53*, 8323–8327.
39. Dubber, M.; Sperling, O.; Lindhorst, T. K. *Org. Biomol. Chem.* **2006**, *4*, 3901–3912.
40. Almant, M.; Moreau, V.; Kovensky, J.; Bouckaert, J.; Gouin, S. G. *Chem. Eur. J.* **2011**, *17*, 10029–10038.
41. Gouin, S. G.; Wellens, A.; Bouckaert, J.; Kovensky, J. *ChemMedChem* **2009**, *4*, 749–755.
42. Touaibia, M.; Wellens, A.; Shiao, T. C.; Wang, Q.; Sirois, S.; Bouckaert, J.; Roy, R. *ChemMedChem* **2007**, *2*, 1190–1201.
43. Bouckaert, J.; Li, Z.; Xavier, C.; Almant, M.; Caveliers, V.; Lahoutte, T.; Weeks, S. D.; Kovensky, J.; Gouin, S. G. *Chem. Eur. J.* **2013**, *19*, 7847–7855.
44. Nierengarten, I.; Buffet, K.; Holler, M.; Vincent, S. P.; Nierengarten, J.-F. *Tetrahedron Lett.* **2013**, *54*, 2398–2402.
45. Obritsch, M. D.; Fish, D. N.; MacLaren, R.; Jung, R. *Pharmacotherapy* **2005**, *25*, 1353–1364.

46. Cecioni, S.; Lalor, R.; Blanchard, B.; Praly, J.-P.; Imberty, A.; Matthews, S. E.; Vidal, S. *Chem. Eur. J.* **2009**, *15*, 13232–13240.
47. Soomro, Z. H.; Cecioni, S.; Blanchard, H.; Praly, J.-P.; Imberty, A.; Vidal, S.; Matthews, S. E. *Org. Biomol. Chem.* **2011**, *9*, 6587–6597.
48. Gening, M. L.; Titov, D. V.; Cecioni, S.; Audfray, A.; Gerbst, A. G.; Tsevetkov, Y. E.; Krylov, V. B.; Imberty, A.; Nifantiev, N. E.; Vidal, S. *Chem. Eur. J.* **2013**, *19*, 9272–9285.
49. Consoli, G. M. L.; Granata, G.; Cafiso, V.; Stefani, S.; Geraci, C. *Tetrahedron Lett.* **2011**, *52*, 5831–5834.
50. Morvan, F.; Meyer, A.; Jochum, A.; Sabin, C.; Chevolot, Y.; Imberty, A.; Praly, J.-P.; Vasseur, J.-J.; Souteyrand, E.; Vidal, S. *Bioconjugate Chem.* **2007**, *18*, 1637–1643.
51. Ligeour, C.; Audfray, A.; Gillon, E.; Meyer, A.; Galanos, N.; Vidal, S.; Vasseur, J.-J.; Imberty, A.; Morvan, F. *RSC Adv.* **2013**, *3*, 19515–19524.
52. Cecioni, S.; Faure, S.; Darbost, U.; Bonnamour, I.; Parrot-Lopez, H.; Roy, O.; Taillefumier, C.; Wimmerová, M.; Praly, J.-P.; Imberty, A.; Vidal, S. *Chem. Eur. J.* **2011**, *17*, 2146–2159.
53. Varga, N.; Sutkeviciute, I.; Ribeiro-Viana, R.; Berzi, A.; Ramdasi, R.; Daghetti, A.; Vettoretti, G.; Amara, A.; Clerici, M.; Rojo, J.; Fieschi, F.; Bernardi, A. *Biomaterials* **2014**, *35*, 4175–4184.
54. André, S.; Sansone, F.; Kaltner, H.; Casnati, A.; Kopitz, J.; Gabius, H.-J.; Ungaro, R. *ChemBioChem* **2008**, *9*, 1649–1661.
55. Consoli, G. M. L.; Cunsolo, F.; Geraci, C.; Sgarlata, V. *Org. Lett.* **2004**, *6*, 4163–4166.
56. André, S.; Grandjean, C.; Gautier, F.-M.; Bernardi, S.; Sansone, F.; Gabius, H.-J.; Ungaro, R. *Chem. Commun.* **2011**, *47*, 6126–6128.
57. André, S.; Liu, B.; Gabius, H.-J.; Roy, R. *Org. Biomol. Chem.* **2003**, *1*, 3909–3916.
58. Wang, G.-N.; André, S.; Gabius, H.-J.; Murphy, P. V. *Org. Biomol. Chem.* **2012**, *10*, 6893–6907.
59. Giguère, D.; Patnam, R.; Bellefleur, M.-A.; St-Pierre, C.; Sato, S.; Roy, R. *Chem. Commun.* **2006**, 2379–2381.
60. Marotte, K.; Préville, C.; Sabin, C.; Moumé-Pymbock, M.; Imberty, A.; Roy, R. *Org. Biomol. Chem.* **2007**, *5*, 2953–2961.
61. Giguère, D.; André, S.; Bonin, M.-A.; Bellefleur, M.-A.; Provencal, A.; Cloutier, P.; Pucci, B.; Roy, R.; Gabius, H.-J. *Bioorg. Med. Chem.* **2011**, *19*, 3280–3287.
62. Astronomo, R. D.; Burton, D. R. *Nat. Rev. Drug Discov.* **2010**, *9*, 308–324.
63. Hakomori, S. *Adv. Cancer Res.* **1989**, *52*, 257–331.
64. Lo-Man, R.; Bay, S.; Vichier-Guerre, S.; Dériaud, E.; Cantacuzène, D.; Leclerc, C. *Cancer Res.* **1999**, *59*, 1520–1524.
65. Lo-Man, R.; Vichier-Guerre, S.; Bay, S.; Dériaud, E.; Cantacuzène, D.; Leclerc, C. *J. Immunol.* **2001**, *166*, 2849–2854.
66. Ragupathi, G.; Coltart, D. M.; Williams, L. J.; Koide, F.; Kagan, E.; Allen, J.; Harris, C.; Glunz, P. W.; Livingston, P. O.; Danishefsky, S. J. *Proc. Natl. Acad. Sci. U.S.A.* **2002**, *99*, 13699–13704.
67. Ragupathi, G.; Koide, F.; Livingston, P. O.; Cho, Y. S.; Endo, A.; Wan, Q.; Spassova, M. K.; Keding, S. J.; Allen, J.; Ouerfelli, O.; Wilson, R. M.; Danishefsky, S. J. *J. Am. Chem. Soc.* **2006**, *128*, 2715–2725.
68. Zhu, J.; Wan, Q.; Lee, D.; Yang, G.; Spassova, M. K.; Ouerfelli, O.; Ragupathi, G.; Damani, P.; Livingston, P. O.; Danishefsky, S. J. *J. Am. Chem. Soc.* **2009**, *131*, 9298–9303.

69. For a recent account, see: Wilson, R. M.; Danishefsky, S. J. *J. Am. Chem. Soc.* **2013**, *135*, 14462–14472.
70. Freire, T.; Lo-Man, R.; Piller, F.; Piller, V.; Leclerc, C.; Bay, S. *Glycobiology* **2006**, *16*, 390–401.
71. Grigalevicius, S.; Chierici, S.; Renaudet, O.; Lo-Man, R.; Dériaud, E.; Leclerc, C.; Dumy, P. *Bioconjugate Chem.* **2005**, *16*, 1149–1159.
72. Renaudet, O.; BenMohamed, L.; Dasgupta, G.; Bettahi, I.; Dumy, P. *ChemMedChem* **2008**, *3*, 737–741.
73. Bettahi, I.; Dasgupta, G.; Renaudet, O.; Chentoufi, A. A.; Zhang, X.; Carpenter, D.; Yoon, S.; Dumy, P.; BenMohamed, L. *Cancer Immunol. Immunother.* **2009**, *58*, 187–200.
74. Fiore, M.; Thomas, B.; Duléry, V.; Dumy, P.; Renaudet, O. *New J. Chem.* **2013**, *37*, 286–289.
75. Richichi, B.; Thomas, B.; Fiore, M.; Bosco, R.; Qureshi, H.; Nativi, C.; Renaudet, O.; BenMohamed, L. *Angew. Chem. Int. Ed.* **2014**, *53*, 11917–11920.
76. Geraci, C.; Consoli, G. M. L.; Galante, E.; Bousquet, E.; Pappalardo, M.; Spadaro, A. *Bioconjugate Chem.* **2008**, *19*, 751–758.
77. Krauss, I. J.; Joyce, J. G.; Finnefrock, A. C.; Song, H. C.; Dudkin, V. Y.; Geng, X.; Warren, J. D.; Chastain, M.; Shiver, J. W.; Danishefsky, S. J. *J. Am. Chem. Soc.* **2007**, *129*, 11042–11044.
78. Joyce, J. G.; Krauss, I. J.; Song H. C.; Opalka, D. W.; Grimm, K. M.; Nahas, D. D.; Esser, M. T.; Hrin, R.; Feng, M.; Dudkin, V. Y.; Chastain, M.; Shiver, J. W.; Danishefsky, S. J. *Proc. Natl. Acad. Sci. U. S. A.* **2008**, *105*, 15684–15689.
79. Wang, J.; Li, H.; Zou, G.; Wang, L.–X. *Org. Biomol. Chem.* **2007**, *5*, 1529–1540.
80. Wang, L.-X.; Ni, J.; Singh, S.; Li, H. *Chem. Biol.* **2004**, *11*, 127–134.
81. Li, H.; Wang, L.-X. *Org. Biomol. Chem.* **2004**, *2*, 483–488.
82. Ni, J.; Song, H.; Wang, Y.; Stamatos, N. M.; Wang, L.-X. *Bioconjugate Chem.* **2006**, *17*, 493–500.
83. Horiya, S.; Bailey, J. K.; Temme, J. S.; Schlippe, Y. V. G.; Krauss, I. J. *J. Am. Chem. Soc.* **2014**, *136*, 5407–5415.
84. Grandjean, C.; Rommens, C.; Gras-Masse, H.; Melnyk, O. *Angew. Chem. Int. Ed.* **2000**, *39*, 1068–1072.
85. Grandjean, C.; Angyalosi, G.; Loing, E.; Adriaenssens, E.; Melnyk, O.; Pancré, V.; Auriault, C.; Gras-Masse, H. *ChemBioChem* **2001**, *2*, 747–757.
86. Adamo, R.; Hu, Q.-Y.; Torosantucci, A.; Crotti, S.; Brogioni, G.; Allan, M.; Chiani, P.; Bromuro, C.; Quinn, D.; Tontini, M.; Berti, F. *Chem. Sci.* **2014**, *5*, 4302–4311.
87. Aoyama, Y. *Chem. Eur. J.* **2004**, *10*, 588–593.
88. Avvakumova, S.; Fezzardi, P.; Pandolfi, L.; Colombo, M.; Sansone, F.; Casnati, A.; Prosperi, D. *Chem. Commun.* **2014**, *50*, 11029–11032.
89. Díaz-Moscoso, A.; Guilloteau, N.; Bienvenu, C.; Méndez-Ardoy, A.; Jiménez Blanco, J. L.; Benito, J. M.; Le Gourriérec, L.; Di Giorgio, C.; Vierling, P.; Defaye, J.; Ortiz Mellet, C.; García Fernández, J. M. *Biomaterials* **2011**, *32*, 7263–7273.
90. Symens, N.; Méndez-Ardoy, A.; Díaz-Moscoso, A.; Sánchez-Fernández, E.; Remaut, K.; Demeester, J.; García Fernández, J. M.; De Smedt, S. C.; Rejman, J. *Bioconjugate Chem.* **2012**, *23*, 1276–1289.
91. Aguilar Moncayo, E. M.; Guilloteau, N.; Bienvenu, C.; Jiménez Blanco, J. L.; Di Giorgio, C.; Vierling, P.; Benito, J. M.; Ortiz Mellet, C.; García Fernández, J. M. *New J. Chem.* **2014**, *38*, 5215–5225.

92. Arima, H.; Yamashita, S.; Mori, Y.; Hayashi, Y.; Motoyama, K.; Hattori, K.; Takeuchi, T.; Jono, H.; Ando, Y.; Hirayama, F.; Uekama, K. J. *Control. Release* **2010**, *146*, 106–117.

93. Motoyama, K.; Mori, Y.; Yamashita, S.; Hayashi, Y.; Jono, H.; Ando, Y.; Hirayama, F.; Uekama, K.; Arima, H. *J. Incl. Phenom. Macrocycl. Chem.* **2011**, *70*, 333–338.

94. Baussanne, I.; Benito, J. M.; Ortiz Mellet, C.; García Fernández, J. M.; Law, H.; Defaye, J. *Chem. Commun.* **2000**, 1489–1490.

95. Benito, J. M.; Gómez-García, M.; Ortiz Mellet, C.; Baussanne, I.; Defaye, J.; García Fernández, J. M. J. Am. *Chem. Soc.* **2004**, *126*, 10355–10363.

96. de Robertis, L.; Lancelon-Pin, C.; Driguez, H. *Bioorg. Med. Chem. Lett.* **1994**, *4*, 1127–1130.

97. Lainé, V.; Coste-Sarguet, A.; Gadelle, A.; Defaye, J.; Perly, B.; Djedaïni-Pilard, F. *J. Chem. Soc., Perkin Trans. 2* **1995**, 1479–1487.

98. Furuike, T.; Aiba, S.; Nishimura, S.-I. *Tetrahedron* **2000**, *56*, 9909–9915.

99. André, S.; Kaltner, H.; Furuike, T.; Nishimura, S.-I.; Gabius, H.-J. *Bioconjugate Chem.* **2004**, *15*, 87–98.

100. Furuike, T.; Sadamoto, R.; Niikura, K.; Monde, K.; Sakairi, N., Nishimura, S.-I. *Tetrahedron* **2005**, *61*, 1737–1742.

101. Yasuda, N.; Aoki, N.; Abe, H.; Hattori, K. *Chem. Lett.* **2000**, *29*, 706–707.

102. García-López, J. J.; Hernández-Mateo, F.; Isac-García, J.; Kim, J. M.; Roy, R.; Santoyo-González, F.; Vargas-Berenguel, A. *J. Org. Chem.* **1998**, *64*, 522–531.

103. García-López, J. J.; Santoyo-González, F.; Vargas-Berenguel, A.; Giménez-Martínez, J. J. *Chem. Eur. J.* **1999**, *5*, 1775–1784.

104. Ortega-Caballero, F.; Giménez-Martínez, J. J.; García-Fuentes, L.; Ortiz-Salmerón, E.; Santoyo-González, F.; Vargas-Berenguel, A. *J. Org. Chem.* **2001**, *66*, 7786–7795.

105. Vargas-Berenguel, A.; Ortega-Caballero, F.; Santoyo-González, F.; García-López, J. J.; Giménez-Martínez, J. J.; García-Fuentes, L.; Ortiz-Salmerón, E. *Chem. Eur. J.* **2002**, *8*, 812–827.

106. Ortiz Mellet, C.; Benito, J. M.; García Fernández, J. M.; Law, H.; Chmurski, K.; Defaye, J.; O'Sullivan, M. L.; Caro, H. N. *Chem. Eur. J.* **1998**, *4*, 2523–2531.

107. Baussanne, I.; Benito, J. M.; Ortiz Mellet, C.; García Fernández, J. M.; Defaye, J. *ChemBiochem* **2001**, *2*, 777–783.

108. Abe, H.; Kenmoku, A.; Yamaguch, N.; Hattori, K. *J. Incl. Phenom. Macrocycl. Chem.* **2002**, *44*, 39–47.

109. Oda, Y.; Yanagisawa, H.; Maruyama, M.; Hattori, K.; Yamanoi, T. *Bioorg. Med. Chem.* **2008**, *16*, 8830–8840.

110. Bernardes, G. J. L.; Kikkeri, R.; Maglinao, M.; Laurino, P.; Collot, M.; Hong, S. Y.; Lepenies, B.; Seeberger, P. H. *Org. Biomol. Chem.* **2010**, *8*, 4987–4996.

111. Zhang, Q.; Su, L.; Collins, J.; Chen, G.; Wallis, R.; Mitchell, D. A.; Haddleton, D. M.; Becer, C. R. *J. Am. Chem. Soc.* **2014**, *136*, 4325–4332.

112. Mazzaglia, A.; Forde, D.; Garozzo, D.; Malvagna, P.; Ravoo, B. J.; Darcy, R. *Org. Biomol. Chem.* **2004**, *2*, 957–960.

113. Mazzaglia, A.; Valerio, A.; Villari, V.; Rencurosi, A.; Lay, L.; Spadaro, S.; Scolaro, L. M.; Micali, N. *New J. Chem.* **2006**, *30*, 1662–1668.

114. Aykaç, A.; Martos-Maldonado, M. C.; Casas-Solvas, J. M.; Quesada-Soriano, I.; García-Maroto, F.; García-Fuentes, L.; Vargas-Berenguel, A. *Langmuir* **2014**, *30*, 234–242.

8

GLYCO-FUNCTIONALIZED LIPOSOMES

JACOB J. WEINGART, PRATIMA VABBILISETTY, AND XUE-LONG SUN

Department of Chemistry, Chemical and Biomedical Engineering and Center for Gene Regulation in Health and Disease (GRHD), Cleveland State University, Cleveland, OH, USA

8.1 INTRODUCTION

Liposomes are self-assembling lipid bilayers having closed colloidal structures in which an outer lipid bilayer surrounds a central aqueous space. Due to the lipid compositions closely resembling that of cell membranes, liposomes have been extensively studied as a model of cell membranes [1]. Reconstitution of membrane-bound proteins within liposome bilayers is one of the most useful techniques in studying the functions of the membrane proteins [2]. During the past decades, liposomes have been extensively studied as drug, gene, image agent, and antigen delivery systems [3]. Liposomes are attractive as delivery system due to their capacity to improve the stability, therapeutic efficiency, and pharmacokinetic properties of drugs while reducing their side effects and have the advantage of being biodegradable and nontoxic as well. In addition, surface chemistry and lipid composition can be easily modified to address specific applications of liposomes.

Liposome surface functionalization facilitates the enormous potential in the application of liposomes. Various biomolecules have been conjugated onto liposome surfaces for a variety of biomedical applications [4]. Cell surface carbohydrates have specific interactions with proteins, which play an important role in various biological recognition processes such as fertilization, metastasis, inflammations, and host–pathogen adhesion [5], thereby serving as attractive molecules for surface modification of liposomes with purpose for specific biomedical applications [6].

Carbohydrate Nanotechnology, First Edition. Edited by Keith J. Stine.

For example, monosialoganglioside (GM1) can enhance circulation lifetimes of liposomes comparable to that observed for PEG [7]. Sialyl Lewis X-decorated liposome has been demonstrated to target delivery of drugs to endothelial cells based on the site-specific expression of E- and P-selectin on blood vessels during inflammation [8]. Mannosylated liposomes revealed superior uptake by macrophages both *in vitro* and *in vivo* after intratracheal administration and show great potential for targeted antigen delivery for immune therapy [9]. On the other hand, carbohydrate-decorated liposomes have been used as a multivalent platform of carbohydrates to inhibit carbohydrate-mediated cell adhesion. For example, sialic acid-decorated liposomes have showed strong inhibitory activity against influenza hemagglutinin and neuraminidase [10]. This book chapter describes recent advances in the preparations, characterization, and evaluation of glycoliposomes and their biomedical research and applications.

8.2 PREPARATION OF GLYCO-FUNCTIONALIZED LIPOSOMES

Many strategies for surface glyco-functionalization of liposomes have been reported in the past decades, of which two strategies/approaches have been commonly used. The first approach, named as direct liposome formulation approach, is to synthesize the glycolipid ligand first, followed by the preparation of liposomes in combination with other principal lipids [11–15]. The second approach, named as postfunctionalization approach, mainly involves grafting the carbohydrate ligand onto the preformed liposomes via various bioconjugation chemistries [16,17].

8.2.1 Direct Liposome Formulation Approach

Various neoglycolipids have been synthesized by either chemical or enzymatic strategies. Traditional NHS coupling reaction is often used for the synthesis of neoglycolipids, such as maltose-DSPE (M2-DSPE) and oligomaltose-DSPE (M15-DSPE) glycolipids, for the preparation of oligosaccharide-coated liposomes in combination with DSPC for use in potential drug delivery applications [18]. ^{BPC}NeuAc-PEGylated lipids for incorporation into doxorubicin-loaded liposomes were synthesized by NHS-based coupling reaction between NHS-activated PEGylated lipids with ^{BPC}NeuAc (sialoside) [19]. Reductive amination is another practical method to synthesize neoglycolipids for glycoliposome preparation. Ikehara *et al.* synthesized the neoglycolipids (Man3-DPPE) by conjugation of DPPE with oligosaccharides (mannotriose) via reductive amination, followed by the preparation of the liposomes with these neoglycolipids by thin-film extrusion method [20, 21]. In addition, chemoenzymatic strategies were used for the synthesis of neoglycolipids such as lactosylceramide mimetics. Harada *et al.* prepared glycoliposomes incorporated with (Lac-DPPA, Lac-DPPC, and Lac-DPPE) neoglycolipid conjugates, which were synthesized by reacting phospholipids and the respective disaccharide glycosides [22].

8.2.2 Post Liposome Functionalization Approach

Though direct liposome formation approach has been widely used for glycoliposome formation, one drawback is that some of the carbohydrate targeting ligands can inevitably face toward the interior aqueous compartment of the formed liposomes and thus become unavailable for the interaction with the respective target molecules. Many post liposome functionalization methods have been developed for glycoliposome preparations. This approach allows for maximum surface functionalization of liposomes with carbohydrates, most importantly on the exterior part allowing for more efficient delivery of drugs/genes/antigens. The most well-known methods are mainly based on the coupling reactions between maleimide and thiol groups [23]. A study done by the Rice group focuses on the chemoenzymatic strategy for the synthesis of tyrosine-linked oligosaccharides, bearing terminal sialyl or GalNac-Le^x along with vinyl sulfonate ending poly(ethylene glycol) spacer arm. This enables for further coupling reactions between thiol-bearing liposomes incorporated with PC, Chol, and phosphatidylthioethanol lipids [24]. Neomannosylated liposomal vesicles were constructed by covalent coupling reaction between preformed vesicles containing 4-(*p*-maleimidophenyl) butyryl phosphatidylethanolamine (MPB-PE) lipids and α-D-thiomannopyranoside residues bearing sulfhydryl group [25]. Chemoselective conjugation between carbohydrate ligands and preformed liposomes requires mild aqueous conditions in order to prevent or minimize any liposomal disruptions in the process. Recent advances led to the development of click chemistry and Staudinger ligation methods that have high specificity, chemoselectivity, and efficiency. Hassane *et al.* employed click chemistry to conjugate an unprotected α-D-mannosyl derivative carrying an azide-armed spacer with the surface of liposomes comprising of a synthetic lipid bearing a terminal alkyne functional group [26]. We demonstrated a chemoselective glyco-functionalization of preformed liposomes via Staudinger ligation, in which azide-functionalized lactose molecules were conjugated onto liposomes incorporated with synthetic anchor lipids bearing triphenylphosphine moiety [27]. Though the aforementioned strategies have been widely used and are highly efficient, many factors need to be considered for glyco-functionalization of liposomes. The affinity of the glycolipids toward their target is mainly based on the type of PEG linker employed that tailors the accessibility and availability of the surface-bound carbohydrate ligand for many interactions [28].

8.3 CHARACTERIZATION AND EVALUATION OF GLYCO-FUNCTIONALIZED LIPOSOMES

Prior to biomedical application of a designed glycoliposome, much effort is required to understand the physical and chemical properties of the liposome as well as the carbohydrate moiety it bears. In the past decades, a tremendous amount of research effort has gone into optimizing the ways liposomes are formulated and characterized to build a fundamental understanding of how various factors, such as lipid size, spacer lengths, and the ligands they bear, affect size distribution, surface charge, and morphology [29, 30]. Apart from the core structure of the liposome, the carbohydrate moieties must also be evaluated toward a particular target to determine the optimal percent range at which it needs to cover the liposome surface in which the "glycoside cluster effect" can take hold enhancing the affinity of the target to that specific carbohydrate moiety for that biomedical application [6,31].

8.3.1 Characterization Methods

As previously mentioned, size distribution and morphology, which can affect the amount of cargo the liposome can carry as well as *in vivo* behaviors such as circulation time in the bloodstream and, consequently, biodistribution, are of specific interest. Due to this, regardless of the method used to build glycoliposome constructs, whether inclusion of the carbohydrate moiety on a synthesized neoglycolipid directly added into the liposome formulation or postfunctionalization of the liposome surface, the types of characterization employed focus heavily on the determination of size and the surface morphology of the liposome. Thus, the characterization of liposome constructs has been heavily dominated by methods that solely address surface morphology. These methods include dynamic light scattering (DLS), transmission electron microscopy (TEM), high-performance liquid chromatography (HPLC), and enzymatic methods [32].

8.3.1.1 DLS DLS has become one of the most widely used methods for particle size determination; this is especially the case for liposomal constructs, due to quick turnaround times in data acquisition at low cost. DLS works based on measuring the intensity fluctuations of light, at an angle, scattered by dispersed particles suspended in solution. These fluctuations result from the continuous motion of the particles within the solution brought about by Brownian motion. Thus, using a correlation function and analyzing the characteristic decay times related to diffusion coefficients allow for the determination of the hydrodynamic or the total diameter of the particle based on the Stokes–Einstein equation [33]. This is very important due to the fact that the size of the liposomes can affect the end application.

For instance, in a seminal study, Liu *et al.* reported that ganglioside GM1-containing liposomes avoid uptake by the mononuclear phagocytic system (MPS), which is mediated by various complement proteins in blood, only if they are between 70 and 200 nm in size [34]. This study was in agreement with the previous findings that reductions in liposome size can reduce complement recognition [35]. Thus, finding the optimal size parameters and ensuring that liposomal constructs fall within that range can ensure successful use of the liposome in *in vivo* application. Others have followed suit in regard to setting an appropriate size distribution ranges for their glycoliposomes, such as Zeisig *et al.* who determined that in the instance of using sialyl Lewis X-glycoliposomes to inhibit E-selectin-mediated tumor cell adhesion, their glycoliposomes must be within 100–120 nm in size with a low polydispersity of 0.20 or lower [36]. Apart from using DLS for looking solely at size distributions of the liposome in the previously mentioned studies, DLS has been used to monitor changes in aggregation states. This is particularly important when dealing with carbohydrate moieties that bear charged sugars or lipid head groups in the surrounding environment. These charged sugars/lipids may induce aggregation among liposomes depending upon the environment and their interactions with buffer salts, other charged macromolecules, and physiologically relevant ions, such as calcium. This induced aggregation event may hinder desired performance of the glycoliposome. For instance, in a study by Nyren-Erickson *et al.*, it was demonstrated that liposomes in the presence of glycosaminoglycans, which are negatively charged, induced liposomal aggregation in the presence of divalent cationic metal species, such as Mg^{2+} [37]. Though DLS is readily used in the characterization of glycoliposomes, and liposomes in general, in the determination of liposomal size distribution, diameter, and polydispersity, there are distinct disadvantages, particularly inaccurate data acquisition due to dust or buffer salt particulates and other aggregates contaminating the sample solution. Thus, to attain similar data in regard to size distribution, diameter, and polydispersity while limiting interferences from particulate contaminants, others have turned to transmission electron spectroscopy.

8.3.1.2 TEM TEM can be used to determine the total particle size, shape, inner core, and surface coating/monolayer as well as the size distribution of many different nanoparticle systems with high resolution. However, the use of traditional TEM is becoming undesirable in its use for characterizing liposomal constructs due to disadvantages that arise from sample preparation. A typical sample preparation requires that a small portion of the liposomal sample be mounted on a coated copper grid and allowed to dry before imaging. This is clearly disadvantageous to the analysis of liposomes, which heavily rely on aqueous solutions to maintain their supramolecular architecture, because the drying process can significantly alter the liposomal characteristics making the TEM image not truly representative of the sample as if it were in aqueous environment. A classic example of this was in the study by Nyren-Erickson *et al.* who investigated aggregations effects of liposomes caused by glycosaminoglycans. In their study, DLS indicated that there were three distinct sizes of POPC- and DSPC-based liposomes, which were approximately 200 nm in diameter. However, upon drying of their sample and TEM analysis, individual liposomes appeared to have diameters that were far less from this value, ranging from 25 to 40 nm in diameter, which is not truly indicative of their actual size when hydrated [37]. Other distinct disadvantages the traditional TEM has are operator bias and improper staining technique resulting in poor contrast. Thus, to limit size variation, and other disadvantages, in TEM micrographs of liposomal constructs, many have turned to the use of cryo-TEM [38–40]. Cryo-TEM involves the flash freezing of hydrated samples in liquid ethane, which is essential for the TEM analysis of samples that are of biological relevance. The flash freezing causes the water in these samples to take on a frozen-hydrated glass-like state so that the structure of the sample is not disrupted by the formation of ice crystals. This technique allows individuals to not have to rely on the use of chemical fixatives, dehydration, or staining, preserving the ultrastructure of the sample as if it were in an aqueous environment. Thus, this technique has become a valuable method in the analysis of liposomal constructs via TEM. Such is the case in the cryo-TEM analysis of liposomes bearing hyaluronic acid (HA), an anionic polysaccharide, targeting CD44 receptor performed by Surace and coworkers [41].

8.3.1.3 HPLC and Enzymatic Methods Apart from characterizing the size and the morphology of the liposome via DLS and TEM, it is equally important to know the molar ratio of the lipid components of the liposome. This is especially the case in dealing with glycoliposomes where a large number of carbohydrates on the surface may be required to achieve the glycoside cluster effect to target/bind a specified receptor for a particular application. HPLC has been readily used to access the molar ratio of the lipid constituents of glycoliposomes, and liposomes in general [42]. Typically, HPLC is used in combination with some type of detector. The most common detector coupled to HPLC for the identification and molar determination of the lipid constituents, especially for glycoliposomes, is a light scattering detector [43]. In addition to using a particular detector, the HPLC is advantageous in the fact that lines can be split allowing for the collection of aliquots of fractionated lipid samples that can be used for other standard molecular characterization techniques, such as nuclear magnetic resonance or mass spectrometry.

Enzymatic analysis of the sugar moieties of the glycoliposomes is often performed in combination, or in addition, to lipid analysis via HPLC. Enzymatic analysis typically requires the use of enzymes to digest specific carbohydrate moieties from the surface of the liposome. The amount of sugar can then be determined by using HPLC analysis. As was the case for DeFrees and coworkers, who determined the percentage of sialyl Lewis X on the surface of their glycoliposomes by first digesting the carbohydrate moiety using the enzyme sialidase and further analyzing the digested sugar residues by HPLC [44], others have relied on more traditional colorimetric techniques for the determination of the amount of carbohydrate on liposomal surfaces such as the resorcinol, phenol–sulfuric acid methods, and the thiobarbituric method [26,27,45].

8.3.2 Evaluation Methods

Once proper analysis has been performed on the prepared glycoliposomes to accurately characterize them, parameters can be set for a particular formulation. These parameters, such as diameter, percent lipid composition, and percent carbohydrate on surface, can then be evaluated and optimized in regard to the effects they have on the liposomal construct binding to its particular target. This ultimately will result in the determination of the absolute formulation for the glycoliposome for its intended application. Two key issues must be addressed when evaluating carbohydrate-bearing liposomes for a specific application, specifically stability and binding affinity. Liposomes as delivery devices, or immunoassay agents, must be stable with good shelf life. Any compromising of the liposomal structure leading to collapse, or unintended aggregation, can lead to leakage of cargo or the inability of targeting ligand to effectively bind to their target prior to their use. If stabilization of the liposomes is not an issue, one then must assess the ability of the carbohydrate moiety on the surface of liposome to bind to a targeted receptor. Thus, evaluating glycoliposome stability and binding affinity is key to success in biomedical applications.

8.3.2.1 ***Liposome Stability*** As previously mentioned, stabilization of liposomes is of utmost importance. Since their first purposed use as drug delivery vehicles by Gregoriadis *et al.* in 1974, much effort has gone into the study of their stabilization to prolong liposomal shelf life and circulation retention [4,46]. One of the most common methods to enhance stability and retention of liposomes is to modify the surface with an even more hydrophilic coating than the already existing head group. A hydrophilic coating allows for the formation of a clathrate of water to form on the exterior surface of the liposome, further stabilizing the supramolecular structure of the bilayer. In addition to enhancing stabilization, this water barrier also inhibits absorption and/or opsonization of proteins and other hydrophobic entities [47]. Hence, using carbohydrate ligands as a means of hydrophilically coating liposomes has become attractive. Due to the diversity of carbohydrate structures, modification of liposome surfaces must be treated on a case-by-case basis. The dominant methods of evaluating liposome stability over time have relied on DLS and fluorescence.

DLS is used to monitor the stability of liposomes over time looking at diameter size and size distribution as well as qualitatively. Liposomes are inherently unstable over time. As the bilayers collapse, they often begin to form smaller unilamellar micelles, which aggregate with existing liposomes to form even larger vesicles. These events can be witnessed, as an increase in size distribution of the liposome sample over time following by splitting of the distribution into two, representing lipid aggregates and large lipid vesicles. An excellent example of this phenomenon occurring is the study we performed by selectively surface modifying liposomes with azido-lactose via Staudinger ligation. Using DLS, a comparison was performed between unmodified liposomes and lactose-modified liposomes demonstrating that the lactose modification enhanced the liposomal stability over time [27]. As seen in Figure 8.1, the unmodified liposomes (A) begin to aggregate by day 13 exhibiting a wide size distribution and have completely collapsed forming lipid aggregates and large vesicles by day 15.

Recently, we set out to investigate how the lipid-anchoring groups the carbohydrate is conjugated to at varying percents of incorporation affect liposomal stability by addressing not only size distribution over time and lectin binding via DLS but also release of fluorophore over time via fluorescence monitoring [48]. To accomplish this, they employed the method developed by Barbet *et al.* for determining stability via monitoring fluorescence leakage [49]. Encapsulating 5,6-carboxyfluorescein, we monitored the leakage of the fluorophore over time from different percent compositions of a lactose-conjugated PEG_{2000}-cholesterol and lactose-conjugated PEG_{2000}-DSPE anchors in liposome formulation determining the optimal percent compositions for the two anchors to enhance the stability permitting fluorescence leakage of only 10% to occur after 2 weeks.

8.3.2.2 ***Binding Affinity*** Apart from stability, the carbohydrates on the surface of the glycoliposome must be able to effectively bind to its target, particularly lectins

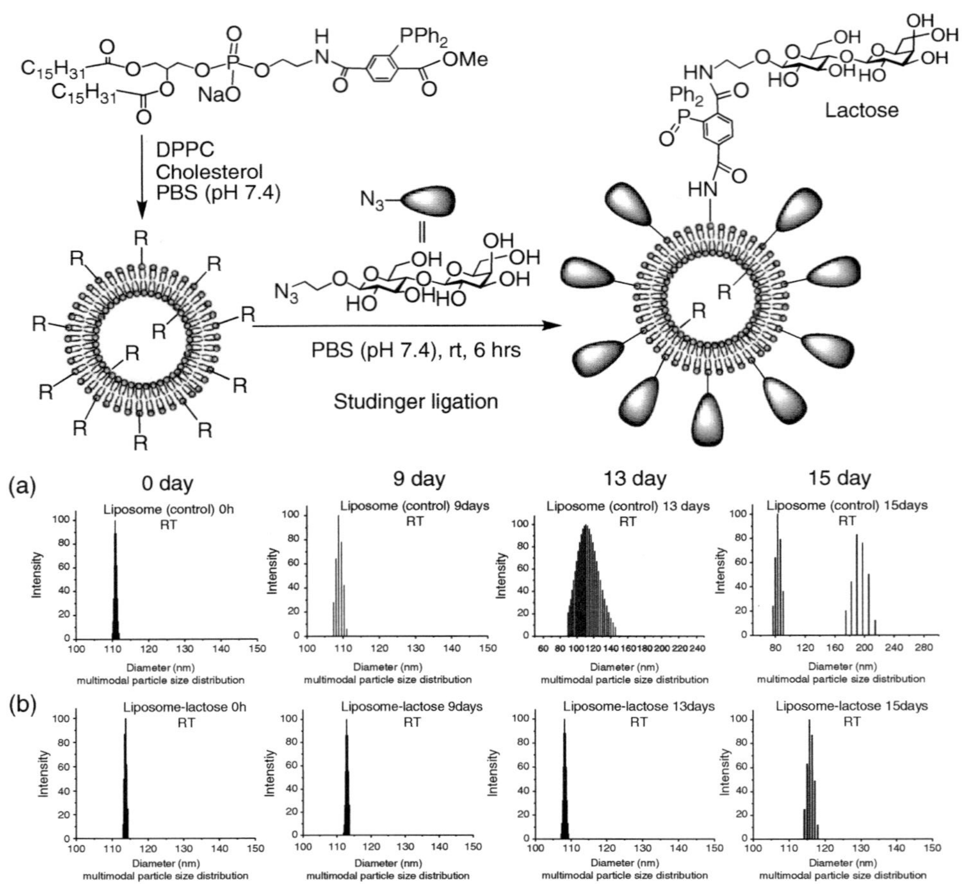

FIGURE 8.1 Azido-lactose modification of liposome and DLS comparison of unmodified (a) and modified liposome (b) over time. *Source*: Reproduced from Ref. 27, with permission from the Royal Society of Chemistry.

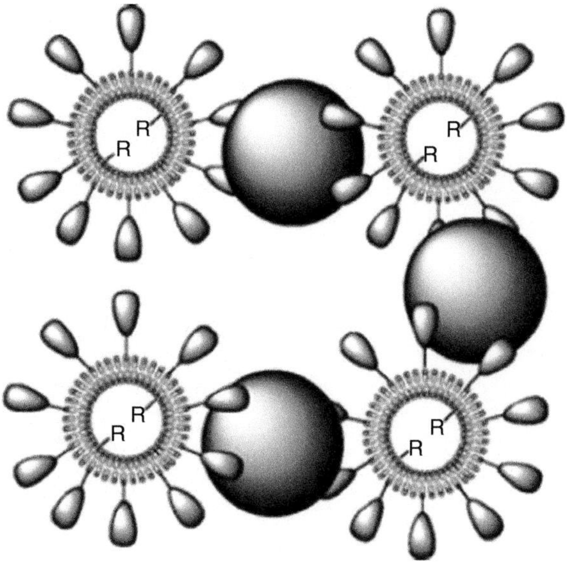

FIGURE 8.2 Glycoliposomes binding to lectins to forming a bound network. *Source*: From Ref. 27, with permission from the Royal Society of Chemistry.

that are sugar-binding proteins expressed by a large number of mammalian cells [4,6]. Thus, many methods, both qualitative and quantitative, are typically employed. Each of these methods varies depending upon the glycoliposome system being investigated. However, these methods can be generalized into two means of measuring lectin binding: agglutination assays and immunoassays.

Agglutination assays rely on the coalescing of the glycoliposomes upon binding to the surface of the targeted lectin. This event is typically measured by changes in light that moves through the sample, whether the light is scattered, blocked, or changed [50]. Prior to the development of DLS, agglutination assays were performed measuring changes in turbidity. Turbidity is the measure of how much light can pass through a sample. Thus, if carbohydrate-bearing liposomes bind their target, they should coalesce forming a bound network of liposomes and lectin, as seen in Figure 8.2, which in turn blocks the passage of light increasing turbidity.

An example of this was the study performed by Slama and Rando investigating the ability of three synthesized mannose-bearing glycolipids with different spacer lengths incorporated into liposomes to bind ricin toxin [12]. Measuring the time-dependent increase in turbidity of the varying glycoliposome formulations upon introduction of the ricin toxin, they were able to determine the optimal spacer length for the synthetic glycolipid based on the kinetics derived from the data. They also were able to demonstrate that by addition of a more competitively binding sugar, in this case lactose, the liposomal aggregates with the ricin can be dispersed again reducing the turbidity. Another method of measuring agglutination is though DLS. As aggregate networks are formed between the glycoliposomes and their target

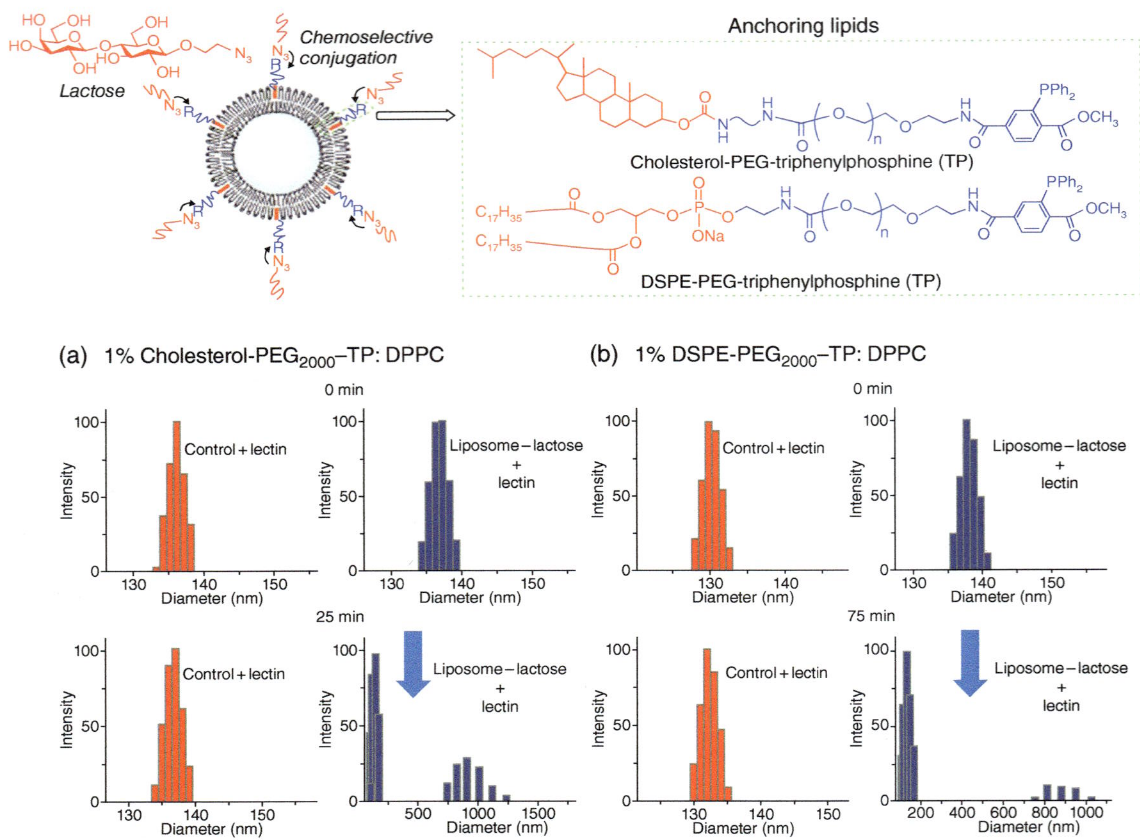

FIGURE 8.3 DLS monitoring of lectin-induced aggregation of glycoliposomes with control. *Source*: Reproduced from Ref. 48, with permission from the Royal Society of Chemistry.

lectins, turbidity increases as well as the way light is scattered in the sample. Thus, by using DLS, one can visualize this aggregation event as a dramatic change in the size distribution compared to that of the original glycoliposome solution. We demonstrated this by monitoring the agglutination of lactose-modified liposomes in the presence of β-galactose-binding lectin in PBS buffer [27]. A similar study using DLS to monitor agglutination was also performed by demonstrating the same phenomena as seen in Figure 8.3 [48]. Others in turn have also used colorimetric methods, whether monitoring changes in fluorescence or UV–Vis spectrum as the result of lectin binding [50]. A seminal example of this is the work performed by Pan and Charych who developed G_{M1} ganglioside-bearing poly(diacetylene) liposomes. These liposomes were found to be very unique in that the assembled and polymerized diacetylene and G_{M1}-bearing lipids give the glycoliposomes a deep blue/purple color while suspended in aqueous solution [51]. Upon the binding of the targeted lectin, cholera toxin, there was a distinct color shift casing the solution to change to an orange color.

In addition to, or aside from, running agglutination assays, others have used a variety of immunoassays to assess the binding ability of designed glycoliposomes to specific lectins. Whether glycoliposomes are incubated with lectins or lectin-bearing cell lines attached to the surface of microtiter plates, these assays often employ the use of a fluorophore that allows for assessment of the binding event [50]. Many of these assays rely on the sandwich method in which an entity's binding activity to its target attached to the microtiter plates is probed with fluorescently tagged antibody. The tagged antibody can be specific to that entity that is bound, which gives fluorescence in event of binding, or to the target. If the tagged antibody is specific to the target, it is vice versa. The bound glycoliposomes will block the antibody, resulting in no fluorescent signal that notes a positive binding event. This type of immunoassay, which is an inhibition assay, is commonly used to assess the binding of glycoliposomes to a specific lectin [6,52]. Such was the work performed by Bruehl *et al.* in investigating liposome assemblies bearing sialyl Lewis X and a sulfo-Lewis X carbohydrate as selectin inhibitors toward binding of the L-selectin toward GlyCAM1, a selectin ligand, bound to microtiter plates. Using different percentage concentrations of glycoliposome solutions, L-selectin was captured and blocked from binding to the sugar adhering in the wells of the plate to varying degrees. In doing so, when the plate was probed with a fluorescently tagged antibody toward L-selectin, in the samples maximally inhibited, there was no fluorescence [53]. In another study, Harada *et al.* designed glycoliposomes with nitrobenzoxadiazole (NBD) fluorescent tags incorporated into the glycolipids of the bilayer. When incubating with targeted lectin immobilized on a microtiter plate, the binding of the liposomes was measured based on their fluorescence intensity as dictated by the amount of glycolipid present in the liposomal formulation [22].

Once a proper assessment of the stability and binding ability of a glycoliposome is performed, it allows individuals to gain more confidence to test their system for the designed application. This is especially true in dealing with biomedical applications, where a glycoliposome may find itself as a targeted delivery device for a variety of payloads to be investigated by using *in vitro* and *in vivo* model systems.

8.4 BIOMEDICAL APPLICATIONS OF GLYCO-FUNCTIONALIZED LIPOSOMES

The use of glyco-functionalized liposomes can present itself in more than one model for the study of glycoconjugate interactions. Glycoliposomes have been used in the study of carbohydrate–carbohydrate recognition events as well as the cluster model, in which many carbohydrate motifs act as ligands [6,54]. However, the predominant use of glycoliposomes still surrounds their application in the study of carbohydrate–protein interactions.

Lectins are carbohydrate-binding proteins serving in conjunction with naturally occurring glycans on lipids and other proteins to moderate a variety of cellular binding events. Binding occurs specifically at the carbohydrate recognition domain (CRD) of lectins, which exist as multiple sites on the protein, where hydrogen bonding between the amino acids and the carbohydrate hydroxyl groups occurs. The affinity of this binding is relatively weak, relying on the multivalent or "cluster effect" in which a specific carbohydrate, in the proper orientation, is clustered together too rapidly to increase affinity and specificity toward that lectin that has several shallow CRDs [55]. Due to the ability of glycoliposomes to display multiple carbohydrate residues across their surface with the fluidity of the bilayer membrane similar to a cell, the carbohydrates can effectively move across the outer membrane layer and cluster to effectively bind a specific lectin. Thus, the carbohydrates conjugated to the liposomal surface can serve as a targeting mechanism. This is very important considering that lectins have been found to be expressed in a variety of protein toxins, viruses, specific cells, and disease states [56]. Apart from the displayed carbohydrate's targeting ability, carbohydrates display ancillary benefits such as biodegradability, low toxicity, and protein-shielding ability that helps to minimize liposome clearance from circulation. The liposomal portion of the glycoliposome itself can serve as a delivery vehicle. Due to the benefits of surface functionalization with carbohydrate and the ability of the liposomal portion to act as a cargo-containing vehicle, glycoliposomes have found themselves applied as glycosylated-targeted delivery devices in a wide variety of biomedical applications [56]. In this section of the chapter, we specifically wish to address recent developments, within the past 5 years, in the use of glycoliposomes in targeted drug, gene, and antigen delivery as well as glycoliposome use as multivalent inhibitors.

8.4.1 Targeted Drug Delivery

Of the many uses of liposomes for targeted delivery, targeted drug delivery is the most widely studied, many of which are geared toward the treatment of many diseases and disease states such as leishmaniasis, fungal infections, cancer, atherosclerosis, and inflammation [56,57]. Though the pharmaceutical industry has successfully developed numerous drugs for the treatment of a variety of diseases, some of these treatments fail or demonstrate severe side effects due to inadequate delivery of the drug to its site of action. Thus, there is an increasing demand for therapeutic systems that can accurately target the diseased cells or tissues, delivering the drugs to a localized area limiting systemic side effects that could possibly result in further illness and even death. Considering that lectins have been found to be expressed on a wide variety of cells, including those only expressed in specific disease states, they have become the driving force for the use of glycoliposomes in drug delivery.

Cancer remains one of the most devastating diseases with more than 10 million new cases every year [4]. Lectins have been found to be expressed on various malignant cancer cells, many of which are involved in metastasis, or spreading of the cancer to other healthy tissues within the body [58,59]. Thus, much effort has gone into using glycoliposomes to target a specific cell/tissue type where the cancer is located or target the cancer itself. The targeting of a specific tissue cell type with a glycoliposome was recently demonstrated by Bernardes and coworkers. In this study, liposomes were functionalized with *N*-acetyl-D-glucosamine (GlcNAc) and *N*-acetyl-D-galactosamine (GalNAc) sugar residues to specifically target asialoglycoprotein receptor (ASGPR), which is a C-type lectin that is expressed primarily by parenchymal hepatocytes in the liver [60]. Using a rhodamine-label within their glycoliposome constructs and flow cytometry, they were able to demonstrate preferential uptake of the liposomes functionalized specifically with GalNAc.

Based on a method developed by Chen *et al.* to deliver chemotherapeutic drugs to B cells for the B-cell-specific cell surface protein CD22, a protein that recognizes sialylated glycans and governs endocytosis, Geert-Jan Boons sought to reverify the internalization of liposomes displaying sialic acid, also known as *N*-acetylneuraminic acid (NeuAc) [19,61]. Sialic acid, or *N*-acetylneuraminic acid, is the terminal sugar linked to galactose, a (2, 6)-linkage, which constitutes the ligand which CD22 specifically recognizes. Using fluorescently labeled glycoliposomes displaying this sugar, it was demonstrated that they bound robustly to CD22-expressing but not to wild-type CHO cells, while unmodified liposomes bound to neither. Taking the study a step further, the NeuAc-glycoliposomes were further tested as a delivery device for the drug doxorubicin. Doxorubicin is a very effective cancer chemotherapeutic that works by interacting with DNA by stopping replication and cellular growth. However, due to its nonspecificity, this drug is very toxic with adverse systemic side effects. Thus, if localized delivery can be controlled by the glycoliposomes, one can lower the dose required and possibly the appearance of side effects that occur with systemic administration of the drug. Such was the case for NeuAc-glycoliposomes, which exhibited a 33-fold higher cytotoxicity toward the targeted B-cell lymphoma compared to naked liposomes, with an inhibitory concentration level of 1.6 μM versus 53 μM, respectively [19]. Apart from Boons work on glycoliposomes, Song *et al.* sought to test the delivery of doxorubicin also using glycoliposome as a targeted delivery device testing of the ability of two common disaccharides, specifically sucrose and maltose, and their effects on intracellular uptake against several cancer cell lines [15]. Upon comparing the disaccharide-modified liposomes against their PEGylated liposome controls, it was determined via fluorescent confocal microscopy and flow cytometry that the glyco-modified liposomes exhibited a significantly higher level of intracellular uptake.

The doxorubicin fluorescence mainly resided both the cytoplasm and nuclei of the cell, whereas the PEGylated liposome carrying doxorubicin did not.

As previously mentioned, the use of glycoliposomes in targeted drug delivery is not just limited to cancer but has been applied to many other disease states. For example, recent studies have been geared toward targeting monocytes and macrophages [62]. Monocytes and macrophages are mononuclear phagocytes that are involved in innate immunity, the primary role of these cells being phagocytosis of apoptotic cells, pathogens, and other targets. The MPS is initiated through a variety of cellular binding events playing a pivotal role in inflammation [63]. It is of no surprise that some of these binding events are governed by a number of protein receptors, including glycan–lectin receptor interactions, making them prime targets for drug delivery to treat diseases where inflammatory processes are the key drivers of a specific disease's progression and pathogenesis. As it turns out, macrophages have a variety of lectin protein receptors expressed on their surface that are able to bind pathogen carbohydrate-associated molecular patterns. These receptors, which are mannose specific, are often overexpressed during times of immune distress caused by infection. Thus, they have been successfully targeted for drug delivery in a number of recent studies [57, 64–66]. For example, many pulmonary disease states cause alveolar macrophages to infiltrate diseased lung tissue and cause inflammation making the delivery of anti-inflammatories a necessity to aid in preserving lung function and ease of breathing for sick individuals. Such study was performed by Chono *et al.* who sought to investigate whether mannose-modified glycoliposome would be a viable drug delivery system via aerosolization [57]. In the study, the mannose-bearing glycoliposomes were tested in both *in vitro* and *in vivo* models, using cultured SD rat alveolar macrophages and male SD rats, respectively. In both models, the mannosylated liposomes exhibited a two- to threefold uptake compared to the unmodified liposome controls as well as noncytotoxicity to the lung tissues and stability in epithelial lung fluid. A recent study by Jeong *et al.* sought to investigate the uptake of mannosylated glycoliposomes by macrophages as well, specifically looking at the effects of various spacer lengths between the sugar and the liposome surface, specifically using bromohexadecane to conjugate di-, tetra-, and hexaethylene glycol spacer lengths [66]. All of the various mannosylated liposomes with varying spacer lengths exhibited superior uptake in the RAW 264.7 macrophage cells monitored by measuring the fluorescence intensity resulting from the NBD used to label the liposomes. Of the varying spacer lengths, the tetra(ethylene glycol) performed the best *in vitro* and demonstrated quick localization over time of the glycoliposome (M-EG4-LP-ICG) to tissues with areas of high macrophage counts *in vivo*, such as the pulmonary system, compared to the liposome control (LP-ICG).

These recent highlights shed light on glycoliposomes as more than just tools to study glycan–lectin interactions. They offer new possibilities in the development of drug delivery platforms with high specificity toward specific lectin-expressing cells with benefits of being biodegradable and low in toxicity and imparting protein-shielding ability resulting in longer biodistribution and increased efficacy.

8.4.2 Targeted Gene Delivery

Cationic liposomes that can electrostatically complex pDNA or mRNA have been proposed as nonviral gene delivery vehicles for the treatment of genetic disorders or disease states [67]. Though they have been able to efficiently bind genetic material to the outer and internal surfaces of the liposomal bilayer and have exhibited transfection capabilities *in vitro*, transfection *in vivo* still remains an issue. This is mostly caused by interactions with plasma proteins and poor targeted delivery. To circumvent this, researchers have began to investigate glycoliposomes in their potential to act as viable targeted gene delivery vehicles. Similar to drug delivery, the sugar residues on the liposome surface act as targeting ligands toward cell types expressing a specific lectin. Once the glycoliposome binds to the lectin on the cell surface, it must support the transfection of its genetic contents to the cell [56, 64]. Recent developments have addressed enhancing the targeting ability by introducing sugars across the surface of cationic liposomes. Naturally, given the knowledge surrounding mannose-binding lectin receptors and their expression on antigen-presenting cells (APCs) of the immune system, researchers have sought to conjugate mannose to the surface of their cationic liposomes [64, 68]. A most recent example is the study performed by Perche *et al.* investigating selective gene delivery in dendritic cells (DCs) with mannosylated liposomes and lipopolyplexes [69]. DCs play a critical role in antigenic presentation within the immune system; thus, they offer a viable way to immunize patients through the delivery of the genetic material encoding for the expression of antigens against diseases like cancer. These antigens are expressed on the DC surface allowing for the interplay of the immune system to illicit a response by producing antibodies against that specific antigen. To demonstrate this, mannosylated cationic liposomes (Man_{11}-LPD100) containing DNA encoding for the luciferase, a fluorescent protein, were used to transfect a murine DC line (DC2.4 cells). After transfection, the cells became visible via confocal microscopy as they began to produce the fluorescent luciferase protein, whereas the control liposomes (LPD100) demonstrated very limited transfection.

In addition to investigating the use of mono- and disaccharides, others have looked toward modifying cationic liposomes with oligo- and polysaccharides for gene delivery [54, 56]. A recent example of this was the work performed by Surace *et al.* in which gene-delivering liposomes were modified with HA to target the CD44 HA receptors of breast cancer cells. Specifically, cationic liposomes complexed with plasmid DNA encoding for luciferase were modified by incorporating HA-bearing lipids [41]. HA is unique in that it is a naturally occurring glycosaminoglycan of high molecular weight as a major component of extracellular matrix. The receptor it targets, CD44, exists at low levels on many cells during normal physiological conditions. However, during uncontrolled replication, such as in cancer, this receptor is overexpressed, which may aid in the metastasis of malignant cells by allowing them to bind to the extracellular matrix, containing HA, of tissues located elsewhere in the body ultimately allowing the cancer to spread [70]. Thus, CD44 represents a marker of cancer and viable target for gene delivery as demonstrated by the uptake and transfection by the HA-conjugated cationic liposomes in MCF-7 breast cancer cells expressing CD44 versus MCF-7 cell that did not express CD44 where uptake and transfection did not occur in the study [41].

Despite these recent advances using glycoliposomes in targeted gene delivery, there remains much opportunity for further exploration of this platform being only limited by what saccharide motifs can be synthesized or isolated. This is especially true given the structural diversity and complexity of carbohydrate moieties involved in the myriad number of cellular binding and adhesion events.

8.4.3 Targeted Antigen Delivery

Liposomes have been widely exploited as antigen delivery systems for treating a variety of diseases [71]. Liposomes show tremendous advantages as antigen delivery system to APCs for immune therapy. First, liposomes have a higher chance of properly activating APCs compared to free antigen as they mimic pathogens such as bacteria and viruses more closely. Second, liposomes protect the antigen against premature degradation, thus reducing the quantity of administered antigen needed to evoke an effective immune response. Moreover, the possibility to incorporate adjuvant along with the antigens will enhance or modulate the immune response more effectively than mixing the components.

Efficient delivery of antigens to immune cells can be facilitated by agents that bind selectively to molecular structures on the surface of the targeted cells. Glycoliposomes have been widely demonstrated as efficient antigen delivery systems for inducing and modifying immune responses [72]. Glycoliposomes with encapsulated or entrapped antigens are suitable as antigen delivery vehicles to APC. Numerous C-type lectin receptors (CLRs) are expressed on APCs that act as phagocytic receptors. The family of DC-SIGNs is of particular interest, since expression of these CLRs is restricted to professional APCs such as DCs. Targeting DC-SIGN by anti-DC-SIGN antibody affords a strategy to deliver antigens or drug into DCs [73]. DC-SIGNs recognize carbohydrate ligands with terminal mannose, fucose, and N-acetylglucosamine residues, such as high-mannose-type glycans [74]. Therefore, mannose residues attached to antigens or coated on antigen-containing particles have the potential to deliver antigens to DCs [75]. Kojima *et al.* have shown that liposomes coated with a mannotriose-containing neoglycolipid (Man3-DPPE) can deliver encapsulated protein antigens to DCs [76], promote presentation of the encapsulated antigen on MHC class I and II molecules [20], and lead to maturation of DCs [77], with resulting induction of antigen-specific immune responses against protozoan infectious diseases, cancer, and allergy [78]. Recently, they have demonstrated that Man3-DPPE-coated liposomes (oligomannose-coated liposomes (OMLs)) could also be used as a vehicle for preferential delivery of lipid antigens such as αGC to DCs to activate iNKT cells via presentation of lipids on CD1d [79]. In this study, they compared the *in vitro* and *in vivo* responses of iNKT cells to αGC-containing liposomes coated with Man3-DPPE (αGC-OMLs) and to those without a Man3-DPPE coating (αGC-BLs).

Sialoadhesin (Sn, Siglec-1, CD169) is a member of the sialic acid-binding Ig-like lectin (siglec) family expressed on macrophages. Its macrophage-specific expression makes it an attractive target for delivering antigens to tissue macrophages via Sn-mediated endocytosis. Paulson *et al.* described a novel approach for delivering antigens to macrophages using liposome displaying high-affinity glycan ligands of Sn [80]. The Sn-targeted liposomes selectively bind to and are internalized by Sn-expressing cells and accumulate intracellularly over time. This research showed that ligand-decorated liposomes are specific for Sn, since they are taken up by bone marrow-derived macrophages that are derived from wild-type but not Sn2/2 mice. Importantly, the Sn-targeted liposomes dramatically enhance the delivery of antigens to macrophages for presentation to and proliferation of antigen-specific T cells. This data provides insights into the potential of cell-specific targeting and delivery of antigens to intracellular organelles of macrophages using Sn-ligand-decorated liposomes.

Glycan-coated liposomes with an encased antigen can induce antigen-specific CTLs from peripheral blood mononuclear cells obtained from patients. Feasibility studies of glycan-coated liposomes have revealed their potential for clinical use as vaccines for diseases where CTLs act as effector cells. Thus, glycan-coated liposomes may provide a new prophylactic strategy for infectious disease and therapeutic approaches for cancer via the induction of T-cell immunity.

8.4.4 Multivalent Inhibitors

Carbohydrate recognition is a crucial event in many biological processes [81], such as in cell–cell signaling, immune recognition events, pathogen–host interactions, tumor metastasis, tissue growth and repair, as well as other cellular events. Therefore, carbohydrate recognition has come to the forefront of biological research aiming to uncover the molecular mechanisms of many physiological and pathological processes and discover potential therapeutic targets or diagnostic mechanisms for various diseases, including viral infections, autoimmune diseases, cancer, and cardiovascular disorders [82]. Consequently, reconstitution of carbohydrate epitopes to mimic authentic compositions and spatial organization has become one of the goals in glycoscience and biomedical science. For example, synthetic glycoligands can act as agonists or antagonists for understanding the molecular mechanisms of many biological processes and also provide tremendous opportunities for therapeutic applications.

Sialic acid-containing oligosaccharides named as sialyloligosaccharide expressed on the host respiratory epithelial cell surface are involved in influenza virus infection in both virus hemagglutinin-based attaching [83] and neuraminidase-based detaching [84] processes. The mechanisms by which influenza virus recognizes sialic acids have therefore been very important targets to design the anti-influenza agents [85]. To date, two effective neuraminidase inhibition-based anti-influenza drugs, zanamivir and oseltamivir, have substantially improved antiviral therapy for influenza infection [86]. On the other hand, site-specific blocking of the viral hemagglutinin binding to host cell receptors has been intensively investigated for anti-influenza drug development [87].

Sialic acid derivative-decorated liposomes have been investigated for influenza inhibition studies [88, 89]. In particular, Sun *et al.* reported a series of sialic acid derivative-decorated liposomes as inhibitors of both influenza virus proteins hemagglutinin and neuraminidase, including those with one of the hydrogen atoms at the C-3 position replaced by either OH or F. These glycoliposomes showed potent and selective inhibitory activities against the hemagglutinin H3 subtype but displayed resistance to the influenza virus neuraminidases N1 and N2 [10]. Further study demonstrated that the compound Neu5Ac3αF-DSPE, in which the C-3 position was modified with an axial fluorine atom, inhibited the catalytic hydrolysis of influenza virus sialidase and the binding activity of hemagglutinin. The inhibitory activities to sialidases were independent of virus isolates examined. With the positive results obtained for inhibition of hemagglutination and hemolysis induced by A/Aichi/2/68 virus, the inhibitory effect of Neu5Ac3αF-DSPE against MDCK cells was examined, and it was found that Neu5Ac3αF-DSPE inhibits the viral infection with IC_{50} value of 5.6 μM based on the cytopathic effects. The experimental results indicate that Neu5Ac3αF-DSPE not only inhibits the attachment of virus to the cell surface receptor but also disturbs the release of the progeny viruses from infected cells by inhibiting both hemagglutinin and sialidase of the influenza viruses. The study suggested that the compound is a new class of bifunctional drug candidates for the future chemotherapy of influenza [90].

These early experiments demonstrate the validity of using SA-functionalized liposomes to inhibit influenza. The SA analogue and its attachment chemistry incorporated into the liposomes can be optimized for stronger binding to HA. HA interacts with the host-glycan motif beyond the terminal SA monosaccharide. Monosaccharides extending beyond the SA critically interact with HA and govern the HA–glycan interaction. A common feature of the previous pandemic strains is their high-affinity binding to α2,6-sialylated glycans that are tetrasaccharides or longer. Wang *et al.* described a sialylneolacto-*N*-tetraose c (LSTc)-decorated liposome as an SA receptor-based therapeutic agent capable of multivalent presentation of SA residues to influenza [91]. They demonstrated that LSTc sialoside-bearing liposomes can present SA in a multivalent fashion to enable high-avidity binding to influenza HA. This research showed that decoy glycoliposomes bind strongly to influenza viral particles and can inhibit IAV infectivity *in vitro* as well as extend survival of mice challenged with a lethal dose of influenza A virus *in vivo*.

8.5 SUMMARY

Cell surface carbohydrates have specific interactions with proteins, which play an important role in various biological recognition processes such as fertilization, metastasis, inflammations, and host–pathogen adhesion, thereby serving as attractive molecules for surface modification of liposomes with purpose for specific biomedical applications. Cell surface carbohydrates are attractive model for liposome surface modification due to their protein-rejecting ability, biodegradability, low toxicity, and, especially, cell targeting ability through specific binding interactions with receptors expressed at the targeted cell surface. Carbohydrate decoration enhances circulation lifetimes of liposomes comparable to that observed for PEG.

Carbohydrate-decorated liposome has been demonstrated to target delivery of drugs, genes, image agents, and antigen delivery for a variety of biomedical applications. On the other hand, carbohydrate-decorated liposomes have been used as multivalent platform of carbohydrates to inhibit carbohydrate-mediated cell adhesion. Recent advances in the preparations, characterization, and evaluation of glyco-functionalized liposomes speed up glycoliposome research and applications. Continued efforts are still needed to facilitate practical biomedical applications of liposomes.

REFERENCES

1. Sessa, G.; Weissmann, G. *J. Lipid Res.* **1968**, *9*, 310–318.
2. Ciancaglini, P.; Simao, A. M. S.; Bolean, M.; Millan, J. L.; Rigos, C. F.; Yoneda, J. S.; Colhone, M. C.; Stabeli, R. G. *Biophys. Rev.* **2012**, *4*, 67–81.
3. Torchilin, V. P. *Nat. Rev. Drug Discov.* **2005**, *4*, 145–160.
4. Jolck, R. I.; Feldborg, L. N.; Andersen, S.; Moghimi, S. M.; Andresen, T. L. *Adv. Biochem. Eng. Biotechnol.* **2011**, *125*, 251–280.
5. Lis, H.; Sharon, N. *Chem. Rev.* **1998**, *98*, 637–674.
6. Jayaraman, N.; Maiti, K.; Narech, K. *Chem. Soc. Rev.* **2013**, *42*, 4640–4656.
7. Mehvar, R. *Curr. Pharm. Biotechnol.* **2003**, *4*, 283–302.
8. Stahn, R.; Grittner, C.; Zeisig, R.; Karsten, U.; Felix, S. B.; Wenzel, K. *Cell. Mol. Life Sci.* **2001**, *58*, 141–147.
9. Irache, J. M.; Salman, H. H.; Gamazo, C.; Espuelas, S. *Expert Opin. Drug. Del.* **2008**, *5*, 703–724.
10. Sun, X.-L.; Kanie, Y.; Guo, C.-T.; Kanie, O.; Suzuki, Y.; Wong, C.-H. *Eur. J. Org. Chem.* **2000**, *14*, 2643–2653.
11. Haywood, A. M.; Boyert, B. P. *Biochemistry* **1986**, *25*, 3925–3929.
12. Slamas, J. S.; Rando, R. R. *Biochemistry* **1980**, *19*, 4595–4600.
13. Zalipsky, S.; Mullah, N.; Dibble, A.; Flaherty, T. *Chem. Commun.* **1999**, 653–654.
14. Zhu, J.; Xue, J.; Guo, Z.; Zhang, L.; Marchant, R. E. *Bioconjugate Chem.* **2007**, *18*, 1366–1369.
15. Song, C. K.; Jung, S. H.; Kim, D.-D.; Jeong, K.-S.; Shin, B. C.; Seong, H. *Int. J. Pharm.* **2009**, *380*, 161–169.
16. Kunimasa, J.-I.; Inui, K.-I.; Hori, R.; Kawamura, Y.; Endo, K. *Chem. Pharm. Bull.* **1992**, *40*, 2565–2567.
17. Vyas, S. P.; Katare, Y. K.; Mishra, V.; Sihorkar, V. *Int. J. Pharm.* **2000**, *210*, 1–14.
18. Xu, Z.; Jayaseharan, J.; Marchant, R. E. *J. Colloid Interface Sci.* **2002**, *252*, 57–65.
19. Boons, G. J. *Expert Rev. Vac.* **2010**, *9*, 1251–1256.
20. Ikehara, Y.; Shiuchi, N.; Ikehara, S. K.; Nakanishi, H.; Yokoyama, N.; Takagi, H.; Nagata, T.; Koide, Y.; Kuzushima, K.; Takahashi, T.; Tsujimura, K.; Kojima, N. *Cancer Lett.* **2008**, *260*, 137–145.
21. Ikehara,Y.; Niwa, T.; Biao, L.; Ikehara, S. K.; Ohashi, N.; Kobayashi, T.; Shimizu, Y.; Kojima,N.; Nakanishi, H. *Cancer Res.* **2006**, *66*, 8740–8748.

22. Harada, Y.; Murata, T.; Totani, K.; Kajimoto, T.; Md. Masum, S.; Tamba, Y.; Yamazaki, M.; Usui, T. *Biosci. Biotechnol., Biochem.* **2005**, *69*, 166–178.
23. Hassane, F. S.; Phalipon, A.; Tanguy, M.; Guerreiro, C.; Belot, F.; Frisch, B.; Mulard, L. A.; Schuber, F. *Vaccine* **2009**, *27*, 5419–5426.
24. Hayden. T. V.; Elhalabi, J.; Rice, K. G. *Carbohydr. Res.* **1998**, *306*, 387–400.
25. Muller, C. D.; Schuber, F. *Biochim. Biophys. Acta* **1989**, *986*, 97–105.
26. Hassane, F. S.; Frisch, B.; Schuber, F. *Bioconjugate Chem.* **2006**, *17*, 849–854.
27. Zhang, H.; Ma, Y.; Sun, X.-L. *Chem. Commun.* **2009**, 3032–3034.
28. Engel, A.; Chatterjee, S. K.; Al-arifi, A.; Riemann, D.; Langner, J.; Nuhn, P. *Pharm. Res.* **2003**, *20*, 51–57.
29. Torchilin, V. P.; Weissing, V. *Liposomes: A Pratical Approach*, 2nd ed.; Oxford University Press: Oxford/New York, **2003**.
30. Jesorka, A.; Orwar, O. *Annu. Rev. Anal. Chem.* **2008**, *1*, 801–832.
31. Mammen, M.; Choi, S.-K.; Whitesides, G.M. *Angew. Chem. Int. Ed.* **1998**, *37*, 2754–2794.
32. Marradi, M.; Chiodo, F.; Garcia, I.; Penades, S. Glycoliposomes and metallic glyconanoparticles in glycoscience, in Olivier Renaudet and Nicolas Spinelli, eds. *Synthesis and. Biological Applications of Glycoconjugates*. Bentham eBooks, Beijing, China **2011**, pp. 164–202.
33. Hupfeld, S.; Holsaeter, A.M.; Skar, M.; Frantzen, C.B.; Barndl, M. *J. Nanosci. Nanotechnol.* **2006**, *6*, 1–7.
34. Liu, D.; Mori, A.; Huang, L. *Biochim. Biophys. Acta* **1992**, *1104*, 95–101.
35. Juliano, R.L.; Stamp, D. *Biochem. Biophys. Res. Commun.* **1975**, *63*, 651–658.
36. Zeisig, R.; Stahn, R.; Wenzel, K.; Behrens, D.; Fichtner, I. *Biohcim. Biophys. Acta* **2004**, *1660*, 31–40.
37. Nyren-Erickson, E.K.; Haldar, M.K.; Totzauer, J.R.; Ceglowski, R.; Patel, D.S.; Friesner, D.L.; Srivastava, D.K.; Mallik, S. *Langmuir* **2012**, *28*, 16115–16125.
38. Coldren, B.; van Zaten, R.; Mackel, M.J.; Zasadzinski, J.A.; Jung, H.T. *Langmuir* **2003**, *19*, 5632–5639.
39. Cui, H.; Hodgton, T.K.; Kaler, E.K.; Abezgauz, L.; Danino, D.; Lubovsky, M.; Talmond, Y.; Pochan, D. J. *Soft Matter* **2007**, *3*, 945–955.
40. Barbeau, J.; Cammas-Marion, S.; Auvray, P.; Benvegnu, T. *J. Drug Deliv.* **2011**, *11*, 396068.
41. Surace, C.; Arpicco, S.; Dufay-Wojcicki, A.; Marsaud, V.; Bouclier, C.; Clay, D.; Cattel, L.; Renoir, J.-M.; Fattal, E. *Mol. Pharm.* **2009**, *6*, 1062–1073.
42. Chen, S. S. H.; Kou, A. Y. *J. Chromatogr. B.*, **1982**, *227*, 25–31.
43. Shimizu, Y.; Nakata, M.; Matsunuma, J.; Mizouchim T. *J. Chromatogr. B.* **2001**, *754*, 127–133.
44. DeFrees, S. A.; Phillips, L.; Guo, L.; Zalipsky, S. *J. Am. Chem. Soc.* **1996**, *188*, 6101–6104.
45. Suroliam A.; Bachhawat, B. K.; Podder, S. K. *Nature* **1975**, *257*, 802–804.
46. Gregoriadis, G.; Wills, E. J.; Swain, C. P.; Tavill, A. S. *Lancet* **1974**, *1*, 1313–1316.
47. Woodle, M. C.; Lasic, D. D. *Biochim. Biophys. Acta* **1992**, *1113*, 171–199.
48. Vabbilisetty, P.; Sun, X.-L. *Org. Biomol. Chem.* **2014**, *12*, 1237–1244.
49. Barbet, J.; Machy, P.; Truneh, A.; Leserman, L. D. *Biochim. Biophys. Acta* **1984**, *772*, 347–356.

50. Jelinek, R.; Kolusheva, S. *Chem. Rev.* **2004**, *104*, 5987–6015.
51. Pan, J. J.; Charych, D. *Langmuir* **1997**, *13*, 1365–1367.
52. Spevak, W.; Foxall, Charych, D. H.; Dasgupta, F., Nagy, J. O. *J. Med. Chem.* **1996**, *36*, 1018–1020.
53. Bruehl, R. E.; Dasgupta, F.; Katsumoto, T. R.; Tan, J. H.; Bertozzi, C. R.; Spevak, W.; Ahn, D. J.; Rosen, S. D.; Nagy, J. O. *Biochemistry* **2001**, *40*, 5964–5974.
54. Davis, B. G. *J. Chem. Soc., Perkin Trans. 1.* **1999**, *1*, 3215–3237.
55. Kiessling, L. L.; Pohl, N. L. *Chem. Biol.* **1996**, *3*, 71–77.
56. Jain, K.; Kesharwani, P.; Gupta, U.; Jain, N. K. *Biomaterials* **2012**, *33*, 4166–4186.
57. Chono, S.; Kaneko, K.; Yamamoto, E.; Togami, K.; Morimoto, K. *Drug Dev. Ind. Pharm.* **2010**, *36*, 102–107.
58. Damodaran, D.; Jeyakani, J.; Chauhan A.; Jeyakani, J.; Chauhan, A.; Kumar, N.; Chandra, N. R.; Surolia, A. *Glycoconjugate J.* **2008**, *25*, 191–198.
59. Gabius, H.-J. *Cancer Invest.* **1987**, *5*, 39–46.
60. Bernardes, G. J. L.; Kikkeri, R.; Maglinao, M.; Laurino, P.; Collot, M.; Hong, S. Y.; Lepenies, B.; Seeberger, P. H. *Org. Biomol. Chem.* **2010**, *8*, 4987–4996.
61. Chen, W. C.; Completo, G. C.; Sigal, D. S.; Crocker, P. R.; Saven, A.; Paulson, J. C. *Blood* **2010**, *115*, 4778–4786.
62. Kelly, C.; Jefferies, C.; Cryan, S.-A. *J. Drug Del.* **2011**, *11*, 727241.
63. Ravichandran, K. S.; Lorenz, U. *Nat. Rev. Immunol.* **2007**, *7*, 964–974.
64. Gupta, A.; Gupta, R. K.; Gupta, G. S. *J. Sci. Indust. Res.* **2009**, *68*, 465–483.
65. Singodia, D.; Verma, A.; Verma, R. K.; Mishra, P. R. *Nanomedicine* **2012**, *8*, 468–477.
66. Jeong, H. S.; Na, K. S.; Hwang, H.; Oh, P. S.; Kim, D. H.; Lim, S. T.; Sohn, M. H.; Jeong, H. J. *J. Biomed. Mater. Res. Part A* **2014**, *102*(12), 4545–4553, doi: 10.1002/jbm.a.35112.
67. Mintzer, M. A.; Simanek, E. E. *Chem. Rev.* **2009**, *109*, 259–302.
68. Kuramoto, Y.; Kawakami, S.; Zhou, S.; Fukuda, K.; Yamshita F.; Hashida, M. *J. Gene Med.* **2008**, *10*, 392–399.
69. Perche, F.; Gosset, D.; Mevel, M.; Miramon, M. L.; Yaouanc, J. J.; Pichon, C.; Benvegnu, T.; Jaffres, P. A.; Midoux, P. *J. Drug Targeting* **2011**, *19*, 315–325.
70. Reyam T.; Morrison, S. J.; Clarke, M. F.; Weissman, I. L. *Nature* **2001**, *414*, 105–111.
71. Kozako, T.; Arima, N.; Yoshimitsu, M.; Honda, S.-I.; Soeda, S. *Int. J. Nanomed.* **2012**, *7*, 4943–4951.
72. Kojima, N.; Ishii, M.; Kawauchi, Y.; Takagi, H. *BioMed Res. Int.* **2013**, *11*, 562924.
73. Pereira, C. F.; Torensma, R.; Hebeda, K.; Kretz-Rommel, A.; Faas, S. J.; Figdor, C. G.; Adema, G. J. *J. Immunother.* **2007**, *30*, 705–714.
74. Kawauchi, Y.; Kuroda, Y.; Kojima, N. *Glycoconjugate J.* **2012**, *29*, 481–90.
75. Singh, S. K.; Stephani, J.; Schaefer, M.; Kalay, H.; Garcia-Vallejo, J. J.; den Haan, J.; Saeland, E.; Sparwasser, T.; van Kooyk, Y. *Mol. Immunol.* **2009**, *47*, 164–174.
76. Ishii, M.; Kato, C.; Hakamada, A.; Kojima, N. *Int. Immunopharmacol.* **2011**, *11*, 164–167.
77. Kojima, N.; Kato, C.; Igarashi, M.; Ishii, M. *Cell Immunol.* **2011**, *271*, 335–341.
78. Nishikawa, Y.; Zhang, H.; Ikehara, Y.; Kojima, N.; Xuan, X.; Yokoyama, N. *Clin. Vac. Immunol.* **2009**, *16*, 792–797.
79. Ishii, M.; Kojima, N. *Int. Immunopharmacol.* **2013**, *15*, 685–692.

80. Chen, W. C.; Kawasaki, N.; Nycholat, C. M.; Han, S.; Pilotte, J.; Crocker, P. R.; Paulson, J. C. *PLoS One* **2012**, *7*, e39039.
81. Wang, B.; Boons, G.-J. *Carbohydrate Recognition, Biological Problems, Methods and Applications*. Wiley, Hoboken, **2011**.
82. McGowan, E. D.; Bowman, K. Background Paper on Glycosciences and Glycomics in the United States. Board on Chemical Sciences and Technology and Board on Life Sciences, National Research Council; **2010**.
83. Skehel, J. J.; Wiley, D. C. *Annu. Rev. Biochem.* **2000**, *69*, 531–569.
84. Palese, P.; Tobita, K.; Ueda M.; Compans, R. W. *Virology* **1974**, *61*, 397–410.
85. Wilson, J. C.; von Itzstein, M. *Curr. Drug Targets* **2003**, *4*, 389–408.
86. Moscona, A. N. *Eng. J. Med.* **2005**, *353*, 1363–1373.
87. Sun, X.-L. *Curr. Med. Chem.* **2007**, *14*, 2304–2313.
88. Kingery-Wood, J. E.; Williams, K. W.; Sigal, G. B.; Whitesides, G. M. *J. Am. Chem. Soc.* **1992**, *114*, 7303–7305.
89. Spevak, W.; Nagy, J. O.; Charych, D. H.; Schaefer, M. E.; Gilbert, J. H.; Bednarski, M. D. *J. Am. Chem. Soc.* **1993**, *115*, 1146–1147.
90. Guo, C.-T.; Sun, X.-L.; Kanie, O.; Shortridge, K. S.; Suzuki, T.; Miyamoto, D.; Hidari, K. I. –P. J.; Wong, C.-H.; Suzuki, Y. *Glycobiology* **2002**, *12*, 183–190.
91. Hendricks, G. L.; Weirich, K. L.; Viswanathan, K.; Li, J.; Shriver, Z. H.; Ashour, J.; Ploegh, H. L.; Kurt-Jones, E. A.; Fygenson, D. K.; Finberg, R. W.; Comolli, J. C.; Wang, J. P. *J. Biological Chem.* **2013**, *288*, 8061–8073.

9

GLYCANS IN MESOPOROUS AND NANOPOROUS MATERIALS

KEITH J. STINE
Department of Chemistry and Biochemistry and Center for Nanoscience, University of Missouri–St. Louis, St. Louis, MO, USA

9.1 INTRODUCTION

The use of porous materials in glycoscience has become an active research area. Porous or "nanoporous" materials have thus far found applications in glycan enrichment, separation, detection, drug delivery, and supported synthesis. In this chapter, a survey of applications of porous materials in glycoscience is presented, including mesoporous silica, mesoporous carbon, porous alumina, porous titania, porous polymer monoliths, and nanoporous gold. The application of these nanomaterials to separations and analysis in glycoproteomics and glycomics and proteomics in general is growing and has recently been reviewed [1–3].

9.2 POROUS MATERIAL CHARACTERISTICS

A range of terms are used to describe porous materials depending on their average pore size. A popular current term is "nanoporous" for pore size features below 1 μm, but earlier terminology is also widely used. Of particular significance are the terms macropore (above 50 nm), mesopore (2–50 nm), and micropore (below 2 nm), with these dimensions defined by IUPAC [4]. A porous material is also characterized by the distribution of pore sizes, by the pore geometry (cylindrical, slit shaped, ink bottle, or funnel shaped), and by the extent of pore interconnectedness as pores may

Carbohydrate Nanotechnology, First Edition. Edited by Keith J. Stine.

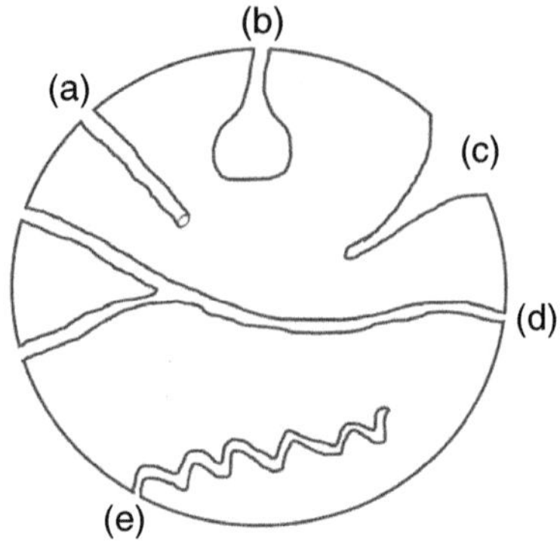

FIGURE 9.1 Common types of pore geometries: (a) cylindrical, (b) ink bottle, (c) funnel shaped, (d) pore open on both ends of the material or particle and also branched, and (e) a pore of high tortuosity.

be closed, only open on one end for fluid flow, or open on both ends of the material. Some of these features of porous materials are depicted in Figure 9.1. Pore size distributions with two dominant and widely separated prevalent diameters are referred to as bimodal or hierarchical and are of special interest since a given size molecule or protein may pass through the large pores but be trapped inside the smaller ones or may be unable to enter the smaller pores and have its diffusion restricted to the larger pores. It is also possible that molecules can flow through the larger pores but must diffuse into the smaller pores in a bimodal structure. The total pore volume is another significant parameter, as is the total porosity defined as the pore volume divided by the total volume, which does not include interparticle voids that may be present. The surface area to volume ratio is also a useful parameter. An additional parameter of interest is the tortuosity, defined as the distance a molecule must travel through the pore structure to move between two points divided by the straight line distance between those two points. The specific surface area given in $m^2\ g^{-1}$ is an essential parameter. Some porous materials are comprised of many agglomerated smaller particles, and others have a bicontinuous structure of pores and solid matter.

9.3 METHODS USEFUL FOR CHARACTERIZING POROUS MATERIALS

Characterization of the porous material being used is desirable for understanding its behavior and suitability for surface modification by glycans or chemical species intended to interact with glycans. Specific surface area is most commonly determined by nitrogen gas adsorption measurements in which the amount of N_2 gas adsorbed on the material at 77 K is measured as a function of P/P_0 (P_0 = vapor pressure over the bulk N_2 liquid) and then analyzed using the Brunauer–Emmett–Teller (BET) isotherm model, typically over the range $0.05 < P/P_0 < 0.30–0.35$ [5].

The BET model allows for multilayer adsorption and conversion of the amount of N_2 adsorbed to specific surface area requires using an established value for the area occupied by each adsorbed N_2 molecule, typically taken as $0.16\,nm^2$. The pore size distribution is most commonly determined from the adsorption or desorption branch using the Barrett–Joyner–Halenda (BJH) theory [6], which is based on the Kelvin equation applied to capillary condensation/evaporation in cylindrical pores as P/P_0 is increased toward 1.0 along the adsorption branch and then decreased along the desorption branch. Different pore size distributions may be obtained upon applying the BJH analysis to the adsorption and desorption branches due to pore network and blockage effects [7]. The isotherms of porous materials generally display hysteresis loops when measured on adsorption followed by desorption that fall into one of four types [8]. For compatible and mechanically stable materials, mercury porosimetry [9] can be used to determine the pore size by putting the material under a gradually increasing pressure of Hg(l) and measuring the entrance of Hg(l) into the pores, subject to the assumption that pores fill with Hg(l) in order of increasing size. Other methods for characterizing macroporous solids have recently been reviewed [10].

When a porous material is surface modified, the amount of material added to the surface can in some cases be readily determined by pyrolysis using thermogravimetric analysis (TGA) in which the sample is heated under a flow of nitrogen gas and the decrease in mass is recorded using a sensitive microbalance [11]. TGA can be used to estimate the extent of surface modification or amount of organic material bound inside a porous material provided that the material is stable over the temperature range over which pyrolysis of the added molecules occurs. The structure of porous materials or the extent of reaction after modifying their surfaces or conjugating molecules to the material can be confirmed using infrared and Raman spectroscopy, with Raman spectra being especially useful for confirming the structure of porous carbons [12]. X-ray photoelectron spectroscopy can be used to probe surface elemental composition, and high-resolution scans on a selected element (such as carbon) can help to assess the bonding states that are present near the surface [13].

Small-angle X-ray scattering is a key method for determining the structure of porous materials [14] and can provide information on the pore geometry. Solid-state NMR can be applied to study the bonding states and dynamics of molecules inside porous materials [15] and to determine the structure of the materials themselves. Small-angle neutron scattering has found application to following the formation mechanism of mesoporous silica [16]. Positron annihilation spectroscopy (PAS) measures the lifetime of positrons or positronium inside a material and detects gamma rays emitted when the positron or positronium annihilates with an electron. PAS can provide information on pore size distribution since the lifetime of a positron or positronium inside a porous material depends on the probability of annihilation with an electron and is a function of pore size [17]. PAS has been applied to characterize mesoporous silica [18].

Electron microscopy is an essential tool for imaging the morphology of porous materials and securing information for image analysis of pore dimensions and shapes, interconnectedness of pores, and agglomeration state for porous materials composed of agglomerates of smaller particles. Electron tomography can be used to obtain 3D reconstructions of porous materials [19–22]. Atomic force microscopy (AFM) can be used to image proteins bound to the surface of a porous material [23–25] or to characterize surface topography [26,27]. Diffusion of fluorescent dye molecules within mesoporous silica monoliths has been followed by single-molecule tracking [28].

9.4 GLYCOPROTEIN ENRICHMENT

Glycosylation is a common posttranslational modification, and over 50% of all mammalian proteins have one or more sites of glycosylation [29]. The role of glycoproteins in cell recognition, immune response, viral infection, cellular differentiation, cancer, and other key biological processes is under investigation and is supported by the intensely active field of glycomics [30]. Selective enrichment of glycans, glycopeptides, or glycoproteins on porous monolithic liquid chromatography columns prior to mass spectrometry analysis is a key enabling technology for glycomics. Current methods for enrichment and separation of glycoproteins and glycopeptides on such columns have been reviewed [31–34]. Given their high surface area and prospect for surface functionalization, porous materials have found growing application as supports for capture and enrichment of glycans, glycopeptides, and glycoproteins. Studies can involve intact glycoproteins, glycoproteins digested by treatment with trypsin to give a glycopeptide mixture, or N-glycans released from glycoproteins or glycopeptides by action of the enzyme PNGase F (peptide N-glycosidase). The release of O-glycans involves use of chemical reactions such as by β-elimination under gentle conditions or use of hydrazine [35–39]. Glycopeptides are usually a minor component of tryptic digests, and hence, their enrichment to improve signal-to-noise (S/N) ratios in mass spectrometry analysis is important in order to detect all the glycans present and allow for their further analysis by MS/MS techniques. Matrix-assisted laser desorption/ionization time of flight (MALDI-TOF) is a widely applied method given its convenience and sensitivity. Electrospray ionization (ESI) is also widely applied together with liquid chromatography. The methods for glycan enrichment that are most widely used include boronate affinity, hydrazide chemistry, lectin affinity, and hydrophilic interaction chromatography (HILIC).

Boronic acid-modified materials can be used to capture glycans via the covalent reaction between boronate and *cis*-diols, shown in Figure 9.2. The reaction to form a cyclic boronate ester under basic conditions, typically pH 8–9, can be used to capture a glycan, and lowering the pH to acidic conditions allows the glycan to be released. The pH should be above the pK_a of the boronic acid (pK_a = 8.9 for phenylboronic acid) in most cases for binding to occur; however, sialic acids are known to bind to boronates at pH < pK_a of the boronic acid [40,41]. In order to assess the stability of cyclic boronate esters, a method was developed using the dye Alizarin Red S, which has a diol in its structure and is nonfluorescent when free but becomes fluorescent when it forms a cyclic boronate ester [42]. The presence of carbohydrate competes with Alizarin Red S for a boronic acid species, and the strength of the diol–boronate association could be followed by changes in fluorescence [43]. The method was

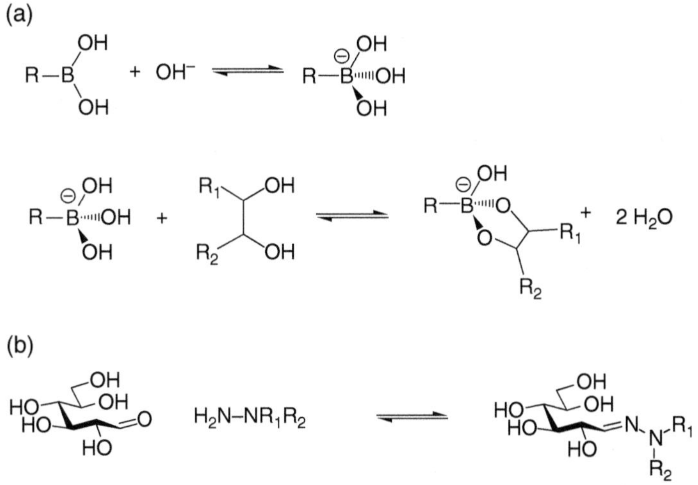

FIGURE 9.2 The two most prevalent covalent schemes for capturing glycans: (a) reversible boronate ester formation, (b) formation of hydrazone bonds by reaction of oxidized sugar with hydrazide groups.

applied to study binding of 25 arylboronic acids (pK_a ranging from 4.2 to 9.0) with sugars including glucose, fructose, and catechol. It was found that boronic acids with lower pK_a did not always give the strongest diol-binding affinity and that the optimal pH for binding was not always above the pK_a, and other steric factors and added buffer components were also relevant. Gold can be used as a surface for MALDI, and gold-coated silicon wafers modified with self-assembled monolayers (SAMs) presenting terminal diboronic acid groups were used to capture glycopeptides from tryptic digests [44]. Enrichment on the SAM-modified gold plate reduced the limit of detection for the three glycoproteins studied by two orders of magnitude relative to that for the digest itself, and the procedure of enrichment involved only sample loading and removal of supernatant. Many of the porous materials developed for glycan enrichment recently have been based on the principle of boronate affinity. Approaches to lowering the pH required for binding glycan to be closer to neutral or to moderately acidic have included use of modified phenylboronic acids substituted with electron-withdrawing groups [45,46], of Wulff-type boronic acids with intramolecular B–O or B–N coordination [47], and the "molecular team" approach as have been reviewed [48]. In the "molecular team" approach, a molecule with an amine is assembled together with the boronic acid to provide B–N coordination so that the combination (team) will perform like a Wulff-type boronic acid. The ability to bind at near-neutral pH or lower is clearly advantageous as it is less likely that biomolecules targeted will become unstable and also to allow binding of glycans over a wider range of pH.

The capture of glycans using hydrazide chemistry involves first oxidizing *cis*-diols on some of the sugars to aldehydes followed by reaction with hydrazide groups on the surface of the porous support to form hydrazone bonds [49], as shown in Figure 9.2. The hydrazone bond is stable under basic conditions, but the glycans can be released by addition of an acidic solution. Examples of this approach include glycan enrichment using hydrazide-modified magnetic nanoparticles [50], hydrazide-modified controlled pore glass (CPG) [51], and hydrazide-coated paramagnetic silica particles [52].

Lectin affinity can provide selectivity toward the glycan structures bound by the selected lectin as many lectins are available with some degree of specificity against a wide range of known glycan structures [53,54]. As such, lectin affinity approaches will not necessarily capture all glycans present in a sample. The glycans captured by immobilized lectins can be released by exposure to free carbohydrate ligand or by change in pH to promote dissociation of the bound glycan. Perhaps the most commonly used lectin is concanavalin A (Con A), isolated from Jack bean, and having affinity for mannose and glucose, the protein is a tetramer above neutral pH and dissociates into dimers at lower pH [55,56]. Concanavalin A can recognize the trimannose core structure of N-glycans, Jacalin has an affinity for D-galactose or β(1-3)-linked *N*-acetylgalactosamine, and the affinity of other lectins has been reviewed [57]. One issue to keep in mind with use of lectin affinity is that the optimal conditions of pH for binding and the need for inclusion of metal ions such as manganese and calcium for concanavalin A are specific to each lectin.

HILIC involves use of a polar stationary phase and nonpolar mobile phase. Analytes partition between a water-rich layer on the surface of the stationary phase and the mobile phase. There are a range of column materials presenting hydrophilic surface chemistries suitable for HILIC, including poly(2-sulfoethyl)aspartamide, poly(2-hydroxyethyl)aspartamide, and TSKgel Amide-80, which presents many pendant amide groups at the ends of spacer chains, a β-cyclodextrin-modified material, and others [31]. HILIC was compared to both use of TiO_2 for binding of sialic acid and use of hydrazide chemistry using a standard glycopeptide mixture and liquid chromatography electrospray ionization tandem mass spectrometry (LC–ESI–MS/MS) [58]. The sulfoalkylbetaine zwitterionic (ZIC-HILIC) material and the TSKgel Amide-80 were found specific for capturing glycopeptides. It was proposed that the sulfobetaine zwitterions formed a particularly stable water layer near the surface of the stationary phase. ZIC-HILIC did not have a bias for particular glycan structures. The enrichment of N-linked glycopeptides by hydrazide chemistry resulted in less efficient glycopeptide recovery and identification of a smaller number of glycopeptides. The standard mixture used was a tryptic digest of bovine fetuin, RNase B, alpha-1-acid glycoprotein, bovine serum albumin (BSA), histone, and human IgG yielding a total of 9 N-linked glycopeptides of known sequence and structure.

Approaches to glycopeptide enrichment using magnetic nanoparticles have also been introduced and proven to be promising. Particles with a magnetic core and a polymethyl methacrylate shell decorated with Ag nanoparticles afforded rapid capture of glycopeptides from rat serum digests treated with PNGase F. The glycans were adsorbed by multivalent hydrophilic interactions with the silver surface [59]. Core–shell nanoparticles with an Fe_3O_4 core and a poly(methacrylic hydrazide) shell were prepared and used to capture glycopeptides from the tryptic digests of fetuin and asialofetuin for MALDI-TOF analysis [60]. Mesoporous nanoparticles of Fe_3O_4 were surface modified with glucose and through hydrophilic interactions used to enrich N-glycans released from human sera by PNGase F and also to enrich glycopeptides from tryptic digests of horseradish peroxidase.

9.5 MESOPOROUS SILICA MATERIALS

9.5.1 Basic Aspects of Mesoporous Silicas

Mesoporous silicas are materials with variable pore sizes in the mesopore range and in many preparations display ordered arrangements of pores. They can be formed by condensation reaction of a silica source such as tetraethyl orthosilicate (TEOS) in the presence of a surfactant that self-assembles into cylindrical micelles that form hexagonal arrays and provide a structural template for pore formation, such as in the formation of the M41S family of mesoporous silicas [61]. The pore size can be varied by changing the surfactant chain length, and the surfactant choice can also give rise to different self-assemblies in solution and hence different pore arrangements with layered, hexagonal, and cubic geometries among the possibilities [62]. Amphiphilic triblock copolymers are also used as structure-directing agents, such as in preparation of SBA-15 materials with larger pore sizes [63, 64]. The surfactant or polymer is burned off by calcination at high temperature to leave the mesoporous silica. Silica surfaces can then be modified by silanes that bear a terminal functional group, with 3-aminopropyltriethoxysilane (3-APTES) or 3-aminopropyltrimethoxysilane (3-APTMS) being examples of widely used surface-modifying agents that provide $-NH_2$ groups that can then be subject to coupling to other species. It is also possible to include cocondensing species bearing the desired functional groups directly during the preparation of the material. Different preparations of mesoporous silica are often referred to by notations such as Mobil Composition of Matter 41 (MCM-41) and Santa Barbara Amorphous 15 (SBA-15), among others.

9.5.2 Enrichment of Peptides Using Mesoporous Silicas

There are many potential biomarkers in the circulating low molecular weight proteome that are of interest for capture and analysis, and mesoporous silicas have proven promising for this application. Mesoporous silica thin films supported on silicon wafers were prepared by spin coating a solution containing TEOS and a polyethylene oxide–polypropylene oxide–polyethylene oxide triblock copolymer (PEO–PPO–PEO) followed by evaporation and calcination. The self-assembly of the triblock copolymer as the solvent evaporates provided the template for the regular pore structure (3.7–3.9 nm) obtained after condensation of TEOS followed by calcination to remove the polymer template. The resulting films were used to capture low molecular weight proteins from serum by incubation for 30 min that were then eluted with acid for MALDI-TOF analysis. 3D honeycomb, 2D hexagonal, or 3D cubic pore nanostructures could be formed by varying the polymer concentration, and these structures exhibited significantly different efficiencies for recovering different peptides, with the 3D structures being the most efficient [65].

Mesoporous silica materials have been used to enrich peptides from human plasma, for analysis of the circulating endogenous peptides referred to as the peptidome, which is in a lower molar mass range than the proteome, generally below 15 kDa. The material MCM-41 with a pore size of 2.05 nm, in which the surface silanol groups were converted to hydrophobic bridging siloxanes, was found to be capable of efficiently binding peptides in the 1–12 kDa range from human plasma. Elution by acetonitrile followed by MALDI-TOF enabled detection of many well-defined mass/charge peaks not seen without enrichment using the material [66]. The small pore size excluded adsorption of larger-size proteins. Adsorption of lysozyme into this material was found to require only 30 min for equilibrium. Subsequently, MCM-41 nanoparticles were modified with either anion or cation exchange groups so that enrichment of endogenous peptides could be carried out near physiological pH. These materials were used to enrich the mouse liver peptidome by suspending the material with the mouse liver extract, centrifuging and eluting using NaCl solutions. The eluted peptides were then analyzed using 2D-nano LC–MS/MS. Use of these cation and anion exchange MCM-41 forms was found to be complimentary and to greatly increase the number of unique peptides that could be captured [67].

9.5.3 Mesoporous Silicas Modified for Boronate Affinity Applications

An increasing number of studies have used boronate-modified mesoporous silicas for glycan affinity applications. In one earlier study [68], boronic acid was grafted onto mesoporous silica (FDU-12) by first reacting 3-glycidyloxypropyltrimethoxysilane and 3-aminophenylboronic acid and then using the product to modify the mesoporous silica surface. The material was found to have average pore size of 18.2 nm with pore entrances of ~3.5 nm diameter after surface modification. The material was used for glycopeptide enrichment from the tryptic digests of horseradish peroxidase, fetuin, asialofetuin, and invertase. For the tryptic digest of horseradish peroxidase, the material was able to bind 5 glycopeptides and leave all nonglycosylated peptides in solution, as detected by MALDI. The limit of detection for glycopeptides by MALDI was found to be improved by factors of 70–125, and interference from addition of tryptic BSA was found to be negligible.

Phenylboronic acid functionalized mesoporous silica was prepared by cocondensation of TEOS and a phenylboronic acid triethoxysilane derivative obtained by reaction of 3-isocyanatopropyltriethoxysilane and 3-aminophenylboronic acid [69]. The boronic acid groups were then reacted with the dye Alizarin Red S through the diol on one of its aromatic rings. The fluorescent dye could be displaced by a sugar and released from the material. The materials were imprinted with D-fructose and D-xylose during the cocondensation of TEOS and cetyltrimethylammonium bromide (CTAB) surfactant and a phenylboronic acid-appended triethoxysilane derivative. Extensive washing with HCl removed the surfactant and the monosaccharide templates. The specific surface area was 576 $m^2 g^{-1}$ for MCM and 269 $m^2 g^{-1}$ for FruIM (D-fructose template material). The pore size was 2.10 nm for MCM and 1.80 nm for

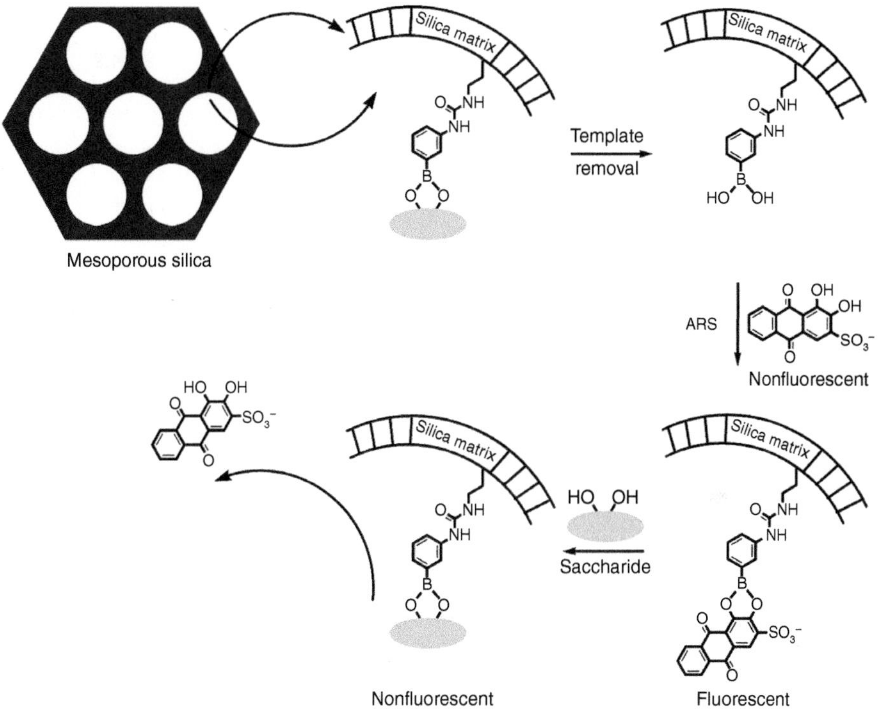

FIGURE 9.3 Scheme indicating the use of molecularly imprinted mesoporous silica modified with boronic acids and used together with the dye Alizarin Red S to create an indicator–displacement sensor for saccharides. *Source*: Reprinted with permission from Ref. 69. Copyright 2009 American Chemical Society.

FruIM. The pore volume was reduced for the FruIM material to $0.191\,cm^3\,g^{-1}$ from $0.264\,cm^3\,g^{-1}$ for MCM. The binding of Alizarin Red S resulted in a change in fluorescence intensity. Titration of NIM, FruIM, and XylIM (D-xylose template material) with eight saccharides (fructose, xylose, arabinose, ribose, sorbose, glucose, mannose, and galactose) was carried out. Decreases in fluorescence were observed, and the response was greatest to fructose for FruIM and greatest to xylose for XylIM, although a response was also seen from other sugars. The general scheme for these sensors is shown in Figure 9.3.

Hierarchically ordered porous materials with bimodal pore size distributions are of interest for enrichment of glycans and glycoproteins due to their high surface area and good mass transport through the larger pores [70]. Hierarchically ordered macro- and mesoporous silica (HOMMS) was prepared using colloidal crystals of polystyrene as a hard template. Following impregnation with TEOS in the presence of an amphiphilic triblock copolymer, and calcination, a structure was obtained with macropores of 1 μm diameter and mesopores of average diameter 4.6 nm. Specific surface area, as determined from BET measurements, was $255\,m^2\,g^{-1}$, and the pore volume was $0.46\,cm^2\,g^{-1}$. Modification of the silica

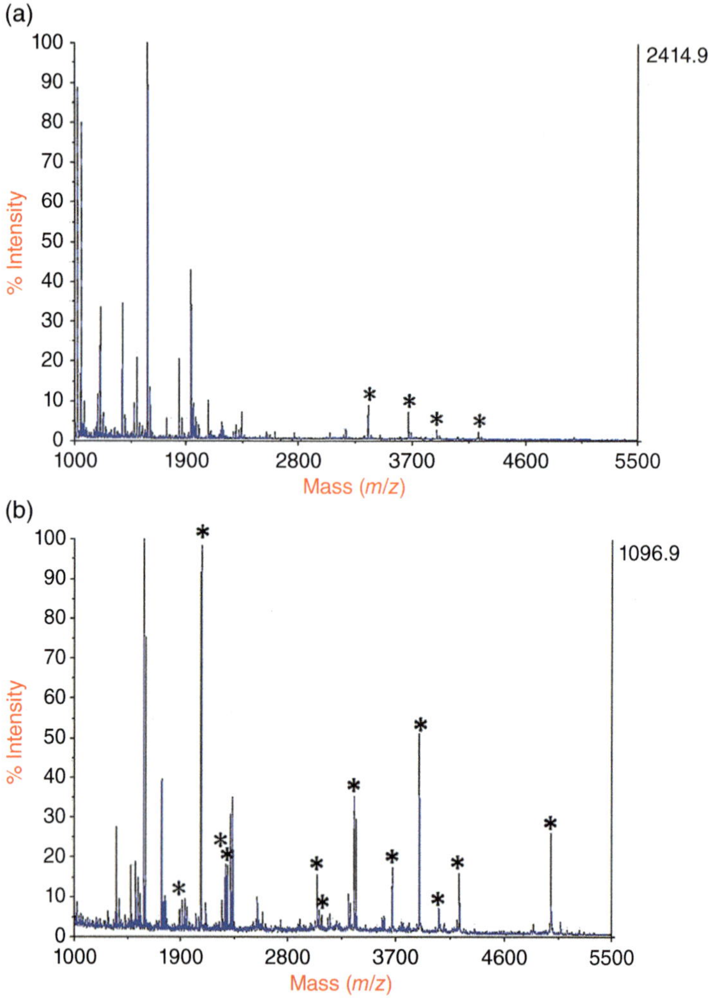

FIGURE 9.4 MALDI-TOF mass spectra of HRP digests (10 ng μl^{-1}). Without (a) and with (b) enrichment by HOMMS@PolyVPBA. The peaks of glycopeptides or their fragments are marked. *Source*: Reprinted with permission from Ref. 70, Copyright 2014 Wiley-VCH Verlag GmbH & Co.

with 3-methacryloxypropyltrimethoxysilane for coupling to 4-vinylphenylboronic acid (VPBA) reduced the surface area and pore volume but resulted in a material with high affinity to glycopeptides. Exposure of the material to the tryptic digest of horseradish peroxidase followed by elution using a mixture of acetonitrile, water, and trifluoroacetic acid (50:49:1 v/v/v) and analysis using MALDI-TOF showed that 12 glycans could be detected, whereas only 4 could be detected from the digest without enrichment, as shown in Figure 9.4. The enrichment process of incubation and elution could be completed in 30 min, an efficiency attributed to good mass transport through the macroporous regions. Enrichment of glycopeptides was demonstrated to be possible even in the presence of large amounts of BSA.

Surface modification of mesoporous silica with 3-APTMS followed by attachment to glucose, fructose, galactose, or mannose by borohydride coupling reactions gave materials of high surface area used to efficiently capture borate from solution. The adsorption capacity was the greatest for the mannose-functionalized material and lowest for that functionalized with galactose [71]. In a subsequent related study, glucose was incorporated into mesoporous silica using click chemistry of the azido-sugar onto acetylene-modified mesoporous silica MCM-41. The material was prepared by cocondensation of TEOS with 3-APTMS to yield an $-NH_2$-functionalized material. Acetylene modification was then achieved by reaction of this material with the widely used combination of 1-ethyl-3-[3-dimethylaminopropyl]carbodiimide hydrochloride (EDC) and *N*-hydroxysuccinimide (NHS) in the presence of 4-pentynoic acid. Specific surface area and pore size distribution were characterized using N_2(g) adsorption–desorption isotherms. The amount of functionalization of the silica surface was determined using TGA. Borate adsorption from aqueous media by the glucose-modified material was determined to be 5.4 mg g^{-1}, noted as twice the amount adsorbed by the unfunctionalized material [72].

9.5.4 Graphene Mesoporous Silica Composites

Graphene mesoporous silica composites have been prepared by first treating graphene with concentrated nitric acid, followed by treating it with TEOS and CTAB, a surfactant that serves as a structure-directing agent since it tends to adsorb and form bilayers [73]. The final step was treatment with sulfuric acid followed by carbonization of CTAB in an N_2(g) environment at high temperature. The resulting product was shown to have uniform pores of about 2.8 nm and a very high surface area of 372 m^2 g^{-1}. Mesoporous channels allowed the material to exclude large proteins. These materials were shown to enhance the S/N ratio in MALDI-TOF analysis for β-cyclodextrin, chosen as a "standard glycan," for the N-linked glycans in a digest of ovalbumin, and from a digest of human serum. In addition to increasing the S/N ratio, the number of glycans that could be detected also increased.

9.5.5 Lectin-Modified Mesoporous Silica

Ultrasonic spray pyrolysis was used to produce 1.6 ± 0.3 μm silica particles of high surface area (203 m^2 g^{-1}) presenting a network of macropores primarily in the 50–150 nm range and suitable for use in liquid chromatography columns under ultrahigh pressure (11,000 psi), as shown in Figure 9.5. The particles were covalently modified with the lectin Con A for use in lectin affinity chromatography. The lectin *Aleuria aurantia* (specific for α-linked fucose residues) was immobilized in a biotinylated form to particles surface modified to present avidin groups. Surface modification was achieved by first generating a highly reactive aldehyde-covered silica surface by 3-glycidyloxypropyltrimethoxysilane. Compared to previously

FIGURE 9.5 (a) Scanning electron micrograph and (b) transmission electron micrograph images of silica particles synthesized by ultrasonic spray pyrolysis/drying using salts as a pore template. (c) Corresponding N_2 adsorption–desorption isotherms of the particles (inset: pore size distributions obtained from BJH analysis of the same samples; surface area based on three-point BET analysis is also denoted). *Source*: Reprinted with permission from Ref. 75, Copyright 2013 American Chemical Society.

developed 10 μm size silica particles modified by Con A [74], the amount of ribonuclease B captured by a column of the smaller microparticles was greater by factors of 37–75. Treatment of the material with NaOH solution to create more hydrophilic silanol groups resulted in higher loading of Con A and a column able to retain analyte longer during elution of glycoprotein (horseradish peroxidase or alpha-1-acid glycoprotein). The new material was tested using diluted blood serum, and using only a 1 μl aliquot, the fractions bound to lectin and also not bound to lectin were subjected to LC–MS/MS analysis [75].

9.5.6 Controlled Release Applications Involving Mesoporous Silica and Glycans

In addition to applications involving enrichment or capture of species, mesoporous silica has also found wide application as a vehicle for controlled release of molecules such as therapeutics contained within its pores, usually using some sort of triggered gating mechanism. Mesoporous silica nanoparticles (MSN) (110 ± 10 nm) were surface modified with alkenyl-terminated silanes, and then thiol-terminated mannose derivatives were attached using thiol–ene click chemistry [76]. The lectin Con A was used to cap the nanoparticle pores. The strongly fluorescing dye molecule rhodamine 6G was trapped inside the particles prior to capping and lowering the pH to 5.5, 3.5, or 1.5 resulting in successively more complete release of the dye from inside the particles. The release was attributed to the removal of the Mn^{2+} and Ca^{2+} cations needed for binding and to dissociation of Con A from tetramers to dimers and monomers. Of potential significance to release of diabetes drugs, addition of glucose was found to compete for Con A bound to the mannose units near the pore exits and result in release of the dye molecule. The particles were fully characterized using solid-state NMR, SEM, TEM, FTIR, small-angle powder X-ray diffraction, and N_2(g) adsorption–desorption isotherm measurements.

MSN were modified with APTES and conjugated to mannose. The mannose modification was designed to target cancer cells that overexpress mannose receptors [77] in two-photon excitation photodynamic therapy. The particles were reported to have a hydrodynamic diameter of 118 nm and contain 6850 porphyrin photosensitizer molecules each, resulting in giant cross sections per particle for two-photon absorption. The photodynamic efficiency against two breast cancer cell lines and a colon cancer cell line was greater for the mannose-modified particles than for the particles not modified by mannose. MSN functionalized with mannose or with galactose were targeted against a retinoblastoma cell line [78], and endocytosis of the particles into the cells was confirmed by confocal microscopy. The particles were also loaded with the anticancer drug camptothecin, and it was found that the combination therapy of one-photon excitation and release of the drug from the particles was significantly more effective than one-photon excitation alone. With two-photon excitation alone, the mannose- or galactose-modified particles were much more effective than the unmodified particles. Cancer cells also have affinity for galactose, and it was shown that galactose-functionalized MSN are routed from the outer cell membrane to endosomes and then to lysosomes [79]. Use of the MSN for dual PDT and delivery of the cancer drug camptothecin to breast, colon, and pancreatic cancer cells was shown to be effective, especially so when both were irradiated and used to deliver camptothecin.

MSN of the MCM-41 type (pore size 2.4 nm, surface area 1200 $m^2 g^{-1}$) also are promising for enzyme-triggered controlled release, demonstrated for lactose-capped material with the release of $[Ru(bipy)_3]^{2+}$ being negligible until action of the enzyme β-D-galactosidase to hydrolyze the glycosidic bond and release the terminal galactose unit [80]. Mesoporous–macroporous silica prepared using an emulsion-based method provided a material with a hierarchical distribution of larger macropores (>50 nm) and mesopores. Variation of the emulsion composition afforded materials with different release profiles for ibuprofen [81]. Release of the dye rhodamine B encapsulated inside magnetic MSN triggered by the presence of aflatoxin B_1 (AFB) was achieved using nanoparticles modified by mannose and whose pores were capped by a biotinylated Con A bound to streptavidin bound also to biotinylated anti-AFB antibody. Gold nanoparticles bound to both invertase and a BSA–AFB conjugate were introduced in the presence of free AFB, and the magnetic mesoporous silica and gold nanoparticles bound together due to the binding of the BSA–AFB conjugate bound to the anti-AFB antibodies were collected using a magnet. Glucose released due to invertase hydrolysis upon addition of sucrose was used to displace mannose on the particle surfaces from Con A, uncap pores, and release rhodamine B such that the fluorescence signal decreased linearly with the AFB concentration [82].

9.6 POROUS ALUMINA AND POROUS TITANIA

9.6.1 Porous Alumina

Porous aluminum oxide was surface modified to present mannose and then found to enhance the adherence of *Lactobacillus plantarum*, a bacterium that binds strongly to mannose [83]. The stability of a number of surface-modifying molecules as a function of time and temperature was explored using measurements of contact angle versus time and of the percent of C on the surface determined by the C1s signal from XPS data. A long-chain phosphonic acid with a terminal alkyne was found to be most stable, and click chemistry was used to attach an azido-terminated mannose with an 11 carbon spacer chain using alkyne–azide cycloaddition catalyzed by Cu^+ ions; from XPS data, it was determined that the efficiency of the click reaction was near 70%. A mixed SAM of 10-undecynynlphosphonic acid and 1-octylphosphonic acid was used to provide spacing between the attached mannose units and reduce steric crowding for improved bacterial binding. A hydroxyl-terminated azide analogue was used as a control. It was found that the bacterium bound significantly to the mannose-modified surfaces but much less so to the hydroxyl-terminated surfaces. The porous aluminum oxide substrates had an average pore size of 200 nm. In a prior study by the same group, porous alumina was modified by 1,15-hexadecadiyne suitable for further modification by click chemistry with 11-azidoundecyl lactoside [84]. The binding of peanut agglutinin to the surface was demonstrated and was stronger for the β-anomer than for the α-anomer. The lactose-modified surfaces enhanced the colonization of *Candida albicans*.

Commercially available "Anopore" membranes of porous alumina were used as supports for the enzyme alliinase assembled together with the lectin Con A [85]. The membrane surfaces were surface modified with 3-APTES and then conjugated to 1,4-butanedioldiglycidyl ether. Binding of mannan, a polysaccharide of mannose, was followed by binding of a layer of Con A subsequently used to bind alliinase (isolated from garlic powder). Alliinase is a glycoprotein with two mannose-rich glycans that catalyzes the conversion of alliin to allylsulfenic acid and pyruvic acid. Enzyme loading was the greatest for an optimal concentration of APTES, either too low or too high of a concentration resulted in less efficient loading (due to pore clogging). The authors proposed that other enzymes presenting mannose-rich glycans could be immobilized using the scheme.

A nanoporous alumina membrane with 2–4 nm pore sizes was used as the basis for a glucose sensor operating between 2 and 12 mM, noted as the most physiologically relevant concentration range [86]. The membrane allowed glucose diffusion while retaining Con A. The sensor contained actuating and sensing piezoelectric diaphragms, a microchannel, and the nanoporous membrane. The response was based on the change in viscosity with glucose concentration and is an example of a "viscometric affinity sensor." The sensitivity of the viscosity to glucose concentration is based on the competition for binding to Con A between dextran and glucose, with displacement of bound dextran from Con A by glucose resulting in a decrease in viscosity. The membrane is needed to retain Con A and dextran in the sensing volume but allow glucose to be at the same concentration as in the testing solution [87]. The strategy for glucose detection based on osmotic pressure, in which a number of porous polymer membranes were studied, also examined use of alumina [88]. Porous alumina membranes have also been surface modified and used for affinity-based separation of proteins, demonstrated by capture of hemoglobin by bound chitosan chelated to Cu^{2+} ions [89].

9.6.2 Porous Titania

Porous titania in the form of microspheres has been applied to enrichment of glycopeptides and phosphopeptides [90]. The TiO_2 surface has an affinity for phosphate and thus can enrich phosphopeptides. The study used monodisperse titania microspheres [91] prepared by a sol–gel method with a surface area of 76.8 $m^2 g^{-1}$ and with a pore diameter of 10.8 nm. The titania surface was found to hydrogen bond to glycopeptides, and glycopeptides from tryptic digests of human serum IgG were able to be captured and eluted for mass spectrometry detection. Elution of the IgG glycopeptides required a more concentrated trifluoroacetic acid solution (5%) than did elution of glycopeptides from a tryptic digest of bovine RNase B, attributed to enhancement of affinity for the TiO_2 surface due to the presence of two successive glutamic acids in the IgG glycopeptides. Elution of the phosphopeptides could be achieved using a basic solution. TiO_2 has a strong affinity for sialic acid, and 5 mm TiO_2 beads have been used to selectively capture and enrich sialic acid containing glycopeptides for subsequent MALDI analysis and applied to detect differences in the sialiome of healthy patients versus those with advanced bladder cancer [92].

9.7 MESOPOROUS CARBON

Mesoporous carbon offers a size exclusion effect whereby large proteins do not enter the pores and selectivity for lower molecular weight peptides is achieved. Oxidized mesoporous carbon (O-CMK-3) was used to selectively enrich N-linked glycans, according to the scheme shown in Figure 9.6 [93]. The material was prepared by taking mesoporous silica (CMK-3) and mixing it with sucrose and sulfuric acid, followed by carbonization at high temperature under N_2(g). The SBA-15 template was removed using HF, and then the resulting carbon material was refluxed with nitric acid, filtered, washed, and dried. O-CMK-3 was characterized by N_2(g) adsorption–desorption isotherms (specific surface area = $1576\,m^2\,g^{-1}$ for O-CMK-3), TEM, and infrared spectroscopy. The pore size distribution was calculated from the desorption branch using the BJH method and found to be fairly narrow and peaked near 3.4 nm. γ-Cyclodextrin was used as a standard glycan to estimate adsorption capacity of the material. Adsorption was complete in less than 10 min. PNGase F was used to release the N-linked glycans from denatured protein. Adsorbed glycans were eluted using 50% acetonitrile solution. The number of N-glycans detected from PNGase F treated ovalbumin increased from 16 to 24, and the MS peak intensities were enhanced. N-glycans released from human serum samples and captured on O-CMK-3 did not show interference from human serum albumin. Comparison of released and captured N-glycans from liver cancer patients showed that compared to samples from healthy patients, the intensities for five core-fucosylated glycans were enhanced.

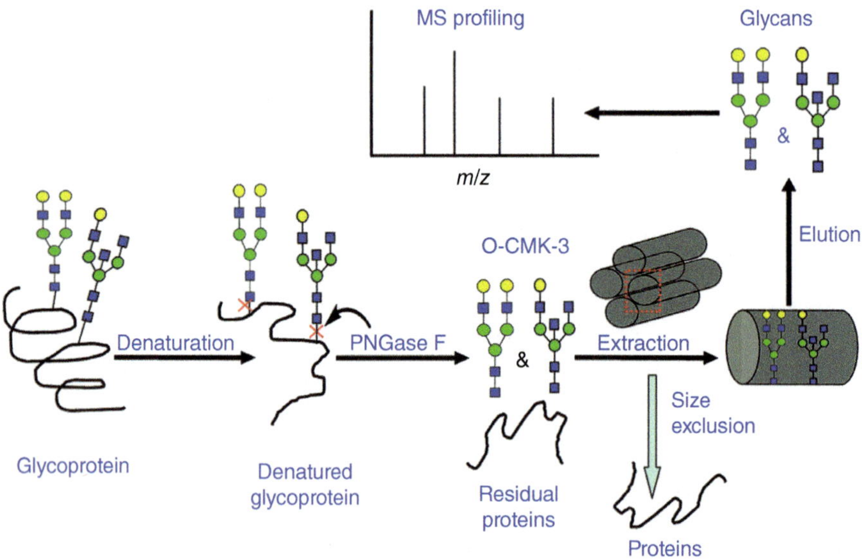

FIGURE 9.6 Scheme for size-selective enrichment of glycans by O-CMK-3 (oxidized mesoporous carbon material). *Source*: Reprinted with permission from Ref. 93, Copyright 2011 American Chemical Society.

In a subsequent study, preparation of silica–carbon composite nanoparticles was achieved using a MCM-41 support and CTAB as the structure-directing agent and carbon source [94]. The particles were of uniform shape, being roughly spherical with diameter of around 70 nm. The regularity of shape was in contrast to O-CMK reported as having irregular shapes and being from microns to millimeters in size. The precursor was carbonized under N_2(g) at 600°C and was a black-colored material showing Raman bands near 1580 and 1360 cm^{-1} indicating the G and D modes of sp^2-hybridized aromatic carbons. Treatment with sulfuric acid prior to thermal carbonization resulted in a greater extent of carbonization, and without this treatment, the material was not effective at glycan enrichment. A surface area of 901 $m^2 g^{-1}$ for the material was obtained. N-glycans were released from ovalbumin by PNGase F and after enrichment and release using the material. Twenty-four N-linked glycans were clearly seen in MALDI-TOF spectra. When applied to glycans released from human serum filtered to remove endogenous peptides, 36 N-linked glycans were detected, 8 more than when using O-CMK. Use of the material denoted as NP-MCM-C gave higher S/N ratios in MALDI-TOF-MS than obtained using the O-CMK material.

Porous graphitic carbon (PGC) has been available as a material for stationary phases in liquid chromatography for over 20 years [95]. The material has a structure of intertwined ribbons of graphite. Molecules are retained on its surface by both hydrophobic interactions and by polar or electrostatic interactions for which the explanation is not simply the presence of oxygen-containing functional groups and is more complex and not fully resolved. In an example of the application of a PGC column to glycomics, application to monkey plasma with detection by triple TOF mass spectrometry resulted in identification of 130 N-glycoproteins demonstrating the value of the mixed mode of retention on a PGC column [96]. PGC retains smaller *N*-glycopeptides better than larger ones, and use of PGC beds in combination with C_{18} reverse phase beads in a column resulted in capture and identification of more *N*-glycopeptides. A PGC HPLC-chip was used together with a PNGase F column for LC–MS and LC–MS/MS analysis of N-glycans released from human serum [97]. Use of graphitized carbon in clinical glycomics using liquid chromatography–mass spectrometry approaches has recently been reviewed [98].

Carbohydrate precursors have been used to form mesoporous carbons for subsequent use in separation of sugars [99]. Alginic acid, a block copolymer of α(1-4)-linked β-D-mannuronic acid and α-L-guluronic acid found in algae or seaweed, was found to adopt a mesoporous form after heating an aqueous solution to form a suspension, cooling to form a gel, exchange of water for ethanol and vacuum or supercritical drying. Heating under argon gas yielded mesoporous carbon materials with a pore diameter near 14 nm and pore volume near 1 $cm^3 g^{-1}$. Limitation of micropore content was a desirable outcome since micropores irreversibly capture analytes by high-energy adsorption. Preparation temperatures were varied from 200 to 1000°C. The material prepared at 1000°C was stable and of low oxygen content. The material was packed into a 50 mm LC column and applied to separate the carbohydrates glucose, sucrose, the trisaccharide raffinose, the tetrasaccharide stachyose, and the pentasaccharide verbascose, with glucose and sucrose giving different retention times and the remaining three sugars having a similar retention time.

9.8 POROUS POLYMER MONOLITHS

9.8.1 Boronate-Modified Porous Polymer Monoliths

Porous polymer monoliths are generally prepared by *in situ* polymerization in a capillary column, resulting in a so-called "fritless" column known to offer a number of advantages in chromatography such as less trouble with occurrence of bubbles and better reproducibility [100,101]. Polymer monoliths containing boronic acid have been used in numerous chromatography applications, including those of modified boronates. An early report of a boronate affinity polymer monolith was by Potter [102]. Here, monoliths of poly(glycidylmethacrylate-*co*-ethylene glycol dimethacrylate) were prepared in fused silica capillaries (75 or 100 μm inner diameter) that had been surface treated so that the polymer would bind to the inner capillary surface. The monoliths were then functionalized with *p*-hydroxyphenylboronic acid. The monolith column retained cytidine over 2-deoxycytidine and adenosine over 2-deoxyadenosine, with retention higher at pH 8 than at pH 6 or pH 7. It is important that boronate affinity monoliths be hydrophilic to limit nonspecific adsorption of proteins (reversed phase retention) by hydrophobic interactions. Liu and coworkers reported the first example of a boronate affinity monolith prepared by *in situ* free radical polymerization [103]. The capillary wall was first modified with γ-methacryloxypropyltrimethoxysilane (γ-MAPS). The active monomer VPBA and cross-linker ethylene glycol dimethacrylate were polymerized in the presence of radical initiator azobisisobutyronitrile (AIBN) in a binary porogenic solvent of ethylene glycol and diethylene glycol. The specific surface area was found to be 47.73 $m^2 g^{-1}$. The monolith exhibited selectivity for catechol over quinol and adenosine over 2-deoxyadenosine and was able to retain flavin adenine dinucleotide. The monolith exhibited some aspects of reversed phase (hydrophobic) and hydrogen-bonding interactions. The same group then prepared a monolith by *in situ* free radical polymerization of VPBA and *N,N'*-methylenebisacrylamide (MBAA), serving as a more hydrophilic cross-linker [104]. The average pore diameter was 11.4 nm, and the specific surface area was 10.6 $m^2 g^{-1}$. SEM showed densely interconnected globules with larger open channels. The monolith was found to bind catechol, but not quinol, and to bind the glycoproteins horseradish peroxidase and lactoferrin while not retaining four nonglycoproteins (BSA, lactoglobulin, myoglobin, and cytochrome *c*).

A boronate affinity monolith was prepared in a capillary by a one-step polymerization procedure inside a silanized capillary tube and applied to the enrichment of glycopeptides and glycoprotein [105]. The polymer was formed by reaction of 3-acrylamidophenylboronic acid and ethylene dimethacrylate in methanol at 60°C for 20–24 h in the presence of 1 wt.% AIBN and poly(ethylene glycol) (PEG, molecular weight 1,000–20,000) as a porogen, which was flushed out with methanol after the reaction. Use of PEG of lower molar mass resulted in more permeable monoliths with a lower HPLC pressure drop, but 20,000 MW PEG was chosen since it gave monoliths of better mechanical strength. From the BET adsorption isotherm from $P/P_0 = 0.05–0.35$, a specific surface area of $20\,m^2\,g^{-1}$ was determined, and BJH analysis of the desorption branch gave an average pore size of 6.9 nm. The columns were shown to retain adenosine or cytidine, but not their 2-deoxy forms. The column retained and enriched 11 glycopeptides from tryptic digests of horseradish peroxidase, as shown by MALDI-TOF, and to enrich horseradish peroxidase in the presence of BSA, as shown by SDS-PAGE applied to the eluate (eluent = 1% TFA solution).

Given that immunoglobulin G antibodies present a glycan at the Fc region, they can be captured by boronate binding leaving the Fv binding domains accessible [106]. Bimodal pore size distribution of macropores and mesopores allows the mesopores to provide size selection. A "restricted access boronate affinity porous monolith" was designed to serve as a mimic of protein A. In this structure, IgG does not enter the smaller nanochannels and is held just at their surface by affinity of the Fc glycan for a "molecular team" on the surface. Boronic acid requires pH 8.5 or higher for covalent binding; however, the combination of 4-mercaptophenylboronic acid and *N,N″*-bis(2-aminoethyl)oxamide immobilized together and serving as a "molecular team" was able to bind the IgG glycan at neutral pH. Polymerization of tris(2,3-epoxypropyl) isocyanurate (TEPIC) cross-linker with *m*-aminophenylboronic acid and 1,6-hexamethylenediamine produced monoliths with bimodal pore size distributions that would bind small molecules with *cis*-diols but would not bind glycoproteins. Coordination of nitrogen to boron in the "molecular team" was confirmed by NMR and provided binding to *cis*-diols at near-neutral pH. Mercury intrusion porosimetry showed that the average size of the macropores was 1.44 μm. From BET isotherm analysis, the mesopores were found to be 3.88 nm on average, and the specific surface area was $13.74\,m^2\,g^{-1}$. A capacity for IgG of $23.73\,mg\,g^{-1}$ was found. Five out of seven glycoproteins tested were not retained. The material was used to fractionate human serum, and capture of the antibody fraction was confirmed by its release and analysis by MALDI-TOF. Captured IgG antibody retained its ability to recognize antigen, confirmed using erythropoietin (EPO) and anti-EPO.

The macropores present in boronate affinity monoliths can promote more rapid flow through a column. A metal–organic gel, formed from Fe^{3+} and 1,3,5-benzenetricarboxylic acid, was found to serve as a porogenic template [107]. *In situ* polymerization in a column of a mixture containing 3-acrylamidophenylboronic acid, ethylene dimethacrylate, AIBN, and the gel precursors was achieved. After polymerization, the metal–organic gel template was washed out using 0.25 M H_2SO_4. The macropores, evident in SEM images as variations in the amount of space between interconnected clusters of globular particles, were approximately 2 μm in size. The extent of macropores influenced the permeability and backpressure of the monolithic column, and these parameters could be adjusted by varying the composition of the initial mixture of components. The better developed distribution of macropores increased the throughput. Specificity for transferrin over BSA was demonstrated in a one-step separation.

A boronate affinity monolithic column selective for one glycoprotein over another was prepared by a molecular imprinting procedure [108]. The interior of a glass capillary was first modified with γ-MAPS and then the monolith formed by polymerization of VPBA and pentaerythritol triacrylate in the binary porogenic solvent of ethylene glycol and cyclohexanol using AIBN initiator. Horseradish peroxidase was then captured on the monolith to full capacity, and then dopamine was polymerized around the glycoprotein before washing out the glycoprotein. The imprinted monolith, used in solid-phase microextraction (SPME), was selective for horseradish peroxidase over transferrin and did not bind nonglycoproteins as confirmed by SDS-PAGE.

A monolith for in-line SPME was prepared by copolymerization of VPBA, ethylene dimethacrylate in the porogenic solvent mixture of diethylene glycol (45 µl), and ethylene glycol (195 µl) using AIBN as initiator. The reaction temperature was 75°C for 12 h followed by flushing with methanol inside a fused silica capillary that had been inner surface treated with γ-MAPS. The capillary had an effective length of 25 cm and had an inner diameter of 100 µm. UV detection at 214 nm was used for following protein elution. A buffer of pH of 8.5 was used for capture (100 mM phosphate buffer with 20% ACN), and elution was carried out using 100 mM acetic acid at pH 2.7. Selectivity for catechol over quinol was demonstrated, for OVA over BSA, and for the glycoproteins from egg white as confirmed by SDS-PAGE [109].

Boronate affinity binding of sialic acids at pH < pK_a of the boronic acid has been exploited to separate sialylated glycoprotein from nonsialylated glycoprotein on a polymer monolith [110]. At pH = pK_a − 1, the sialylated glycoproteins were favored for capture, while at pH = pK_a + 1, the nonsialylated glycoproteins could be captured. The monolith was prepared from VPBA and MBAA in dodecanol with AIBN initiator in a column that had first been vinyl silanized on the inner wall. The monolith was selective for sialic acid over mannose at pH 7.4 with the selectivity greatly improved by increasing the NaCl concentration from 100 to 500 mM in order to reduce electrostatic repulsion between negatively charged sialic acid and boronic acid that is slightly negatively charged at this pH. The sialylated glycoprotein EPO could be extracted from spiked human serum samples at pH 7.4 along with a minor amount of human serum albumin and α2–HS-glycoprotein that is sialylated.

A porous polymer coating on a silver electrode formed by electrodeposition of 3-aminophenylboronic acid using cyclic voltammetry was found to be useful for SPME coupled to HPLC [111]. The electrodeposited coating was 7 µm thick and found to be useful for boronate affinity extraction of riboflavin, including from beer samples.

Another general strategy as an alternate to *in situ* polymerization is to modify a monolith after it has been prepared. In one recent example, monoliths of a copolymer were amine modified and then coupled to 3-carboxy-benzoboroxole, which bound adenosine efficiently as low as pH 5. SEM imaging showed a porous agglomeration of polymer microparticles each approximately 1 µm in diameter. The monoliths were used to capture glycopeptides from a tryptic digest of horseradish peroxidase and also to capture glycoproteins from human saliva samples. The process of sample loading, washing, and then elution of the captured glycopeptides or glycoproteins was followed by MALDI-TOF. Selectivity was demonstrated by showing no capture of 2-deoxyadenosine while adenosine was captured, and capture of RNase B that has a glycan chain while RNase A that has no glycans was not retained. The column was able to capture 22 glycopeptides from the tryptic digest, noted as the most reported to date, and also to have captured 6 intact glycoproteins, the most reported to date [112]. A prior example used 3-carboxy-benzoboroxole to modify a polymer monolith prepared by copolymerization of glycidyl methacrylate (GMA) and methylenebisacrylamide [113]. For this monolith, the average pore size was 6.12 nm, and the specific surface area by BET analysis was $11.40\,m^2\,g^{-1}$. The monolith was shown to exhibit selectivity for binding glycoproteins and nucleosides. This monolith was also employed in a scheme for systematic evolution of ligands by exponential enrichment (SELEX) for identifying DNA aptamers that would strongly bind to glycoproteins captured on the monolith. The members of the ssDNA library bound to the glycoprotein (horseradish peroxidase) were eluted for amplification by PCR and then introduced again for another round of selection. After six rounds of selection, K_d values of $\sim 10^{-8}$ were achieved for aptamer–glycoprotein binding [114].

Preparation of a monolith using 4-(3-butenylsulfonyl) phenylboronic acid as the functional monomer and MBAA as cross-linker gave a material of specific surface area $5.0\,m^2\,g^{-1}$ that was able to bind adenosine, RNase B, and ovalbumin at neutral pH, with elution accomplished using 100 mM acetic acid at pH 2.7 [45]. A secondary separation mechanism involving hydrogen bonding was also identified [46]. A Wulff-type boronate affinity monolith was prepared by synthesizing 3-(dimethylaminomethyl)aniline-4-pinacol boronate and then using it to modify a monolith that had been prepared by copolymerization of GMA and poly(ethylene glycol) diacrylate (PEGDA) in the presence of a binary porogenic solvent of dimethyl sulfoxide (DMSO) and 1,4-butanediol. The monolith was shown to be able to bind adenosine at pH 5.5 and to enrich nucleosides from human urine samples of pH 6.5. While in most reported cases, the boronic acid is incorporated in the functional monomer prior to polymerization, postmodification of monoliths is also widely used and has been reviewed [115].

A monolith capable of binding at lower pH was prepared by coordinating hexamethylenediamine and *m*-aminophenylboronic acid to form a complex with B–N Wulff-type coordination [116]. The complex was used to cure an epoxy resin TEPIC and thereby form a macroporous monolith using PEG200 as a porogen. The material was capable of capturing adenosine or catechol under neutral pH, the release at acidic conditions occurring when the nitrogen is protonated and the B–N association is broken. Use of the functional monomer 3-acrylamidophenyl boronic acid, which has a lower pK_a than vinylphenylboronic acid, was used to produce a more hydrophilic monolith with greater binding capacity that was also found to capture nucleosides at near-neutral pH [117].

Gold nanoparticles containing porous polymeric monoliths have been reported [118–120], for the capture and separation of peptides and also for functioning in boronate affinity mode. A polymer monolith was first prepared by polymerization in a capillary of GMA and PEGDA in cyclohexanol with 1-propanol serving as a porogen [121]. Reaction with cystamine generated free thiol groups throughout the monolith used to capture and fix in place gold nanoparticles (13 nm average diameter) onto which a "molecular team" of 4-mercaptophenylboronic acid and 2-mercaptoethylamine were immobilized. The monolith showed excellent selectivity for horseradish peroxidase over BSA and was able to recover 84.8 ± 1.9% of horseradish peroxidase from a solution containing 4 μg. The monolith was also shown to successfully enrich glycoproteins from human plasma.

Click chemistry was used to modify an azide-functionalized monolith with an alkyne boronate derivative. The monolith was first prepared in a stainless steel column by reacting tetramethoxysilane, 3-(chloropropyl)-trimethoxysilane, and PEG (10,000 average molar mass) porogen followed by reaction with sodium azide to replace Cl [122]. The hybrid inorganic–organic monolith could also be prepared in one step with the click reaction happening at the same time as the polymerization. The monolith was selective for binding of glycoproteins over nonglycoproteins. The binding capacity increased with pH over the range 7–9; however, the monolith was less stable above pH 8. A hybrid inorganic–organic monolith was prepared by one-pot reaction of tetramethyloxysilane with 3-methacryloxypropyltrimethoxysilane and VPBA in a porogenic solvent diethylene glycol using AIBN as initiator. Adjustment of reagent ratios and reaction temperature was needed in order to successfully produce the monolith. The specific surface area was found to be $35.2\,m^2\,g^{-1}$, and the average pore diameter was 15.2 nm. The monolith column passed nonglycoproteins but retained the glycoproteins ovalbumin, transferrin, and HRP. Passage of egg white showed retention of ovotransferrin and ovalbumin but passage of lysozyme by SDS-PAGE analysis [123].

Cryogels are formed by lowering the temperature of the polymer and cross-linker solution to below the freezing point, resulting in lowering the concentration required for gelation and increasing the rate of gelation as part of the solution remains unfrozen. The result is a "supermacroporous" gel possessing pores in the range of several to hundreds of microns. Such cryogel formed from agarose and chitosan with glutaraldehyde as the cross-linker was subsequently activated with epoxy groups and then reacted with 3-aminophenylboronic acid. The gel was formed in a column and was shown to be effective for the high-efficiency capture and separation of ovalbumin and ovotransferrin from egg white [124].

9.8.2 Hydrophilic Interactions on Porous Polymer Monoliths

Hydrophilic interactions on porous polymer monoliths have also been exploited. Porous polyacrylamide monoliths for use in capillary electrochromatography were prepared by polymerization of acrylamide in the presence of a cross-linker and a porogen in a capillary [125]. Oligosaccharides of glucose (4–10 units) and ribonuclease B were used to test the performance of the column. Mercury porosimetry was used to determine the pore volume and BET analysis to determine specific surface area. Hydrophilic ligands were included during the copolymerization such as tris-acrylamide (*N*-[trishydroxymethylmethyl])acrylamide, MBAA, 2-cyanoethyl acrylate, *N,N′*-diacryloylpiperazine, and ethanesulfonic acid. Phase separation takes place during polymerization. The best separation of glycans released from ribonuclease B for MALDI-TOF analysis occurred in the absence of any copolymerizing ligand, and the interaction site was identified as the amides in the polyacrylamide backbone. The retention time increased with number of glucose units in oligoglucose. SEM of the monolith showed aggregated spherical features. In the absence of ligand, a 34.4 nm mean pore radius, 62.3% total porosity, and specific surface area of 73.9 $m^2 g^{-1}$ were determined. The observed electroosmotic flow indicated that larger pore sizes formed when the monolith was swelled in solvent.

When monoliths are prepared in a capillary, the inner capillary wall must be silanized so the monolith will bond to the capillary wall during polymerization. A porous polymer layer bonded to the inner walls of a capillary and leaving an open path in the center was prepared by reacting vinylbenzene chloride and divinylbenzene in methanol with added AIBN at 70°C for 16 h [126]. The reactive chlorides were then displaced by ethylenediamine to give $-NH_2$-terminated surfaces. The column allowed for low flow rates and high sensitivity, demonstrated by LC–MS analysis of 28 N-linked glycans released from ovalbumin by PNGase F and using only 3 ng of digest. Referred to as an HILIC-porous layer open tubular (PLOT) column, SEM confirmed the open path in the center. The column was used with ESI–MS/MS and allowed detection limits for glycans in the low femtomole to sub-femtomole level. The PLOT column allowed flow rates down to 20 nl min^{-1} that corresponded to higher ion intensities.

9.8.3 Glycan Release by Enzyme-Modified Porous Polymer Monoliths

The release of glycans by PNGase F was accomplished by immobilizing the enzyme in a porous polymer support inside a silica capillary so that the released glycans could flow through a column for subsequent separation by HILIC and then be analyzed by ESI/TOF-MS. The deglycosylated protein was captured on a hydrophobic monolith trap column placed after the PNGase F reactor. The porous polymer monolith for the reactor column was prepared using GMA, ethylene dimethacrylate, dodecanol, cyclohexanol, and AIBN followed by washing out the porogens. Photografting of ethoxylated hydroxyethyl methacrylate was used to create the hydrophilic column, and further photografting of vinyl azlactone was done to prepare the column for immobilization of PNGase F. The set of three columns (PNGase F reactor, hydrophobic trap column, and hydrophilic interaction column) were used to deglycosylate glycoproteins, trap the deglycosylated protein, and separate the glycans for analysis by ESI/TOF-MS. The method was able to deglycosylate human IgG much faster than solution-phase enzyme and was also applied to analysis of N-glycans released from ribonuclease B, chicken egg white albumin [127]. Palm and Novotny [128] first demonstrated the preparation of a polyacrylamide monolith in a column into which PNGase F was immobilized concurrently with polymerization. N-glycans were released from ribonuclease B, asialofetuin, alpha-1-acid glycoprotein, and ovalbumin flowed through the column followed by MALDI-TOF analysis. The column afforded rapid deglycosylation within 3.5 min, which was comparable to that achieved in 18 h in solution with free enzyme and using only 0.1–20 μg glycoprotein. The column was found to be stable for at least 8 weeks at room temperature. PNGase F is a precious enzyme, and hence, its efficient use in immobilized form is desirable. A related column was previously developed for use with trypsin [129] with polymerization and immobilization accomplished in one step within 1 h.

9.8.4 Porous Polymer Monoliths Modified by or Interacting with Lectins

A nanospray emitter for MS/MS was prepared in microcolumns by copolymerization of GMA and ethylene glycol dimethacrylate in the presence of porogenic solvents including methanol and mixtures of methanol and other fatty alcohols [130]. The inner wall of the fused silica capillary was first silanized using 3-(trimethoxysilyl) propyl methacrylate. Photopatterning was achieved by masking off the parts of the capillary where polymerization was not desired. Con A was immobilized after activation by cyanoborohydride, which oxidized ethylene glycol groups on the column to aldehydes. A tryptic digest of ribonuclease B, which has a single site of high-mannose Man5–Man9, was preconcentrated on the lectin affinity sprayer, and enhanced detection of the glycan peaks by MS and MS/MS was observed. Methanol as the porogen gave the lowest back pressure and largest pores. Use of the porous polymer monolith in the nanospray emitter for preconcentration provided better stable electrospray of ions over a range of flow rates.

A polymer monolith was prepared using GMA and ethylene dimethacrylate in cyclohexanol and dodecanol as porogenic binary solvent and AIBN initiator at 50°C for 12 h in a 200 μm inner diameter column that was 10 cm in length and that had been first surface treated with γ-MAPS. The monolith was then activated with iminodiacetic acid (IDA) and a sandwich complex of IDA–Cu(II)–Con A was formed. The Con A was able to recognize more *N*-glycoproteins from human urine that Con A loaded onto agarose. The glycoproteins were detected by ESI–MS with linear ion trap detection [131].

A polymer monolith was prepared by electron beam-induced free radical polymerization of pentaethyleneglycolmethacrylate and triethyleneglycoldimethacrylate using 2-propanol and 1-dodecanol and "macroporogen" and toluene as "microporogen." The monolith was more stable and less prone to shrinkage when also loaded with hydrophilic fumed silica particles. The monoliths were prepared in 200 μl pipette tips suitable for use in 96 well plates. The monolith was supermacroporous with an average pore diameter of 14.3 ± 0.7 μm. Con A was immobilized by glutaraldehyde cross-linking. Nonspecific BSA adsorption was very low. The monolith was highly selective and efficient for binding ovalbumin over BSA [132].

An ethylene dimethacrylate porous polymer monolith was formed within 20 μl polypropylene pipette tips [133]. Using azlactone chemistry, Au nanoparticles of 20 nm size were immobilized. The lectin *Erythrina cristagalli* (ECL) was conjugated to the surfaces using dithiobis(succinimidyl propionate) (DTSP linker). The monolith was used for enrichment of galactose-presenting glycoproteins such as desialylated transferrin prepared by treatment with neuraminidase. The performance of the nanoparticle-decorated monolith was better than that of a monolith prepared without gold nanoparticles to which ECL was attached covalently to grafted chains of poly(vinyl azlactone). SEM revealed the gold nanoparticles homogenously decorating a globular macroporous structure. RNase B did not bind ECL in the monolith since it is mannose terminated but desialylated transferrin, which presents galactose, was shown to be retained by ECL in the monolith. The monolith was also selective for desialylated transferrin over transferrin and over transferrin treated with PNGase F. Selectivity was also seen in a test mixture of nine proteins, of which six were nonglycoproteins.

Adsorption of Con A on macroporous chitosan and chitin membranes was examined. Poly(2-hydroxyethyl methacrylate) UV grafted onto microporous polypropylene membranes (MPPM) was modified by reaction of acetylated sugars (glucose pentaacetate and galactose pentaacetate) and used in lectin affinity separation [134]. After reaction, the acetates were removed by sodium methoxide. The MPPM membranes were commercially available and had a 200 nm average pore size and were 160 μM thick. Using labeled Con A and confocal fluorescence spectroscopy, it was possible to image the depth profile of the Con A in the membrane. The membrane was found to reach near monolayer capacity for Con A, which was strongly bound due to multivalent binding, requiring use of 1 M acetic acid for full elution that could not be achieved by glucose or methyl α-mannopyranoside, noted as not denaturing the Con A. This approach to preparing affinity membranes was subsequently extended to include the disaccharides lactose and maltose and also galactose [135]. Selectivity between binding of Con A and PNA was demonstrated and enhanced binding of PNA by lactose over galactose, but not of Con A, upon comparing glucose and maltose.

A MPPM was functionalized with acrylic acid by UV-induced graft polymerization and then EDC/NHS coupling with propargylamine gave an alkyne-modified MPPM, ready to be glycosylated by thiol–ene click with 2,3,4,4,6-tetra-*O*-acetyl-β-D-glucopyranosyl thiol solution in the presence of 2% benzophenone while irradiated at 365 nm [136]. The acetate-protecting groups were removed and the modified MPPM characterized using FTIR and XPS. Efficiency of the click reaction was determined from XPS of the ratio of the S2p/N1s peaks to be 56.5%. Recognition of Con A was demonstrated using FITC-labeled Con A and fluorescence microscopy. Less nonspecific binding of PNA occurred than was seen for the azide-clicked glucose. Con A-binding equilibrium took about 5 h, and there was some nonspecific binding of BSA.

Glycopolymer brushes were prepared by UV graft polymerization of α-D-allyl glucose on polypropylene microporous membranes with average pore diameter 200 nm and 80% porosity. The grafting density was determined from the mass increase. Modification with glucose was confirmed by IR and XPS and by decrease in contact angle of water. The binding of Con A by the membrane in phosphate-buffered saline required the presence of Mn^{2+} and Ca^{2+} and a sugar density above about 90 μg cm^{-2} where the Con A binding increased significantly due to multivalent interactions [137].

A macroporous chitosan membrane was prepared from a solution of chitosan using silica particles as a porogen that could be removed by treatment with a basic solution. The chitosan membranes could be converted to chitin membranes by acetylation. Using intrinsic protein fluorescence, these membranes were found to have a strong affinity for Con A [138].

Protein A or G columns are often used for removal of IgG from human serum; however, the carbohydrate region of IgG can also be targeted. A macroporous cryogel modified with Con A was prepared for this purpose [139]. Copolymerization of acrylamide and allyl glycidyl ether was achieved by free radical polymerization in frozen aqueous solution at −12°C, creating monolithic cryogels described as "opaque, spongelike, and elastic." The cryogel had large interconnected pores 10–100 μm in diameter and thin nonporous walls. The monolithic cryogel was shown to adsorb Con A and then could be used to adsorb IgG from aqueous solutions and also from human plasma with a capacity ($25.6\,mg\,g^{-1}$) that compared favorably to other available adsorbents and was reusable for at least 10 cycles with only a 2% decrease in capacity. Inside PEEK columns, a polymer monolith was prepared by ring opening metathesis reaction to yield a polymer with many pendant amino groups for immobilization of *Lens culinaris* hemagglutinin used in the extraction of glucose oxidase from buffer. The polymer monolith morphology was one of agglomerated microglobules 4–5 μm in size with flow through voids near 12 μm in size [140].

Bombardment of polymer membranes by accelerated ions can create nanopores that can be surface modified and used as nanosensors as their ion transport properties will vary with molecular recognition events. Conical nanopores can be made by applying an etching solution on one side of the membrane and cylindrical nanopores made by applying the 2 M NaOH etching solution on both sides of the membrane. Twelve microns thick terephthalate membranes were bombarded with 11.4 MeV Pb, U, or Au ions to create nanopores at a density of $10^7\,cm^{-2}$. The surface of the pores will present –COOH groups from the etching, and HRP was linked to the pore walls using EDC/NHS coupling reaction, and then Con A was bound by its affinity to the sugar residues of HRP [141]. Measurements of current were made using a picoammeter as voltages of −2 to +2 V was applied. The Poisson–Nernst–Planck equations were used to determine the pore diameters from the I-V data at pH 7.6. The cylindrical pore diameter was determined to be 8.5 nm and to decrease to 7 nm upon conjugation to horseradish peroxidase (44 kDa, $6.0 \times 7.0 \times 7.0$ nm) indicating that horseradish peroxidase did not cover the entire pore surface. Binding of Con A gave an s-shaped I-V dependence, indicating inhomogeneous charge distribution and reduction of the pore diameter to the 4–5 nm range. The resistance of the nanopores, normalized to that of the unmodified pores, decreased upon addition of free glucose relative to that for the horseradish peroxidase–Con A-modified pores.

9.9 NANOPOROUS GOLD

Nanoporous gold [142–144], prepared by the dealloying process in which less noble metals are removed from alloys of 20–50% gold, can produce a material typically with pores in the 50–200 nm range and comprise of interconnected pores and ligaments through which liquid flow is possible. The pore size obtained depends upon the method of dealloying, with acid dealloying at room temperature producing larger pores than electrochemical dealloying. Dealloying at low temperature can produce smaller pores sizes [145], and postdealloying annealing at high temperatures can produce larger pore sizes [146, 147]. The pore sizes provide a material of high surface area suitable for modification with SAMs [148]. The material presents opportunities for the supported synthesis of glycans, for the development of sensors and assays for glycans or glycoproteins, and potentially for the capture, enrichment, and release of glycans. The material can be regenerated back to bare gold and used many times, or the gold material can be recovered and processed back into a new alloy.

The application of NPG to supported synthesis of oligosaccharides was first reported by Stine and Demchenko with their introduction of the surface-tethered iterative carbohydrate synthesis (STICS) scheme [149]. In this study, NPG is introduced as an alternative to synthesis on polymer support and to overcome some of the drawbacks of supported synthesis on polymer beads, such as eliminating the need for time-consuming swelling and drying, reducing the generation of large volumes of waste solvent, poisoning of reagents inside resin beads, and lacking of compatibility with molecular sieves. Presenting a high surface area of gold does not involve any of the issues that arise if a porous material with many surface hydroxyls were to be used. CPG, a material with a somewhat similar porous morphology, was used in one report by Schmidt and coworkers for the preparation of a dimannoside and a trimannoside, prepared after modification of the CPG with 3-mercaptopropyl trimethoxysilane [150]. A SEM image of the NPG used for supported synthesis of carbohydrates is shown in Figure 9.7, together with the scheme for the STICS procedure. The NPG surface is first modified by a multidentate thiol or sulfide species that has a base-cleavable linker terminated by a glycosyl acceptor. The NPG is in the form of plates of dimensions 8.0 mm × 8.0 mm × 0.25 mm. The loading achieved on these plates was 0.9–1.1 μmol. Glycosylation reactions of three different immobilized acceptors with four different donor saccharides resulted in yields ranging from 31 to 65% as determined by HPLC analysis of the protected product disaccharide cleaved off of the NPG. Synthesis of a trisaccharide with 52% yield was also demonstrated. In a follow-up study [151], instead of NPG plates moved manually between reaction vessels, NPG chips were packed into a small HPLC column for use in a new HPLC-assisted approach to supported carbohydrate synthesis [152]. The focus of the study was on improvement of the yield of the glycosylation reaction achieved by increasing the length of the linker chain between the acceptor sugar and the lipoic acid "anchor." Increasing the length of the linker to C_8H_{16}–O–C_8H_{16} from the anomeric carbon to the ester linkage to lipoic acid and carrying out the reaction under flow in the NPG packed HPLC column achieved a yield of 90% for the glycosylation reaction. It was found beneficial to increase the reaction space around the acceptor sugar both by increase in the linker length and by using mixed SAMs with the acceptor sugar to diluting lipoic acid methyl ester of 2:3 in solution.

The high surface area of NPG has recently been demonstrated as the basis for new electrochemical assays for glycoproteins [153]. The high surface area allowed detection of 4-aminophenol, the product of the reaction of alkaline phosphatase on the substrate *p*-aminophenyl phosphate. Immobilization of glycoproteins to SAMs on NPG electrodes followed by incubation with Con A–ALP conjugates and then incubation with substrate resulted in square-wave voltammetry peaks related to the affinity of the glycoprotein for the lectin–enzyme conjugate. NPG electrodes modified with lipoic acid SAMs conjugated to the lectin–enzyme conjugates enabled a kinetic difference assay in which the rate of ALP conversion of substrate to product as assessed by the square-wave peak difference current was different

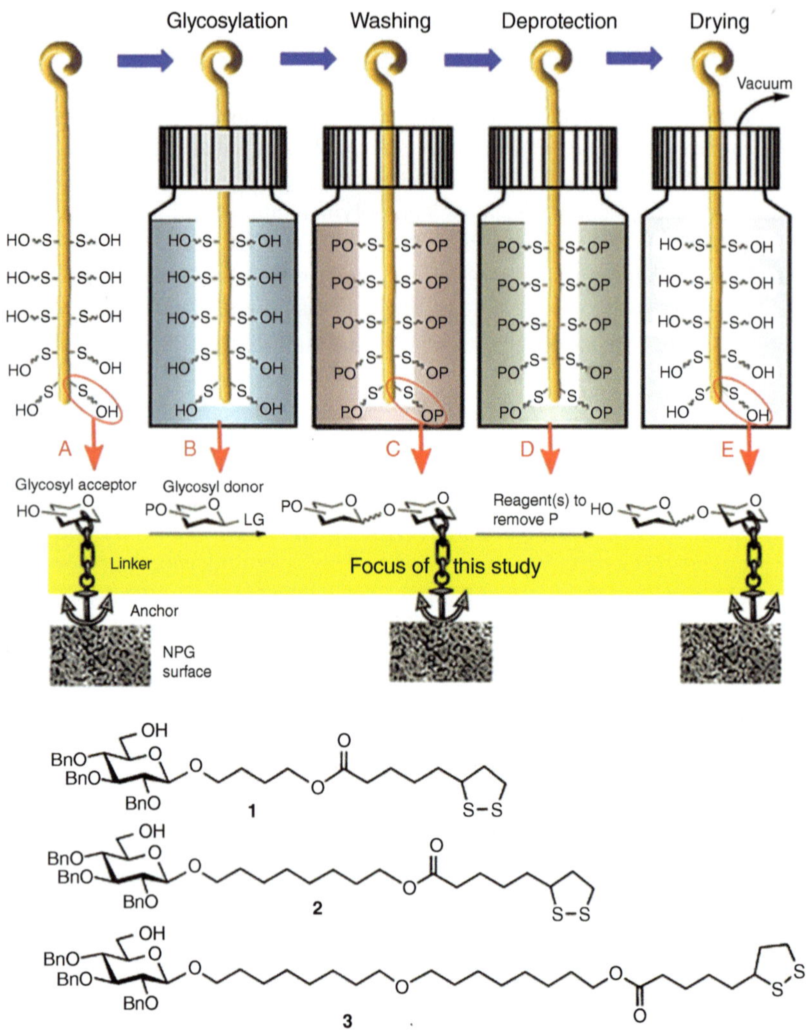

FIGURE 9.7 Surface-tethered iterative carbohydrate synthesis (STICS) procedure that is carried out on nanoporous gold monoliths and a series of lipoic acid anchor/extended linker/glycosyl acceptor scaffold molecules **1–3**. *Source*: Reprinted with permission from Ref. 151, Copyright 2013 American Chemical Society.

before and after glycoprotein binding and varied with glycoprotein concentration. Additionally, competitive assays with immobilized glycoprotein and glycoprotein in solution competed for the lectin–enzyme conjugate were also demonstrated. Electrochemical impedance spectroscopy (EIS) studies suggest that the structure of NPG affects the apparent binding affinity of lectin for carbohydrates in SAMs on the NPG surface, possibly due to the effect of NPG structure on the structure of the SAMs themselves [154,155].

NPG monoliths of 1.0 cm × 1.0 cm × 0.25 mm were used in a flow cell and modified with SAMs presenting mannose. The monoliths were demonstrated to capture Con A, which could be eluted by flowing through methyl-α-D-mannopyranoside [156]. By use of TGA, the monolayer coverage was found to be near the theoretical expectation of 4.7×10^{10} molecules cm^{-2} for SAMs of octadecanethiol and lower as expected for the more sterically bulky mannose derivative. Modification of the NPG monoliths with a short-chain thiolated PEG was found to make nonspecific protein adsorption negligible. AFM examination of the exterior and interior surfaces (exposed by cleaving the material) confirmed binding of Con A to mannose presenting SAMs on NPG.

9.10 FUTURE

The fields of glycoscience and nanoscience are both advancing rapidly, and there is a significant role for nanoporous materials to play in the advancement of glycan analysis, synthesis, and recognition. Nanoporous materials also clearly have great promise to continue to find new applications in assays and biosensors for glycoproteins. Nanoporous materials that can be conveniently surface modified with complex molecular architectures such as mixed SAMs, bound glycopolymers, and reactive functional groups are especially of interest. The integration of such modified nanomaterials with analytical systems in chromatography and mass spectroscopy is also primed for further advancement, as are applications in the separation, purification, and production systems for glycoproteins and other glycoconjugates. A field not addressed in this chapter, but of emerging interest, will be the use of glycans and related species as the basis for creating new nanomaterials, including new porous structures that can be surface modified and used for the above, and other applications such as in cellular engineering and biomaterials development.

REFERENCES

1. Li, L.; Xu, L.; Li, Z.; Bai, Y.; Liu, H. *Anal. Bioanal. Chem.* **2014**, *406*, 35–47.
2. Zhao, H.; Li, Y.; Hu, Y. *Clin. Proteom.* **2014**, *11*, 21.
3. Connolly, D.; Currivan, D.; Paull, B. *Proteomics* **2012**, *12*, 2904–2917.
4. Rouquerol, J.; Avnir, D.; Fairbridge, C. W.; Everett, D. H.; Haynes, J. H.; Pernicone, N.; Ramsay, J. E. F.; Sing, K. S. W.; Unger, K. K. *Pure Appl. Chem.* **1994**, *66*, 1739–1758.
5. Sing, K. S. W. *Colloids Surf. A* **2001**, *187–188*, 3–9.
6. Barrett, E. P.; Joyner, L. G.; Halenda, P. P. *J. Am. Chem. Soc.* **1951**, *73*, 373–380.
7. Sing, K. S. W. *Pure Appl. Chem.* **1982**, *54*, 2201–2218.
8. Gregg, S. J.; K. S. W. Sing, *Adsorption, Surface Area and Porosity*, 2nd edition. Academic Press, New York, **1982**.
9. Giesche, H.; *Part. Part. Syst. Charact.* **2006**, *23*, 9–19.
10. Rouquerol, J.; Baron, G. V.; Denoyel, R.; Giesche, H.; Groen, J.; Klobes, P.; Levitz, P.; Neimark, A. V.; Rigby, S.; Skudas, R.; Sing, K.; Thommes, M.; Unger, K.. *Microporous Mesoporous Mater.* **2012**, *154*, 2–6.

11. Keattch, C. J.; Dollimore, D. *An Introduction to Thermogravimetry*, 2nd edition. Heyden & Son Ltd. England, **1975**.
12. Dresselhaus, M. S.; Jorio, Ado; Hofmann, Mario; Dresselhaus, G. Saito, R. *Nano Lett.* **2010** *10*, 751–758.
13. Fadley, C. S. *J. Electron. Spectrosc. Relat. Phenom.* **2010**, *178–179*, 2–32.
14. Schmidt, P. W. *Stud. Surf. Sci. Catal.* **1988**, *39*, 35–48.
15. Buntkowsky, G.; Breitzke, H.; Adamczyk, A.; Roelofs, F.; Emmler, T.; Gedat, E.; Grünberg, B.; Xy, Y.; Limbach, H.-H.; Shenderovich, I.; Vyalikh, A.; Findenegg, G. *Phys. Chem. Chem. Phys.* **2007**, *9*, 4843–4853.
16. Hollamby, M. J.; Borisova, D.; Brown, P.; Eastoe, J.; Grillo, I.; Shchukin, D. *Langmuir* **2012**, *28*, 4425–4433.
17. Gidley, David W.; Peng, Hua-Gen; Vallery, Richard S. *Annu. Rev. Mater. Res.* **2006**, *36*, 49–79.
18. Boskovic, S.; Hill, A. J.; Turney, T. W.; Gee, M. L.; Stevens, G. W.; O'Connor, A. J. *Prog. Solid State Chem.* **2006**, *34*, 67–75.
19. Gommes, C. J.; Friedrich, H.; Wolters, M.; de Jongh P. E; De Jong, K. P. Q. *Chem. Mater.* **2009**, *21*, 1311–1317.
20. Stoeckel, D.; Kübel, C.; Hormann, K.; Höltzel, A.; Smarsly, B. M.; Tallarek, U. *Langmuir* **2014**, *30*, 9022–9027.
21. Leary, R.; Midgley, P. A.; Thomas, J. M. *Acc. Chem. Res.* **2012**, *45*, 1782–1791.
22. Yao, Y.; Czymmek, K. J.; Pazhianur, R.; Lenhoff, A. *Langmuir* **2006**, *22*, 11148–11157.
23. Prestidge, C. A.; Barnes, T. J. Mierczynska-Vasilev, A.; Kempson, I.; Peddie, F.; Barnett, C. *Phys. Status Solidi* **2008**, *205*, 311–315.
24. Xu, Tongwen; Fu, Rongqiang; Yan, Lifeng. *J. Colloid Interface Sci.* **2003**, *262*, 342–350.
25. Tan, Y. H.; Schallom, J. R.; Ganesh, N. V.; Fujikawa, K.; Demchenko, A. V.; Stine, K. J. *Nanoscale* **2011**, *3*, 3395–3407.
26. Ocelli, M. L.; Gould, S. A. C.; Stucky, G. D. *Stud. Surf. Sci. Catal.* **1994**, *84*, 485–492.
27. Shulga, O. V.; Jefferson, K.; Khan, A. R.; D'Souza, V. T.; Liu, J.; Demchenko, A. V.; Stine, K. *J. Chem. Mater.* **2007**, *19*, 3902–3911.
28. Park, S. C.; Ito, T.; Higgins, D. *J. Phys. Chem. B* **2012**, *117*, 4222–4230.
29. Varki, A. *Essentials of Glycobiology*, 2nd edition, Cold Spring Harbor Laboratory Press, Cold Spring Harbor, October 3, **2008**.
30. Cummings, R. D.; J. Michael Pierce. *Handbook of Glycomics*. Academic Press, New York, **2009**.
31. Ongay, S.; Boichenko, A.; Govorukhina, N.; Bischoff, R. J. *Sep. Sci.* **2012**, *35*, 2341–2372.
32. Li, H.; Liu, Z. *Trends Anal. Chem.* **2012**, *37*, 148–161.
33. Sproß, J.; Sinz, A. *J. Sep. Sci.* **2011**, *34*, 1958–1973.
34. Huang, B.-Y.; Yang, C.-K.; Liu, C.-P.; Liu, C.-Y. *Electrophoresis* **2014**, *35*, 2091–2107.
35. Goetz, J. A.; Novotny, M. V.; Mechref, Y. *Anal. Chem.* **2009**, *81*, 9546–9552.
36. Miura, Y.; Kato, K.; Takegawa, Y.; Kurogochi, M.; Furukawa, J.; Shinohara, Y.; Nagahori, N.; Amano, M.; Hinou, H.; Nishimura, S. *Anal. Chem.* **2010**, *82*, 10021–10029.
37. Zauner, G.; Koeleman, C. A. M.; Deelder, A. M.; Wuhrer, M. *Biochim. Biophys. Acta* **2012**, *1820*, 1420–1428.

38. Patel, T. P.; Parekh, R. B. *Methods Enzymol.* **1994**, *230*, 57–66.
39. Kozak, R. P.; Royle, L.; Gardner, R. A.; Bondt, A.; Fernandes, D. L.; Wuhrer, M. *Anal. Biochem.* **2014**, *453*, 29–37.
40. Djanashvili, K.; Frullano, L., Peters, J. A. *Chem. Eur. J.* **2005**, *11*, 4010–8.
41. Otsuka, H.; Uchimura, E.; Koshino, H.; Okano, T.; Kataoka, K. *J. Am. Chem. Soc.* **2003**, *125*, 3493–3502.
42. Yan, J.; Springsteen, G.; Deeter, S.; Wang, B. *Tetrahedron* **2004**, *60*, 11205–11209.
43. Springsteen, G.; Wang, B. *Tetrahedron* **2002**, *58*, 5291–5300.
44. Xu, Y.; Zhang, L.; Lu, H.; Yang, P. *Proteomics* **2010**, *10*, 1079–1086.
45. Liu, Y.; Ren, L.; Liu, Z. *Chem. Commun.* **2011**, *47*, 5067–5069.
46. Li, H.; Liu, Y.; Liu, J.; Liu, Z. *Chem. Commun.* **2011**, *47*, 8169–8171.
47. Liang, L.; Liu, Z. *Chem. Commun.* **2011**, *47*, 2255–2257.
48. Wu, X.; Li, Z.; Chen, X.-X.; Fossey, J. S.; James, T. D.; Jiang, Y.-B. *Chem. Soc. Rev.* **2013**, *42*, 8032–8048.
49. El Khadema, H. S.; Fatiadi, A. J. *Adv. Carbohydr. Chem. Biochem.* **2000**, *55*, 175–263.
50. Liu, L.; Yu, M.; Zhang, Y.; Wang, C.; Lu, H. *ACS Appl. Mater. Interfaces* **2014**, *6*, 7823–7832.
51. Klement, E.; Lipinski, Z.; Kupihár, Z.; Udvardy, A.; Medziharadsky, K. F. *J. Proteome Res.* **2010**, *9*, 2200–2206.
52. Yang, S. J.; Zhang, H. *Anal. Chem.* **2012**, *84*, 2232–2238.
53. Sharon, N. *J. Biol. Chem.* **2007**, *282*, 2753–2764.
54. Sharon, N.; Lis, H. *Lectins*, 2nd edition. Springer New York, **2003**.
55. Senear, D. F.; Teller, D. C. *Biochemistry* **1981**, *20*, 3076–3083.
56. Mandal, D.; Brewer, C. F. *Biochemistry* **1993**, *32*, 5116–5120.
57. Fanayan, S.; Hincapie, M.; Hancock, W. S. *Electrophoresis* **2012**, *33*, 1746–1754.
58. Wohlgemuth, J.; Karas, M.; Eichorn, T.; Hendriks, R.; Andrecht, S. *Anal. Biochem.* **2009**, *395*, 178–188.
59. Ma, W.-F.; Li, L.-L.; Zhang, Y.; An, Q.; You, L.-J.; Li J.-M.; Zhang, Y.-T.; Xu, S.; Yu, M.; Guo, J.; Lu, H.-J.; Wang, C.-C.. *J. Mater. Chem.* **2012**, *22*, 23981–23988.
60. Zheng, J.; Xiao, Y.; Wang, L.; Lin, Z.; Yang, H.; Zhang, L.; Chen, G. *J. Chromatogr. A* **2014**, *1358*, 29–38.
61. Asefa, T.; Tao, Z. *Can. J. Chem.* **2012**, *90*, 1015–1031.
62. ALOthman, Z. A. *Materials* **2012**, *5*, 2874–2902.
63. Rahmat, N.; Abdullah, A. Z.; Mohamed, A. R. *Am. J. Appl. Sci.* **2012**, *7*, 1579–1586.
64. Zhao, D.; Sun, J.; Li, Q.; Stucky, G. D. *Chem. Mater.* **2000**, *12*, 275–279.
65. Hu, Y.; Bouamrani, A.; Tasciotti, E.; Li, L.; Liu, X.; Ferrari, M. *ACS Nano* **2010**, *4*, 439–451.
66. Tian, R.; Zhang, H.; Ye, M.; Jiang, X.; Hu, L.; Li, X.; Bo, X.; Zou, H. *Angew. Chem. Int. Ed.* **2007**, *46*, 962–965.
67. Tian, R.; Ren, L.; Ma, H.; Li, X.; Hu, L.; Ye, M.; Wu, R.; Tian, Z.; Liu, Z.; Zou, H. *J. Chromatogr. A.* **2009**, *1216*, 1270–1278.
68. Xu, Y.; Wu, Z.; Zhang, L.; Lu, H.; Yang, P.; Webley, P. A.; Zhao, D. *Anal. Chem.* **2009**, *81*, 503–508.

69. Tan, J.; Wang, H.-F.; Yan, X.-P. *Anal. Chem.* **2009**, *81*, 5273–5280.
70. Yan, Y.; Deng, C.; Zheng, Z.; Zhang, X.; Yang, P. *Chempluschem* **2014**, *79*, 31–34.
71. Rodríguez-López, G.; Dolores M. M.; Martínez-Máñez, R.; Sancenón, F.; Soto, J.; Villaescusa, L. A.; Beltrán, D.; Amorós, P. *Chem. Commun.* **2004**, 2198–2199.
72. Fried, D. I.; Schlossbauer, A.; Bein, T. *Microporous Mesoporous Mater.* **2012**, *147*, 5–9.
73. Sun, N.; Deng, C.; Li, Y.; Zhang, X. *Anal. Chem.* **2014**, *86*, 2246–2250.
74. Madera, M.; Mechref, Y.; Novotny, M. V. *Anal. Chem.* **2005**, *77*, 4081–4090.
75. Mann, B. F.; Mann, A. K. P.; Skrabalak, S. E.; Novotny, M. V. *Anal. Chem.* **2013**, *85*, 1905–1912.
76. Wu, S.; Huang, X.; Du, X. *Angew. Chem. Int. Ed.* **2013**, *52*, 5580–5584.
77. Gary-Bobo, M.; Mir, Y.; Rouxel, C.; Brevet, D.; Basile, I.; Maynadier, M.; Vaillant, O.; Mongin, O.; Blanchard-Desce, M.; Morère, A.; Garcia, M.; Durand, J.-O.; Raehm, L. *Angew. Chemie* **2011**, *50*, 11425–11429.
78. Gary-Bobo, M.; Mir, Y.; Rouxel, C.; Brevet, D.; Hocine, O.; Maynadier, M.; Gallud, A.; Da Silva, A.; Mongin, O.; Blanchard-Desce, M.; Richeter, S.; Loock, B.; Maillard, P.; Morère, A.; Garcia, M.; Raehm, L., Durand, J.-O. *Int. J. Pharm.* **2012**, *432*, 99–104.
79. Gary-Bobo, M.; Hocine, O.; Brevet, D.; Maynadier, M.; Raehm, L.; Richeter, S.; Charasson, V.; Loock, B.; Morère, A.; Maillard, P.; Garcia, M.; Durand, J.-O. *Int. J. Pharm.* **2012**, *423*, 509–15.
80. Bernados, A.; Aznar, E.; Marcos, M. D.; Martínez-Máñez, R. Sancenón, F.; Soto, J.; Barat, J. M.; Amorós, P. *Angew. Chem. Intl. Ed.* **2009**, *48*, 5884–5887.
81. Santamaría, E.; Maestro, A.; Porras, M.; Gutiérrez, J. M.; González, C. J. *Solid State Chem.* **2014**, *210*, 242–250.
82. Tang, D.; Liu, B.; Niessner, R.; Li, P.; Knopp, D. *Anal. Chem.* **2013**, *85*, 10589–10596.
83. Debrassi, A.; Ribbera, A.; de Vos, W. M.; Wennekes, T.; Zuilhof, H. *Langmuir* **2014**, *30*, 1311–1320.
84. Ter Maat, J.; Regeling, R.; Ingham, C. J.; Weijers, C. A. G. M.; Giesbers, M.; de Vos, W. M.; Zuilhof, H. *Langmuir* **2011**, *27*, 13606–13617.
85. Milka, P.; Krest, I.; Keusgen, M. *Biotechnol. Bioeng.* **2000**, *69*, 344–348.
86. Boss, C.; Meurville, E.; Sallese, J.-M.; Ryser, P. *J. Memb. Sci.* **2012**, *401–402*, 217–221.
87. Boss, C.; Meurville, E.; Sallese, J.-M.; Ryser, P. *Biosens. Bioelect.* **2011**, *30*, 223–228.
88. Krushinitskaya, O.; Tonnessen, T. I.; Jakobsen, J. E. A. *J. Memb. Sci.* **2011**, *376*, 153–161.
89. Shi, W.; Shen, Y.; Ge, D.; Xue, M.; Cao, H.; Huang, S.; Wang, J.; Zhang, G.; Zhang, F. *J. Memb. Sci.* **2008**, *325*, 801–808.
90. Yan, J.; Li, X.; Yu, L. Jin, Y.; Zhang, X.; Xue, X., Ke, Y.; Liang, X. *Chem. Commun.* **2010**, *46*, 5488–5490.
91. Yan, J.; Li, X.; Cheng, S.; Ke, Y.; Liang, X. *Chem. Commun.* **2009**, 2929–2931.
92. Larsen, M. R.; Jensen, S. S.; Jakobsen, L. A.; Heegaard, N. H. H. *Molec. Cell. Proteom.* **2007**, *6*, 1778–1787.
93. Qin, H.; Zhao, L.; Li, R.; Wu, R.; Zou, H. *Anal. Chem.* **2011**, *83*, 7721–7728.
94. Qin, H.; Hu, Z.; Wang, F.; Zhang, Y.; Zhao, L.; Xu, G.; Wu, R.; Zou, H. *Chem. Commun.* **2013**, *49*, 5612–5164.
95. West, C.; Elfakir, C.; Lafosse, M. *J. Chromatogr. A*, **2010**, *1217*, 3201–3216.

96. Zhao, Y.; Szeto, S. S. W.; Kong, R. P. W.; Law, C. H.; Li, G.; Quan, Q.; Zhang, Z.; Wang, Y.; Chu, I. K. *Anal. Chem.* **2014**, *86*, 12172–12179.

97. Jmeian, Y.; Hammad, L. A.; Mechref, Y. *Anal. Chem.* **2012**, *84*, 8790–8796.

98. Stavenhagen, K.; Kolarich, D.; Wuhrer, M. *Chromatographia*, **2015**, *78*, 307–320. doi:10.1007/s10337-014-2813-7

99. White, R. J.; Antonio, C.; Budarin, V. L.; Bergstrom, E.; Thomas-Oates, J.; Clark, J. H. *Adv. Funct. Mater.* **2010**, *20*, 1834–1841.

100. Chirica, G. S.; Remcho, V. T. *Electrophoresis* **2000**, *21*, 3093–3101.

101. Fujimoto, C. *Chromatography* **2001**, *22*, 145–150.

102. Potter, O. G.; Breadmore, M. C.; Hilder, E. F. *Analyst* **2006**, *131*, 1094.

103. Ren, L.; Liu, Z.; Dong, M.; Ye, M.; Zou, H. *J. Chromatogr. A* **2009**, *1216*, 4768–4774.

104. Ren, L.; Liu, Y.; Dong, M.; Liu, Z. *J. Chromatogr. A* **2009**, *1216*, 8421–8425.

105. Chen, M.; Lu, Y.; Ma, Q.; Guo, L.; Feng, Y.Q. *Analyst* **2009**, *134*, 2158–2164.

106. Liu, Y.; Lu, Y.; Liu, Z. *Chem. Sci.* **2012**, *3*, 1467–1471.

107. Yang, F.; Lin, Z.; He, X.; Chen, L.; Zhang, Y. *J. Chromatogr. A* **2011**, *1218*, 9194–9201.

108. Lin, Z.; Wang, J.; Tan, X.; Sun, L.; Yu, R.; Yang, H.; Chen, G. *J. Chromatogr. A.* **2013**, *1319*, 141–147.

109. Lin, A.; Pang, J. L.; Lin, Y.; Huang, H.; Cai, C. W.; Zhang, L.; Chen, G. N. *Analyst* **2011**, *136*, 3281–3288.

110. Lu, Y.; Bie, Z.; Liu, Y.; Liu, Z. *Analyst* **2013**, *138*, 290–298.

111. He, J.; Liu, Z.; Dou, P.; Liu, J.; Ren, L.; Chen, H.-Y. *Talanta* **2009**, *79*, 746–751.

112. Bie, Z.; Chen, Y.; Li, H.; Wu, R.; Liu, Z. *Anal. Chim. Acta* **2014**, *834*, 1–8.

113. Li, H.; Wang, H.; Liu, Y.; Liu, Z. *Chem. Commun.* **2012**, *48*, 4115–4117.

114. Nie, H.; Chen, Y.; Lu, C.; Liu Z. *Anal. Chem.* **2013**, *85*, 8277–8283,

115. Currivan, S.; Jandera, P. *Chromatography* **2014**, *1*, 24–53.

116. Wang, X.; Liu, Y.; Ren, L.; Li, H.; Liu, Z. *Anal. Methods* **2013**, *5*, 5444–5449.

117. Ren, L.; Liu, Z.; Liu, Y.; Dou, P.; Chen, H.-Y. *Angew. Chem. Int. Ed.* **2009**, *48*, 6704–6707.

118. Xu, Y.; Cao, Q.; Svec, F.; Fréchet, J. M. *Anal. Chem.* **2010**, *82*, 3352–3358.

119. Cao, Q.; Xu, Y.; Liu, F.; Svec, F.; Fréchet, J. M. *Anal. Chem.* **2010**, *82*, 7416–7421.

120. Lv, Y.; Alejandro, F. M.; Fréchet, J. M.; Svec, F. *J. Chromatogr. A* **2012**, *1261*, 121–128.

121. Wi, C.; Liang, Y.; Zhao, Q.; Qu, Y.; Zhang, S.; Wu, Q.; Liang, Z.; Zhang, L.; Zhang, Y. *Chem. Eur. J.* **2014**, *20*, 8737–8743,

122. Yang, F.; Mao, J.; He, X. W.; Chen, L. X.; Zhang, Y. K. *Anal. Bioanal. Chem.* **2013**, *405*, 5321–5331.

123. Lin, Z.; Pang, J.; Yang, H.; Cai, Z.; Zhang, L.; Chen, G. *Chem. Commun.* **2011** *47*, 9675–9677.

124. Sun, S., Tang, Y., Fu, Q., Liu, X., Du, W., Guo, L., Zhao, Y. *J. Sep. Sci.* **2012**, *35*, 893–900.

125. Guryča, V.; Mechref, Y.; Palm, A. K.; Micahalek, J.; Pacáková, V.; Novotný, M. *J. Biochem. Biophys. Methods* **2007**, *70*, 3–13.

126. Luo, Q.; Rejtar, T.; Wu, S.-L.; Karger, B. L. *J. Chromatogr. A* **2009**, *1216*, 1233–1231.

127. Krenkova, J.; Lacher, N. A.; Svec, F. *J. Chromatogr. A* **2009**, *1216*, 3252–3259.

128. Palm, A. K.; Novotny, M. *Rapid Commun. Mass. Spectrom.* **2005**, *19*, 1730–1738.
129. Palm, A. K.; Novotny, M. *Rapid Commun. Mass Spectrom.* **2004**, *18*, 1374–82.
130. Bedair, M.; Oleschuk, R. D.; *Analyst* **2006**, *131*, 1316–1321.
131. Feng, S.; Yang, N.; Pennathur, S.; Goodison, S.; Lubman, D. M. *Anal. Chem.* **2009**, *81*, 3776–3783.
132. Reichelt, S.; Elsner, C.; Prager, A.; Naumov, A.; Kuballa, J.; Buchmeiser, M. R. *Analyst* **2012**, *137*, 2600–2607.
133. Alwael, H.; Connolly, D.; Clarke, P.; Thompson, R.; Twamley, B.; O'Conner, B.; Paull, B. *Analyst* **2011**, *136*, 2619–2628.
134. Hu, M.-X.; Wan, L.-S.; Xu, Z.-K. *J. Membr. Sci.* **2009**, *335*, 111–117.
135. Hu, M.-X.; Xu, Z.-K. *Colloids Surf. B* **2011**, *85*, 19–25.
136. Wang, C.; Ren, P.-F.; Huang, X.-J.; Wu, J.; Xu, Z.-K. *Chem. Commun.* **2011**, *47*, 3930–3932.
137. Yang, Q.; Hu, M.-X.; Dai, Z.-W.; Tian, J.; Xu, Z.-K. *Langmuir* **2006**, *22*, 9345–9349.
138. Gümüşderelioğlu; M.; Agi, P. *React. Funct. Polym.* **2004**, *61*, 211–220.
139. Babac, C.; Yavuz, H.; Galaev, I. Y.; Pişkin, E.; Denizli, A. *React. Funct. Polym.* **2006**, *66*, 1263–1271.
140. Bandari, R.; Kuballa, J.; Buchmeiser, M. R. *J. Sep. Sci.* **2013**, *36*, 1169–1175.
141. Ali, M.; Ramirez, P.; Tahir, M. N.; Mafe, S.; Siwy, Z.; Neumann, R.; Tremel, W.; Ensinger, W. *Nanoscale* **2011**, *3*, 1894–1903.
142. Ding, Y.; Kim, Y.-J.; Erlebacher, J. *Adv. Mater*, **2004**, *16*, 1897–1900.
143. Seker, E.; Reed, M. L.; Begley, M. R. *Materials* **2009**, *2*, 2188–2215.
144. Wittstock, A.; Biener, J.; Bäumer, M. *Phys. Chem. Chem. Phys.* **2010**, *12*, 12919–12930.
145. Qian, L. H.; Chen, M. W. *Appl. Phys. Lett.* **2007**, *91*, 083105.
146. Li, R.; K. Sieradzki, K. *Phys. Rev. Lett.* **1992**, *68*, 1168–1171.
147. Tan, Y. H.; Davis, J. A.; Fujikawa, K.; Ganesh, N. V.; Demchenko, A. V.; Stine, K. J. *J. Mater. Chem.* **2012**, *22*, 6733–6745.
148. Collinson, M. M. *IRSN Anal. Chem.* **2013**, 21692484.
149. Pornsuriyasak, P.; Ranade, S.; Li, A.; Parlato, M. C.; Sims, C. R.; Shulga, O. V.; Stine, K. J.; Demchenko, A. V. *Chem. Commun.* **2009**, 1834–1835.
150. Heckel, A.; Mross, E.; Jung, K.-H.; Schmidt, R. R. *Synlett* **1998**, 171–173.
151. Ganesh, N. V.; Fujikawa, K.; Tan, Y.-H.; Stine, K. J.; Demchenko, A. V. J. *Org. Chem.* **2013**, *78*, 6849–6857.
152. Ganesh, N. V.; Fujikawa, K.; Tan, Y.-H.; Stine, K. J.; Demchenko, A. V. *Org. Lett.* **2012**, *14*, 3036–3039.
153. Pandey, B.; Bhattarai, J. K.; Pornsuriyasak, P.; Fujikawa, K.; Catania, R.; Demchenko, A. V.; Stine, K. J. *J. Electroanal. Chem.* **2014**, *717–718*, 47–60.
154. Pandey, B. P.; Tan, Y.-H.; Fujikawa, K.; Demchenko, A. V.; Stine, K. J. *J. Carbohydr. Chem.* **2012**, *31*, 466–503.
155. Pandey, B.; Tan, Y. H.; Parameswar, A. R; Pornsuriyasak, P.; Demchenko, A. V.; Stine, K. J. *Carbohydr. Res.* **2013**, *373*, 9–17.
156. Tan, Y. H.; Fujikawa, K.; Pornsuriyasak, P.; Alla, A. J.; Demchenko, A. V.; Stine, K. J. *New J. Chem.* **2013**, *37*, 2150–2165.

10

APPLICATIONS OF NANOTECHNOLOGY IN ARRAY-BASED CARBOHYDRATE ANALYSIS AND PROFILING

JARED Q. GERLACH[1], MICHELLE KILCOYNE[1,2], AND LOKESH JOSHI[1]

[1] *Glycoscience Group, National Centre for Biomedical Engineering Science, National University of Ireland Galway, Galway, Ireland*

[2] *Microbiology, School of Natural Sciences, National University of Ireland Galway, Galway, Ireland*

10.1 INTRODUCTION

10.1.1 The Rise of the Multiplexed Affinity Array in Glycoanalysis

Glycomics, the study of carbohydrates to unravel the roles of their structures throughout biology, has moved from being a side note to a mainstream focus over the past quarter century. As a result, the field of glycobiology and the associated fields of glycoproteomics, glycodiagnostics, and recombinant biotechnology have become ever more dependent upon high-throughput (HTP) glycomic analysis strategies to gain further insight and advance understanding of the roles of carbohydrates in biological processes [1].

Within the scope of better defining structural details and functional relationships, carbohydrates (popularly referred to as "glycans") present specific analytical challenges not common to other areas of biological analysis. Carbohydrates, specifically in the form of polysaccharides, proteoglycans, glycosaminoglycans (GAGs), and also the oligosaccharide and monosaccharide modifications of proteins and lipids,

Carbohydrate Nanotechnology, First Edition. Edited by Keith J. Stine.

are biological products built and regulated largely through mechanisms outside of core nucleic acid transcription and protein translation [2]. Even after over a century of study, the controlling mechanisms for the production of many carbohydrate structures, in particular those that are posttranslational modifications (PTMs), are still poorly understood. In stark contrast to the examination of mRNA levels that may allow the direct extrapolation of the relative abundance of protein expression, few situations exist by which even indirect inferences can be made in the analysis of glycans and events relating to their synthesis, attachment, and/or function. Thus, detailed structural and functional analyses within glycobiology still rely predominantly on classical biochemical and chemical methods acting directly on the carbohydrate structures and their receptor molecules instead of their precursor or mediating molecules.

Carbohydrate analysis has long been accomplished primarily through qualitative and quantitative chemical (destructive) methods (e.g., benchtop colorimetric assays using resorcinol, phenol, etc.) [3,4] or more recently through high-end analytical approaches such as liquid chromatography (LC) [5], nuclear magnetic resonance (NMR) [6–8], or mass spectrometry (MS) [5,9]. LC and LC-hyphenated methods (e.g., LC–MS) remain the highest-throughput tools for mammalian oligosaccharide structural analysis and quantification, and MS and NMR remain the most utilized for detailed structural information, especially of nonmammalian and unknown carbohydrate structures [5,10]. These techniques have undergone incremental advancement and improvement in sensitivity and resolution over time and have also gained some ground for nonspecialist accessibility during the past decade. However, their high capital costs along with time-consuming sample analysis and the requirement for a high level of expertise for both instrument operation and data interpretation still tend to make these methods ill suited for many applications where ease of use, speed, cost, and durability are primary concerns (e.g., clinical settings, production lines, and fully automated environmental or process monitoring) [11]. Furthermore, carbohydrate research at the biological scale is difficult in part because available sample and reagent quantities are often very small (mere picograms to micrograms), thus restricting the use of some types of analyses (e.g., NMR) altogether.

The advent of microarray-based, multiplexed, affinity analysis methods is making it possible for glycoanalysis to move from the realm of the specialized carbohydrate laboratory to the general biological laboratory and, at the same time, is also helping to pave the road for a variety of different field-deployed glycoanalytical tools applicable to clinical, environmental, production, and research pursuits. Glycan- and carbohydrate-binding protein (lectin) microarrays are the two major examples of multiplexed, affinity-based carbohydrate analysis tools that have begun to see wide distribution and use [12–14]. Lectins are nonenzymatic proteins that selectively and reversibly bind to glycan structures [15]. Lectin recognition of carbohydrates is usually restricted to mono- and disaccharides or the outermost residues of oligosaccharides and polysaccharides. During the first decade of glycomic microarray development, variations of the method have been used to profile carbohydrates, glycoproteins, glycolipids, complex membranes, intact microorganisms, and even mammalian cells.

10.1.2 Characteristics of Microarrays Used for Glycomics

Current glycomic microarray variants are roughly based on DNA microarrays previously developed for genomics. Most glycomic microarrays consist of a standardized 1 in. × 3in. glass microscope slide functionalized to facilitate attachment of derivatized glycans [14,16], natural and synthetic glycosylated macromolecules of various types [13,17–19], or the protein receptors that bind to these sugar structures (lectins and antibodies) [20–22]. Typical arrangements include multiple identical subarrays of multiplexed items on the same slide to allow the side-by-side comparison of multiple analytes or experimental conditions (Fig. 10.1).

Microarray construction and use in glycomics is still dominated by the academic laboratory. However, commercial offerings have begun to emerge after slightly more than a decade of development. Lectin microarrays and accompanying data analysis services are being or have been marketed by several companies including Procognia, Plexera, GP Biosciences, and GlycoTechnica, among others. These lectin microarray formats typically consist of 50–100 unique lectins with varying binding preferences. Typically, lectins comprising these arrays are natural proteins, usually best characterized for interactions with mammalian glycans and sourced primarily from plants,

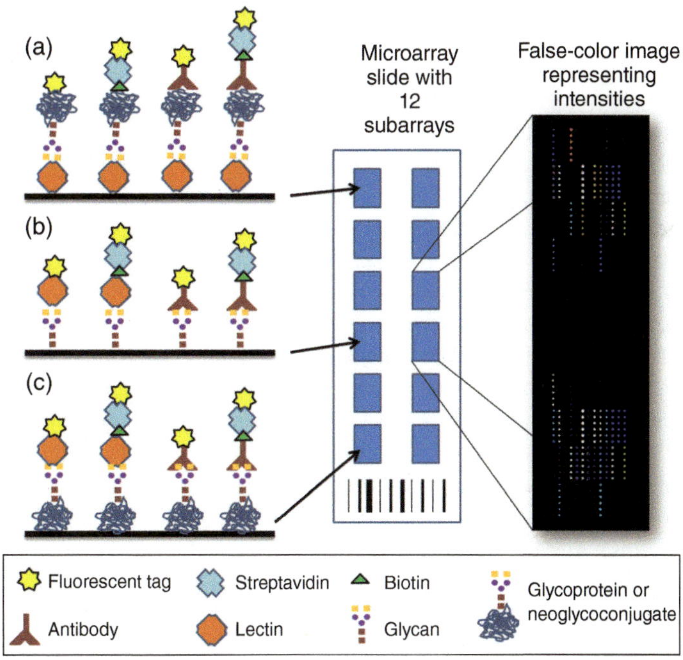

FIGURE 10.1 Representatives of the most frequently encountered microarray variants used in glycomics. Molecules printed are shown for (a) lectin microarray, (b) glycan microarray, (c) and glycoprotein or neoglycoconjugate microarray. Each microarray variation is shown with direct analyte label and potential sandwich detection strategy combinations.

fungi, and bacteria due to their relative abundance and stability [20,23]. Microarrays may also include recombinant versions of lectins incorporating additional peptide domains useful for oriented attachment [24]. Select animal lectins, most frequently from snails, slugs, and crabs, have also been applied to lectin microarrays. However, maintaining activity of many animal lectins after conjugation to microarray substrates has presented additional challenges which need to be overcome before the widespread adoption of more complex animal lectins for this purpose. The carbohydrate structures used for glycan microarrays may be either natural or synthetic. Glycan arrays typically consist of modified (e.g., aminated) monosaccharides and oligosaccharides or glycoconjugates consisting of such structures chemically attached to proteins (e.g., bovine or human serum albumin (BSA or HSA), polyacrylamide (PAA), or synthetic dendrimeric scaffolds). Access to a high-content glycan microarray is available through the Consortium for Functional Glycomics (CFG) Core H now based at Emory University, Georgia, United States. The CFG glycan microarray is the most diverse array reported to date with over 600 unique carbohydrate structures [13]. At least one company, Z Biotech, has begun offering custom glycan microarrays as commercial products.

10.1.3 The Advantages of Going Small in Glycoanalysis

The application of microarray technologies to carbohydrate-focused biological questions has been highly advantageous for a number of reasons. The first, and perhaps most obvious, is that microarray technologies have enabled researchers to interrogate many carbohydrate-mediated interactions simultaneously. This has been especially useful for profiling in situations where signatures of multiple binding events are as important or are more important to the application than the actual structures involved (e.g., often encountered in many biosensing situations) or where exact binding partners or preferences are completely unknown. Second, it has enabled many times more research to be carried out per unit volume of biological sample or substrate than older multiplexed systems such as plate-based assays, even at the 384-well density level. Third, unlike with other methods, the analysis of many types of glycosylated molecules by microarray profiling does not require release of the carbohydrate portion. Finally, and very importantly, microarray technology has also helped amplify sensitivity and lower the limits of detection (LOD) through the use of high-gain microarray imaging and reporting approaches with very high dynamic range capabilities.

Since the first glycomics-based macroarrays (essentially multiplexed versions of ELISA-type plate assays [25,26]) were described, the power of multiplexed glycoanalytical techniques has become clearer. In turn, there has been a rapid succession of developments during the last decade, which led to the appearance of glycan microarrays in many forms [13,16,27–29], antibody and antibody-sandwich hybrid microarrays aimed at glycoanalysis [21], and variations of lectin microarrays [23,29,30]. Microarrays have become outstanding platforms by which glycomic hypotheses can be developed. Understanding the broad potential roles and relationships of many carbohydrates and their binding partners throughout nature has definitively benefited from simply being able to identify a range of glycan structures and receptors that have an ability to interact. These may be intraorganismal interactions (e.g., cell–cell communications) or interorganismal interactions (e.g., pathogenic and commensal interactions, xenotransplantation issues, and adverse immunological events).

One case in point that demonstrates the potential for great gains through the effective use of microarray technology is the study of functional lectins from higher animals [31]. The roles and binding preferences of many of these lectins are only now being elucidated although they are ubiquitous throughout Animalia. Currently, only a handful of fully functional mammalian lectins are available as recombinant proteins. Many that are available to the research community are subject to strict handling requirements and short lifetime constraints, which make them difficult to work with and handle. They are also generally very expensive with very high initial loss rates. Employing microarray technology has the potential to vastly increase the amount of information available per unit of these fragile molecules [32,33].

10.1.4 The Glycomic Microarray's Gateway to Nanotechnology

Nanoscale materials frequently have unique attributes that bulk materials lack, including novel magnetic, optical, catalytic, electronic, or photonic properties [34]. Because of these attributes, it may have been inevitable for "micro" to shift toward "nano" for the next wave of major improvements to array-based glycomic research. To date, a variety of nanomaterials have been examined as potentially useful for biomolecular detection and biosensing applications. Examples of nanomaterials tested for these purposes include nanowires, nanoparticles, and a number of carbon-based nanostructures including carbon quantum dots, graphene, and carbon nanotubes [35].

Initially, the primary emphasis of the application of nanostructures to glycomic microarrays was in the area of detection. Fluorescence is currently the most popular method of detection employed in microarray-based glycoanalysis. Historically, single-molecule fluorophores (e.g., cyanine, fluorescein, rhodamine, tetramethylrhodamine, Alexa Fluor®, and DyLight®) that have seen wide use in flow cytometry, immunohistochemistry, and DNA-based microarray analysis have offered good performance for affinity-based glycoanalysis. However, many such fluorescent agents photobleach easily and are also sensitive to pH variation and environmental degradation (e.g., ozone degradation). The advent of nanostructured fluorescent agents (e.g., quantum dots) has helped to circumvent some of these issues and, at the same time, allowed a new avenue for multivalent presentation of carbohydrates to receptors or vice versa [36,37].

Typically, monovalent interactions of lectins are quite weak, and the dissociation constants (K_d) are usually in the mM range [38]. Because of this fact, many lectins rely on the compounding of affinities to achieve an avidity that is great enough to "stay put." Prolonging on-times (k_{on}) is the key to reliable detection of minute amounts of weaker binding activity. Multivalency frequently mediates biologically important binding and recognition, and the exact spatial arrangement of the carbohydrates is of great importance for biological recognition and interaction [39]. Therefore, the spatial presentation of carbohydrates in model binding systems is very important and biologically relevant. In models or analytical systems, carbohydrate linker flexibility, array surface chemistry, ligand density, orientation, and the molecular architecture of the ligand scaffolding all affect the presentation, accessibility, and therefore biological recognition and binding of the ligands [13,19,38,40,41]. Indeed, different neighboring structures and their physical arrangements have even been shown to modulate antibody binding to specific glycans [42]. Thus, glycodendrimeric structures consisting of multiple carbohydrate determinants in controlled density and spacing are often much better for use as biologically relevant test molecules due to the need for multivalent interaction of many carbohydrate-binding proteins to increase avidity [40,43,44]. These areas are where nanoscale structural enhancements have already begun to be employed in glycomics and are also the obvious points of inclusion of nanoscale engineering destined for more efficient array-based glycoanalysis.

This chapter examines some of the major contributions to the ongoing development of nanostructural technologies reported from 2010 through 2014, which are either currently incorporated into array-based affinity study of carbohydrate interactions or are technologies with great potential for future integration into such systems.

10.2 APPLICATIONS OF NANOTECHNOLOGY FOR GLYCOMIC ARRAYS

10.2.1 Nanostructures on Printed Microarray Surfaces

The direct application of nanostructures to the surface of glycomic microarrays is a promising prospect for increasing surface area and spatial distribution or simply ensuring suitable dimensions for targets to interact optimally with arrays. One application that may benefit greatly from nanoengineered surfaces is the study of carbohydrate-mediated interactions between pathogenic and commensal bacterial species and their hosts. The capture and identification of bacteria by lectin or carbohydrate microarray are of interest for a variety of applications including strain discrimination, disease diagnosis, and identification of potential antiadhesive therapeutics [17,45,46]. Current approaches are challenged with providing a suitable environment for bacterial adhesion that often requires working around the inherent limitations of a two-dimensional (2D) surface, and this can be enhanced by the application of three-dimensional (3D) structures (Fig. 10.2).

Aiming to increase sensitivity and expand the range of potential applications, Liu and colleagues have described a method of building enhanced, 3D PAA structures for the capture and identification of bacteria on microarray surfaces [47]. The microarray was initialized by assembling their 3D glutaraldehyde-activated PAA surfaces by photopolymerization directly on the surface of the slide substrate. The resulting 80 (±20) nm diameter porous structures were further functionalized by the covalent addition of lectins or monosaccharide structures specifically recognizing or recognized by select bacteria [47]. The density effects imparted by attachment of the carbohydrate ligand/receptor to the 3D structure of the cross-linked PAA resulted in an improvement for capture of both Gram-negative and

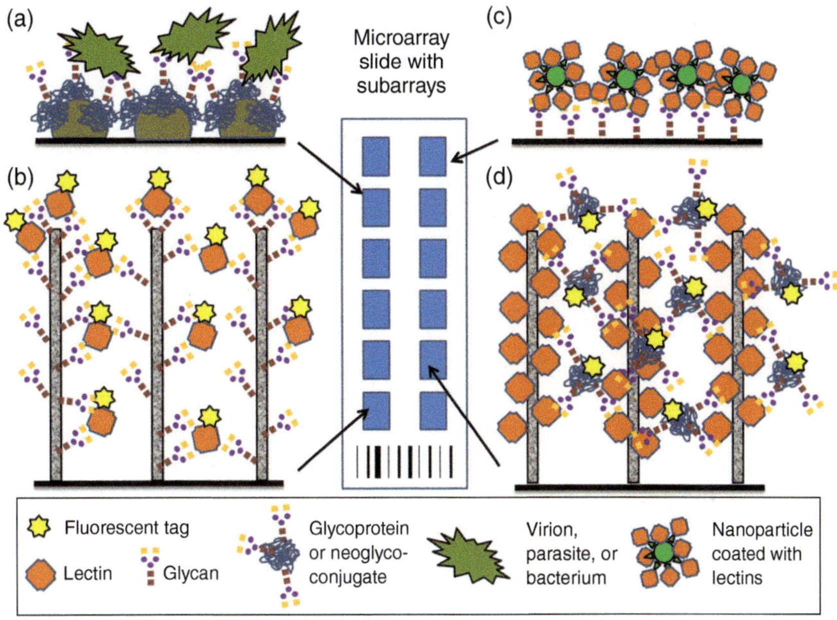

FIGURE 10.2 Examples of schemes designed to increase carbohydrate-mediated binding avidity by inclusion of nanostructures within glycomic microarrays and detection methods. (a) Printed 3D nanostructures with glycoconjugates allow better attachment of microbes with surface lectins. (b) Glycan microarray with carbon nanotube scaffolds to increase cross-ligand interaction with analytes. (c) Lectin-coated nanoparticles interacting with standard 2D glycan array surface. (d) Lectin microarray with 3D enhancement by carbon nanotubes to help facilitate cross-ligand interaction with complex glycoconjugates.

Gram-positive species compared to more standard 2D microarray surfaces. The LOD for *E. coli* strain BL21 was reported as ~50 cells per microarray feature or spot [47].

In a bid to develop a lectin nanoarray of sufficient sensitivity for detection of the generally weak monovalent lectin–glycan interactions, a fluorescent single-walled carbon nanotube (SWCT) was employed [48]. SWCTs are nanoscale empty cylinders composed of rolled graphite sheets [34]. To enable lectin attachment, the SWCTs were first wrapped with chitosan (partially deacetylated chitin, which is a carbohydrate polymer composed of a repeating unit of β-linked *N*-acetylglucosamine (GlcNAc)), which was subsequently functionalized with carboxylic acids to which nitrilotriacetic acid (NTA) groups were then attached. Ni^{2+} was then chelated with the NTA, and finally, a histidine (His)-tagged recombinant lectin was bound to the coated and functionalized SWCT. A custom near-infrared microscope was then employed to probe the fluorescence emissions of the microarrayed SWCT-lectins upon carbohydrate binding. Carbohydrates were presented as either biotinylated monosaccharides or as streptavidin-tethered monosaccharides. This setup allowed dissociation constants (K_d) to be calculated for the lectin–glycan interactions. For example, K_d values of 106 and 142 μM were generated for the binding of free and tethered biotinylated fucose (Fuc) to PA-IIL, a Fuc-specific lectin from the pathogenic bacterium *Pseudomonas aeruginosa*. This was significantly weaker compared to the K_d of 3 μM obtained by isothermal titration microcalorimetry (ITC) that was expected as the lectin was surface immobilized compared to the solution-phase ITC approach. Remarkably, kinetic information could also be generated from only a single SWCT in this format [48]. Although at present a laborious process, the extension of this approach to construct a "nanoarray" may prove an exciting advancement to current technology.

10.2.2 Printed Polymeric and Dendrimeric Structures

Precision synthesis and attachment of glycans with defined density to better model natural carbohydrate presentation have been pursued for some time, but not without many technical and engineering obstacles [49,50]. One of the toughest challenges is the control of orientation and density throughout the several changes in hydration and environmental conditions required to attach carbohydrates to any sort of planar surface.

The exploitation of boronic acid (BA) for use in many glycoanalysis strategies follows from the utility of complexes formed between *cis*-diols of sugars and the boron ions [51]. Recently, a microarray constructed with "macroligands" was reported, which demonstrated controlled density and glycan orientation in which BA was used in conjunction with various sizes of molecular spacers (e.g., lysozyme, BSA, and PAA) as a temporary scaffold to maintain the precise spatial arrangement of printed materials during the assembly process [44]. By varying the size of the molecules associated with the BA, the ability to carefully space the density of 3D glycopolymers was demonstrated on the 2D slide surface. The print process described relied on two different sets of conditions during the assembly of the macroligands. Initially, the pH was maintained above 10.0 to facilitate the isourea bond formation between the synthetic glycopolymer structures and the functionalized slide surface during assembly. Afterward, the pH was shifted to 7.4 to allow the release of the BA scaffolds, which had been holding the glycopolymers in their optimum 3D configuration during attachment.

Prior to microarraying the macroligands, an initial interrogation of a range of the substitution-variable macroligands (1:18 to 1:51 of lactose (Lac, Gal-β-(1,4)-Glc) incorporation onto a PAA backbone) with fluorescein isothiocyanate (FITC)-labeled *Arachis hypogaea* (peanut) lectin (PNA) showed that the lectin preferentially bound to structures with a 1:30 substitution [44]. The density of this macroligand was then modulated on the 2D surface with the BA scaffold inclusions as described earlier. FITC-PNA and another fluorescently labeled Gal-binding lectin, RCA-I from *Ricinus communis*, were used to determine the effects of changes in the lateral distribution. Ultimately, the authors demonstrated that it was possible to optimize the presentation of carbohydrate ligands in a 3D format for microarray applications. While not strictly "nano" structures in their physical size, the 3D structures optimized in two dimensions do fit into the technological niche of the modified surfaces required for next-generation array-based glycomics.

10.2.3 Nanostructures Assisting in Glycan-Mediated Binding Detection

Optical methods of detection, namely, fluorescence and chemiluminescence, are currently the most widely used for microarrays destined for glycomics. While the labeling of analytes or secondary detection molecules has been necessary for glycomic microarrays since their invention, it requires samples that are free of contaminants that can block or compete with the labeling process. Labeling also requires conditions that are favorable to the labeling process but sometimes not favorable for the analyte (e.g., pH, temperature, ionic strength), and most labeling requires the removal of excess labeling reagent prior to analysis. Furthermore, labeling processes can be lengthy and may often not be suitable for some classes of molecules (e.g., polysaccharides) without further chemical modification. Therefore, label-free technologies are highly desirable for glycomic microarray analyses, but often are not practical in many current situations due to the physical requirements of label-free detection (e.g., dependence upon molecular mass in plasmon resonance). However, some new approaches have emerged to alleviate some of the inherent problems associated with labeled glycan–lectin detection.

Light scattering combined with gold nanoparticle sandwich detection is one approach that has garnered attention as a means to reduce the need for primary analyte labeling. A sandwich-assay system consisting of printed glycodendrimers in which the detection method relies upon plasmon resonance has been described [52]. Using glycodendrimeric structures consisting of NH_2-terminated poly(amidoamine) (PAMAM) branches capped with aminated carbohydrates through glutaraldehyde cross-linking, a 3D microarray was constructed with better-positioned ligands for lectin binding. Interactions between carbohydrates and biotinylated lectins were detected using streptavidin-coated gold nanoparticles (13 nm in diameter) and a resonance light scattering (RLS) scanner. Signal-to-noise ratios and other qualitative analyses were conducted with two lectins—RCA-I/120, which preferentially binds to β-linked terminal galactose (Gal), and Con A, which binds mannose (Man)—along with three variants of glycodendrimeric structures terminating in α-linked Man, β-linked Gal, or α-linked glucose (Glc) to show that the approach provided selectivity. Finally, the assay was tested for quantitative performance using the same lectins and sugar structures. For the quantitative assessments, the concentration of α-linked Man, β-linked Gal, or α-linked Glc was modulated during the print process, while the lectin concentration was held constant during the subsequent interaction experiments. Ultimately, the LOD for both lectins was determined to be approximately 50 ng/mL with a dynamic range of two orders of magnitude [52]. In this report, biotinylated lectins were used as a proof of concept (POC); however, it is conceivable that this system could be extended to a sandwich assay including only a gold nanoparticle-labeled immunodetection molecule (e.g., scFv or immunoglobulin), thus completely removing the need for the biotin–avidin pairing in the detection system.

Biophotonic imaging has been employed in a microarray that allowed the label-free, real-time detection of interactions between glycoconjugates and their binding partners [53]. Gold nanoparticles of 4 nm diameter were printed on microarray substrates in 200 μm diameter spots as seed particles that were used to grow truncated polyhedral structures *in situ* with a final diameter of approximately 120 nm. These polyhedral nanoparticles functioned as the light scattering centers for biophotonic imaging-based detection. Modified glycans and proteins were attached to the polyhedral nanoparticles through carbodiimide/*N*-hydroxysuccinimide (EDC/NHS) carboxylic acid coupling. Varying concentrations of monosaccharides, controlled through the ratio of the functional linker to inert spacer included in the self-assembled monolayer (SAM), were attached to the polyhedral structures to test the effects of density on binding behavior. The LOD for the interaction between covalently linked Man and Con A was determined to be just 100 ng/mL (0.93 nM) when the Man was conjugated with a linker/spacer ratio of 1:1000 [53]. A comparative analysis utilizing a panel of soluble lectins was carried out on surface-immobilized human and porcine versions of fibrinogen, a protein reported to have both N- and O-linked glycosylation PTMs [54,55]. In the lectin analysis, the method was sensitive enough to detect the relative increase in binding of the lectins AAL and LcH to the porcine fibrinogen, which indicated the presence of fucosylation, but no binding was noted with the human fibrinogen [53]. In comparison to the lectin responses with only tethered monosaccharides, enhanced binding created by the multivalent presentation of glycans on the fibrinogen was noted. As a negative control, each lectin was also tested with serum albumin, which is not glycosylated. Noise was generally very low, which suggested that the system had very good selectivity and minimal nonspecific binding.

Dye-doped silica (DDS) has been proposed as an alternative to both traditional organic dyes and quantum dots. Some of the attributes of organic dyes and also quantum dots have been examined, and these molecules have recently been effectively combined into a nanoparticle that has positive characteristics of each and fewer of their inherent drawbacks [56]. Using a modified version of the Stöber synthesis method, FITC was silanized and then copolymerized with tetraethyl orthosilicate to build DDS particles with a diameter of approximately 100 nm. Perfluorophenyl azide (PFPA) molecules were covalently bonded to the DDS particles, each serving as both a linker and chemical attachment point for photolinking of carbohydrates to the particles. Using a simple microarray with only two lectins printed at multiple concentrations, selective binding of mannose-coated DDS particles to Con A was demonstrated, although the researchers did not perform a comparison to existing dye molecules [56]. In any case, the DDS nanostructures provide an intriguing combination of robust detection with a vehicle for multivalent presentation through photocoupling.

Along similar lines, a glycan microarray was fabricated from glyconanoparticles (GNPs) by printing the GNPs on a poly(allyl amine) perfluorophenyl azide ((PAAm)-PFPA) surface and then photochemically covalently binding the GNPs to the surface. The immobilized GNPs were probed with both FITC and FITC-doped silica nanoparticle (FSNP)-labeled lectins. After printing, the GNP microarray was coated or blocked with a film of either poly(2-ethyl-2-oxazoline) (PEOX), a hydrophilic polymer, or polystyrene (PS), a hydrophobic polymer. Interestingly, FSNP-labeled Con A adsorbed strongly on PEOX and minimally on PS, while the opposite case was observed with FITC-labeled Con A. Higher signal was also observed for FSNP-Con A compared to the FITC-labeled lectin. A comparison of GNP microarrays with a carbohydrate microarray constructed by printing-free, nonderivatized carbohydrates on the same surface as the GNP microarray was performed [57]. The comparison showed that Man 3 and Man 2 structures displayed higher signals with the GNP microarray than with the plain carbohydrate array, while Man and Glc produced lower signals in the GNP microarray [57].

Although this effect was not noted or discussed in the article [57], it is interesting that the figure depicting FITC-Con A binding to the printed GNPs demonstrated "doughnutting" of the printed GNPs, while FSNP-Con A showed almost complete and evenly distributed feature coverage for the Man-GNPs, with decreasing intensity for decreasing Man density. This observation may be the result of an altered avidity of binding, a reduction in surface charge, or simply better steric access of Con A induced by the different labels. To our knowledge, a systematic study of the effect of different labels on the binding and specificity of lectins has yet to be carried out.

As a POC, a glycan array variant has been reported in which positively charged nanospherical surfactant vesicles were used to control the density of carbohydrates on nitrocellulose slides [58]. Unilamellar vesicles consisting of sodium dodecylbenzene sulfonate (SDBS) and cetyltrimethylammonium tosylate (CTAT) incorporating long-chain hydrocarbon (*n*-dodecyl-β-) glycosides terminating in Glc or maltose (Glc-α(1,4)-Glc) were assembled by mixing in an aqueous suspension. Finished vesicles were analyzed by particle sizing, which revealed that the diameter of the Glc-incorporating vesicles was well controlled with a distribution between 115 and 142 nm despite a mole fraction (mf) of glycolipid incorporation ranging from 0.005 to 0.300 mf [58]. The effects of the carbohydrate density variation were evaluated using plant lectins. Con A, a plant lectin with strongest affinity for Man and some affinity for Glc, demonstrated decreased binding to vesicles proportional to increasing density of incorporated Glc. Maximal Con A binding was observed at 0.005 mf with a statistically significant reduction in binding corresponding with the increased Glc density of vesicles of 0.050 mf and beyond [58].

Density of presented carbohydrates is particularly important in host–pathogen interactions. Viruses first interact with hosts most often via glycan-mediated interactions, and influenza viruses are known to depend upon terminal sialylation of host oligosaccharides for their attachment [59]. Carbohydrates were attached in nanoscale membrane arrangements modeling cell surface presentation of glycan structures on microarrays with potential relevance to influenza viruses [60]. The system employed was, in effect, a hybrid of techniques used for neoglycolipids [18] and those used for PAA neoglyconjugates [61]. Spatial distribution of glycans was facilitated through the combined attachment of biotinylated PAA glycoconjugates at varying concentrations with streptavidin on the surface of aldehyde-functionalized polyethersulfone (PES) membrane. A nitro blue tetrazolium and 5-bromo-4-chloro-3′-indolyl phosphate (NBT/BCIP) reporting system for densitometric analysis of binding events was used; however, a variety of detection methods could conceivably be employed with this system to detect interactions at the array surface. Using a GlcNAc-specific lectin from wheat, WGA, the authors showed that the glycan membrane array demonstrated high linearity from 1 ng/mL to 1 μg/mL of PAA–chitotriose glycoconjugates [60]. The full membrane microarray, consisting of 88 distinct PAA glycoconjugates, was profiled with an assortment of lectins and anti-Lewis antibodies to demonstrate the utility of the spatial presentation provided by this approach. Ultimately, a subset of 30 PAA neoglycoconjugates was used to construct a membrane microarray on which a recombinant influenza hemagglutinin (HA) and five individual strains of intact influenza B (IB) particles were profiled. The subtle differences between glycan preferences for each of the five virus strains were revealed in their profiles. Most profoundly, it was demonstrated that while IB strain Taiwan/966/2001 bound almost exclusively to structures terminating with 5-*N*-acetylneuraminic acid (Neu5Ac) in an α-(2,6) linkage, the recombinant HA and also the four additional IB strains under test also bound to structures terminating with α-(2,3)-linked Neu5Ac, reflecting the possibility of greater cross-species transfer capability for those strains [60].

10.2.4 Nanostructures for Next-Generation Detection of Carbohydrate-Mediated Interactions

Newer methods of detection are now being developed that span the gap between the conventional multiplexed microarray and the label-free, multiplexed biosensor [62,63]. Multiplexed systems incorporating impedimetric biosensor, cantilever, quartz microbalance, and plasmon resonance, among other techniques, are all being tested as approaches for the next generation of array technologies. These will not require fluorescent labels or secondary (sandwich) detection for analytes, and many of these approaches incorporate novel nanoscale architecture [64]. In some cases, the surface of a microarray itself is composed of nanostructures capable of detecting modification and binding events. The solid microarray surface itself may also function as a 3D multivalent scaffold for optimal ligand presentation to the lectin, thereby facilitating an increase in ligand density, avidity, and improvement of overall signal-to-noise ratio.

SWCTs have been recently employed as surfaces for noncovalently immobilized glycosylated porphyrins. The electronic properties of the SWCTs enable detection by field-effect transistor (FET) devices of unlabeled lectins upon binding to specific carbohydrates [65], and noncovalent immobilization maintains the electrical conductivity necessary for FET measurements [66]. While SWCTs are very sensitive to adsorption, after functionalization with the glycoconjugate, no nonspecific adsorption of lectins on the surface was observed [65]. The SWCT-FET device had excellent sensitivity with the response measurable for lectin at 2 nM, which is comparable to optical microarray sensitivity. This configuration also enabled kinetic measurements of lectin–glycoconjugate interaction. For example, a K_d calculated for galactosylated porphyrin and PA-IL, a galactose-specific lectin from *P. aeruginosa*, was one order of magnitude less than that obtained using ITC (~6.8 μM compared to 133 nM). This difference was ascribed to the porphyrin glycoconjugate ligands being adsorbed on the SWCT surface compared to being free in solution in the ITC configuration [65].

This work has been more recently expanded by the same group to usage of chemically converted graphene (CCG) as a surface with immobilized pyrene-based glycoconjugates [66]. For both pyrene- and porphyrin-based glycoconjugates, the SWCT devices had greater response and selectivity compared to the CCG-FET devices, which may have been a result of greater accessibility and flexibility of the carbohydrates on the SWCT surfaces. A K_d value of 147.7 nM was obtained using the SWCT configuration for galactosylated porphyrin and PA-IL, though the one magnitude improvement in K_d measurement in ostensibly the same system as previously assessed was not explained [66].

10.3 WHEREVER MICROARRAYS AND NANOTECHNOLOGY MEET, THERE WILL BE PROGRESS

Array-based carbohydrate analysis has enjoyed considerable attention during the first decade of its existence, and this attention continues to expand. As a maturing technology, microarrays have now moved from the one-off experimental POC to a routine analysis technique utilized for screening and hypothesis building. The incorporation of ever-smaller materials and structures, those at the nanoscale and below, is enhancing affinity approaches in carbohydrate research and will ultimately bring a new level of refinement to glycomic arrays. The examples included in this short review illustrate the current trends in nanostructure incorporation relevant to affinity glycoanalysis and afford a glimpse into the future of upcoming developments.

REFERENCES

1. Rakus, J.F.; Mahal, L.K. New Technologies for Glycomic Analysis: Toward a Systematic Understanding of the Glycome. *Annu. Rev. Anal. Chem.*, **2011**, *4*, 367–392.
2. Varki, A.; Sharon, N. Historical Background and Overview. In *Essentials of Glycobiology*; Varki, A.; Cummings, R.D.; Esko, J.D.; Freeze, H.H.; Stanley, P.; Bertozzi, C.R.; Hart, G.W.; Etzler, M.E., Eds.; Cold Spring Harbor Laboratory Press: Cold Spring Harbor, NY, **2009**.
3. DuBois, M.; Gilles, K.A.; Hamilton, J.K.; Rebers, P.A.; Smith, F. Colorimetric Method for Determination of Sugars and Related Substances. *Anal. Chem.*, **1956**, *28*, 350–356.
4. Kilcoyne, M.; Gerlach, J.Q.; Farrell, M.P.; Bhavanandan, V.P.; Joshi, L. Periodic acid–Schiff's Reagent Assay for Carbohydrates in a Microtiter Plate Format. *Anal. Biochem.*, **2011**, *416*, 18–26.
5. Mariño, K.; Bones, J.; Kattla, J.J.; Rudd, P.M. A Systematic Approach to Protein Glycosylation Analysis: A Path through the Maze. *Nat. Chem. Biol.*, **2010**, *6*, 713–723.
6. Lundborg, M.; Widmalm, G. Structural Analysis of Glycans by NMR Chemical Shift Prediction. *Anal. Chem.*, **2011**, *83*, 1514–1517.
7. Roldós, V.; Cañada, F.J.; Jiménez-Barbero, J. Carbohydrate–Protein Interactions: A 3D View by NMR. *ChemBioChem*, **2011**, *12*, 990–1005.
8. Nazarenko, E.L.; Komandrova, N.A.; Gorshkova, R.P.; Tomshich, S.V.; Zubkov, V.A.; Kilcoyne, M.; Savage, A.V. Structures of Polysaccharides and Oligosaccharides of Some Gram-Negative Marine Proteobacteria. *Carbohydr. Res.*, **2003**, *338*, 2449–2457.
9. Leymarie, N.; Zaia, J. Effective Use of Mass Spectrometry for Glycan and Glycopeptide Structural Analysis. *Anal. Chem.*, **2012**, *84*, 3040–3048.

10. Pabst, M.; Altmann, F. Glycan Analysis by Modern Instrumental Methods. *Proteomics*, **2011**, *11*, 631–643.

11. Gerlach, J.Q.; Cunningham, S.; Kane, M.; Joshi, L. Glycobiomimics and Glycobiosensors. *Biochem. Soc. Trans.*, **2010**, *38*, 1333–1336.

12. Liu, Y.; Palma, A.S.; Feizi, T. Carbohydrate Microarrays: Key Developments in Glycobiology. *Biol. Chem.*, **2009**, *390*, 647–656.

13. Wang, L.; Cummings, R.D.; Smith, D.F.; Huflejt, M.; Campbell, C.T.; Gildersleeve, J.C.; Gerlach, J.Q.; Kilcoyne, M.; Joshi, L.; Serna, S.; Reichardt, N.-C.; Parera Pera, N.; Pieters, R.J.; Eng, W.; Mahal, L.K. Cross-Platform Comparison of Glycan Microarray Formats. *Glycobiology*, **2014**, *24*, 507–517.

14. Rillahan, C.D.; Paulson, J.C. Glycan Microarrays for Decoding the Glycome. *Annu. Rev. Biochem.*, **2011**, *80*, 797–823.

15. Sharon, N.; Lis, H. History of Lectins: From Hemagglutinins to Biological Recognition Molecules. *Glycobiology*, **2004**, *14*, 53R–62R.

16. Blixt, O.; Head, S.; Mondala, T.; Scanlan, C.; Huflejt, M.E.; Alvarez, R.; Bryan, M.C.; Fazio, F.; Calarese, D.; Stevens, J.; Razi, N.; Stevens, D.J.; Skehel, J.J.; van Die, I.; Burton, D.R.; Wilson, I.A.; Cummings, R.; Bovin, N.; Wong, C.-H.; Paulson, J.C. Printed Covalent Glycan Array for Ligand Profiling of Diverse Glycan Binding Proteins. *Proc. Natl. Acad. Sci. U. S. A.*, **2004**, *101*, 17033–17038.

17. Kilcoyne, M.; Gerlach, J.Q.; Gough, R.; Gallagher, M.E.; Kane, M.; Carrington, S.D.; Joshi, L. Construction of a Natural Mucin Microarray and Interrogation for Biologically Relevant Glyco-Epitopes. *Anal. Chem.*, **2012**, *84*, 3330–3338.

18. Palma, A.S.; Feizi, T.; Childs, R.A.; Chai, W.; Liu, Y. The Neoglycolipid (NGL)-Based Oligosaccharide Microarray System Poised to Decipher the Meta-Glycome. *Curr. Opin. Chem. Biol.*, **2014**, *18*, 87–94.

19. Oyelaran, O.; Li, Q.; Farnsworth, D.; Gildersleeve, J.C. Microarrays with Varying Carbohydrate Density Reveal Distinct Subpopulations of Serum Antibodies. *J. Proteome Res.*, **2009**, *8*, 3529–3538.

20. Gerlach, J.Q.; Kilcoyne, M.; Joshi, L. Microarray Evaluation of the Effects of Lectin and Glycoprotein Orientation and Data Filtering on Glycoform Discrimination. *Anal. Methods*, **2013**, *6*, 440–449.

21. Chen, S.; LaRoche, T.; Hamelinck, D.; Bergsma, D.; Brenner, D.; Simeone, D.; Brand, R.E.; Haab, B.B. Multiplexed Analysis of Glycan Variation on Native Proteins Captured by Antibody Microarrays. *Nat. Methods*, **2007**, *4*, 437–444.

22. Haab, B.B.; Yue, T. High-Throughput Studies of Protein Glycoforms Using Antibody-Lectin Sandwich Arrays. *Methods Mol. Biol. (Clifton, NJ)*, **2011**, *785*, 223–236.

23. Pilobello, K.T.; Krishnamoorthy, L.; Slawek, D.; Mahal, L.K. Development of a Lectin Microarray for the Rapid Analysis of Protein Glycopatterns. *ChemBioChem*, **2005**, *6*, 985–989.

24. Propheter, D.C.; Mahal, L.K. Orientation of GST-Tagged Lectins via in Situ Surface Modification to Create an Expanded Lectin Microarray for Glycomic Analysis. *Mol. Biosyst.*, **2011**, *7*, 2114–2117.

25. Knibbs, R.N.; Goldstein, I.J.; Ratcliffe, R.M.; Shibuya, N. Characterization of the Carbohydrate Binding Specificity of the Leukoagglutinating Lectin from Maackia Amurensis. Comparison with Other Sialic Acid-Specific Lectins. *J. Biol. Chem.*, **1991**, *266*, 83–88.

26. Najjam, S.; Gibbs, R.V.; Gordon, M.Y.; Rider, C.C. Characterization of Human Recombinant Interleukin 2 Binding to Heparin and Heparan Sulfate Using an ELISA Approach. *Cytokine*, **1997**, *9*, 1013–1022.
27. Feizi, T.; Fazio, F.; Chai, W.; Wong, C.H. Carbohydrate Microarrays—A New Set of Technologies at the Frontiers of Glycomics. *Curr. Opin. Struct. Biol.*, **2003**, *13*, 637–645.
28. Schwarz, M.; Spector, L.; Gargir, A.; Shtevi, A.; Gortler, M.; Altstock, R.T.; Dukler, A.A.; Dotan, N. A New Kind of Carbohydrate Array, Its Use for Profiling Antiglycan Antibodies, and the Discovery of a Novel Human Cellulose-Binding Antibody. *Glycobiology*, **2003**, *13*, 749–754.
29. Angeloni, S.; Ridet, J.L.; Kusy, N.; Gao, H.; Crevoisier, F.; Guinchard, S.; Kochhar, S.; Sigrist, H.; Sprenger, N. Glycoprofiling with Micro-Arrays of Glycoconjugates and Lectins. *Glycobiology*, **2005**, *15*, 31–41.
30. Kuno, A.; Uchiyama, N.; Koseki-Kuno, S.; Ebe, Y.; Takashima, S.; Yamada, M.; Hirabayashi, J. Evanescent-Field Fluorescence-Assisted Lectin Microarray: A New Strategy for Glycan Profiling. *Nat. Methods*, **2005**, *2*, 851–856.
31. Kilpatrick, D.C. Animal Lectins: A Historical Introduction and Overview. *Biochim. Biophys. Acta BBA Gen. Subj.*, **2002**, *1572*, 187–197.
32. Blixt, O.; Han, S.; Liao, L.; Zeng, Y.; Hoffmann, J.; Futakawa, S.; Paulson, J.C. Sialoside Analogue Arrays for Rapid Identification of High Affinity Siglec Ligands. *J. Am. Chem. Soc.*, **2008**, *130*, 6680–6681.
33. Guo, Y.; Feinberg, H.; Conroy, E.; Mitchell, D.A.; Alvarez, R.; Blixt, O.; Taylor, M.E.; Weis, W.I.; Drickamer, K. Structural Basis for Distinct Ligand-Binding and Targeting Properties of the Receptors DC-SIGN and DC-SIGNR. *Nat. Struct. Mol. Biol.*, **2004**, *11*, 591–598.
34. Reichardt, N.C.; Martín-Lomas, M.; Penadés, S. Glyconanotechnology. *Chem. Soc. Rev.*, **2013**, *42*, 4358–4376.
35. Mu, B.; Zhang, J.; McNicholas, T.P.; Reuel, N.F.; Kruss, S.; Strano, M.S. Recent Advances in Molecular Recognition Based on Nanoengineered Platforms. *Acc. Chem. Res.*, **2014**, *47*, 979–988.
36. Adak, A.K.; Lin, H.-J.; Lin, C.-C. Multivalent Glycosylated Nanoparticles for Studying Carbohydrate–protein Interactions. *Org. Biomol. Chem.*, **2014**, *12*, 5563–5573.
37. Zhao, J.; Liu, Y.; Park, H.-J.; Boggs, J.M.; Basu, A. Carbohydrate-Coated Fluorescent Silica Nanoparticles as Probes for the Galactose/3-Sulfogalactose Carbohydrate–Carbohydrate Interaction Using Model Systems and Cellular Binding Studies. *Bioconjug. Chem.*, **2012**, *23*, 1166–1173.
38. Gening, M.L.; Titov, D.V.; Cecioni, S.; Audfray, A.; Gerbst, A.G.; Tsvetkov, Y.E.; Krylov, V.B.; Imberty, A.; Nifantiev, N.E.; Vidal, S. Synthesis of Multivalent Carbohydrate-Centered Glycoclusters as Nanomolar Ligands of the Bacterial Lectin LecA from *Pseudomonas Aeruginosa*. *Chem. Eur. J.*, **2013**, *19*, 9272–9285.
39. Hall, M.K.; Weidner, D.A.; Chen, J. M.; Bernetski, C.J.; Schwalbe, R.A. Glycan Structures Contain Information for the Spatial Arrangement of Glycoproteins in the Plasma Membrane. *PLoS One*, **2013**, *8*, e75013.
40. Wehner, J.W.; Hartmann, M.; Lindhorst, T.K. Are Multivalent Cluster Glycosides a Means of Controlling Ligand Density of Glycoarrays? *Carbohydr. Res.*, **2013**, *371*, 22–31.

41. Kilcoyne, M.; Gerlach, J.Q.; Kane, M.; Joshi, L. Surface Chemistry and Linker Effects on Lectin–carbohydrate Recognition for Glycan Microarrays. *Anal. Methods*, **2012**, *4*, 2721–2728.

42. Liang, C.-H.; Wang, S.-K.; Lin, C.-W.; Wang, C.-C.; Wong, C.-H.; Wu, C.-Y. Effects of Neighboring Glycans on Antibody–Carbohydrate Interaction. *Angew. Chem. Int. Ed.*, **2011**, *50*, 1608–1612.

43. Godula, K.; Bertozzi, C.R. Density Variant Glycan Microarray for Evaluating Cross-Linking of Mucin-like Glycoconjugates by Lectins. *J. Am. Chem. Soc.*, **2012**, *134*, 15732–15742.

44. Narla, S.N.; Sun, X.-L. Glyco-Macroligand Microarray with Controlled Orientation and Glycan Density. *Lab. Chip*, **2012**, *12*, 1656–1663.

45. Kilcoyne, M.; Twomey, M.E.; Gerlach, J.Q.; Kane, M.; Moran, A.P.; Joshi, L. Campylobacter Jejuni Strain Discrimination and Temperature-Dependent Glycome Expression Profiling by Lectin Microarray. *Carbohydr. Res.*, **2014**, *389*, 123–133.

46. Naughton, J.A.; Mariño, K.; Dolan, B.; Reid, C.; Gough, R.; Gallagher, M.E.; Kilcoyne, M.; Gerlach, J.Q.; Joshi, L.; Rudd, P.; Carrington, S.; Bourke, B.; Clyne, M. Divergent Mechanisms of Interaction of Helicobacter Pylori and Campylobacter Jejuni with Mucus and Mucins. *Infect. Immun.*, **2013**, *81*, 2838–2850.

47. Liu, X.; Lei, Z.; Liu, F.; Liu, D.; Wang, Z. Fabricating Three-Dimensional Carbohydrate Hydrogel Microarray for Lectin-Mediated Bacterium Capturing. *Biosens. Bioelectron.*, **2014**, *58*, 92–100.

48. Reuel, N.F.; Ahn, J.-H.; Kim, J.-H.; Zhang, J.; Boghossian, A.A.; Mahal, L.K.; Strano, M.S. Transduction of Glycan–Lectin Binding Using Near-Infrared Fluorescent Single-Walled Carbon Nanotubes for Glycan Profiling. *J. Am. Chem. Soc.*, **2011**, *133*, 17923–17933.

49. Roy, R.; Shiao, T.C.; Rittenhouse-Olson, K. Glycodendrimers: Versatile Tools for Nanotechnology. *Braz. J. Pharm. Sci.*, **2013**, *49*, 85–108.

50. Chiodo, F.; Marradi, M.; Tefsen, B.; Snippe, H.; van Die, I.; Penadés, S. High Sensitive Detection of Carbohydrate Binding Proteins in an ELISA-Solid Phase Assay Based on Multivalent Glyconanoparticles. *PLoS One*, **2013**, *8*, e73027.

51. Mu, B.; McNicholas, T.P.; Zhang, J.; Hilmer, A.J.; Jin, Z.; Reuel, N.F.; Kim, J.-H.; Yum, K.; Strano, M.S. A Structure–Function Relationship for the Optical Modulation of Phenyl Boronic Acid-Grafted, Polyethylene Glycol-Wrapped Single-Walled Carbon Nanotubes. *J. Am. Chem. Soc.*, **2012**, *134*, 17620–17627.

52. Li, X.; Gao, J.; Liu, D.; Wang, Z. Studying the Interaction of Carbohydrate–Protein on the Dendrimer-Modified Solid Support by Microarray-Based Plasmon Resonance Light Scattering Assay. *Analyst*, **2011**, *136*, 4301–4307.

53. Olkhov, R.V.; Weissenborn, M.J.; Flitsch, S.L.; Shaw, A.M. Glycosylation Characterization of Human and Porcine Fibrinogen Proteins by Lectin-Binding Biophotonic Microarray Imaging. *Anal. Chem.*, **2014**, *86*, 621–628.

54. Zauner, G.; Hoffmann, M.; Rapp, E.; Koeleman, C.A.M.; Dragan, I.; Deelder, A.M.; Wuhrer, M.; Hensbergen, P.J. Glycoproteomic Analysis of Human Fibrinogen Reveals Novel Regions of O-Glycosylation. *J. Proteome Res.*, **2012**, *11*, 5804–5814.

55. Adamczyk, B.; Struwe, W.B.; Ercan, A.; Nigrovic, P.A.; Rudd, P.M. Characterization of Fibrinogen Glycosylation and Its Importance for Serum/Plasma N-Glycome Analysis. *J. Proteome Res.*, **2013**, *12*, 444–454.

56. Wang, X.; Ramström, O.; Yan, M. Dye-Doped Silica Nanoparticles as Efficient Labels for Glycans. *Chem. Commun.*, **2011**, *47*, 4261–4263.

57. Tong, Q.; Wang, X.; Wang, H.; Kubo, T.; Yan, M. Fabrication of Glyconanoparticle Microarrays. *Anal. Chem.*, **2012**, *84*, 3049–3052.

58. Pond, M.A.; Zangmeister, R.A. Carbohydrate-Functionalized Surfactant Vesicles for Controlling the Density of Glycan Arrays. *Talanta*, **2012**, *91*, 134–139.

59. Liao, H.-Y.; Hsu, C.-H.; Wang, S.-C.; Liang, C.-H.; Yen, H.-Y.; Su, C.-Y.; Chen, C.-H.; Jan, J.-T.; Ren, C.-T.; Chen, C.-H.; Cheng, T.-J.R.; Wu, C.-Y.; Wong, C.-H. Differential Receptor Binding Affinities of Influenza Hemagglutinins on Glycan Arrays. *J. Am. Chem. Soc.*, **2010**, *132*, 14849–14856.

60. Lao, W.-I.; Wang, Y.-F.; Kuo, Y.-D.; Lin, C.-H.; Chang, T.-C.; Su, I.-J.; Wang, J.-R.; Chang, C.-F. Profiling of Influenza Viruses by High-Throughput Carbohydrate Membrane Array. *Future Med. Chem.*, **2011**, *3*, 283–296.

61. Dyukova, V.I.; Dementieva, E.I.; Zubtsov, D.A.; Galanina, O.E.; Bovin, N.V.; Rubina, A.Y. *Hydrogel Glycan Microarrays. Anal. Biochem.*, **2005**, *347*, 94–105.

62. Hushegyi, A.; Tkac, J. Are Glycan Biosensors an Alternative to Glycan Microarrays? *Anal. Methods*, **2014**, *6*, 6610–6620.

63. Reuel, N.F.; Mu, B.; Zhang, J.; Hinckley, A.; Strano, M.S. Nanoengineered Glycan Sensors Enabling Native Glycoprofiling for Medicinal Applications: Towards Profiling Glycoproteins without Labeling or Liberation Steps. *Chem. Soc. Rev.*, **2012**, *41*, 5744–5779.

64. Kluková, L.; Bertok, T.; Kasák, P.; Tkac, J. Nanoscale-Controlled Architecture for the Development of Ultrasensitive Lectin Biosensors Applicable in Glycomics. *Anal. Methods*, **2014**, *6*, 4922–4931.

65. Vedala, H.; Chen, Y.; Cecioni, S.; Imberty, A.; Vidal, S.; Star, A. Nanoelectronic Detection of Lectin-Carbohydrate Interactions Using Carbon Nanotubes. *Nano Lett.*, **2011**, *11*, 170–175.

66. Chen, Y.; Vedala, H.; Kotchey, G.P.; Audfray, A.; Cecioni, S.; Imberty, A.; Vidal, S.; Star, A. Electronic Detection of Lectins Using Carbohydrate-Functionalized Nanostructures: Graphene versus Carbon Nanotubes. *ACS Nano*, **2012**, *6*, 760–770.

11

SCANNING PROBE MICROSCOPY FOR THE STUDY OF INTERACTIONS INVOLVING GLYCOPROTEINS AND CARBOHYDRATES

YIH HORNG TAN

Department of Chemistry and Biochemistry and Center for Nanoscience, University of Missouri–St. Louis, St. Louis, MO, USA

11.1 INTRODUCTION

The great diversity of carbohydrates, in combination with different glycoproteins and other molecules (collectively known as the glycocalyx), that cover the apical surface of many epithelia and other cells determines a variety of biological processes. These processes include cell–cell signaling, initiation of infection, promotion of symbiosis, control of differentiation, and organ formation [1–7]. Lectins are carbohydrate-binding proteins that are known for their selectivity and recognition of specific carbohydrate moieties. Various studies have indicated that lectin–carbohydrate binding is generally weak in nature, especially for monovalent binding pairs [8], and the lectin-binding pocket for interacting with the carbohydrate is usually shallow and solvent exposed [9–11]. The weak association between a typical lectin–carbohydrate pair is on the order of hundred-to-thousandfold weaker than protein–protein interactions and can be compensated by multivalent or cross-linking interactions to achieve high binding affinity [12]. Owing to their importance in many biological processes, in-depth understanding and exploitation of the selectivity of lectin–carbohydrate interactions are crucial for developing many relevant bioanalytical detection and biomedical diagnosis applications. Good examples of the importance of distinguishing

Carbohydrate Nanotechnology, First Edition. Edited by Keith J. Stine.

complex carbohydrate structures/interactions on cell surfaces are their application from blood-typing to clinical diagnostics [13,14].

Knowledge of the molecular recognitions/interactions between any given pair of lectins and carbohydrates is a key to understanding many biological events and to developing next-generation bioanalytical applications. Many advanced analytical methods have been used to observe/study the dynamics and distributions of sugar chains on different model membranes via supported monolayer systems (sugar scaffolds) [15,16]. Modern analytical techniques, such as isothermal titration calorimetry (ITC) [17,18], affinity capillary electrophoresis [19], surface plasmon resonance (SPR) [20,21]. X-ray photoelectron spectroscopy (XPS) [21], mass spectrometry (MS) [22], fluorescence correlation spectroscopy [23], and many more, all contribute to significant advances and are prominent tools in the emerging field of glycoscience. However, even with promising detection sensitivity for some techniques mentioned above, they are often limited by either the amount of available materials or lack *in situ* single molecule interrogation capability.

In order to circumvent this limitation, atomic force microscope (AFM) was introduced as a high-resolution imaging tool to examine the topography and mechanical properties of glyco-SAM (carbohydrate-decorated self-assembled monolayer) scaffolds on surfaces without additional labeling or complicated sample treatment. Many researchers have taken advantage of the capability of AFM to extend the characterization of the distribution of carbohydrate moieties on a substrate surface, from in-air to under-liquid environments. Furthermore, the *in situ* imaging method renders AFM as a powerful tool to interrogate the interaction force between individual and specific lectin–carbohydrate complexes. The attogram-level mass and picoNewton force detection sensitivities achieved by the AFM cantilever-based sensing technique can be applied to specific targeted carbohydrate moieties by selectively modifying/functionalizing the AFM cantilever with complementary probe molecules or lectins [24,25].

Given the advance in carbohydrate-related research and interdisciplinary collaboration efforts, many groups have reported novel and bioinspired carbohydrate scaffolds to study lectin–carbohydrate interactions at molecular level. Among several types of scaffold model systems available, self-assembled monolayer of alkanethiolates on gold surfaces (specifically glyco-SAMs) offers one of the most effective strategies in preparing monolayer surfaces. The SAM system is promising, due to the high fidelity in the formation of large monolayer domains and great control over the ligand presentation, geometry, density, and orientation of carbohydrate moieties. When approached using AFM, eliciting structural activity relationship among lectin–carbohydrate pairs and exploring/exploiting mono- or multivalent lectin–carbohydrate interactions are all possible.

A brief introduction of the fundamental components and the basic concepts of the working principles of AFM will be discussed at the beginning of this chapter. This will be followed by an overview of a variety of recent AFM imaging modes and some considerations about image resolution and acquisition conditions. The method of exploiting force curves to investigate molecular interaction between complementary ligand–receptor pairs (e.g., lectin–carbohydrate) will be introduced. The chemistries for the preparation of glyco-SAM scaffolds and *in situ* acquisition of the glyco-SAM morphology and phase behavior will be briefly mentioned. Promising integration of AFM with other modern analytical tools available for probing the physical, chemical, and biological nature of lectin–carbohydrate interactions will be summarized at the end of the chapter.

11.2 FUNDAMENTAL ELEMENTS OF THE ATOMIC FORCE MICROSCOPE

Atomic force microscopy (AFM) belongs to the broad assortment of scanning probe techniques capable of measuring local properties, such as height, friction, and magnetism, with an atomically sharp tip/probe. In 1986, Binnig, Quate, and Gerber demonstrated for the first time the implementation of AFM, using an ultrasmall probe tip supported at the end of a flexible cantilever as a new type of microscope, capable of investigating surfaces of insulators on an atomic scale [26]. The ideal of this marvelous technique, which borrowed the technology of scanning tunneling microscopy (STM), by employing piezoelectric transducers for subangstrom-scale motion over sample surfaces, and implementation of various feedback schemes adapted from STM to map out or provide information on surface topography, with minimum damage to the sample surface. The development of AFM extended the capability of STM to investigate electrically nonconductive materials, including proteins and other biological molecules. Since the inception of AFM, a large number of related techniques were developed in response to different characterization needs. Many excellent reviews can be found in the literature pertaining to different imaging modes or instrumental aspects of AFM [27–32]. In this chapter, we will briefly recall the fundamental operating principles of an AFM and a few key imaging modes that are of main interest for the study of interactions involving glycoproteins and carbohydrates.

11.2.1 How Does AFM Work?

In order for AFM to form an image, the sharp probe utilized by AFM is brought into close proximity of the sample surface. The microscope raster scans the probe over the sample while measuring the local property in question. The raster scanning pattern (a series of rows in a zigzag pattern covering the designated scanning area) is controlled and performed by using a piezoelectric actuator (usually a tube, sometimes a tuning fork) with subangstrom-scale accuracy [29,33]. Upon interacting with the sample, the probe deflects, due to the sample–tip interaction, and such deflection is normally monitored via a laser beam impinging on the back of the probe. The reflection of the beam on the split photodetector then configures an optical lever, which amplifies the probe deflections, shown in Figure 11.1a. A feedback circuit connected to the probe deflection sensor keeps the tip–sample at a fixed interaction force, this force value is then used to control the tip–sample distance. The interaction force is

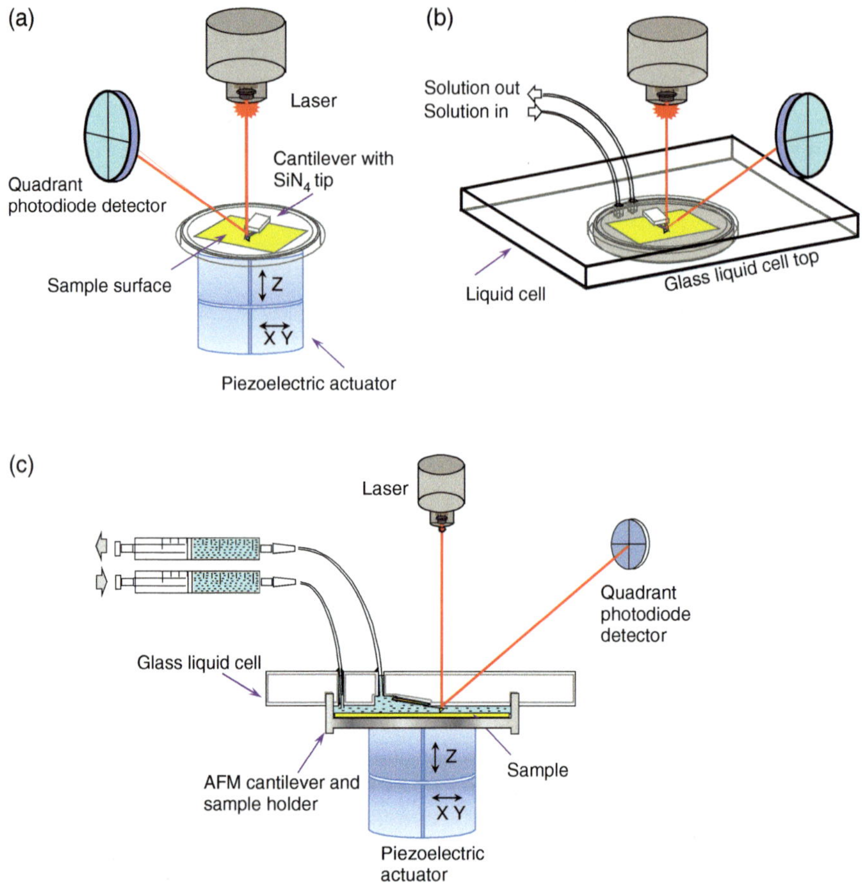

FIGURE 11.1 Schematic diagram of laser deflection-type AFM. (a) The AFM consists of a cantilever (probe) with a sharp tip at its end that is used to scan the specimen surface. The vertical deflection and lateral twisting of the cantilever as the sample rasters against the tip are measured using a laser spot reflected from the top of the cantilever onto an array of photodiodes. When the focused laser spot changes position, the differences between the top and bottom half, along with the signal change between the left and right halves of the position-sensitive detector (4-quadrant photodiode) are translated into a Z signal and a friction signal. By corresponding the coordinates and the piezo movement or lateral signal, a three-dimensional map of the sample surface can be obtained. (b) and (c) Schematic diagrams of the liquid cell setup for AFM instrument for easy insertion or exchange of solutions during an experiment without creating much disturbance, thus allowing *in situ* experiments to be possible. (Please refer to ProQuest/UMI Publication No. 3401508 or 304849342 for further details.)

thereby straightforwardly obtained from Hooke's law [34]. A three-dimensional surface topography $z(x,y, F_c = \text{constant force})$ of the sample surface can be mapped, where z is the height and the sample surface stays in the x–y plane. Atomic resolution of a periodic lattice of NaCl(001) reported by Amer et al. is a good example of this technique [35].

11.2.2 AFM Data Acquisition Environment

Unlike other high-resolution characterization techniques, such as transmission electron microscopy (TEM), scanning electron microscope (SEM), STM, and others, AFM is capable of operating in various environments such as in air, in liquid, and under vacuum. Because of this unique ability, AFM has been used to image biomaterials (such as DNA, protein, glycoprotein, bacteria flagella, cell, etc.) under near-physiological conditions and even to monitor biological processes in real time [1,27,36–40].

To date, many commercially available AFM instruments on the market can enable researchers to acquire data, at molecular resolution in air or liquid environment, from the localized interaction between the sharp probe and surface structure. Typical lateral and vertical resolutions of AFM for a crystal surface can reach 1.0 and 0.1 Å, respectively [35]. A schematic diagram of a typical liquid cell setup for the AFM instrument to acquire data in liquid environment is shown in Figure 11.1b. This type of setup allows for easy insertion or exchange of solutions during an experiment without creating much disturbance, thus allowing *in situ* data acquisition to be possible. Figure 11.1c shows the side view of an AFM liquid cell in liquid exchange setting.

11.2.3 AFM Imaging Modes

Contact mode is the most common method for the operation of AFM. During imaging, the tip and sample surface are always in contact. Typically, images are acquired in the repulsive interaction regime, where the cantilever bends upward. During the scan, the tip deformation is maintained constant by regulating the height of the sample. The output of the Z-piezo voltage manifests itself in the topographic image of the sample, and the imaging mode under this configuration is defined as "constant force mode."

Tapping mode is the next most common mode used in AFM. In tapping mode, an AFM tip is oscillated by a piezo-disk actuator. A modulated current is applied to the piezo actuator at the resonance frequency of the cantilever, causing it to rapidly "tap" the surface during scans [41,42]. The oscillation amplitude decreases when the tip "taps" the surface. A feedback system moves the piezo up or down to maintain the oscillation amplitude. The output Z-piezo voltage is used to generate the topographic image of the sample surface. Since the tip only comes in contact with the surface for a small portion of its oscillation period, the shear force is minimized. When imaging biological samples, tapping mode can greatly improve resolution and minimize damage to surfaces [43,44].

Noncontact mode is yet another operation method for AFM. During noncontact imaging the tip is also oscillated at the resonance frequency of the cantilever. In contrast to tapping mode, in noncontact mode the tip does not come in contact with the surface but is very close to the surface. When the tip is close enough to the surface, it will bend toward the surface due to an attractive force between the tip and surface. Thus, the oscillation amplitude also changes. Similar to tapping mode, the feedback loop maintains the amplitude by moving the piezo up or down. The movement of the piezo is used to create a topographic image. Because noncontact mode works in the attractive regime of the tip–surface interaction, this mode is difficult to operate in ambient conditions.

11.2.4 Cantilevers and AFM Tips

To acquire a high-resolution AFM image of a sample in question, the AFM probe plays a crucial role when it rasters across the sample surface. The dimensions of a typical cantilever are in the scale of micrometer. The tip at the end of the lever typically has a radius in the scale of a few nanometers. The high quality and dimension consistency of the AFM probe can be manufactured by exploiting microelectromechanical system (MEMS) technology [45–48]. Figure 11.2a shows a scanning electron microscopy image of a typical AFM tip microfabricated on a cantilever chip. Depending on the material and/or application, the nanometer size AFM tip can be shaped like a pyramid or a longer cone shaped. The cantilever itself is normally either triangular or rectangular in shape, as shown in Figure 11.2a and b. It is very difficult to dictate which cantilever shape is preferable for any given type of sample to be interrogated, because sample treatment and the intermediate imaging environment are also influential factors. If tip availability is not an issue, a trial and error approach

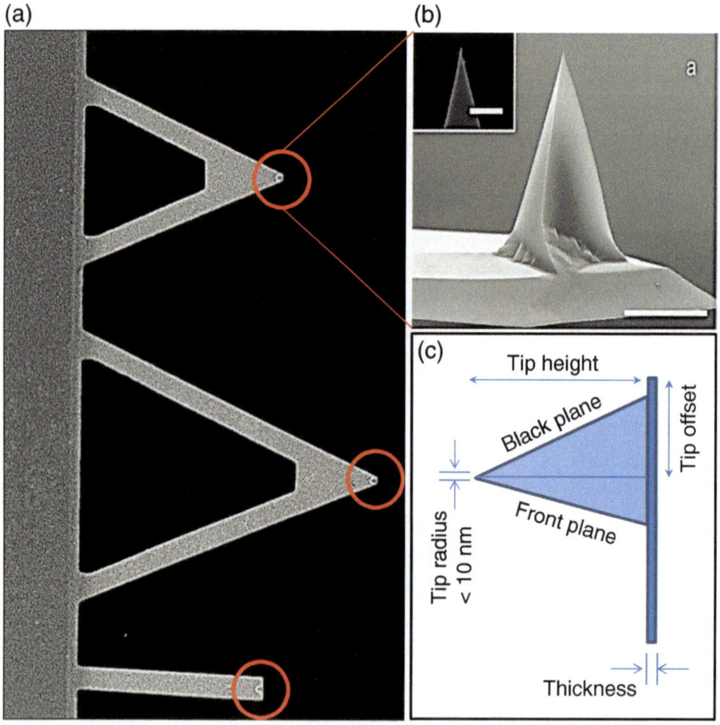

FIGURE 11.2 Morphology of typical silicon nitride AFM cantilever. (a) Scanning electron microscope image of rectangular (aka diving board) and triangular-shaped cantilevers. (b) Higher magnificent rendering of the sharp pyramidal shaped tip at the end of cantilever [51]. Scale bars represent 5 μm and 500 nm (inset). *Source*: Reprinted with permission from Ref. 51. (c) Typical geometry of an AFM cantilever, with a sharp tip radius of less than 10 nm gives high-resolution or high-quality imaging for all standard AFM.

is a good way to find a suitable tip for image acquisition. However, in general, a triangular-shaped cantilever is better than a springboard cantilever for intermittent-contact mode imaging for cells. Additionally, a triangular-shaped lever is known for its lateral stability to keep the tip position stable during force measurement; alternatively, rectangular cantilever is better for lateral force measurement studies.

Cantilevers with a spring constant between 0.01 and 0.1 N/m can be used for contact mode imaging of fixed cells or proteins [39,49]. For contact mode imaging of living cells, it is better that spring constant be as low as possible (0.01–0.03 N/m) to minimize the lateral shear forces that may alter (or dull) the tip or rearrange (or damage) the sample surface during acquisition [50]. Regardless, a wide range of AFM tips with different spring constants, 0.01–100 N/m, are available to suit researchers need to visualize mechanically different components of multicomponent samples. The typical geometry of an AFM cantilever is shown in Figure 11.2c.

11.3 VISUALIZATION AND NANOSCALE IMAGING OF CARBOHYDRATE DISTRIBUTION IN SAMS PREPARED ON GOLD VIA NATURAL ASSEMBLY OR NANOFABRICATION APPROACHES

AFM is best known for its nanometer-scale resolution imaging capabilities and a unique capacity for imaging biological molecules and subsequent protein interactions in liquid environments. Many researchers have since taken advantage of AFM to acquire high-resolution images of carbohydrate self-assembled monolayers previously assembled onto solid substrate surfaces [7,16,38,52,53]. For example, Dufrêne and coworkers used thiol-terminated hexasaccharides to study their interactions with a plant lectin, concanavalin A (Con A), by means of AFM imaging and quantification of the interaction forces via AFM force–distance measurements [53]. Using these approaches, information such as the unbinding forces that originate from specific lectin–carbohydrate interactions and the rupture lengths of corresponding glycomolecules from Con A can be obtained. Alternatively, investigation of lectin interactions with carbohydrate self-assembled monolayers on solid substrate surfaces is also promising for screening specific lectin, carbohydrate-binding proteins, or even potential application in high-throughput characterization of carbohydrate–protein interactions [16]. However, the studies of carbohydrate–protein interactions are a daunting task because of the (i) high heterogeneity of carbohydrate molecules (both naturally available and via synthesis routes), (ii) when coupling with the high complexity of multivalent protein and carbohydrate recognition domains on receptor surface, the inherent challenge required many different fields of study and resources to answer a wide range of questions.

In order to mimic the cellular carbohydrate–protein interactions occurring through glycoproteins, glycolipids, or polysaccharides displayed on the cell surface with lectins, many researches have creatively employed various molecular, surface strategies and technologies to directly quantify these complex glycobiological processes [8,52,54,55]. Some of the popular and conventional approaches, including enzyme-linked lectin assay (ELLA), SPR, hemagglutination inhibition assay, ITC, MS, and other techniques, have been successfully used to evaluate carbohydrate–protein recognition events [18,22,56–58]. Although these approaches have been successfully applied to evaluate carbohydrate–protein interactions, they often suffer from the limitation of small quantity of carbohydrate samples, being labor-intensive, and limited availability of state-of-the-art equipment for sensitive, rapid, or reliable characterizations.

Utilization of self-assembled monolayer scaffolds of carbohydrate-containing molecules on a solid support has recently emerged as one of the promising model systems to quantify molecular interactions between glycomolecules and their respective receptors on proteins in near-physiological conditions. When coupling glyco-SAMs and the proven ability of AFM for nanometer-scale resolution imaging and probing surfaces, many suitable strategies can be developed to probe specific lectin–carbohydrate interactions. The first critical aspect to be taken into account prior to AFM imaging is immobilization of the relevant glycomolecules on a solid substrate. Ideally, the best approach to immobilize the targeted molecule should be able to provide a bond between the molecule and the surface of choice, strong enough to withstand the dragging effects induced by the AFM tip during the raster scanning process. Furthermore, the molecule packed on the surface should retain its native structural integrity, functionality, and flexibility to be accessed by the binding sites on the protein. Among several classes of self-assembled monolayer formation strategies, the formation of alkanethiolates on clean gold surfaces currently holds the role of the best and most convenient system [59]. This strategy has numerous advantages, including the following: (i) the monolayer formation is a spontaneous process, governed by inter- and intramolecular forces that drive the molecules into a stable, low-energy state; (ii) the gold–sulfur (Au–S) bonds are readily formed and stable near-neutral pH and moderate temperatures; (iii) the adsorption of alkanethiol molecules from thiol solution onto gold surfaces will result in crystalline-like monolayers formed from millimolar concentrations of thiol molecules in ethanolic solution; and (iv) it is also known for its convenience of preparation in common research laboratory settings.

Despite the advantages of the SAM strategy, limitations and challenges of generating carbohydrate SAMs exist, especially during the synthetic effort of thiolated carbohydrate preparation steps. Furthermore, the complexity of molecule presentation, density, and chain length of spacer molecule tethered to the carbohydrate moiety increases as the preparation parameters vary [60,61]. It is not guaranteed that the thiolated carbohydrate molecules will form a structurally well-defined monolayer, and the orientation or tilt angle of the carbohydrate moiety is variable. Therefore, other strategies are in place for anchoring carbohydrate moieties on solid supports including Diels–Alder reaction, addition of thiol molecule to activated maleimide [62], 1,3-dipolar cycloaddition reactions ("click chemistry") [63], disulfide exchange [20], and others [59]. Other popular carbohydrate immobilization strategies (covalent and/or site specific) on derivatized surface are out of the scope of this article; readers are encouraged to refer to the review article by Shin and coworkers (Fig. 11.3) [59].

Besides the natural assembly of carbohydrate moieties onto solid supports, micro- and nanofabrication approaches have demonstrated the ability to transfer molecules onto surfaces at micro- to nanometer-scale precision. Some of the common examples

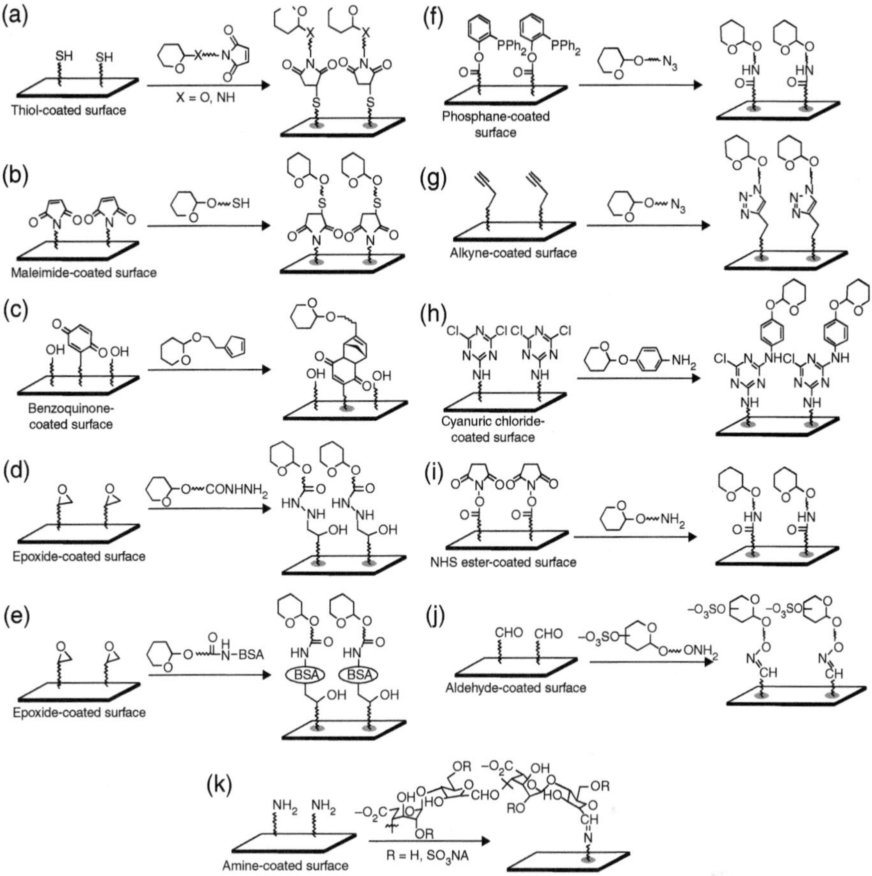

FIGURE 11.3 Strategies for covalent and site-specific display of carbohydrates on the derivatized surface for interrogating or mapping the carbohydrate–protein interactions and used for microarray construction for high-throughput screening. (See original article, Shin et al. 2008.) (a) attachment of maleimide-linked sugars to the thiol-coated surface, (b) attachment of thiol-linked sugars to the maleimide-coated surface, (c) attachment of cyclopentadiene-linked sugars to the benzophenone-coated surface, (d) attachment of hydrazide-linked sugars to the epoxide-coated surface, (e) attachment of glycan-linked BSA to the epoxide-coated surface, (f) attachment of azide-linked carbohydrates to the phosphane-coated surface, (g) attachment of azide-linked glycans to the alkyne-coated surface, (h) attachment of p-aminophenyl-linked sugars to the cyanuric chloride-coated surface, (i) attachment of amine-linked sugars to the NHS ester-coated surface, (j) attachment of aminooxy-linked sugars to the aldehyde-coated surface and (k) attachment of aldehyde-linked sugars to the amine-coated surface. *Source*: Reprinted with permission from Ref. 59.

of this strategy include the use of alkane thiolates to imprint onto a gold surface via microcontact printing [64,65], dip-pen lithography [66], and AFM tip-based grafting techniques [1,53,67]. All of these techniques enable direct deposition of millimolar quantities (carbohydrate moieties included) to supported surfaces to form a two-dimensional pattern or feature at micro- to nanometer-scale dimensions for understanding the specific molecular interactions between biologically related molecules [68,69].

11.3.1 Topographic Imaging and Mapping of Carbohydrate Self-Assembled Monolayers

The sulfur–gold bond, which is often exploited to immobilize biomolecules on surfaces, was demonstrated by Salmeron et al. in 1994 to be able to withstand AFM lateral imaging forces up to hundreds of nanoNewtons without any evidence of monolayer damage [70]. Therefore, along with the high-resolution capability, the stability of the thiol–gold bond under the general lateral forces of a scanning AFM tip has drawn many researchers to explore a wide variety of glyco-SAMs in an attempt to elucidate the binding pattern, affinity (either weak or strong dissociation forces of single ligand–receptor complexes in the picoNewton range), and specificity of lectins at a molecular level [53,71–74].

The preparation and intrinsic tunability of the glyco-SAM on the gold surface resemble those of alkanethiolate SAMs on gold. Frequently, optimization of carbohydrate's availability, presentation, and density is a crucial initial step toward gaining insight into the behavior of the single component or mixed component glyco-SAMs on different gold substrate surfaces [1,38,75,76]. Recent work done by Ratner and coworkers, using quantitative XPS and SPR imaging methods, has demonstrated that surface carbohydrate density and the nature of spacer molecule (oligoethylene glycols, OEG) dictate the lectin-binding behaviors [21]. In another published work, their in-depth investigation of the mixed SAMs of sugar and OEG thiols, using AFM, illustrated and confirmed the formation of nanoscale segregated domains in glyco-SAM over time, as seen in Figure 11.4 [52]. This strongly implies the importance of the bioavailability, molecular stability, and presentation of carbohydrate moieties on surface before using it for investigating carbohydrate–protein interactions. These parameters ultimately determine the bioactivity of surface-bound biomolecules, in order to reduce the nonspecific adsorption of proteins or understand the specific molecular interactions between lectins and carbohydrates for future application and development of new bioanalytical devices.

The importance of the local structure and density of carbohydrate molecules for subsequent glycoprotein recognition was demonstrated by Yu et al. using recombinant HIV-1 surface protein gp120 and thiolated galactosylceramide (GalCer, a primary receptor for HIV viral envelope glycoprotein on epithelial cells). In his investigation, the influence of GalCer local structure (spatially and geometrically arranged using an AFM-based nanofabrication technique) and recombinant gp120 protein immobilization and biorecognition was demonstrated, as shown in Figure 11.5 [1]. Multidisciplinary techniques and tools have been successfully implemented for studying protein–carbohydrate interactions in order to gain understanding of biological function and roles that these interactions play in disease states. Surface-based carbohydrate arrays provide another approach that has been used to facilitate the study of lectin recognition events by precisely presenting carbohydrate moieties in micro- and nanometer-scale arrangement to simulate multivalent-to-polyvalent interactions. Work done by Kiessling and Corn et al. has demonstrated that arrays of carbohydrate on gold films prepared via μCP can be used to assess carbohydrate-binding selectivity to specific lectins [20]. Furthermore, the strength of protein–carbohydrate

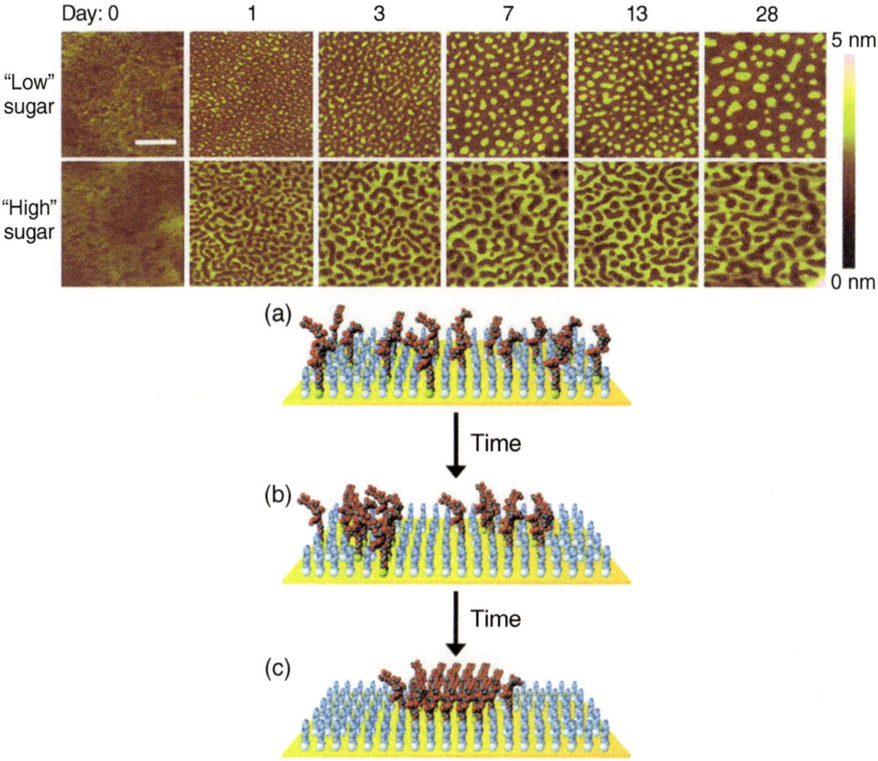

FIGURE 11.4 Time-course morphological changes of nanometer-scale clustering of thiolated carbohydrate in mixed self-assembled monolayers on gold. The clusters observed on the aged mixed glyco-SAM surfaces on polycrystalline Au substrates. The low-sugar SAM was made up of 2:3 molar ratio of sugar to OEG, and the high-sugar SAM was a 4:1 sugar/OEG molar ratio. Schematic shows the proposed time-dependent clustering mechanism in mixed SAMs of sugar (red/black) and OEG (light blue/gray) thiols (sugar/OEG ratio in the illustration is roughly 1:3). *Source*: Reprinted (adapted) with permission from Ref. 52. Copyright 2014 American Chemical Society.

interactions can be quantified using SPR techniques to determine the significance of ligand density on gold films, monitor protein binding to surfaces presenting different compositions of the immobilized carbohydrates, and investigate multivalent protein–carbohydrate interactions [20].

11.3.2 AFM Imaging of Supported Mixed and Hybrid Bilayers

Among all the available surface analysis tools and techniques mentioned in previous sections or other chapters, atomic force microscopy has been applied to investigate amphiphilic ganglioside monolayer films generated via the Langmuir–Blodgett (LB) technique. Together with glycoproteins, glycosaminoglycans, and glycosphingolipids

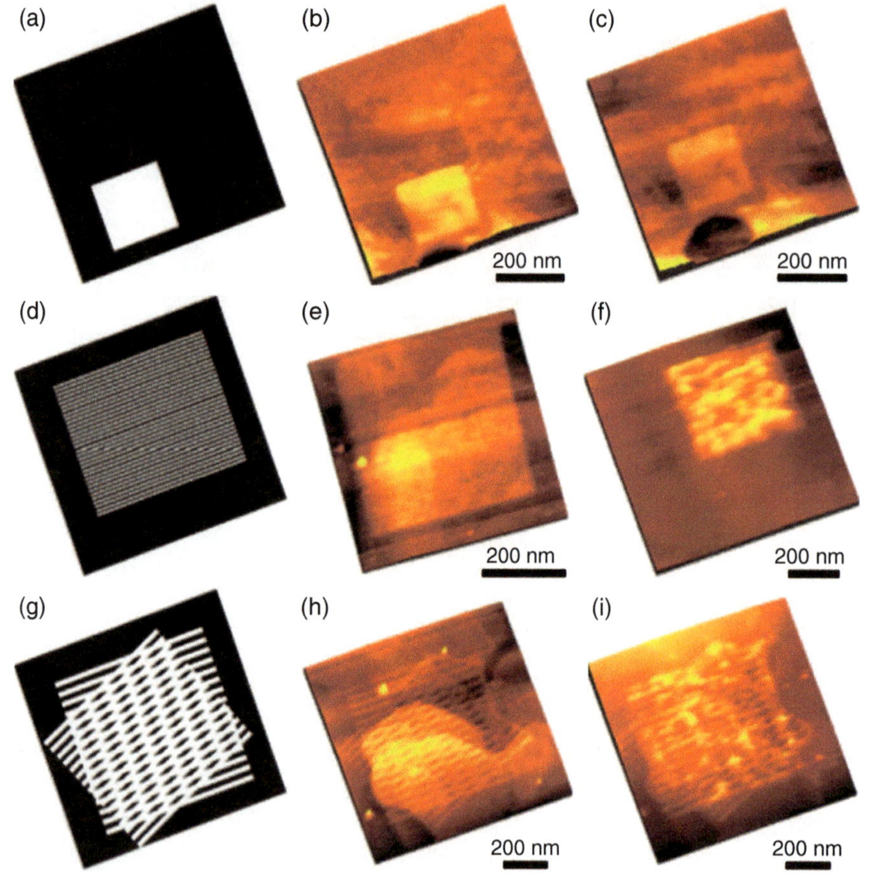

FIGURE 11.5 Polyvalent interactions of HIV-gp120 protein and nanostructures of carbohydrate ligands. A series of nanostructures of galactosylceramide (GalCer) are designed and produced using an AFM-based lithography method for guiding polyvalent interactions of HIV viral envelope glycoprotein gp120 to GalCer. The geometry dependence of recombinant gp120 binding to these nanostructures is monitored using high-resolution AFM imaging in liquid setting. (a) The schematic of a 200 nm × 200 nm square inlaid in C8 SAM. (b) AFM topograph of a 200 nm × 200 nm GalCer square pattern produced using nanografting: 256 lines per frame, scanning speed 500 nm/s. (c) Same area as (b) after 60 min immersion in a 25 μg/mL rgp120 solution. (d) The schematic of an array of lines with designed separation. (e) AFM topograph of an array of 32 lines covering a 400 nm × 400 nm area, produced using nanografting: 32 lines per frame under a shaving speed of 500 nm/s. (f) Same area as (e) after 60 min immersion in 25 μg/mL rgp120 solution. (g) A logo of an array of crossed lines with designed line width, separation, and angle. (h) AFM topograph of two arrays of lines crossing in space. Each line array contains 16 lines homogeneously distributed within a 600 nm × 600 nm area, and the two arrays are rotated 30° with respect to each other. Nanografting conditions are 16 lines per frame with a shaving speed of 500 nm/s. (i) Same area as (h) after 60 min immersion in 25 μg/mL rgp120 solution followed by rinsing with 1% Tween 20 solution. *Source*: Reprinted with permission from Ref. 1.

(GSLs), gangliosides are sialic acid-containing GSLs that are known to contribute to the glycocalyx that covers eukaryotic cell surfaces and are abundantly expressed in the nervous system [77,78]. The increase in research focus on gangliosides is partly due to the biological importance of gangliosides in neurological abnormalities [79], onset of hearing loss to deafness [80], degeneration of organs [80], and many other disorders [78].

In cells, gangliosides are primarily, but not exclusively, localized in the outer leaflets of plasma membranes and are known to play an important role in cell signaling processes [78]. Besides their role as receptors for numerous biological active agents, gangliosides are also believed to play a role in the formation of rafts [81], because the physical start of the lipid rafts (or organized microdomains of GSLs and protein receptors) determines their membrane functions [82,83]. Supported phospholipid monolayers and bilayers incorporating gangliosides prepared by the LB technique have been extensively studied as membrane models and provided a wealth of information on lipid properties and organization [84,85]. Recently, AFM and other related techniques have proved to be ideally suited to visualizing the morphological characteristic of these monolayer and bilayers. Detailed topographical information such as phase separation in supported lipid layers, roughness, height, and fluidity of supported membrane systems have been revealed [81,86–92].

Many theoretical and experimental results revealed by various groups of researchers have demonstrated the distribution of ganglioside GM1 among other components such as phosphatidylcholines, which plays an important role in the formation and stabilization of specific lipid membrane domains [93–95]. The evidence for nanometer-scale distributions of low mole fractions of GM1 between mixed dipalmitoylphosphatidylcholine (DPPC) and dioleoylphosphatidylcholine (DOPC) phase-separated LB monolayers was demonstrated by Grimellec et al. [96] and Johnston et al. [97] using AFM. The results indicated the diversity and complexity of GM1 partitioning and the advantage of high-spatial-resolution AFM for direct detection of ganglioside domains at nanometer scales that are too small for study with most other techniques, as seen in Figure 11.6.

11.4 PROBING SPECIFIC LECTIN–CARBOHYDRATE INTERACTIONS USING FORCE MEASUREMENT

11.4.1 Force Measurement and Data Analysis

The use of AFM is not limited to the high-resolution imaging of the technique. The capability of AFM as a force measurement tool has drawn interest in the glycoscience community because of its potential use to specifically interrogate intra- and intermolecular forces among selected/specific lectin–carbohydrate combinations. Since the AFM tip is capable of probing a small interaction area, highly dependent on AFM tip diameter (normally ranges from 10 to 50 nm), this makes AFM highly sensitive to small forces. In AFM, an interaction force is acquired by collecting a force curve. A typical force curve consists of a plot of cantilever deflection, as a function of the sample position along the (±) *Z*-axis. The force–distance curve collected is often displayed as a simple *XY* plot, as shown in Figure 11.7a and b, where

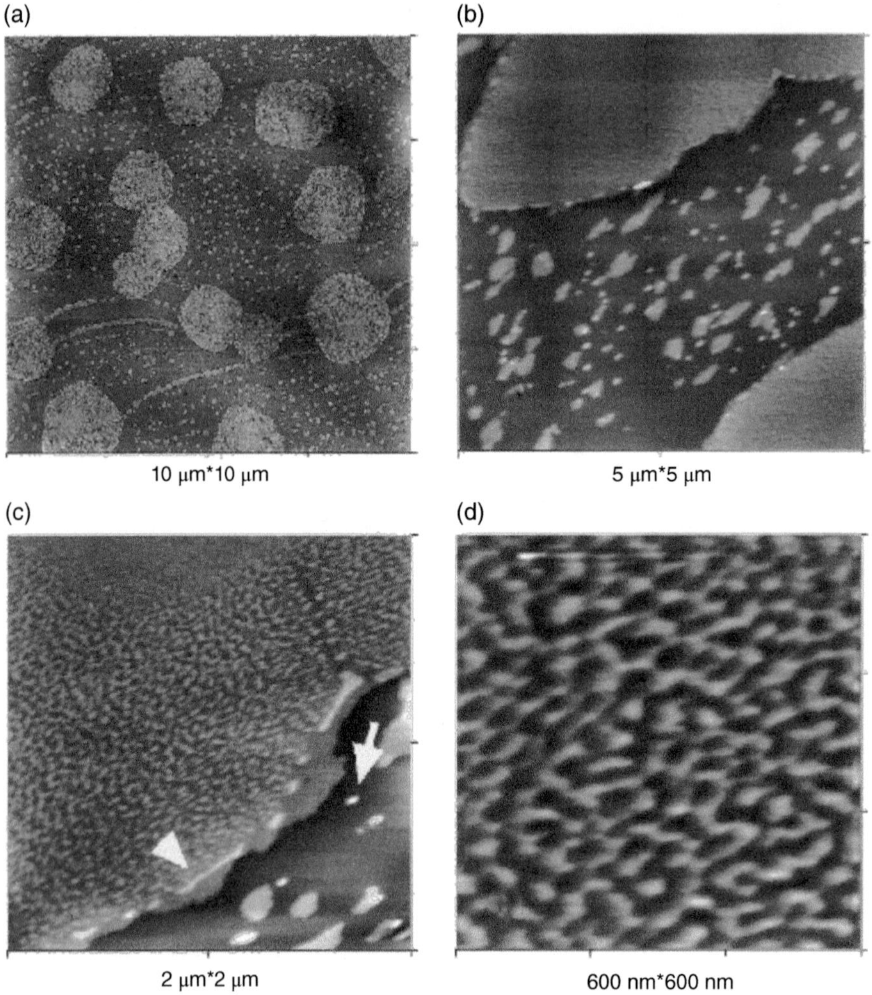

FIGURE 11.6 Topography of DPPC/DOPC LB films containing 3% GM1. Detail of the organization in DPPC macrodomains. Height scales: (a) 10 nm; (b), (c), 6 nm; (d), 3 nm. Scan rates: (a), (b), (c), 2 Hz; (d), 3 Hz. *Source*: Reprinted (adapted) with permission from Ref. 96. Copyright 2014 American Chemical Society.

the approach and retraction distance traveled or height of an AFM tip often referred to the *X*-axis and the vertical deflection of the cantilever is plotted on the *Y*-axis to give a direct measure of interaction force [98].

The basics of a force spectroscopy curve can be understood by dividing tip–surface interactions into several regions. As shown in Figure 11.7c (region I), a cantilever is in air approaching a sample surface. As the cantilever approaches the surface, the initial forces are too small or no interaction is detected to give a measurable deflection of the cantilever, and the undisturbed cantilever holds its deflection position in air. As the cantilever moves closer to the sample surface, the attractive

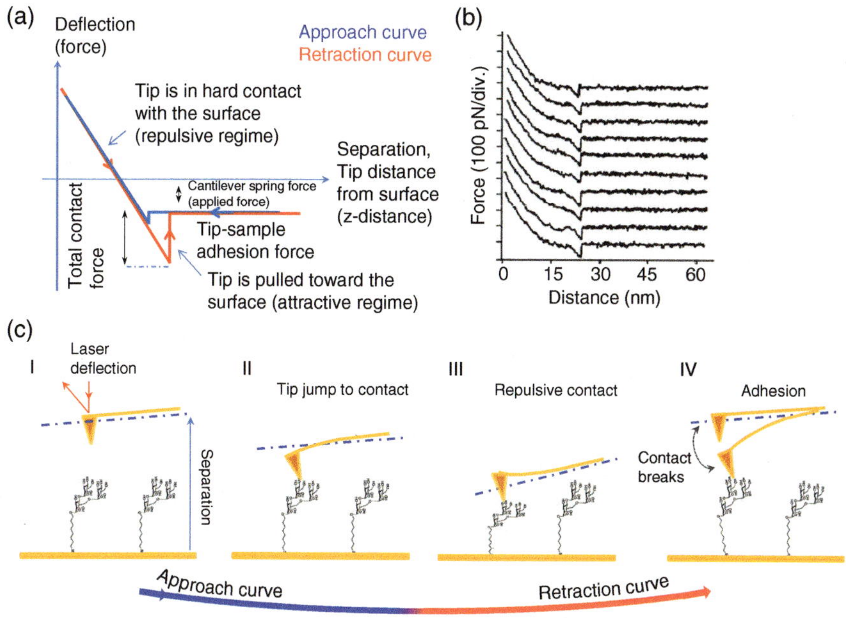

FIGURE 11.7 Ideal force–distance curve showing attraction and retraction regions between the tip and the sample. (a) Two different lines and the position of cantilever at different regions of the approach and withdrawal portions of an AFM tip–surface interaction. (b) An example of typical interaction force curves recorded on the specific interaction between purified glycophorin A and *Psathyrella velutina* lectin (PVL) with an AFM tip functionalized with PVL [99]. *Source*: Reprinted with permission from Ref. 99. (c) Schematic of a single molecule force spectroscopy experiment showing the approach half-cycle and retraction half-cycle of a single tip and sample adhesions events.

forces (van der Waals and capillary forces) overcome the cantilever spring constant and cause the tip to deflect toward the sample, as shown in Figure 11.7c (region II). It is worth noting that, at each position in the force curve, the cantilever remains deflected until it is restoring the constant imaging force to the preset force value (tip interaction force with tip), causing the system to be always at equilibrium. Once the tip is in contact with the sample, it remains on the surface and continues to approach closer to the surface as the separation between the base of the sample and the outer surface of the sample decreases further, causing an accumulation of deflection of the tip and an increase in the repulsive contact force. This region III is a repulsive regime (or positive deflection) in the force curve, in which sample elastic properties can be measured. The cantilever continues to travel further into the sample until a preset force threshold is reached, and the tip movement direction is reversed, until the tip returns to the initial starting position (region I). At the beginning of the tip retraction movement, the tip–sample interaction remains unchanged or similar to that in the approaching process due to some adhesion, and the cantilever is deflected downward. This process continues until the adhesion force between the tip and sample is overcome by the cantilever restoring force and the contact breaks (region IV) [100,101].

In a typical force–distance curve cycle, the deflection of the cantilever spring is directly proportional to the tip–sample interaction force. However, in order to extract the force value from the force curve (conversion of photodetector signal to force value), a calibration of the photodetector sensitivity and cantilever spring constant is required. The sensitivity of the cantilever actually deflects for a certain measured change in photodetector voltage that is required to convert the deflection (in volts) into the deflection in nanometer. Similarly, depending of the type or condition of a given cantilever, a known spring constant will permit the conversion of deflection (in nanometers) into force in nanoNewtons, using Hooke's law, $F = -kX$ [101,102].

11.4.2 Tip Modification

In a typical force measurement experiment, the immediate environment/setting plays a crucial role in the force measurement. The large adhesion caused by the capillary forces arising from the thin water film wetting both the tip and sample can limit the range of information that can be obtained from a force curve. However, the capillary forces can be significantly reduced by performing the force measurement in a liquid setting, a near-physiological liquid environment that permits both imaging analysis and force–distance curve acquisition.

For quantitative force measurement, a complication arises when nonspecific interactions contributing to the adhesive force convolute the actual force measurement. Nevertheless, such issues can be minimized by selectively modifying either the tip–surface or sample with an adlayer of appropriate chemical groups to quantify the adhesion force between different chemical functionalities. By subtracting the cantilever deflection from the calibrated deflection value, the actual force versus tip–sample distance can be recovered and further interpreted based on force law to obtain the force as a function of the probe sample separation. In general, different molecules or even the regions of the sample may adhere and interact with the tip differently, and a variety of adhesion events are possible. Thus, the successive force curves can show very different responses. Therefore, force measurement is often repeated a number of times within the sampling point to give a meaningful statistical understanding of the interaction or at different locations to accrue a map of the tip–sample interactions [34].

For application in glycoscience, an AFM tip or substrate surface can be selectively modified and used to investigate recognition phenomena of biomolecules. In recent years, many reports, including ligand–receptor binding [1,39,103–105], carbohydrate–lectin recognition [37,38,53,106,107], antigen–antibody interaction [39,108,109], protein unfolding [110–112], protein–cell interaction [7,40,113,114], and protein–cellulose fibrils [115,116], have been investigated by AFM either via topographical analysis or force measurement. A good example investigating lectin–carbohydrate interactions was demonstrated by Dufrêne et al. in 2003. The selectivity of lectin Con A and mannose interactions was exploited to investigate the specific interaction force between this pair, by performing AFM force measurements using hexasaccharide-terminated AFM probes and Con A-terminated substrates blocked with mannose [53].

Furthermore, AFM force spectroscopy analysis has widely been used in various receptor–ligand interactions that involved glycoproteins that are known to decorate on

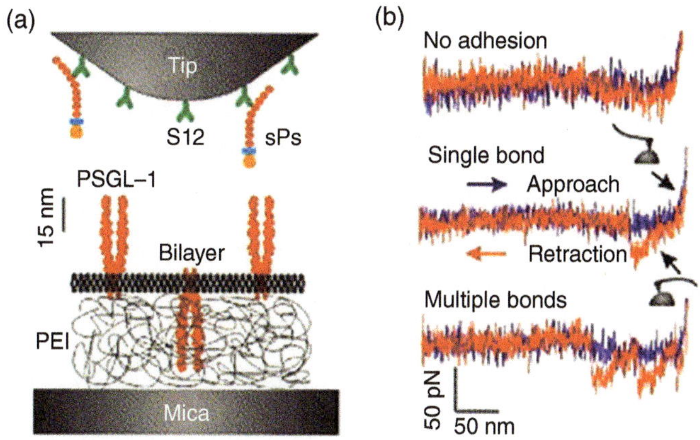

FIGURE 11.8 Quantifying the effects of contact duration, loading rate, and approach velocity on P-selectin–PSGL-1 interactions using AFM. (a) Soluble P-selectin was captured by its nonblocking mAb S12 preadsorbed on the functionalized AFM tip. P-selectin glycoprotein ligand 1 (PSGL-1) was reconstituted in a lipid bilayer supported by a polyethylenimine-cushioned mica. (b) Force–displacement curves were generated when PSGL-1-incorporated bilayer was driven to approach to (from left to right), contact with, and retract from (from right to left) soluble P-selectin-coated AFM tip. Adhesion was visualized from the cantilever deflection and rupture force was measured from the force–displacement curve (middle and lower panels). *Source*: Reprinted with permission from Ref. 118.

the extracellular surface and participate in signal transduction pathway from the outside to the inside of a live cell [117]. By using the glycoprotein P-selectin modified tip and P-selectin glycoprotein ligand 1 (PSGL-1) bilayer previously assembled on mica as a model system, Long and coworkers have shown some of the different biophysical factors (protein–ligand contact duration, loading rate, approach velocity, and bond strength) that influence the kinetics of receptor–ligand interactions can be obtained, as shown in Figure 11.8 [118]. For ligand–protein interaction, bonds have specific lifetimes and affinities. Long and coworkers found that a contact time greater than 500 ms, for P-selectin–PSGL-1 interactions, does not shift the peak force, which indicated that contact time is not as critical a factor with high-affinity bonds as other parameters, for example, loading rate [118]. The parameters focused on by the Long group, and many other factors worked on and reported by other groups, will continue to provide new insights into how physical factors regulate receptor–ligand interactions and further enhance our understanding of complex materials and cellular responses [119–122].

It is accepted concept that, in a typical ligand–receptor interaction, there are direct molecular interactions whose specificity and strength can be dictated by the chemical, physical, and mechanical properties of the participating molecules [123]. Thus, elucidating the biophysical basis of the ligand–receptor interaction, as well as the related biological structures, could provide interesting hints for new therapeutic approaches. An example of investigating the lectin–carbohydrate interactions was made by functionalizing an AFM tip with different types of lectins for the possibility of differentiating the oligosaccharide moieties on the surface of the tumor and healthy cells by using a dynamic force spectroscopy measurement at single-molecule level [7,99].

The application of modified AFM tip is not limited to acquiring picoNewton range forces for the interactions in individual ligand–receptor complexes. Recently, Wu et al. have demonstrated the modified AFM tip as a potential sensing tool (based on gravimetric detection principles) to detect the interaction between Con A and glucose, by monitoring the frequency shift of the AFM cantilever. Their creative approach in modifying the AFM tip and substrate platform with glucose-specific lectin Con A enabled them to detect the glucose from a mixture of carbohydrate solution. The changes in cantilever resonance frequency upon glucose recognition enabled them to detect the glucose binding at attogram mass sensitivity [25].

11.5 PERSPECTIVES

In the past decade, AFM has become an established technique that can be used to study single molecules. In addition to imaging surfaces with nanometer-scale resolution, AFM has gradually evolved and been implemented in many applications related to biophysical research, for determining surface roughness, probing local changes in friction, accessing inter- and intramolecular forces between ligand–receptor pairs, and quantifying the changes in local elasticity over a sample surface [71,124].

Different aspects of development in imaging modes such as lateral force microscopy [125], chemical force microscopy [104], and phase-contrast imaging [126] have been developed for specific tip–sample interactions to acquire various chemical information. However, many known limitations of these imaging modes render these techniques difficult for distinctively identifying multiple surface species reliably. Other known limitations for AFM include lateral convolution (or profile broadening effect) [127], elastic deformation (height lowering) of studied objects [128,129], scan speed limitation for contact mode imaging that potentially damage or degrade studied objects due to the increase in dragging force that occurs during acquisition [130], and the interplay between thermal, electrical, optical, and mechanical noises that can greatly hamper force sensitivity [131].

When surveying the literature from different fields, many promising technical modifications and developments have been reported by combining other established optical microscopy and spectroscopy techniques to compensate the shortcomings of AFM. The combination includes but is not limited to Raman microscopy [132,133], fluorescence/confocal microscopy [134], scanning ion conductance microscope [134], and many more that are constantly being improved and developed to provide spatially correlated physical, chemical, and biological information on the studied object during a single experimental setting. Such combinations of techniques can greatly minimize the discrepancy in sample preparation process and time, add cost saving, and reduce the variability caused by multiplatform approaches arising from technical errors and instrumentation factors.

11.6 CONCLUSIONS

Carbohydrate–protein interactions occur in a wide variety of biological events, particularly those involving cell surface recognition events that are crucial for regulating diverse cellular processes. Many advanced synthetic strategies, biophysical methods, and modern analytical tools were successfully interweaved and demonstrated through original research to help understand the mechanism of how complex glycoconjugates (carbohydrates covalently linked to proteins and lipids) establish and stabilize the conformation of proteins, influence protein secretion from cells, control the turnover process, and protect proteins from degradation. These ultimately regulate the diverse roles in the development of the cell, including regulation of cell growth, recognition, and differentiation.

In the last decade, atomic force microscopy (AFM) has quickly became one of the most versatile and effective scanning probe techniques to investigate the topographic images of carbohydrate scaffold model surfaces at submicrometer length scale, as well as quantification of forces acting between a probe and the sample in the pico- to nanoNewton regime. Once the carbohydrate moiety functionalized on to the solid support, a variety of investigations can be carried out by using *in situ* AFM imaging setup to monitor the lectin-binding behavior of two or more disparate types of carbohydrate moieties, protein screening, or multivalent lectin–carbohydrate interactions. The information gained certainly helps in promoting future experimental design for advancing next-generation biomedical diagnostic kits or pushing the frontiers of creating nanomaterials for nanomedicine applications.

The rapid growing interest in glycoscience field in conjunction with many collaborative interdisciplinary researches provides promising prospects for better insights into the difference in multivalent carbohydrate–protein interactions for different ligand–receptor pairs. Therefore, many venues of opportunity await for creative scientists to elucidate and manipulate a diversity of cellular processes.

REFERENCES

1. Yu, J.-J.; Nolting, B.; Tan, Y.; L, X.; Grvay-Hague, J.; Lu, G.-Y. *Nanobiotechnology* **2005**, *1*, 201.
2. McReynolds, K. D.; Gervay-Hague, J. *Chemical Reviews* **2007**, *107*, 1533.
3. Sharon, N. *Biochimica et Biophysica Acta (BBA) - General Subjects* **2006**, *1760*, 527.
4. Sharon, N.; Lis, H. *Science* **1989**, *246*, 227.
5. Singh, R. S.; Tiwary, A. K.; Kennedy, J. F. *Critical Reviews in Biotechnology* **1999**, *19*, 145.
6. Hamid, R.; Masood, A.; Wani, I. H.; Rafiq, S. *Journal of Applied Pharmaceutical Science* **2013**, *3*, S93.
7. Zhao, W.; Liu, S.; Cai, M.; Xu, H.; Jiang, J.; Wang, H. *Chemical Communications* **2013**, *49*, 2980.
8. Liang, P.-H.; Wang, S.-K.; Wong, C.-H. *Journal of the American Chemical Society* **2007**, *129*, 11177.
9. Kogelberg, H.; Solís, D.; Jiménez-Barbero, J. *Current Opinion in Structural Biology* **2003**, *13*, 646.
10. Woods, R. J.; Tessier, M. B. *Current Opinion in Structural Biology* **2010**, *20*, 575.
11. Sharon, N.; Lis, H. In *The Molecular Immunology of Complex Carbohydrates*, 2nd edn.; Wu, A., Ed.; Springer: New York, **2001**; Vol. *491*, p 1.

12. Dam, T. K.; Brewer, C. F. In *Advances in Carbohydrate Chemistry and Biochemistry*; Derek, H., Ed.; Academic Press: Boston, **2010**; Vol. *63*, p 139.
13. Sontheimer, R. D.; Bergstresser, P. R. *Journal of Investigative Dermatology* **1982**, *79*, 237.
14. Watanabe, K.; Hakomori, S. I. *The Journal of Experimental Medicine* **1976**, *144*, 644.
15. Kiessling, L. L.; Strong, L. E.; Gestwicki, J. E. In *Annual Reports in Medicinal Chemistry*; Academic Press: Boston, **2000**; Vol. *35*, p 321.
16. Zhang, Y.; Luo, S.; Tang, Y.; Yu, L.; Hou, K.-Y.; Cheng, J.-P.; Zeng, X.; Wang, P. G. *Analytical Chemistry* **2006**, *78*, 2001.
17. Dam, T. K.; Gerken, T. A.; Brewer, C. F. *Biochemistry* **2009**, *48*, 3822.
18. Dam, T. K.; Brewer, C. F. *Chemical Reviews* **2002**, *102*, 387.
19. Heegaard, N. H. H.; Nilsson, S.; Guzman, N. A. *Journal of Chromatography B: Biomedical Sciences and Applications* **1998**, *715*, 29.
20. Smith, E. A.; Thomas, W. D.; Kiessling, L. L.; Corn, R. M. *Journal of the American Chemical Society* **2003**, *125*, 6140.
21. Dhayal, M.; Ratner, D. M. *Langmuir* **2009**, *25*, 2181.
22. Jiménez-Castells, C.; Defaus, S.; Moise, A.; Przbylski, M.; Andreu, D.; Gutiérrez-Gallego, R. *Analytical Chemistry* **2012**, *84*, 6515.
23. Chiantia, S.; Ries, J.; Chwastek, G.; Carrer, D.; Li, Z.; Bittman, R.; Schwille, P. *Biochimica et Biophysica Acta (BBA): Biomembranes* **2008**, *1778*, 1356.
24. Müller, D. J.; Krieg, M.; Alsteens, D.; Dufrêne, Y. F. *Current Opinion in Biotechnology* **2009**, *20*, 4.
25. Hsieh, S.; Hsieh, S.-L.; Hsieh, C.-W.; Lin, P.-C.; Wu, C.-H. *Journal of Analytical Methods in Chemistry* **2013**, *2013*, 5.
26. Binnig, G.; Quate, C. F.; Gerber, C. *Physical Review Letters* **1986**, *56*, 930.
27. Alsteens, D.; Dufrêne, Y. In *Super-Resolution Microscopy Techniques in the Neurosciences*; Fornasiero, E. F., Rizzoli, S. O., Eds.; Humana Press: New York, **2014**; Vol. *86*, p 225.
28. Bhushan, B. *Scanning Probe Microscopy in Nanoscience and Nanotechnology*, 2nd edn.; Springer: Berlin, **2011**.
29. Giessibl, F. J. *Reviews of Modern Physics* **2003**, *75*, 949.
30. Muller, D.; Dufrene, Y. *Nature Nanotechnology* **2008**, *3*, 261.
31. Sarid, D. *Scanning Force Microscopy*, 2nd edn.; Oxford University Press: New York, **1994**.
32. Takeyasu, K. *Atomic Force Microscopy in Nanobiology*; Pan Stanford: Singapore, **2014**.
33. Patil, N. G.; Levy, J. *Review of Scientific Instruments* **2002**, *73*, 486.
34. Butt, H.-J.; Cappella, B.; Kappl, M. *Surface Science Reports* **2005**, *59*, 1.
35. Meyer, G.; Amer, N. M. *Applied Physics Letters* **1990**, *56*, 2100.
36. Shao, Z.; Mou, J.; Czajkowsky, D. M.; Yang, J.; Yuan, J.-Y. *Advances in Physics* **1996**, *45*, 1.
37. Tan, Y. H.; Davis, J. A.; Fujikawa, K.; Ganesh, N. V.; Demchenko, A. V.; Stine, K. J. *Journal of Materials Chemistry* **2012**, *22*, 6733.
38. Tan, Y. H.; Fujikawa, K.; Pornsuriyasak, P.; Alla, A. J.; Vijaya Ganesh, N.; Demchenko, A. V.; Stine, K. J. *New Journal of Chemistry* **2013**, *37*, 2150.
39. Tan, Y. H.; Liu, M.; Nolting, B.; Go, J. G.; Gervay-Hague, J.; Liu, G.-Y. *ACS Nano* **2008**, *2*, 2374.
40. Grandbois, M.; Dettmann, W.; Benoit, M.; Gaub, H. E. *Journal of Histochemistry & Cytochemistry* **2000**, *48*, 719.

41. Zhong, Q.; Inniss, D.; Kjoller, K.; Elings, V. B. *Surface Science* **1993**, *290*, L688.
42. Hansma, P. K.; Cleveland, J. P.; Radmacher, M.; Walters, D. A.; Hillner, P. E.; Bezanilla, M.; Fritz, M.; Vie, D.; Hansma, H. G.; Prater, C. B.; Massie, J.; Fukunaga, L.; Gurley, J.; Elings, V. *Applied Physics Letters* **1994**, *64*, 1738.
43. Möller, C.; Allen, M.; Elings, V.; Engel, A.; Müller, D. J. *Biophysical Journal* **1999**, *77*, 1150.
44. San Paulo, A.; García, R. *Biophysical Journal* **2000**, *78*, 1599.
45. Albrecht, T. R.; Akamine, S.; Carver, T. E.; Quate, C. F. *Journal of Vacuum Science & Technology A: Vacuum, Surfaces, and Films* **1990**, *8*, 3386.
46. Akamine, S.; Barrett, R. C.; Quate, C. F. *Applied Physics Letters* **1990**, *57*, 316.
47. Wolter, O.; Bayer, T.; Greschner, J. *Journal of Vacuum Science & Technology B* **1991**, *9*, 1353.
48. Algre, E.; Legrand, B.; Faucher, M.; Walter, B.; Buchaillot, L. In Micro Electro Mechanical Systems (MEMS), **2010** IEEE 23rd International Conference on 2010, p 292.
49. Bowers, C. M.; Carlson, D. A.; Rivera, M.; Clark, R. L.; Toone, E. J. *The Journal of Physical Chemistry B* **2013**, *117*, 4755.
50. Ahimou, F.; Denis, F. A.; Touhami, A.; Dufrêne, Y. F. *Langmuir* **2002**, *18*, 9937.
51. Koehne, J. E.; Stevens, R. M.; Zink, T.; Deng, Z.; Chen, H.; Weng, I. C.; Liu, F. T.; Liu, G. Y. *Ultramicroscopy* **2011**, *111*, 1155.
52. Tantakitti, F.; Burk-Rafel, J.; Cheng, F.; Egnatchik, R.; Owen, T.; Hoffman, M.; Weiss, D. N.; Ratner, D. M. *Langmuir* **2012**, *28*, 6950.
53. Touhami, A.; Hoffmann, B.; Vasella, A.; Denis, F. A.; Dufrêne, Y. F. *Langmuir* **2003**, *19*, 1745.
54. Lim, Y.-B.; Lee, M. *Organic & Biomolecular Chemistry* **2007**, *5*, 401.
55. Hirabayashi, J. *Journal of Biochemistry* **2008**, *144*, 139.
56. McCoy Jr, J. P.; Varani, J.; Goldstein, I. J. *Analytical Biochemistry* **1983**, *130*, 437.
57. Duverger, E.; Frison, N.; Roche, A.-C.; Monsigny, M. *Biochimie* **2003**, *85*, 167.
58. Vornholt, W.; Hartmann, M.; Keusgen, M. *Biosensors and Bioelectronics* **2007**, *22*, 2983.
59. Park, S.; Lee, M.-R.; Shin, I. *Chemical Communications* **2008**, *37*, 4389.
60. Fujikawa, K.; Ganesh, N. V.; Tan, Y. H.; Stine, K. J.; Demchenko, A. V. *Chemical Communications* **2011**, *47*, 10602.
61. Ganesh, N. V.; Fujikawa, K.; Tan, Y. H.; Nigudkar, S. S.; Stine, K. J.; Demchenko, A. V. *The Journal of Organic Chemistry* **2013**, *78*, 6849.
62. Houseman, B. T.; Gawalt, E. S.; Mrksich, M. *Langmuir* **2002**, *19*, 1522.
63. Aragão-Leoneti, V.; Campo, V. L.; Gomes, A. S.; Field, R. A.; Carvalho, I. *Tetrahedron* **2010**, *66*, 9475.
64. Xia, Y.; Whitesides, G. M. *Annual Review of Materials Science* **1998**, *28*, 153.
65. Peramo, A.; Albritton, A.; Matthews, G. *Langmuir* **2006**, *22*, 3228.
66. Piner, R. D.; Zhu, J.; Xu, F.; Hong, S.; Mirkin, C. A. *Science* **1999**, *283*, 661.
67. Liu, M.; Amro, N. A.; Liu, G.-y. *Annual Review of Physical Chemistry* **2008**, *59*, 367.
68. Prats-Alfonso, E.; Albericio, F. *Journal of Material Science* **2011**, *46*, 7643.
69. Voskuhl, J.; Brinkmann, J.; Jonkheijm, P. *Current Opinion in Chemical Biology* **2014**, *18*, 1.
70. Liu, G.-y.; Salmeron, M. B. *Langmuir* **1994**, *10*, 367.

71. Hinterdorfer, P.; Dufrene, Y. F. *Nature Methods* **2006**, *3*, 347.
72. Jelinek, R.; Kolusheva, S. *Chemical Reviews* **2004**, *104*, 5987.
73. Becer, C. R. *Macromolecular Rapid Communications* **2012**, *33*, 742.
74. Dupres, V.; Verbelen, C.; Dufrêne, Y. F. *Biomaterials* **2007**, *28*, 2393.
75. de la Fuente, J. M.; Penadés, S. *Biochimica et Biophysica Acta (BBA): General Subjects* **2006**, *1760*, 636.
76. Marradi, M.; Martín-Lomas, M.; PenadÉs, S. In *Advances in Carbohydrate Chemistry and Biochemistry*; Derek, H., Ed.; Academic Press: Boston, **2010**; Vol. *64*, p 211.
77. Sonnino, S.; Mauri, L.; Ciampa, M. G.; Prinetti, A. *Journal of Neurochemistry* **2013**, *124*, 432.
78. Yu, R. K.; Tsai, Y.-T.; Ariga, T.; Yanagisawa, M. *Journal of Oleo Science* **2011**, *60*, 537.
79. Okada, M.; Itoh, M.-i.; Haraguchi, M.; Okajima, T.; Inoue, M.; Oishi, H.; Matsuda, Y.; Iwamoto, T.; Kawano, T.; Fukumoto, S.; Miyazaki, H.; Furukawa, K.; Aizawa, S.; Furukawa, K. *Journal of Biological Chemistry* **2002**, *277*, 1633.
80. Yoshikawa, M.; Go, S.; Takasaki, K.; Kakazu, Y.; Ohashi, M.; Nagafuku, M.; Kabayama, K.; Sekimoto, J.; Suzuki, S.-i.; Takaiwa, K.; Kimitsuki, T.; Matsumoto, N.; Komune, S.; Kamei, D.; Saito, M.; Fujiwara, M.; Iwasaki, K.; Inokuchi, J.-I. *Proceedings of the National Academy of Sciences* **2009**, *106*, 9483.
81. Armendariz, K. P.; Dunn, R. C. *The Journal of Physical Chemistry B* **2013**, *117*, 7959.
82. Lingwood, D.; Simons, K. *Science* **2010**, *327*, 46.
83. Simons, K.; Sampaio, J. L. *Cold Spring Harbor Perspectives in Biology* **2011**, *3*, 1.
84. Picas, L.; Milhiet, P.-E.; Hernández-Borrell, J. *Chemistry and Physics of Lipids* **2012**, *165*, 845.
85. Tan, Y. H.; Stine, K. J. In *Langmuir Monolayers in Thin Film Technology*; Sherwin, J. A., Ed.; Nova Science Publishers: New York, **2011**, p 101.
86. Boxer, S. G. *Current Opinion in Chemical Biology* **2000**, *4*, 704.
87. Bassereau, P.; Pincet, F. *Langmuir* **1997**, *13*, 7003.
88. Dufrêne, Y. F.; Barger, W. R.; Green, J.-B. D.; Lee, G. U. *Langmuir* **1997**, *13*, 4779.
89. McKiernan, A. E.; Ratto, T. V.; Longo, M. L. *Biophysical Journal* **2000**, *79*, 2605.
90. Yuan, C.; Johnston, L. J. *Biophysical Journal* **2001**, *81*, 1059.
91. Milhiet, P. E.; Vié, V.; Giocondi, M.-C.; Le Grimellec, C. *Single Molecules* **2001**, *2*, 109.
92. Khatun, U. L.; Gayen, A.; Mukhopadhyay, C. *Chemistry and Physics of Lipids* **2013**, *170–171*, 8.
93. Dietrich, C.; Volovyk, Z. N.; Levi, M.; Thompson, N. L.; Jacobson, K. *Proceedings of the National Academy of Sciences* **2001**, *98*, 10642.
94. Samsonov, A. V.; Mihalyov, I.; Cohen, F. S. *Biophysical Journal* **2001**, *81*, 1486.
95. Dietrich, C.; Bagatolli, L. A.; Volovyk, Z. N.; Thompson, N. L.; Levi, M.; Jacobson, K.; Gratton, E. *Biophysical Journal* **2001**, *80*, 1417.
96. Vié, V.; Van Mau, N.; Lesniewska, E.; Goudonnet, J. P.; Heitz, F.; Le Grimellec, C. *Langmuir* **1998**, *14*, 4574.
97. Coban, O.; Burger, M.; Laliberte, M.; Ianoul, A.; Johnston, L. J. *Langmuir* **2007**, *23*, 6704.
98. Cappella, B.; Dietler, G. *Surface Science Reports* **1999**, *34*, 1.

99. Yan, C.; Yersin, A.; Afrin, R.; Sekiguchi, H.; Ikai, A. *Biophysical Chemistry* **2009**, *144*, 72.
100. García, R.; San Paulo, A. *Physical Review B* **1999**, *60*, 4961.
101. Devendra, K.; Kanta, L. S. *The Himalayan Physics* **2011**, *2*, 80.
102. Domke, J.; Radmacher, M. *Langmuir* **1998**, *14*, 3320.
103. Allen, S.; Davies, J.; Dawkes, A. C.; Davies, M. C.; Edwards, J. C.; Parker, M. C.; Roberts, C. J.; Sefton, J.; Tendler, S. J. B.; Williams, P. M. *FEBS Letters* **1996**, *390*, 161.
104. Wong, S. S.; Joselevich, E.; Woolley, A. T.; Cheung, C. L.; Lieber, C. M. *Nature* **1998**, *394*, 52.
105. Yuan, C.; Chen, A.; Kolb, P.; Moy, V. T. *Biochemistry* **2000**, *39*, 10219.
106. Tan, Y. H.; Schallom, J. R.; Ganesh, N. V.; Fujikawa, K.; Demchenko, A. V.; Stine, K. J. *Nanoscale* **2011**, *3*, 3395.
107. Pandey, B.; Tan, Y. H.; Fujikawa, K.; Demchenko, A. V.; Stine, K. J. *Journal of Carbohydrate Chemistry* **2012**, *31*, 466.
108. Dammer, U.; Hegner, M.; Anselmetti, D.; Wagner, P.; Dreier, M.; Huber, W.; Güntherodt, H. J. *Biophysical Journal* **1996**, *70*, 2437.
109. Harada, Y.; Kuroda, M.; Ishida, A. *Langmuir* **1999**, *16*, 708.
110. Rief, M.; Gautel, M.; Oesterhelt, F.; Fernandez, J. M.; Gaub, H. E. *Science* **1997**, *276*, 1109.
111. Fisher, T. E.; Marszalek, P. E.; Fernandez, J. M. *Nature Structural & Molecular Biology* **2000**, *7*, 719.
112. Oberhauser, A. F.; Carrión-Vázquez, M. *Journal of Biological Chemistry* **2008**, *283*, 6617.
113. Oh, Y. J.; Sekot, G.; Duman, M.; Chtcheglova, L.; Messner, P.; Peterlik, H.; Schäffer, C.; Hinterdorfer, P. *Journal of Molecular Recognition* **2013**, *26*, 542.
114. Afrin, R.; Nakaji, M.; Sekiguchi, H.; Lee, D.; Kishimoto, K.; Ikai, A. *Cytoskeleton* **2012**, *69*, 101.
115. Zhang, M.; Chen, G.; Kumar, R.; Xu, B. *Biotechnology for Biofuels* **2013**, *6*, 147.
116. Zhang, M.; Wang, B.; Xu, B. *Physical Chemistry Chemical Physics* **2013**, *15*, 6508.
117. Benoit, M.; Gabriel, D.; Gerisch, G.; Gaub, H. E. *Nature Cell Biology* **2000**, *2*, 313.
118. Lü, S.; Ye, Z.; Zhu, C.; Long, M. *Polymer* **2006**, *47*, 2539.
119. Dos Santos Ferreira, O.; Gelinck, E.; de Graaf, D.; Fischer, H. *Applied Surface Science* **2010**, *257*, 48.
120. Somyot, C.; Adam, J. E. *Journal of Physics: Condensed Matter* **2010**, *22*, 194102.
121. Xu, L.-C.; Siedlecki, C. A. *Langmuir* **2009**, *25*, 3675.
122. Getfert, S.; Reimann, P. *Biophysical Journal* **2012**, *102*, 1184.
123. Prezhdo, O. V.; Pereverzev, Y. V. *Accounts of Chemical Research* **2009**, *42*, 693.
124. Santos, N. C.; Castanho, M. A. R. B. *Biophysical Chemistry* **2004**, *107*, 133.
125. Overney, R. M.; Meyer, E.; Frommer, J.; Guentherodt, H. J.; Fujihira, M.; Takano, H.; Gotoh, Y. *Langmuir* **1994**, *10*, 1281.
126. Cleveland, J. P.; Anczykowski, B.; Schmid, A. E.; Elings, V. B. *Applied Physics Letters* **1998**, *72*, 2613.
127. Chen, L.; Cheung, C. L.; Ashby, P. D.; Lieber, C. M. *Nano Letters* **2004**, *4*, 1725.
128. Heuberger, M.; Dietler, G.; Schlapbach, L. *Journal of Vacuum Science & Technology B* **1996**, *14*, 1250.

129. Vinckier, A.; Semenza, G. *FEBS Letters* **1998**, *430*, 12.
130. Butt, H. J.; Siedle, P.; Seifert, K.; Fendler, K.; Seeger, T.; Bamberg, E.; Weisenhorn, A. L.; Goldie, K.; Engel, A. *Journal of Microscopy* **1993**, *169*, 75.
131. Smith, D. P. E. *Review of Scientific Instruments* **1995**, *66*, 3191.
132. Dorozhkin, P.; Kuznetsov, E.; Schokin, A.; Timofeev, S.; Bykov, V. *Microscopy Today* **2010**, *18*, 28.
133. Moretti, M.; Proietti Zaccaria, R.; Descrovi, E.; Das, G.; Leoncini, M.; Liberale, C.; De Angelis, F.; Di Fabrizio, E. *Plasmonics* **2013**, *8*, 25.
134. Kassies, R.; Van Der Werf, K. O.; Lenferink, A.; Hunter, C. N.; Olsen, J. D.; Subramaniam, V.; Otto, C. *Journal of Microscopy* **2005**, *217*, 109.

12

SIALIC ACID-MODIFIED NANOPARTICLES FOR β-AMYLOID STUDIES

HOVIG KOUYOUMDJIAN AND XUEFEI HUANG

Department of Chemistry, Michigan State University, East Lansing, MI, USA

12.1 INTRODUCTION

Alzheimer's disease (AD) is a progressive, irreversible neurodegenerative disorder characterized by the destruction of brain cells and neural connections. In 2014, 5.2 million people in the United States had AD with an estimated cost of $214 billion in medical care [1]. It is forecasted that the number of AD patients and the associated medical care cost will triple by 2050 [2]. This trajectory suggests the urgent need for research into the diagnosis, treatment, and possible prevention of AD. For early detection of AD before any clinical manifestation, biomarker detection-based strategies are gaining momentum.

While the root cause or the exact etiology of AD is still not clear, there are two main histopathological hallmarks of the disease [3]. These include the extracellular amyloid plaques formed through aggregation of β-amyloid (Aβ) peptides and the intracellular neurofibrillary tangles consisting mainly of tau protein. With its extracellular nature, Aβ has become a prominent target for AD diagnosis and treatment studies.

Aβ is formed through the cleavage of the transmembrane amyloid precursor protein expressed in the central nervous system by β- and γ-secretases [4,5]. Due to the relaxed specificity of γ-secretase, Aβ can exist in a variable length ranging from 36 to 43 amino acid residues. The most common forms of Aβ contain 40 and 42 residues, which are commonly referred to as Aβ(1–40) and Aβ(1–42) [6,7]. The sequences of

Carbohydrate Nanotechnology, First Edition. Edited by Keith J. Stine.

	1	10	20	30	40
Aβ(1–40)	DAEFRHDSGYEVHHQKLVFFAEDVGSNKGAIIGLMVGGVV				
	1	10	20	30	40
Aβ(1–42)	DAEFRHDSGYEVHHQKLVFFAEDVGSNKGAIIGLMVGGVVIA				

FIGURE 12.1 Sequences of Aβ(1–40) and Aβ(1–42).

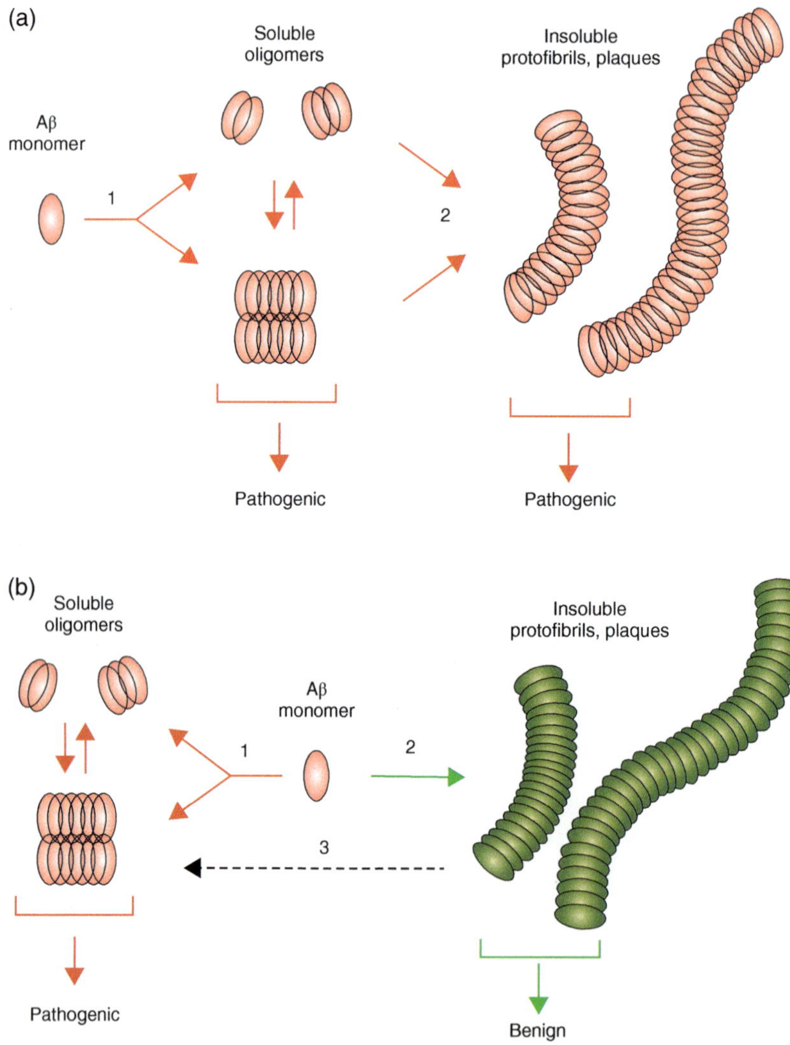

SCHEME 12.1 Potential pathways of Aβ aggregation. *Source*: Adapted from Ref. 17 with permission from Elsevier.

Aβ(1–40) and Aβ(1–42) are shown in Figure 12.1. The two extra hydrophobic residues in the C-terminus of Aβ(1–42) render it capable of aggregating faster than Aβ(1–40) with higher cytotoxicity [7,8].

Aβ can accumulate in the brain due to an imbalance between its production and clearance mechanisms [9]. With the increased concentration, Aβ is prone to aggregate forming higher-order structures, including oligomers, protofibrils, and fibrils. The fibrillogenesis process involves conformation changes from random coil to an α-helix-containing intermediate and finally to extended β-sheet structures [10].

While the deposits of polymeric fibril form of Aβ are one of the most prominent features of AD brains, it is generally accepted that the Aβ oligomers are the most toxic form of Aβ [11–15]. The Aβ oligomers isolated from human AD brains potently inhibited long-term potentiation with IC_{50} values in the picomolar range [16]. Disruption of memory of a learned behavior in normal rats was also shown to be caused by soluble oligomeric Aβ obtained from AD-like rat brains [12].

The potential polymerization pathways of Aβ are summarized in Scheme 12.1 [17]. Scheme 12.1a depicts a linear approach with Aβ monomers initially forming low molecular weight soluble oligomers. Subsequently, aggregation into insoluble protofibrils and fibrils occurs. In Scheme 12.1b, two distinct pathways were envisioned. In the first path, the pathogenic soluble oligomers are formed from the monomers, which are then converted to insoluble fibrils. In the second pathway, the direct formation of benign insoluble aggregates and plaques is proposed (step 2), which can deaggregate and transform to the pathogenic soluble oligomers (step 3) [18].

12.2 CARBOHYDRATES PLAY IMPORTANT ROLES IN Aβ AGGREGATION

On the surface of mammalian cells, there are many carbohydrate structures in the form of glycolipids and glycoproteins [19]. Multiple carbohydrates including sialic acid-containing gangliosides [20–23] and glycosaminoglycans (GAGs) [24,25] have been shown to bind with Aβ. These cell surface glycans have been proposed to serve as nucleating sites for Aβ peptides to further aggregate leading to the formation of plaques on cell surface [26,27]. In the following, we will summarize the interactions of ganglioside GM1 and GAGs with Aβ.

12.2.1 Sialic Acid-Containing Ganglioside GM1 and Aβ

Sialic acids (*N*-acetylneuraminic acid or Neu5Ac) are a family of monosaccharides with a nine-carbon backbone that are typically attached to the nonreducing terminal ends of glycan chains. This location bestows sialic acid the opportunity to influence numerous physiological and pathological processes. For instance, sialic acids serve as components of binding sites for various pathogens and toxins [28–32], including influenza viruses, *Vibrio cholera*, *Plasmodium falciparum*, *Clostridium botulinum*, and *Helicobacter pylori*.

Sialic acid terminates glycolipids such as gangliosides. On neuronal cells, one of the most abundant sialic acid-containing glycolipid is GM1, which constitutes 20% of brain gangliosides (Fig. 12.2) [33]. Aβ can bind to GM1 on cell membrane forming a GM1–Aβ complex [34–36], which has been found in brains of patients with AD as well as Down syndrome [37–39]. Micelles coated with GM1 aggregated with Aβ with a dissociation constant on the order of 10^{-6} M [40]. It is noteworthy that Aβ

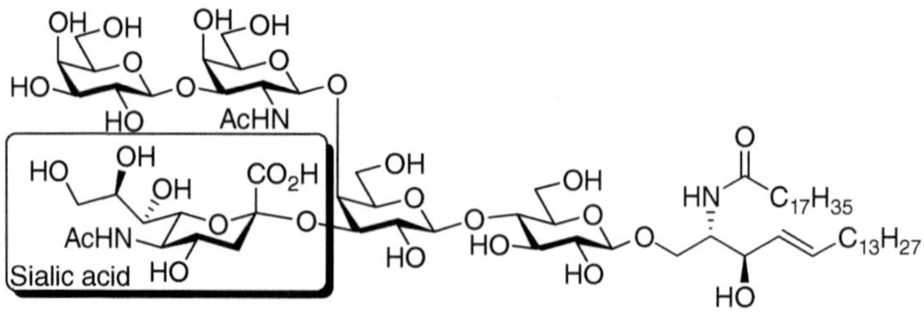

FIGURE 12.2 The representative structure of GM1 ganglioside with the sialic acid moiety shown in the box.

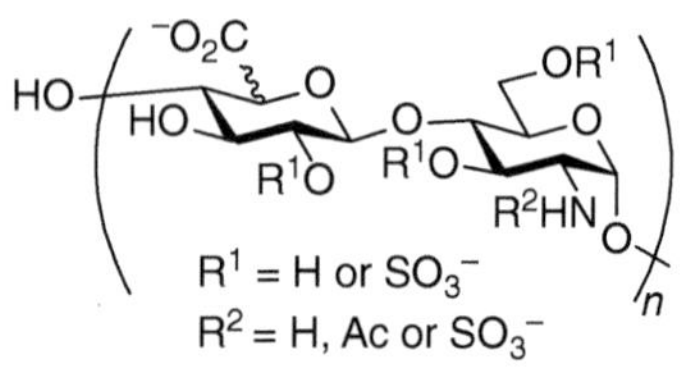

FIGURE 12.3 Structure of heparan sulfate, a member of the GAG family.

aggregation on the neurons was enhanced by the presence of high-density clusters of gangliosides encountered in lipid raft-like microdomains or by increased degree of sialylation [26,41,42]. The amyloid fibrils formed in the presence of GM1 were found to be much more neurotoxic compared to those without GM1 [24,34,36,43]. This is presumably because of the larger hydrophobic surface of Aβ exposed in GM1–Aβ complexes [43].

The acidic character of GM1 can be attributed to the sialic acid moiety, which is believed to have an imperative role in Aβ binding [20,21,23,39]. Removal of cell surface sialic acid through treatment of sialidase almost completely abolished the Aβ-induced cytotoxicity in cell culture [44]. The majority of interactions with sialic acid are from the N-terminal sequence of Aβ, while its C-terminus does not contribute much. The binding between the carbohydrate ring and the two aromatic residues of Aβ (Phe-4, Tyr-10) in addition to the hydrogen bonding of carbohydrate hydroxyls with residues such as His-13 is believed to be most important [45].

12.2.2 Proteoglycans and Aβ

Besides GM1, another class of glycoconjugates that can interact with Aβ is proteoglycans. Proteoglycans are a diverse family of large molecules that are composed of GAGs bonded to the protein backbone [46,47]. They are mostly found in the extracellular matrix (ECM) and on the cell surface. The GAGs are highly negatively charged molecules, an example of which is heparan sulfate-containing uronic acid in the backbone and multiple *O*- and *N*-sulfates (Fig. 12.3).

GAGs such as heparan sulfate were found to be associated with amyloid plaques isolated from AD patients [48–51]. Several studies demonstrated that GAGs can accelerate the aggregation of Aβ to form β-sheeted Aβ fibrils [52–55]. The structures of GAGs are important for Aβ binding with the minimum binding motif being an *N*-sulfated hexasaccharide containing 2-*O*-sulfated iduronic acid [56,57]. Residues 13–16 of the Aβ N-terminal region are particularly important for heparan sulfate/Aβ interactions [58–61].

12.3 NANOTECHNOLOGY IN Aβ RESEARCH

Nanotechnology deals with the fabrication and functionalization of nanoparticles (1–100 nm) for many applications including disease detection and therapeutics. One of the most important characteristics of nanoparticles is their high surface area-to-volume ratio. This enables the installation of high-density ligands on nanoparticle surface through a variety of surface functional groups (e.g., carboxylic and amino groups).

Nanoparticles have been shown to have a profound impact on Aβ aggregation. Through binding with Aβ, nanoparticles could inhibit Aβ fibrillation presumably by quenching nucleation and elongation processes. The nanoparticle platforms studied include inorganic nanoparticles such as gold [62–72], iron oxide [73–77], and quantum dots [63,78,79]; polymeric nanoparticles including dendrimers [80–85], biomolecular [86–91], and copolymeric nanoparticles [92–94]; as well as carbon-based nanomaterials such as graphenes [95,96], fullerenes [97–100], and carbon nanotubes [101–103]. A variety of ligands including anionic phospholipids [88], amino acids [79], peptides [76,77,104,105], and antibodies [72,87,106] have been utilized to functionalize the nanoparticles to facilitate Aβ interactions. Among these, carbohydrate ligands are attractive due to their biomimetic nature.

12.4 SIALIC ACID-FUNCTIONALIZED NANOPARTICLES FOR Aβ STUDIES

Compared to the full structures of GM1 and GAGs, sialic acid is more accessible synthetically. As a result, a variety of sialic acid-conjugated nanomaterials have been prepared. In the following, we will review the results on the interactions of these nanoparticles with Aβ and their effects on Aβ binding and cytotoxicity.

12.4.1 Sialic Acid-Functionalized Dendrimers

As cell surface sialic acid is critical for Aβ binding to cells [44], it is envisioned that exogenous sialic acid-coated nanoparticles could potentially compete with the endogenous sialic acid on cells for Aβ binding, thus reducing Aβ-induced cytotoxicities. Good's group performed pioneering research using sialic acid-functionalized dendrimer platforms to study Aβ [80,81]. Their first-generation approach was to couple sialic acid with amine-bearing PAMAM dendrimers through amide bonds (Scheme 12.2a). Second, third, and fourth generation of PAMAM dendrimers were

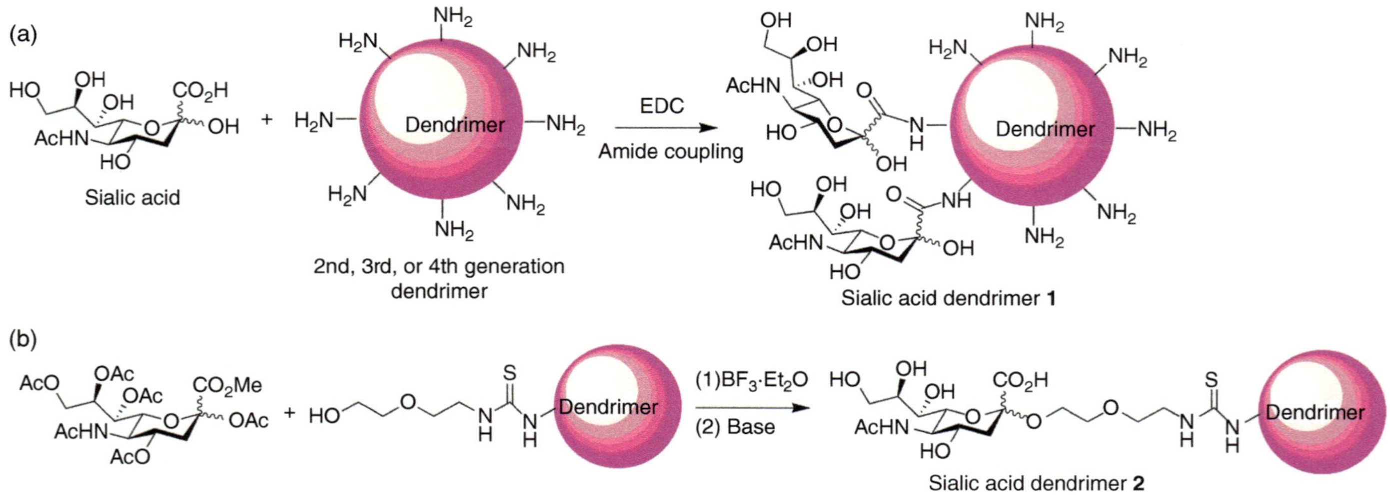

SCHEME 12.2 Sialic acid-conjugated dendrimers used for Aβ binding studies.

TABLE 12.1 Equilibrium Dissociation Constants of Aβ Binding To Different Generations of Dendrimers With Sialic Acid Modification

Generation	$K_{A\beta}$ (μM)
2	0.25 ± 0.02
3	0.02 ± 0.01
4	0.009 ± 0.005

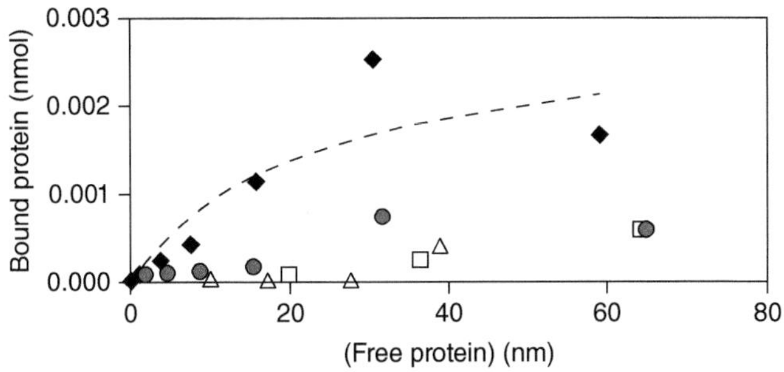

FIGURE 12.4 Equilibrium binding isotherms of Aβ (solid diamonds), insulin (open squares), and lysozyme (open triangles) to sialic acid-bearing third-generation dendrimer. The solid circles represent the binding isotherm of unmodified dendrimer to Aβ. *Source*: Adapted from Ref. 80 with permission from Elsevier.

utilized with 16, 41, and 22% of the amines on the dendrimers functionalized with sialic acid, respectively. Although the amide bond between the sialic acid and the dendrimers was not a native linkage (naturally existing sialosides are linked through its hydroxyl group at the anomeric center with α stereochemistry), the sialic acid dendrimers showed strong binding to Aβ with dissociation constants around 20 nM for the third- and fourth-generation dendrimers (Table 12.1). The affinities of the sialic acid-functionalized dendrimers to Aβ were orders of magnitude higher than those to either soluble sialic acid or GM1 ganglioside [82, 107], which was presumably due to the multivalent effects from clustering sialic acid on the dendrimers and/or the combined effects of interactions with dendrimer backbones and sialic acid.

The selectivity of the dendrimers was examined next using insulin and lysozyme. Although these two proteins are known to form amyloid-like fibrils under the experimental conditions, they exhibited much less binding to sialic acid dendrimers compared to Aβ (Fig. 12.4), demonstrating high selectivity of these dendrimers toward Aβ binding.

The effects of the sialic acid dendrimers on cell viability were determined via a cell viability assay using SH-SY5Y neuroblastoma cells, a model of neuronal cells [80]. Treatment of the cells with Aβ (50 μM) for 24 h (dashed line in Fig. 12.5) led to significant cell death with about 55% of the cells remaining viable. When the second generation of sialic acid dendrimer was added in the presence of Aβ, it reduced Aβ toxicity to the cells (filled diamonds in Fig. 12.5a) in a dose-dependent manner,

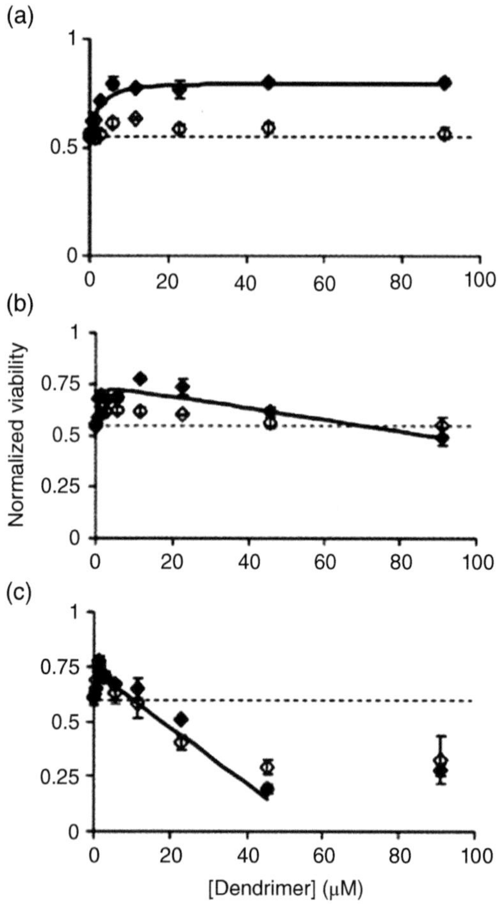

FIGURE 12.5 Normalized cell viability upon treatment of SH-SY5Y cells with sialic acid-conjugated dendrimers (filled diamonds) and unmodified dendrimers (open diamonds) when treated with Aβ. (a) Second-generation, (b) third-generation, and (c) fourth-generation dendrimers. *Source*: Adapted from Ref. 80 with permission from Elsevier.

while the unfunctionalized dendrimer had little effects (open diamonds in Fig. 12.5a). With the higher generation of dendrimers, protective effects were also observed at low dendrimer concentration (Fig. 12.5b and c). However, further increase in dendrimer concentrations led to very high cytotoxicities, which were presumably due to the intrinsic toxicity of these dendrimers. These results suggest that sialic acid functionalization can enhance recognition of Aβ by the dendrimers, although there were serious toxicity issues related to the platform.

To enhance the binding of the dendrimers with Aβ, the effect of the linkage site between sialic acid and the dendrimers was examined. Sialic acid dendrimer **2** was synthesized by linking sialic acid through the anomeric center, leaving the carboxylic acid moiety intact as in naturally existing sialosides (Scheme 12.2b)

[81]. For the fourth-generation dendrimer, higher loading of sialic acid was achieved compared to the corresponding dendrimer **1** (61% vs. 22%). However, despite the higher sialic acid loading and the native glycosidic linkage, the binding affinity of dendrimer **2** to Aβ did not improve. The exact reason for the lack of binding enhancement was not clear. The cytotoxicity of dendrimer **2** toward SH-SY5Y cells was lowered presumably due to the reduction of the amount of free amine on the dendrimer.

12.4.2 Sialic Acid-Functionalized Gold Nanoshells

Gold nanoparticles have been extensively used in nanoscience. Taking advantage of their large surface areas and conductance properties, Chikae et al. investigated the utility of sialic acid-functionalized gold nanoparticles in electrochemical sensing of Aβ [108]. Gold nanoparticles were formed on a carbon electrode through electrodeposition and subsequently coated with alkyne-terminated disulfide. Azide-functionalized sialic acid was immobilized onto the gold nanoparticle surface through a triazole linkage formed by the 3+2 copper-catalyzed azide–alkyne cycloaddition reaction (Fig. 12.6).

The binding of Aβ onto the sialic acid-functionalized gold nanoshell was visualized using atomic force microscopy (AFM) (Fig. 12.7). Upon introduction of sialic acid, a slight roughness of the surface was observed by AFM. Incubation of the resulting surface with either Aβ(1–40) or Aβ(1–42) led to a significant increase in surface roughness, indicating binding of Aβ.

Besides AFM, the binding of Aβ (5 μM) on the sialic acid-modified gold nanoshell electrode was detected electrochemically. Since Aβ contains a tyrosine residue, strong oxidation signals were observed in differential pulse voltammetry. When another protein such as insulin prone to fibril formation was tested using this system, little oxidation signals were observed despite the fact that insulin contains four tyrosine residues (Fig. 12.8). This suggests the selectivity of the electrode toward Aβ.

Raman spectroscopy is an optical technique for biological detection. The detection sensitivity can be drastically enhanced (on the order of 10^6–10^{14} times) by absorbing the reporter molecules on the surfaces of metal nanoparticles or nanostructures, which is referred to as surface-enhanced Raman spectroscopy (SERS). In order to apply SERS for Aβ detection, sialic acid was functionalized with cysteamine, which was then absorbed onto gold nanoshell forming a self-assembled monolayer (Fig. 12.9) [109]. The sialic acid-immobilized surface was capable of binding Aβ, which was ascertained by performing equilibrium binding isotherms using I^{125}-labeled Aβ. The equilibrium dissociation constant of the Aβ with the surfaces was found to be 14 ± 8 nM (Fig. 12.10a), which was comparable to the affinity values obtained in sialic acid-conjugated dendrimer study [80].

SERS detection of Aβ was aided by Congo red, a dye capable of recognizing aggregated Aβ in β-sheet form. Upon addition of Congo red to sialic acid gold nanoshell incubated with Aβ, SERS signals from Congo red were observed. The detection limit of Aβ was at a remarkable level of 1 pM (Fig. 12.10b).

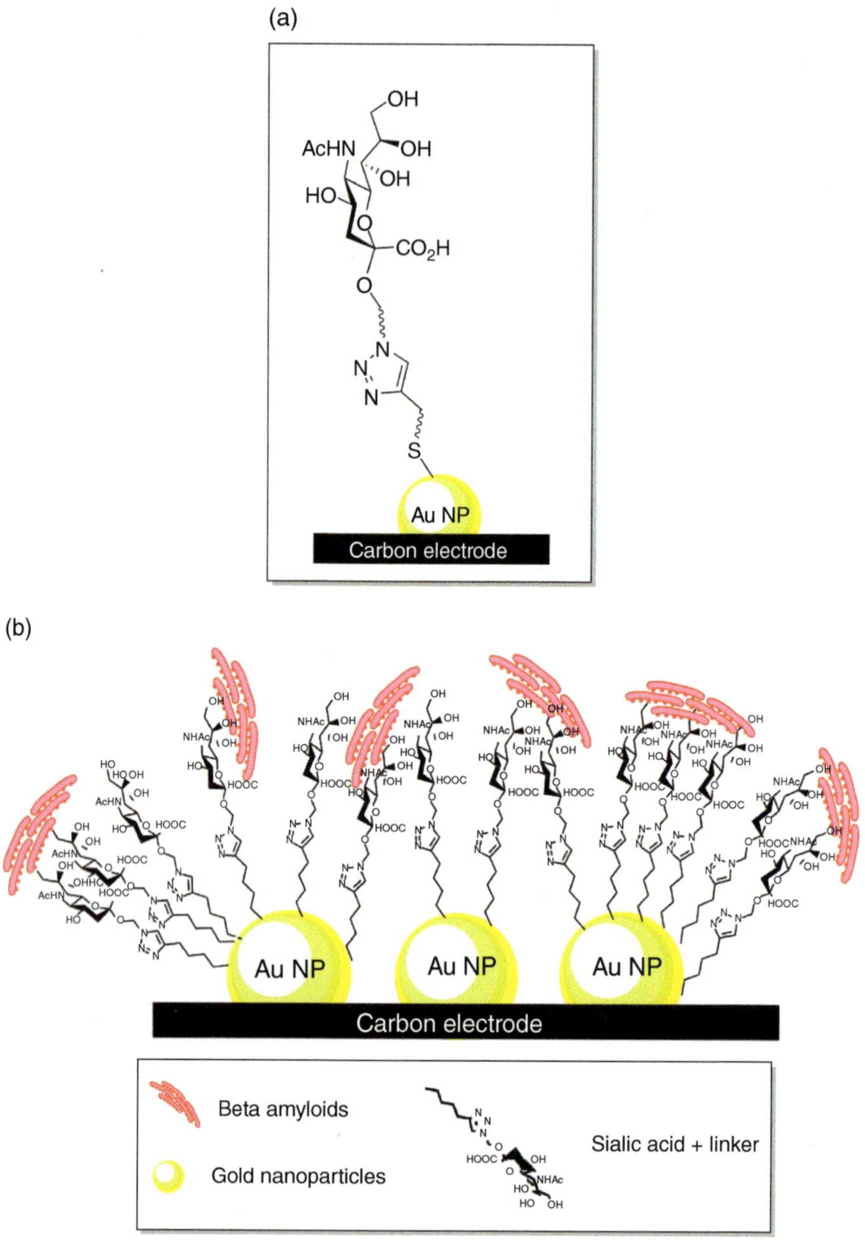

FIGURE 12.6 (a) Sialic acid-conjugated gold nanoshell deposited on a carbon electrode for Aβ binding. (b) Schematic demonstration of Aβ binding to sialic acid-conjugated gold nanoshell.

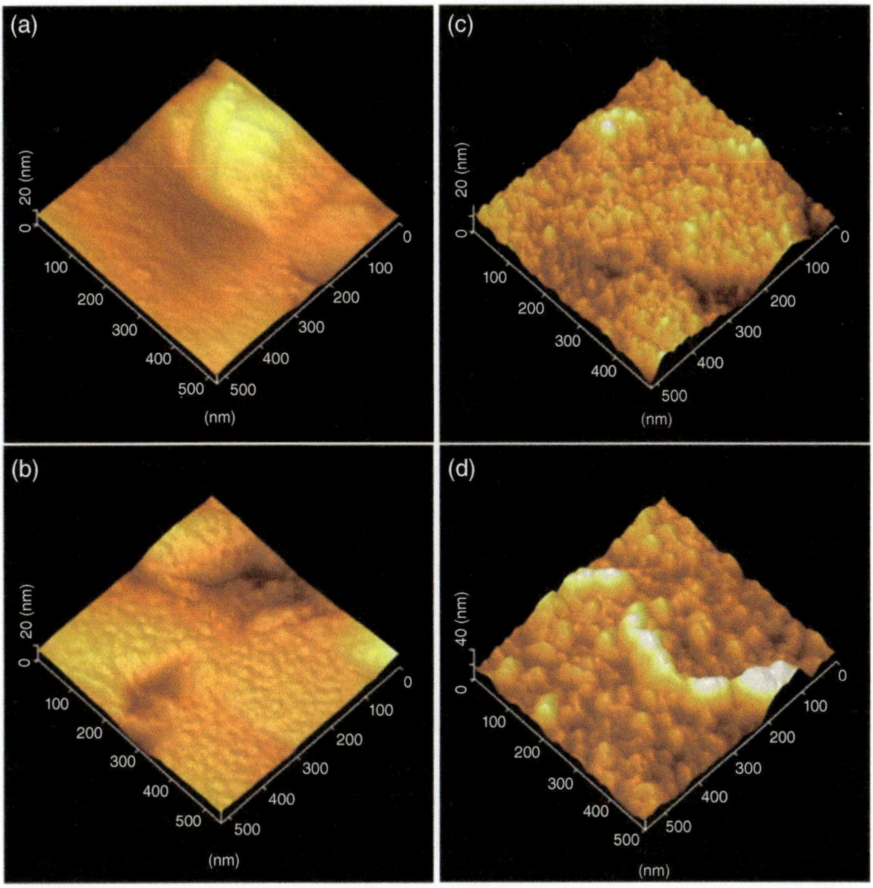

FIGURE 12.7 AFM images of (a) the bare gold substrate, (b) after sialic acid conjugation, (c) after incubation of Aβ(1–40), and (d) after incubation of Aβ(1–42). *Source*: Adapted from Ref. 108 with permission from Elsevier.

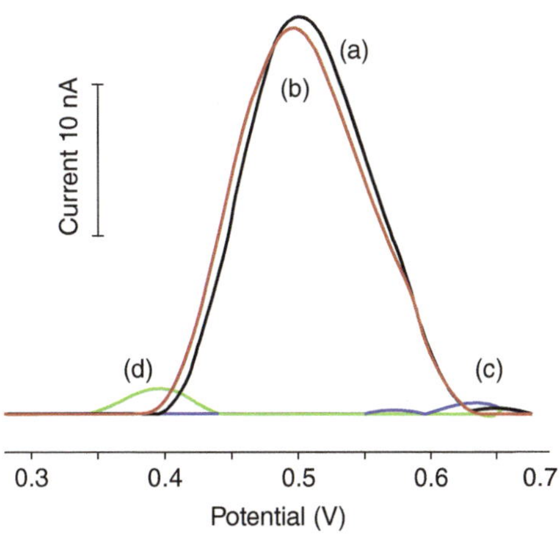

FIGURE 12.8 Electrochemical detection of Aβ using sialic acid-functionalized gold nanoshell-bearing carbon electrode. (a) Aβ(1–40), (b) Aβ(1–42), (c) insulin (5 μM), and (d) PBS only. *Source*: Adapted from Ref. 108 with permission from Elsevier.

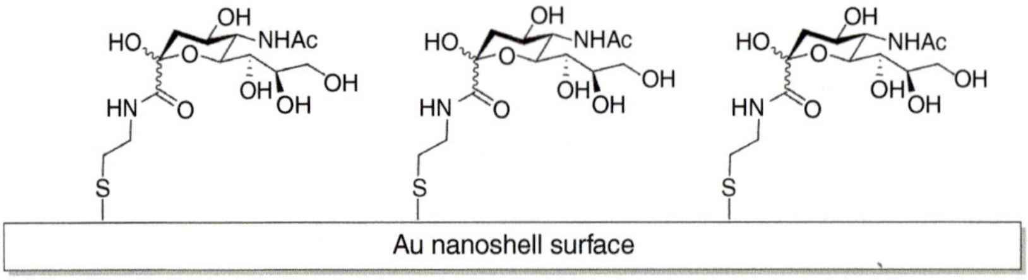

FIGURE 12.9 Self-assembled monolayer of sialic acid on gold nanoshells.

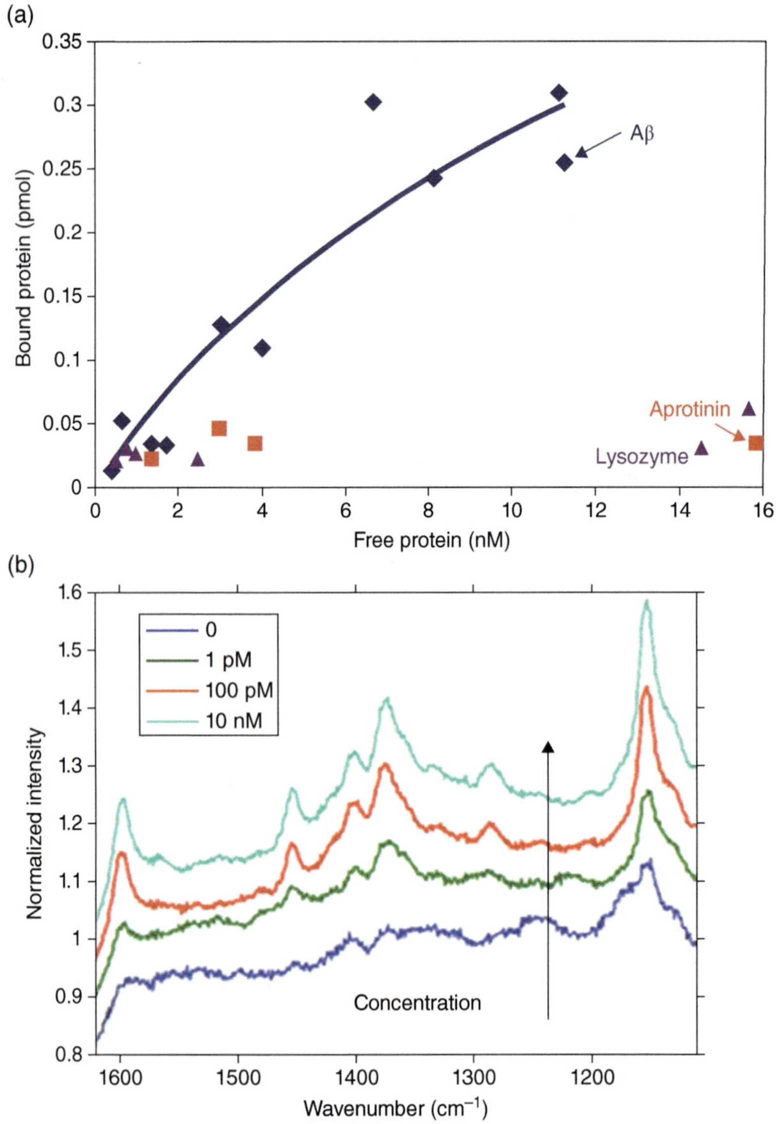

FIGURE 12.10 (a) Equilibrium binding isotherm of I^{125}-labeled Aβ (diamond) bound to the sialic acid gold nanoshell. Aprotinin (square) and lysozyme (triangle) did not exhibit significant binding to the nanoshell under the experimental condition. (b) SERS spectra showing sensitive detection of Aβ using the sialic acid-functionalized gold nanoshell. *Source*: Adapted from Ref. 109 with permission from Springer.

12.4.3 Sialic Acid–Chitosan Conjugate

The cationic polysaccharide chitosan is another platform that was used as carriers of sialic acid for Aβ study. Dhavale et al. conjugated sialic acid with chitosan polysaccharides and tested their abilities to attenuate Aβ toxicity [86]. In order to decipher the importance of various structural components of sialic acid, four analogs representing partial structures of sialic acid, that is, cyclohexanecarboxylic acid (CHC), tetrahydropyran-2-carboxylic acid (pyran), ketodeoxynonulosonic acid (KDN), and D-(+)-galacturonic acid (GA), were acquired and conjugated onto chitosan through amide bonds in addition to sialic acid (Fig. 12.11) [86].

The Aβ toxicity attenuation assays were performed to assess the protective abilities of the four analogs either alone or as chitosan conjugates. The compounds were added to a mixture of Aβ incubated with neuroblastoma SH-SY5Y cells followed by cell viability measurements. All analogs alone had no or very weak activities in reducing the toxicity of Aβ. In contrast, at lower concentrations, chitosan-conjugated analogs showed pronounced attenuation of Aβ toxicity on the SH-SY5Y cells, which was attributed to the clustering effect bestowed by the polysaccharide backbone. KDN chitosan conjugate was the most potent in the group. As KDN is the only analog tested containing the side chain with multiple hydroxyl groups, it suggests the side chain may be an important element for Aβ binding (Fig. 12.12). The pyran chitosan conjugate exhibited higher protective effects than the corresponding CHC adduct indicating the oxygen moiety in the pyranosyl ring may be important for higher protective effects. It should be pointed out that the observed Aβ toxicity

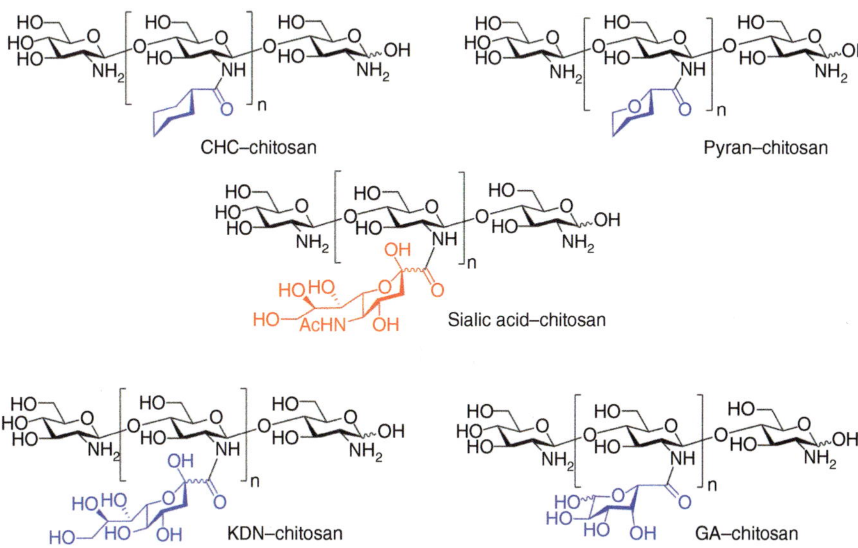

FIGURE 12.11 Structures of chitosan conjugates with sialic acid, and four other analogs: cyclohexanecarboxylic acid (CHC), tetrahydropyran-2-carboxylic acid (pyran), ketodeoxynonulosonic acid (KDN) and D-(+)-galacturonic acid (GA).

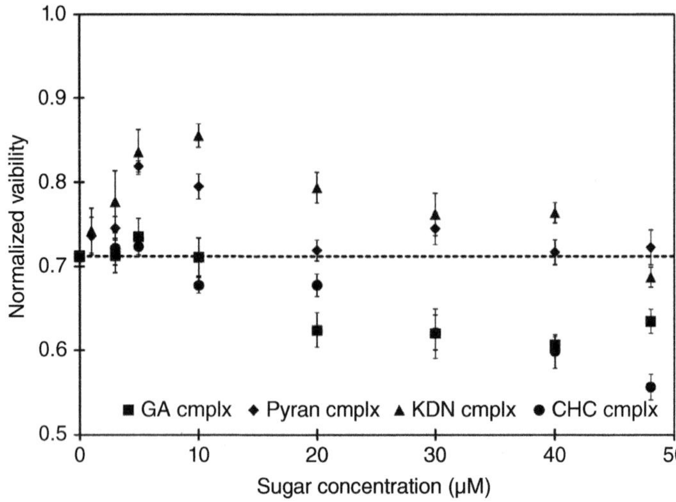

FIGURE 12.12 Attenuation of Aβ toxicity by sialic acid analog–chitosan conjugates. Conjugates were incubated with 50 μM Aβ followed by cell viability measurements. KDN complex, triangle; pyran complex, diamond; CHC complex, circle; and GA complex, square. The dashed line represents the cell viability with 50 μM Aβ in the absence of any sialic acid analogs. *Source*: Adapted from Ref. 86 with permission from Elsevier.

attenuation effects of these sialic acid analogs were modest. Furthermore, with higher concentrations of the conjugates, cell viabilities decreased presumably due to the intrinsic cytotoxicities of these constructs, thus posting a serious challenge for *in vivo* applications.

12.4.4 Iron Oxide Nanoparticles Functionalized with Sialic Acid

With the recognition of the importance of Aβ in AD, imaging methods allowing the noninvasive evaluation of Aβ load are being actively pursued. Among a variety of technologies used including positron emission tomography [110–113] and near-infrared fluorescence imaging [114], magnetic resonance imaging (MRI) [104, 105, 115–117] is a highly attractive approach due to its high spatial and temporal resolution. Furthermore, it does not require ionizing radiation rendering it suitable for longitudinal analysis.

Contrast agents are compounds that can improve target visibility in MRI. Due to their high magnetic relaxivity, the superparamagnetic magnetite nanoparticles (SPIONs) are one of the promising contrast agents used in MRI [118,119]. To explore their applications in Aβ imaging, Huang and coworkers designed sialic acid-functionalized SPIONs [73]. SPIONs were synthesized using the coprecipitation method, which were subsequently modified with amines on the external surface (SPION-NH_2). A sialic acid derivative bearing a carboxylic acid at the reducing end was coupled to SPION-NH_2 yielding the sialic acid-functionalized magnetic

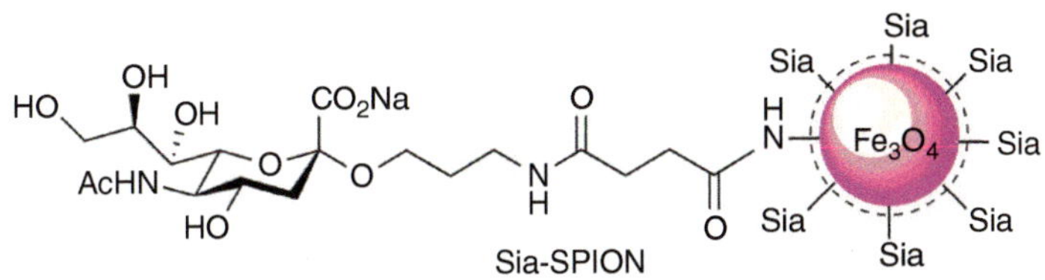

FIGURE 12.13 Sialic acid-functionalized iron oxide magnetic nanoparticles (Sia-SPION).

nanoparticles (Sia-SPION) (Fig. 12.13). The Sia-SPION could bind with Aβ fibril with an apparent K_d of 0.7 μM as determined via an enzyme-linked immunosorbent assay (ELISA).

Another interesting property of the magnetic nanoparticles is that their transverse relaxivities can be significantly enhanced when they form larger clusters due to target binding [120–124]. This magnetic relaxation switching phenomenon has been utilized for biological detections [120–124]. Clusters of NPs create larger local magnetic field inhomogeneities and thus are more efficient in dephasing the spins of surrounding water protons and lowering T2* relaxation time.

The detection of Aβ fibril binding with Sia-SPION was explored by MRI [73]. After incubation of Aβ with Sia-SPION, T2*-weighted magnetic resonance (MR) images were acquired, showing that Aβ led to many more darkened spots (Fig. 12.14a vs. Fig. 12.14c). Quantification of the images demonstrated that the T2* relaxation time of Sia-SPION was significantly reduced in the presence of Aβ (from 23 to 16 ms) (Fig. 12.14e). The reduction of T2* relaxation time was due to the enhancement of the transverse relaxivities of SPIONs when they formed larger clusters upon target binding [120–124]. Addition of 10,000-fold higher concentration of free sialic acid competed with Sia-SPION for Aβ binding and reduced the number of dark spots in T2*-weighted images (Fig. 12.14b and e). This suggested the dependence of sialic acid on Aβ binding and the high avidity of Sia-SPION with Aβ.

To lay the groundwork for future *in vivo* detection, *ex vivo* binding of Sia-SPION to Aβ was explored. Mouse brains were coated with Aβ on the external surface. Upon incubation with Sia-SPION, the presence of the nanoparticle was detected by MRI as evident from the darkened perimeter of the brain in T2*-weighted MR image (Fig. 12.15a). Addition of free sialic acid together with Sia-SPION significantly reduced darkening of the surface of the brain in the T2*-weighted MR image (Fig. 12.15b). Mouse brains without Aβ did not exhibit much darkening (Fig. 12.15c). The selectivity in binding of Sia-SPION with Aβ was also confirmed by Prussian blue staining, a staining method specific for ferric ion. Only the Aβ brain incubated with Sia-SPION exhibited the characteristic blue color upon Prussian blue staining (Fig. 12.15e), whereas the other brains (Aβ brain + Sia-SPION + free sialic acid, normal brain + Sia-SPION, Aβ brain) showed little blue coloration (Fig. 12.15f–h), thus supporting the MRI results.

The Sia-SPION was highly biocompatible exhibiting no toxicity to SH-SY5Y cells at 0.3 mg/mL. Upon incubation of Sia-SPION with Aβ and SH-SY5Y cells, Sia-SPION was found to mitigate the toxicity induced by Aβ in a dose-dependent

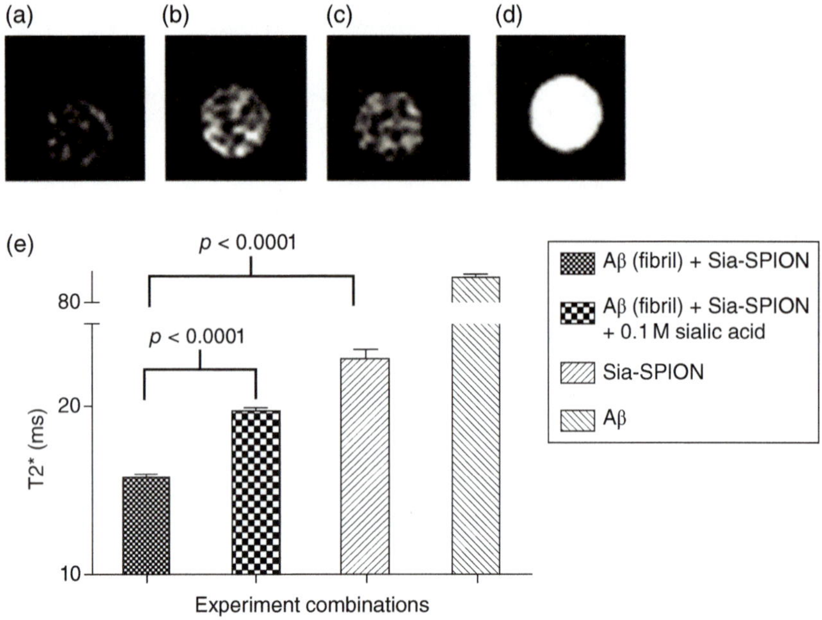

FIGURE 12.14 T2*-weighted MR images of (a)–(d) incubation of Sia-SPION (0.1 mg/ml) with Aβ fibrils and their controls. (e) Quantification of T2* (ms) of images in (a)–(d). *Source*: Adapted from Ref. 73 with permission from the American Chemical Society.

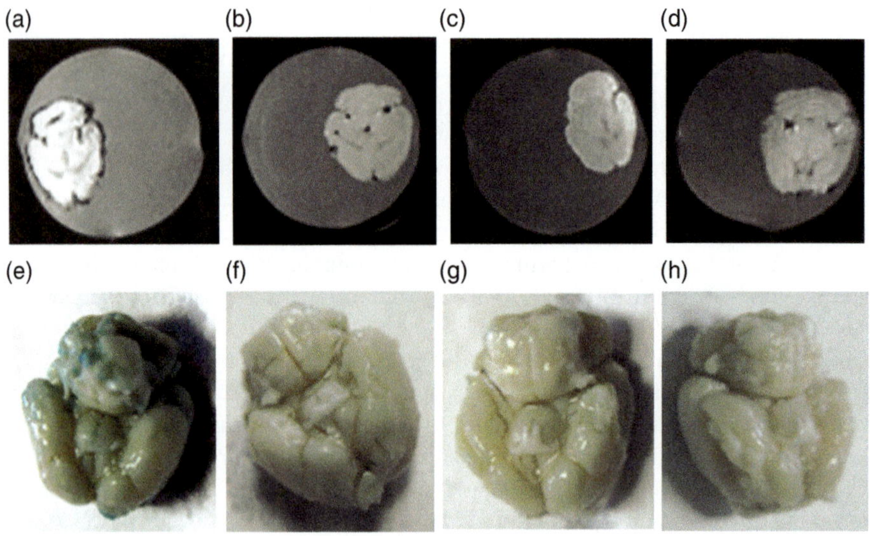

FIGURE 12.15 T2*-weighted MR images of (a) Aβ mouse brain incubated with Sia-SPION, (b) Aβ brain incubated with Sia-SPION in the presence of free sialic acid, (c) normal brain incubated with Sia-SPION, and (d) Aβ brain; Prussian blue staining of the previous mouse brains (e–h). *Source*: Adapted from Ref. 73 with permission from the American Chemical Society.

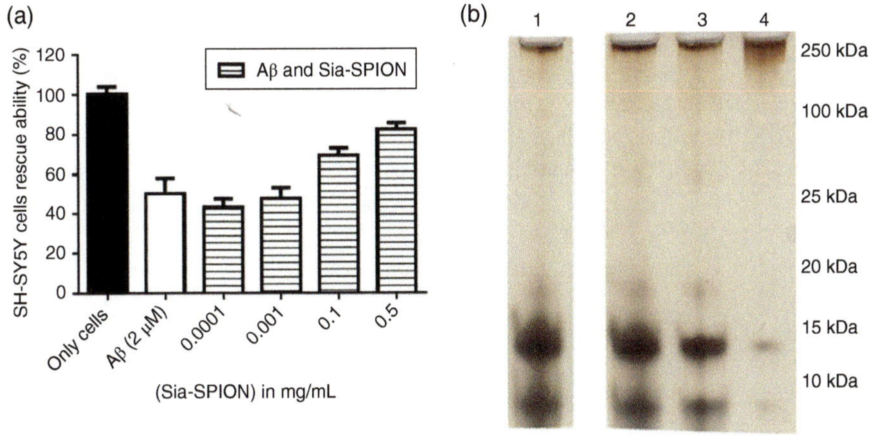

FIGURE 12.16 (a) Addition of Sia-SPION to SH-SY5Y neuroblastoma cells incubated with Aβ (2 μM) reduced Aβ cytotoxicity and (b) PAGE gel of Aβ. Aβ only (lane 1), Aβ incubated with 0.02 mg/ml (lane 2), 0.2 mg/ml (lane 3), and 2 mg/ml (lane 4) of Sia-SPION. *Source*: Adapted from Ref. 73 with permission from the American Chemical Society.

manner (Fig. 12.16a). This was presumably due to the binding of Sia-SPION with Aβ, thus reducing the amount of free Aβ available to interact with the cells. Furthermore, Sia-SPION could affect the Aβ equilibrium (Fig. 12.16b). When Aβ was induced to aggregate in the presence of Sia-SPION, the formation of Aβ oligomers was completely suppressed favoring the higher molecular weight fibril. The ability to shift Aβ equilibrium away from the more toxic oligomeric form [11–15] can be another potential reason for the cytoprotective effect of Sia-SPION.

12.5 CONCLUSION AND FUTURE OUTLOOK

In this chapter, we summarized the results on the biomimetic approach using various sialic acid conjugates to study Aβ binding. The nanomaterials provide a multivalent platform to display sialic acid leading to significantly enhanced avidity with Aβ. Furthermore, several constructs have shown abilities to protect cells from Aβ toxicity. The sialic acid-coated iron oxide nanoparticles enabled imaging of Aβ via MRI. Coupled with its high biocompatibility and ability to reduce Aβ toxicity, the iron oxide nanoparticle platform holds great promises for *in vivo* applications.

The results obtained to date have demonstrated the potential of using sialic acid to study Aβ. As sialic acid is only part of GM1 ganglioside structure, to further improve the binding affinity and selectivity, future studies should explore the usage of full GM1 structure as affinity ligand. In addition, other carbohydrates such as heparan sulfate can be examined. Further in-depth studies on characterization of glycan nanoparticle–Aβ complexes, detailed understanding of how the glycans interact with various forms of Aβ, and how carbohydrate binding impacts toxicity of Aβ are needed to realize the full potential of glyconanotechnology in Aβ research.

REFERENCES

1. Alzheimer's Association *Alzheimer's disease facts and figures*; Alzheimer Foundation: Washington, DC, **2014**.
2. Hebert, L. E.; Weuve, J.; Scherr, P. A.; Evans, D. A., Alzheimer disease in the United States (2010-2050) estimated using the 2010 census. *Neurology* **2013**, *80*, 1778–1783.
3. Serrano-Pozo, A.; Frosch, M. P.; Masliah, E.; Hyman, B. T., Neuropathological alterations in Alzheimer disease. *Cold Spring Harb. Perspect. Med.* **2011**, *24*, a006189.
4. Mattson, M. P., Pathways towards and away from Alzheimer's disease. *Nature* **2004**, *431*, 107.
5. Selkoe, D. J., Cell biology of protein misfolding: The examples of Alzheimer's and Parkinson's diseases. *Nat. Cell Biol.* **2004**, *6*, 1054–1061.
6. Hartmann, T.; Bieger, S. C.; Bruhl, B.; Tienari, P. J.; Ida, N.; Allsop, D.; Roberts, G. W.; Masters, C. L.; Dotti, C. G.; Unsicker, K.; Beyreuther, K., Distinct sites of intracellular production for Alzheimer's disease Aβ 40/42 amyloid peptides. *Nat. Med.* **1997**, *3*, 1016–1020.
7. Bitan, G.; Kirkitadze, M. D.; Lomakin, A.; Vollers, S. S.; Benedek, G. B.; Teplow, D. B., Amyloid β-protein (Aβ) assembly: Aβ40 and Aβ42 oligomerize through distinct pathways. *Proc. Natl. Acad. Sci. U.S.A.* **2003**, *100*, 330–335.
8. Klein, A. M.; Kowall, N. W.; Ferrante, R. J., Neurotoxicity and oxidative damage of beta amyloid 1-42 versus beta amyloid 1-40 in the mouse cerebral cortex. *Ann. NY Acad. Sci.* **1999**, *893*, 314–320.
9. Masters, C. L.; Cappai, R.; Barnham, K. J.; Villemagne, V. L., Molecular mechanisms for Alzheimer's disease: implications for neuroimaging and therapeutics. *J. Neurochem.* **2006**, *97*, 1700–1725.
10. Kirkitadze, M. D.; Condron, M. M.; Teplow, D. B., Identification and characterization of key kinetic intermediates in amyloid β-protein fibrillogenesis. *J. Mol. Biol.* **2001**, *312*, 1103–1119.
11. Benilova, I.; Karran, E.; De Strooper, B., The toxic Aβ oligomer and Alzheimer's disease: an emperor in need of clothes. *Nat. Neurosci.* **2012**, *15*, 349–357.
12. Shankar, G. M.; Li, S.; Mehta, T. H.; Garcia-Munoz, A.; Shepardson, N. E.; Smith, I.; Brett, F. M.; Farrell, M. A.; Rowan, M. J.; Lemere, C. A.; Regan, C. M.; Walsh, D. M.; Sabatini, B. L.; Selkoe, D. J., Amyloid-beta protein dimers isolated directly from Alzheimer's brains impair synaptic plasticity and memory. *Nat. Med.* **2008**, *14*, 837–842.
13. Lesne, S.; Koh, M. T.; Kotilinek, L.; Kayed, R.; Glabe, C. G.; Yang, A.; Gallagher, M.; Ashe, K. H., A specific amyloid-beta protein assembly in the brain impairs memory. *Nature* **2006**, *440*, 352–357.
14. Bitan, G.; Teplow, D. B., Preparation of aggregate-free, low molecular weight amyloid-beta for assembly and toxicity assays. *Methods Mol. Biol.* **2005**, *299*, 3–9.
15. Klein, W. L.; Krafft, G. A.; Finch, C. E., Targeting small Abeta oligomers: the solution to an Alzheimer's disease conundrum? *Trends Neurosci.* **2001**, *24*, 219–224.
16. Puzzo, D.; Privitera, L.; Leznik, E.; Fa, M.; Staniszewski, A.; Palmeri, A.; Arancio, O., Picomolar amyloid-beta positively modulates synaptic plasticity and memory in hippocampus. *J. Neurosci.* **2008**, *28*, 14537–14545.
17. Pimplikar, S. W., Reassessing the amyloid cascade hypothesis of Alzheimer's disease. *Int. J. Biochem. Cell Biol.* **2009**, *41*, 1261–1268.

18. Meyer-Luehmann, M.; Spires-Jones, T. L.; Prada, C.; Garcia-Alloza, M.; de Calignon, A.; Rozkalne, A.; Koenigsknecht-Talboo, J.; Holtzman, D. M.; Bacskai, B. J.; Hyman, B. T., Rapid appearance and local toxicity of amyloid-beta plaques in a mouse model of Alzheimer's disease. *Nature* **2008**, *451*, 720–724.

19. Brandley, B. K.; Schnaar, R. L., Cell-surface carbohydrates in cell recognition and response. *J. Leukocyte Biol.* **1986**, *40*, 97–111.

20. Wakabayashi, M.; Okada, T.; Kozutsumi, Y.; Matsuzaki, K., GM1 ganglioside-mediated accumulation of amyloid beta-protein on cell membranes. *Biochem. Biophys. Res. Commun.* **2005**, *328*, 1019–1023.

21. Ariga, T.; Kobayashi, K.; Hasegawa, A.; Kiso, M.; Ishida, H.; Miyatake, T., Characterization of high-affinity binding between gangliosides and amyloid beta-protein. *Arch. Biochem. Biophys.* **2001**, *388*, 225–230.

22. Yanagisawa, K., GM1 ganglioside and the seeding of amyloid in Alzheimer's disease: Endogenous seed for Alzheimer amyloid. *Neuroscientist* **2005**, *11*, 250–260.

23. Matsuzaki, K.; Horikiri, C., Interactions of amyloid beta-peptide (1-40) with ganglioside-containing membranes. *Biochemistry* **1999**, *38*, 4137–4142.

24. Ariga, T.; Miyatake, T.; Yu, R. K., Role of proteoglycans and glycosaminoglycans in the pathogenesis of Alzheimer's disease and related disorders: amyloidogenesis and therapeutic strategies: A review. *J. Neurosci. Res.* **2010**, *88*, 2303–2315.

25. Fraser, P. E.; Darabie, A. A.; McLaurin, J., Amyloid-beta interactions with chondroitin sulfate-derived monosaccharides and disaccharides—Implications for drug development. *J. Biol. Chem.* **2001**, *276*, 6412–6419.

26. Kakio, A.; Nishimoto, S. I.; Yanagisawa, K.; Kozutsumi, Y.; Matsuzaki, K., Cholesterol-dependent formation of GM1 ganglioside-bound amyloid beta-protein, an endogenous seed for Alzheimer amyloid. *J. Biol. Chem.* **2001**, *276*, 24985–24990.

27. Choo-Smith, L.-P. I.; Garzon-Rodriguez, W.; Glabe, C. G.; Surewicz, W. K., Acceleration of amyloid fibril formation by specific binding of A-beta-(1-40) peptide to ganglioside-containing membrane vesicles. *J. Biol. Chem.* **1997**, *272*, 22987–22990.

28. Varki, A., Sialic acids in human health and disease. *Trends Mol. Med.* **2008**, *14*, 351–360.

29. Schauer, R., Achievements and challenges of sialic acid research. *Glycoconjugate J.* **2000**, *17*, 485–499.

30. Angata, T.; Varki, A., Chemical diversity in the sialic acids and related alpha-keto acids: An evolutionary perspective. *Chem. Rev.* **2002**, *102*, 439–469.

31. Ilver, D.; Johansson, P.; Miller-Podraza, H.; Nyholm, P. G.; Teneberg, S.; Karlsson, K. A., Bacterium-host protein-carbohydrate interactions. *Methods Enzymol.*, **2003**, *363*, 134–157.

32. Lehmann, F.; Tiralongo, E.; Tiralongo, J., Sialic acid-specific lectins: Occurrence, specificity and function. *Cell. Mol. Life Sci.* **2006**, *63*, 1331–1354.

33. Kracun, I.; Rosner, H.; Cosovic, C.; Stavljenic, A., Topographical atlas of the gangliosides of the adult human brain. *J. Neurochem.* **1984**, *43*, 979-989.

34. Yanagisawa, K., Role of gangliosides in Alzheimer's disease. *Biochim. Biophys. Acta-Biomembr.* **2007**, *1768*, 1943–1951.

35. Ariga, T.; McDonald, M. P.; Yu, R. K., Role of ganglioside metabolism in the pathogenesis of Alzheimer's disease: A review. *J. Lipid Res.* **2008**, *49*, 1157–1175.

36. Yanagisawa, K., Pathological significance of ganglioside clusters in Alzheimer's disease. *J. Neurochem.* **2011**, *116*, 806–812.

37. Pitto, M.; Raimondo, F.; Zoia, C.; Brighina, L.; Ferrarese, C.; Masserini, M., Enhanced GM1 ganglioside catabolism in cultured fibroblasts from Alzheimer patients. *Neurobiol. Aging* **2005**, *26*, 833–838.
38. Yanagisawa, K.; Ihara, Y., GM1 ganglioside-bound amyloid beta-protein in Alzheimer's disease brain. *Neurobiol. Aging* **1998**, *19*, S65–S67.
39. Yanagisawa, K.; Odaka, A.; Suzuki, N.; Ihara, Y., GM1 ganglioside-bound amyloid beta-protein (Abeta): A possible form of preamyloid in Alzheimer's disease. *Nat. Med.* **1995**, *1*, 1062–1066.
40. Yamamoto, N.; Hasegawa, K.; Matsuzaki, K.; Naiki, H.; Yanagisawa, K., Environment- and mutation-dependent aggregation behavior of Alzheimer amyloid β-protein. *J. Neurochem.* **2004**, *90*, 62–69.
41. Yamamoto, N.; Matsubara, T.; Sato, T.; Yanagisawa, K., Age-dependent high-density clustering of GM1 ganglioside at presynaptic neuritic terminals promotes amyloid β-protein fibrillogenesis. *Biochim. Biophys. Acta* **2009**, *1778*, 2717–2726.
42. McLaurin, J.; Chakrabartty, A., Membrane disruption by Alzheimer beta-amyloid peptides mediated through specific finding to either phospholipids or gangliosides—Implications for neurotoxicity. *J. Biol. Chem.* **1996**, *271*, 26482–26489.
43. Matsuzaki, K.; Kato, K.; Yanagisawa, K., Aβ polymerization through interaction with membrane gangliosides. *Biochim. Biophys. Acta* **2010**, *1801*, 868–877.
44. Wang, S. S. S.; Rymer, D. L.; Good, T. A., Reduction in cholesterol and sialic acid content protects cells from the toxic effects of beta-amyloid peptides. *J. Biol. Chem.* **2001**, *276*, 42027–42034.
45. Williamson, M. P.; Suzuki, Y.; Bourne, N. T.; Asakura, T., Binding of amyloid β-peptide to ganglioside micelles is dependent on histidine-13. *Biochemical J.* **2006**, *397*, 483–490.
46. Tumova, S.; Woods, A.; Couchman, J. R., Heparan sulfate proteoglycans on the cell surface: Versatile coordinators of cellular functions. *Int. J. Biochem. Cell Biol.* **2000**, *32*, 269–288.
47. Bernfield, M.; Götte, M.; Park, P. W.; Reizes, O.; Fitzgerald, M. L.; Lincecum, J.; Zako, M., Functions of cell surface heparan sulfate proteoglycans. *Annu. Rev. Biochem.* **1999**, *68*, 729–777.
48. Snow, A. D.; Mar, H.; Nochlin, D.; Kimata, K.; Kato, M.; Suzuki, S.; Hassell, J.; Wight, T. N., The presence of heparan sulfate proteoglycans in the neuritic plaques and congophilic angiopathy in Alzheimer's disease. *Am. J. Pathol.* **1988**, *133*, 456–463.
49. Snow, A. D.; Mar, H.; Nochlin, D.; Sekiguchi, R. T.; Kimata, K.; Koike, Y.; Wight, T. N., Early accumulation of heparan sulfate in neurons and in the beta-amyloid protein-containing lesions of Alzheimer's disease and Down's syndrome. *Am. J. Pathol.* **1990**, *137*, 1253–1270.
50. Perlmutter, L. S.; Chui, H. C.; Saperia, D.; Athanikar, J., Microangiopathy and the colocalization of heparan sulfate proteoglycan with amyloid in senile plaques of Alzheimer's disease. *Brain Res.* **1990**, *508*, 13–19.
51. Snow, A. D.; Sekiguchi, R. T.; Nochlin, D.; Kalaria, R. N.; Kimata, K., Heparan sulfate proteoglycan in diffuse plaques of hippocampus but not of cerebellum in Alzheimer's disease brain. *Am. J. Pathol.* **1994**, *144*, 337–347.
52. McLaurin, J.; Franklin, T.; Zhang, X. Q.; Deng, J. P.; Fraser, P. E., Interactions of Alzheimer amyloid-beta peptides with glycosaminoglycans—Effects on fibril nucleation and growth. *Eur. J. Biochem.* **1999**, *266*, 1101–1110.

53. Snow, A. D.; Sekiguchi, R. T.; Nochlin, D.; Fraser, P.; Kimata, K.; Mizutani, A.; Arai, M.; Schreier, W. A.; Morgan, D. G., An important role of heparan sulfate proteoglycan (perlecan) in a model system for the deposition and persistence of fibrillar Aβ-amyloid in rat brain. *Neuron* **1994**, *12*, 219–234.

54. Castillo, G. M.; Ngo, C.; Cummings, J.; Wight, T. N.; Snow, A. D., Perlecan binds to the beta-amyloid proteins (A beta) of Alzheimer's disease, accelerates A beta fibril formation, and maintains A beta fibril stability. *J. Neurochem.* **1997**, *69*, 2452–2465.

55. Sandwall, E.; O'Callaghan, P.; Zhang, X.; Lindahl, U.; Lannfelt, L.; Li, J. P., Heparan sulfate mediates amyloid-beta internalization and cytotoxicity. *Glycobiology* **2010**, *20*, 533–541.

56. Lindahl, B.; Westling, C.; Giménez-Gallego, G.; Lindahl, U.; Salmivirta, M., Common binding sites for β-amyloid fibrils and fibroblast growth factor-2 in heparan sulfate from human cerebral cortex. *J. Biol. Chem.* **1999**, *274*, 30631–30635.

57. Smits, N. C.; Kurup, S.; Rops, A. L.; Dam, G. B.; Massuger, L. F.; Hafmans, T.; Turnbull, J. E.; Spillmann, D.; Li, J. P.; Kennel, S. J.; Wall, J. S.; Shworak, N. W.; Dekhuijzen, P. N. R.; ven der Vlag, J.; Van Kuppevelt, T. H., The heparan sulfate motif $(GlcNS6S\text{-}IdoA2S)_3$, common in heparin, has a strict topography and is involved in cell behavior and disease. *J. Biol. Chem.* **2010**, *285*, 41143–41151.

58. Buee, L.; Ding, W.; Delacourte, A.; Fillit, H., Binding of secreted human neuroblastoma proteoglycans to the Alzheimer's amyloid A4 peptide. *Brain Res.* **1993**, *601*, 154–163.

59. Buee, L.; Ding, W. H.; Anderson, J. P.; Narindrasorasak, S.; Kisilevsky, R.; Boyle, N. J.; Robakis, N. K.; Delacourte, A.; Greenberg, B.; Fillit, H. M., Binding of vascular heparan sulfate proteoglycan to Alzheimer's amyloid precursor protein is mediated in part by the N-terminal region of A4 peptide. *Brain Res.* **1993**, *627*, 199–204.

60. Fraser, P. E.; Nguyen, J. T.; Chin, D. T.; Kirschner, D. A., Effects of sulfate ions on Alzheimer beta/A4 peptide assemblies—Implications for amyloid fibril proteoglycan interactions. *J. Neurochem.* **1992**, *59*, 1531–1540.

61. McLaurin, J.; Fraser, P. E., Effect of amino-acid substitutions on Alzheimer's amyloid-beta peptide-glycosaminoglycan interactions. *Eur. J. Biochem.* **2000**, *267*, 6353–6361.

62. Shaw, C. P.; Middleton, D. A.; Volk, M.; Levy, R., Amyloid-derived peptide forms self-assembled mono layers on gold nanoparticle with a curvature-dependent beta-sheet structure. *ACS Nano* **2012**, *6*, 1416–1426.

63. Chan, H.-M.; Xiao, L.; Yeung, K.-M.; Ho, S.-L.; Zhao, D.; Chan, W.-H.; Li, H.-W., Effect of surface-functionalized nanoparticles on the elongation phase of beta-amyloid (1-40) fibrillogenesis. *Biomaterials* **2012**, *33*, 4443–4450.

64. Wang, C.; Liu, D.; Wang, Z., Gold nanoparticle based dot-blot immunoassay for sensitively detecting Alzheimer's disease related beta-amyloid peptide. *Chem. Commun.* **2012**, *48*, 8392–8394.

65. Triulzi, R. C.; Dai, Q.; Zou, J.; Leblanc, R. M.; Gu, Q.; Orbulescu, J.; Huo, Q., Photothermal ablation of amyloid aggregates by gold nanoparticles. *Colloids Surf., B* **2008**, *63*, 200–208.

66. Bastus, N. G.; Kogan, M. J.; Amigo, R.; Grillo-Bosch, D.; Araya, E.; Turiel, A.; Labarta, A.; Giralt, E.; Puntes, V. F., Gold nanoparticles for selective and remote heating of beta-amyloid protein aggregates. *Mater. Sci. Eng., C* **2007**, *27*, 1236–1240.

67. Antosova, A.; Gazova, Z.; Fedunova, D.; Valusova, E.; Bystrenova, E.; Valle, F.; Daxnerova, Z.; Biscarini, F.; Antalik, M., Anti-amyloidogenic activity of glutathione-covered gold nanoparticles. *Mater. Sci. Eng., C* **2012**, *32*, 2529–2535.

68. Bastus, N. G.; Sanchez-Tillo, E.; Pujals, S.; Farrera, C.; Kogan, M. J.; Giralt, E.; Celada, A.; Lloberas, J.; Puntes, V., Peptides conjugated to gold nanoparticles induce macrophage activation. *Mol. Immunol.* **2009**, *46*, 743–748.

69. Zhang, D.; Neumann, O.; Wang, H.; Yuwono, V. M.; Barhoumi, A.; Perham, M.; Hartgerink, J. D.; Wittung-Stafshede, P.; Halas, N. J., Gold nanoparticles can induce the formation of protein-based aggregates at physiological pH. *Nano Lett.* **2009**, *9*, 666–671.

70. Araya, E.; Olmedo, I.; Bastus, N. G.; Guerrero, S.; Puntes, V. F.; Giralt, E.; Kogan, M. J., Gold nanoparticles and microwave irradiation inhibit beta-amyloid amyloidogenesis. *Nano. Res. Lett.* **2008**, *3*, 435–443.

71. Liao, Y.-H.; Chang, Y.-J.; Yoshiike, Y.; Chang, Y.-C.; Chen, Y.-R., Negatively charged gold nanoparticles inhibit Alzheimer's amyloid-beta fibrillization, induce fibril dissociation, and mitigate neurotoxicity. *Small* **2012**, *8*, 3631–3639.

72. Sakono, M.; Zako, T.; Maeda, M., Naked-eye detection of amyloid aggregates using gold nanoparticles modified with amyloid beta antibody. *Anal. Sci.* **2012**, *28*, 73–76.

73. Kouyoumdjian, H.; Zhu, D. C.; El-Dakdouki, M. H.; Lorenz, K.; Chen, J.; Li, W.; Huang, X., Glyconanoparticle aided detection of β-amyloid by magnetic resonance imaging and attenuation of β-amyloid induced cytotoxicity. *ACS Chem. Neurosci.* **2013**, *4*, 575–584.

74. Mahmoudi, M.; Quinlan-Pluck, F.; Monopo, M. P.; Sheibani, S.; Vali, H.; Dawson, K. A.; Lynch, I., Influence of the physiochemical properties of superparamagnetic iron oxide nanoparticles on amyloid beta protein fibrillation in solution. *ACS Chem. Neurosci.* **2013**, *4*, 475–485.

75. Skaat, H.; Margel, S., Synthesis of fluorescent-maghemite nanoparticles as multimodal imaging agents for amyloid-beta fibrils detection and removal by a magnetic field. *Biochem. Biophys. Res. Commun.* **2009**, *386*, 645–649.

76. Skaat, H.; Shafir, G.; Margel, S., Acceleration and inhibition of amyloid-beta fibril formation by peptide-conjugated fluorescent-maghemite nanoparticles. *J. Nanopart. Res.* **2011**, *13*, 3521–3534.

77. Wadghiri, Y. Z.; Li, J.; Wang, J.; Hoang, D. M.; Sun, Y.; Xu, H.; Tsui, W.; Li, Y.; Boutajangout, A.; Wang, A.; de Leon, M.; Wisniewski, T., Detection of amyloid plaques targeted by bifunctional USPIO in Alzheimer's disease transgenic mice using magnetic resonance microimaging. *PLoS One* **2013**, *8*, e57097.

78. Yoo, S. I.; Yang, M.; Brender, J. R.; Subramanian, V.; Sun, K.; Joo, N. E.; Jeong, S.-H.; Ramamoorthy, A.; Kotov, N. A., Inhibition of amyloid peptide fibrillation by inorganic nanoparticles: Functional similarities with proteins. *Angew. Chem. Int. Ed.* **2011**, *50*, 5110–5115.

79. Xiao, L.; Zhao, D.; Chan, W.-H.; Choi, M. M. F.; Li, H.-W., Inhibition of beta 1-40 amyloid fibrillation with N-acetyl-L-cysteine capped quantum dots. *Biomaterials* **2010**, *31*, 91–98.

80. Patel, D.; Henry, J.; Good, T., Attenuation of beta-amyloid induced toxicity by sialic acid-conjugated dendrimeric polymers. *Biochim. Biophys. Acta* **2006**, *1760*, 1802–1809.

81. Patel, D. A.; Henry, J. E.; Good, T. A., Attenuation of beta-amyloid-induced toxicity by sialic-acid-conjugated dendrimers: Role of sialic acid attachment. *Brain Res.* **2007**, *1161*, 95–105.

82. Cowan, C. B.; Cote, G. L.; Good, T. A., Development of photocrosslinked sialic acid containing polymers for use in Aβ toxicity attenuation. *Biomaterials* **2008**, *29*, 3408–3414.

83. Klajnert, B.; Cortijo-Arellano, M.; Bryszewska, M.; Cladera, J., Influence of heparin and dendrimers on the aggregation of two amyloid peptides related to Alzheimer's and prion diseases. *Biochem. Biophys. Res. Commun.* **2006**, *339*, 577–582.

84. Chafekar, S. M.; Malda, H.; Merkx, M.; Meijer, E. W.; Viertl, D.; Lashuel, H. A.; Baas, F.; Scheper, W., Branched KLVFF tetramers strongly potentiate inhibition of beta-amyloid aggregation. *Chembiochem* **2007**, *8*, 1857–1864.

85. Klajnert, B.; Wasiak, T.; Ionov, M.; Fernandez-Villamarin, M.; Sousa-Herves, A.; Correa, J.; Riguera, R.; Fernandez-Megia, E., Dendrimers reduce toxicity of A beta 1-28 peptide during aggregation and accelerate fibril formation. *Nanomed.-Nanotechnol.* **2012**, *8*, 1372–1378.

86. Dhavale, D.; Henry, J. E., Evaluation of sialic acid-analogs for the attenuation of amyloid-beta toxicity. *Biochim. Biophys. Acta Gen. Subj.* **2012**, *1820*, 1475–1480.

87. Canovi, M.; Markoutsa, E.; Lazar, A. N.; Pampalakis, G.; Clemente, C.; Re, F.; Sesana, S.; Masserini, M.; Salmona, M.; Duyckaerts, C.; Flores, O.; Gobbi, M.; Antimisiaris, S. G., The binding affinity of anti-Aβ 1-42 mAb-decorated nanoliposomes to Aβ 1-42 peptides in vitro and to amyloid deposits in post-mortem tissue. *Biomaterials* **2011**, *32*, 5489–5497.

88. Gobbi, M.; Re, F.; Canovi, M.; Beeg, M.; Gregori, M.; Sesana, S.; Sonnino, S.; Brogioli, D.; Musicanti, C.; Gasco, P.; Salmona, M.; Masserini, M. E., Lipid-based nanoparticles with high binding affinity for amyloid-beta(1-42) peptide. *Biomaterials* **2010**, *31*, 6519–6529.

89. Mourtas, S.; Canovi, M.; Zona, C.; Aurilia, D.; Niarakis, A.; La Ferla, B.; Salmona, M.; Nicotra, F.; Gobbi, M.; Antimisiaris, S. G., Curcumin-decorated nanoliposomes with very high affinity for amyloid-beta 1-42 peptide. *Biomaterials* **2011**, *32*, 1635–1645.

90. Ikeda, K.; Okada, T.; Sawada, S.-i.; Akiyoshi, K.; Matsuzaki, K., Inhibition of the formation of amyloid beta-protein fibrils using biocompatible nanogels as artificial chaperones. *FEBS Lett.* **2006**, *580*, 6587–6595.

91. Pai, A. S.; Rubinstein, I.; Onyuksel, H., PEGylated phospholipid nanomicelles interact with beta-amyloid((1-42)) and mitigate its beta-sheet formation, aggregation and neurotoxicity in vitro. *Peptides* **2006**, *27*, 2858–2866.

92. Le Droumaguet, B.; Nicolas, J.; Brambilla, D.; Mura, S.; Maksimenko, A.; De Kimpe, L.; Salvati, E.; Zona, C.; Airoldi, C.; Canovi, M.; Gobbi, M.; Noiray, M.; La Ferla, B.; Nicotra, F.; Scheper, W.; Flores, O.; Masserini, M.; Andrieux, K.; Couvreur, P., Versatile and efficient targeting using a single nanoparticulate platform: application to cancer and Alzheimer's disease. *ACS Nano* **2012**, *6*, 5866–5879.

93. Brambilla, D.; Verpillot, R.; Taverna, M.; De Kimpe, L.; Le Droumaguet, B.; Nicolas, J.; Canovi, M.; Gobbi, M.; Mantegazza, F.; Salmona, M.; Nicolas, V.; Scheper, W.; Couvreur, P.; Andrieux, K., New method based on capillary electrophoresis with laser-induced fluorescence detection (CE-LIF) to monitor interaction between nanoparticles and the amyloid-beta peptide. *Anal. Chem.* **2010**, *82*, 10083–10089.

94. Cabaleiro-Lago, C.; Quinlan-Pluck, F.; Lynch, I.; Lindman, S.; Minogue, A. M.; Thulin, E.; Walsh, D. M.; Dawson, K. A.; Linse, S., Inhibition of amyloid beta protein fibrillation by polymeric nanoparticles. *J. Am. Chem. Soc.* **2008**, *130*, 15437–15443.

95. Li, M.; Yang, X.; Ren, J.; Qu, K.; Qu, X., Using graphene oxide high near-infrared absorbance for photothermal treatment of Alzheimer's disease. *Adv. Mater.* **2012**, *24*, 1722–1728.

96. Mahmoudi, M.; Akhavan, O.; Ghavami, M.; Rezaee, F.; Ghiasi, S. M. A. Graphene oxide strongly inhibits amyloid beta fibrillation. *Nanoscale* **2012**, *4*, 7322–7325.

97. Kim, J. E.; Lee, M., Fullerene inhibits beta-amyloid peptide aggregation. *Biochem. Biophys. Res. Commun.* **2003**, *303*, 576–579.

98. Huang, H. M.; Ou, H. C.; Hsieh, S. J.; Chiang, L. Y., Blockage of amyloid beta peptide-induced cytosolic free calcium by fullerenol-1, carboxylate C60 in PC12 cells. *Life Sci.* **2000**, *66*, 1525–1533.

99. Lee, C.-M.; Huang, S.-T.; Huang, S.-H.; Lin, H.-W.; Tsai, H.-P.; Wu, J.-Y.; Lin, C.-M.; Chen, C.-T., C-60 fullerene-pentoxifylline dyad nanoparticles enhance autophagy to avoid cytotoxic effects caused by the beta-amyloid peptide. *Nanomed. Nanotechnol.* **2011**, *7*, 107–114.

100. Bobylev, A. G.; Kornev, A. B.; Bobyleva, L. G.; Shpagina, M. D.; Fadeeva, I. S.; Fadeev, R. S.; Deryabin, D. G.; Balzarini, J.; Troshin, P. A.; Podlubnaya, Z. A., Fullerenolates: Metallated polyhydroxylated fullerenes with potent anti-amyloid activity. *Org. Biomol. Chem.* **2011**, *9*, 5714–5719.

101. Fu, Z.; Luo, Y.; Derreumaux, P.; Wei, G., Induced beta-barrel formation of the Alzheimer's Abeta 25-35 oligomers on carbon nanotube surfaces: Implication for amyloid fibril inhibition. *Biophysical J.* **2009**, *97*, 1795–1803.

102. Li, H.; Luo, Y.; Derreumaux, P.; Wei, G., Carbon nanotube inhibits the formation of beta-sheet-rich oligomers of the Alzheimer's amyloid-beta(16-22) peptide. *Biophys. J.* **2011**, *101*, 2267–2276.

103. de la Fuente, E.; Adura, C.; Kogan, M. J.; Bollo, S., Carbon nanotubes electrochemistry allows the in situ evaluation of the effect of β-sheet breakers on the aggregation process of β-amyloid. *Electroanalysis* **2012**, *24*, 938–944.

104. Siegemund, T.; Paulke, B. R.; Schmiedel, H.; Bordag, N.; Hoffmann, A.; Harkany, T.; Tanila, H.; Kacza, J.; Hartig, W., Thioflavins released from nanoparticles target fibrillar amyloid beta in the hippocampus of APP/PS1 transgenic mice. *Int. J. Dev. Neurosci.* **2006**, *24*, 195–201.

105. Poduslo, J. F.; Wengenack, T. M.; Curran, G. L.; Wisniewski, T.; Sigurdsson, E. M.; Macura, S. I.; Borowski, B. J.; Jack, C. R., Molecular targeting of Alzheimer's amyloid plaques for contrast-enhanced magnetic resonance imaging. *Neurobiol. Dis.* **2002**, *11*, 315–329.

106. Skaat, H.; Corem-Slakmon, E.; Grinberg, I.; Last, D.; Goez, D.; Mardor, Y.; Margel, S., Antibody-conjugated, dual-modal, near-infrared fluorescent iron oxide nanoparticles for antiamyloidgenic activity and specific detection of amyloid-β fibrils. *Int. J. Nanomed.* **2013**, *8*, 4063–4076.

107. Cowan, C. B.; Patel, D. A.; Good, T. A., Exploring the mechanism of beta-amyloid toxicity attenuation by multivalent sialic acid polymers through the use of mathematical models. *J. Theor. Biol.* **2009**, *258*, 189–197.

108. Chikae, M.; Fukuda, T.; Kerman, K.; Idegami, K.; Miura, Y.; Tamiya, E., Amyloid-β detection with saccharide immobilized gold nanoparticle on carbon electrode. *Bioelectrochemistry* **2008**, *74*, 118–123.

109. Beier, H. T.; Cowan, C. B.; Chou, I. H.; Pallikal, J.; Henry, J. E.; Benford, M. E.; Jackson, J. B.; Good, T. A.; Cote, G. L., Application of surface-enhanced Raman spectroscopy for detection of beta amyloid using nanoshells. *Plasmonics* **2007**, *2*, 55–64.

110. Barthel, H.; Gertz, H.-J.; Dresel, S.; Peters, O.; Bartenstein, P.; Buerger, K.; Hiemeyer, F.; Wittemer-Rump, S. M.; Seibyl, J.; Reininger, C.; Sabri, O.; Florbetaben Study, G., Cerebral amyloid-beta PET with Florbetaben (F-18) in patients with Alzheimer's disease and healthy controls: A multicentre phase 2 diagnostic study. *Lancet Neurol.* **2011**, *10*, 424–435.

111. Villemagne, V. L.; Ong, K.; Mulligan, R. S.; Holl, G.; Pejoska, S.; Jones, G.; O'Keefe, G.; Ackerman, U.; Tochon-Danguy, H.; Chan, J. G.; Reininger, C. B.; Fels, L.; Putz, B.; Rohde, B.; Masters, C. L.; Rowe, C. C., Amyloid imaging with F-18-Florbetaben in Alzheimer disease and other dementias. *J. Nucl. Med.* **2011**, *52*, 1210–1217.

112. Wong, D. F.; Rosenberg, P. B.; Zhou, Y.; Kumar, A.; Raymont, V.; Ravert, H. T.; Dannals, R. F.; Nandi, A.; Brasic, J. R.; Ye, W.; Hilton, J.; Lyketsos, C.; Kung, H. F.; Joshi, A. D.; Skovronsky, D. M.; Pontecorvo, M. J., In vivo imaging of amyloid deposition in Alzheimer disease using the radioligand F-18-AV-45 (Florbetapir F 18). *J. Nucl. Med.* **2010**, *51*, 913–920.

113. Clark, C. M.; Schneider, J. A.; Bedell, B. J.; Beach, T. G.; Bilker, W. B.; Mintun, M. A.; Pontecorvo, M. J.; Hefti, F.; Carpenter, A. P.; Flitter, M. L.; Krautkramer, M. J.; Kung, H. F.; Coleman, R. E.; Doraiswamy, P. M.; Fleisher, A. S.; Sabbagh, M. N.; Sadowsky, C. H.; Reiman, P. E. M.; Zehntner, S. P.; Skovronsky, D. M.; Grp, A. A. S., Use of Florbetapir-PET for imaging beta-amyloid pathology. *J. Am. Med. Assoc.* **2011**, *305*, 275–283.

114. Hintersteiner, M.; Enz, A.; Frey, P.; Jaton, A.-L.; Kinzy, W.; Kneuer, R.; Neumann, U.; Rudin, M.; Staufenbiel, M.; Stoeckli, M.; Wiederhold, K.-H.; Gremlich, H.-U., In vivo detection of amyloid-β deposits by near-infrared imaging using an oxazine-derivative probe. *Nat. Biotechnol.* **2005**, *23*, 577–583.

115. Wadghiri, Y. Z.; Sigurdsson, E. M.; Sadowski, M.; Elliott, J. I.; Li, Y. S.; Scholtzova, H.; Tang, C. Y.; Aguinaldo, G.; Pappolla, M.; Duff, K.; Wisniewski, T.; Turnbull, D. H., Detection of Alzheimer's amyloid in transgenic mice using magnetic resonance micro-imaging. *Magn. Reson. Med.* **2003**, *50*, 293–302.

116. Petiet, A.; Santina, M.; Bertrand, A.; Wiggins, C. J.; Petit, F.; Houitte, D.; Hantraye, P.; Benavides, J.; Debeir, T.; Rooney, T.; Dhenain, M., Gadolinium-staining reveals amyloid plaques in the brain of Alzheimer's transgenic mice. *Neurobiol. Aging* **2012**, *33*, 1533–1544.

117. Higuchi, M.; Iwata, N.; Matsuba, Y.; Sato, K.; Sasamoto, K.; Saido, T. C., ^{19}F and ^{1}H MRI detection of amyloid beta plaques in vivo. *Nat. Neurosci.* **2005**, *8*, 527–533.

118. Jun, Y. W.; Lee, J. H.; Cheon, J., Chemical design of nanoparticle probes for high-performance magnetic resonance imaging. *Angew. Chem. Int. Ed.* **2008**, *47*, 5122–5135.

119. Na, H. B.; Song, I. C.; Hyeon, T., Inorganic nanoparticles for MRI contrast agents. *Adv. Mater.* **2009**, *21*, 2133–2148.

120. El-Boubbou, K.; Zhu, D. C.; Vasileiou, C.; Borhan, B.; Prosperi, D.; Li, W.; Huang, X., Magnetic glyco-nanoparticles: A tool to detect, differentiate, and unlock the glyco-codes of cancer via magnetic resonance imaging. *J. Am. Chem. Soc.* **2010**, *132*, 4490–4499.

121. Roch, A.; Gossuin, Y.; Muller, R. N.; Gillis, P., Superparamagnetic colloid suspensions: Water magnetic relaxation and clustering. *J. Magn. Magn. Mater.* **2005**, *293*, 532–539.

122. Yung, K. T., Empirical models of transverse relaxation for spherical magnetic perturbers. *Magn. Reson. Imaging* **2003**, *21*, 451–463.

123. Lee, H.; Sun, E.; Ham, D.; Weissleder, R., Chip-NMR biosensor for detection and molecular analysis of cells. *Nat. Med.* **2008**, *14*, 869–874.

124. Koh, I.; Hong, R.; Weissleder, R.; Josephson, L., Sensitive NMR sensors detect antibodies to influenza. *Angew. Chem. Int. Ed.* **2008**, *47*, 4119–4121.

13

CARBOHYDRATE NANOTECHNOLOGY AND ITS APPLICATIONS FOR THE TREATMENT OF CANCER

Shailesh G. Ambre and Joseph J. Barchi, Jr.

Glycoconjugate and NMR Section, Chemical Biology Laboratory, Center for Cancer Research, National Cancer Institute at Frederick, Frederick, MD, USA

13.1 INTRODUCTION

Cancer is a group of more than 200 disease states, which develop due to unrestrained growth of cells resulting from mutations in their genetic code. Major contributors to the accumulation of these genetic changes leading to DNA damage include lifestyle patterns (smoking, alcohol consumption, obesity, and inactivity) and environmental factors (UV radiation, pollution, and microbial exposure). All cells have mechanisms that can repair such damage or initiate one of various self-destructive processes. However, most tumors have adapted ways to subvert both the cellular repair machinery and the mechanisms that facilitate programmed cell death (apoptosis).

It is well known that cancer is one of the main causes of disease-related mortality in the United States, and even with the major discoveries and advances in the diagnosis and treatment of cancer, there remain an alarming number of yearly deaths reported. According to the statistics available with the National Cancer Institute (NCI), about 1.7 million new cases have been reported for cancer, and nearly 600,000 deaths are predicted for the current year (2014). Breast cancer remains the most commonly diagnosed cancer and the cause of cancer-related death in women, while prostate cancer is the prime culprit in men. Each type of cancer is unique in its

Carbohydrate Nanotechnology, First Edition. Edited by Keith J. Stine.

manifested characteristics and molecular complexities. Thus, there is no common treatment to cancer. Clinical treatment of cancer can follow a variety of therapies. Conventionally, localized tumors are removed by surgery, followed by chemotherapy or radiation therapy. Chemotherapeutic intervention to control and eliminate cancer is quite common and employs anticancer agents that act through different cellular mechanisms to interrupt growth and proliferation of cells or induce apoptotic mechanisms to kill these cells. Although anticancer drugs are intended to stop and destroy abnormally growing cells, they can similarly interact with healthy cells and are the cause of major toxic side effects. Additionally, most drugs suffer from poor solubility and pharmacokinetics, making them less able to access the tumor cells leading to increased nonspecific toxicities. Decades of research have been focused on addressing these issues, and nanomedicine [1,2] has emerged as one of the solutions. Nanomedicine [3] is nanotechnology applied to either the diagnosis or treatment of disease, usually through the novel use of nanoscale delivery systems. In recent years, the number of reports using nanoparticles for effective delivery of cytotoxic drugs has increased logarithmically [4].

In this chapter, we briefly introduce the biological features of cancer and the challenges they present for successful cancer therapy. Next, we discuss the merits of nanomedicine and how they help alleviate some of the difficulties in cancer therapy. Later, we introduce the role of sugars in cancer biology and their utility in current or prospective carbohydrate-based nanomedicinal therapies.

13.2 CANCER BIOLOGY AND CONSIDERATIONS FOR EFFECTIVE NANOTHERAPY

13.2.1 Tumor Heterogeneity

Cancer arises from the abnormal growth of cells. Since a particular tissue is comprised of multiple cell types, it is logical that cancers occurring in different types of cells carry significant molecular disparities. However, the functional heterogeneity is not limited to intertumoral tissue. Intratumoral heterogeneity, both biological and functional, occurs due to the diverse genetic mutations acquired by different cells within the tumor [5]. Two different models had been proposed for the development of tumor heterogeneity—the clonal evolution model and the cancer stem cell (CSC) model [6]. Both models provide valuable insight into the design of anticancer treatments. Lately, these models are considered to be less mutually exclusive. The clonal evolution model hypothesizes that the cancer cells randomly acquire genetic mutations, triggered by either internal or external stimuli, and develop heterogeneous clones [6,7]. Alternatively, the CSC model proposes that CSCs possess the ability to self-renew and differentiate. Thus, they are able to generate heterogeneous phenotypes of the tumor population [6,7]. The CSC model has gained considerable attention in the last two decades; however, its origins can be traced as early as 1937. At that time, Kahn and coworkers showed that a single mouse leukemia cell was able to generate new tumors in recipient mice [8]. Early experiments were hampered by the technical limitations in isolating and studying CSCs; these limitations were largely overcome with the availability of new techniques such as the fluorescence-activated cell sorting (FACS). By combining FACS with dilution transplantations into highly immunodeficient mice, Bonnet *et al.* were able to isolate the first leukemic CSCs [9]. The first isolated CSCs from solid tumors were from a specific breast cancer [10]. Building on the initial success, CSCs from other types of cancers were also isolated. CSCs share a variety of similar functional properties with normal tissue stem cells. Thus, it is logical to hypothesize that they are resistant to therapies designed to target rapidly dividing cancer cells [11,12]. An example of this would be the resistance to the drug imatinib, an inhibitor of a tyrosine kinase that is highly overexpressed in chronic myeloid leukemia patients [13]. Hence, the past decade has seen a renewed focus on deciphering the mechanism(s) of drug resistance in cancer cells. These mechanisms include but are not limited to the quiescent status of CSCs, overexpression of drug efflux pumps, reduced DNA damage repair, and activation of antiapoptotic machinery [12,14,15]. The issue of tumor heterogeneity presents the obvious problems with the design of "global" inhibitors, but the ability to target all cancer cells, both bulk tumors and CSCs, will be desired in new therapies to improve prognosis of the disease and prevent reoccurrences.

13.2.2 Tumor Microenvironment and Enhanced Permeability and Retention Effect

A tumor can be seen as a complex organ-like structure. Cancer cells are embedded in unique extracellular matrix (ECM) and are surrounded by various tumor stromal cells. The tumor undergoes constant remodeling through reciprocal communications between the cells and the surrounding microenvironment [16]. A detailed understanding of the tumor microenvironment is critical for the design of therapies that can specifically target cancer cells leaving the surrounding normal tissue unharmed. The tumor microenvironment is highly complex, and here, we will focus on the aspects that are most essential to successful design of nanomedicine.

The formation of new blood vessels (angiogenesis) is essential to the survival of any tissue, and in particular developing tumors, as these vessels act as the carrier of oxygen and nutrients for cellular growth and development. Once tumors reach a size of about 2 mm^3, a state of hypoxia develops, and thus, there becomes a distinct need for new blood vessels [17]. New blood vessels are developed from existing blood vessels and by recruiting specialized cells from the bone marrow. CSCs may also differentiate into endothelial cells toward the same goal [18–21]. Angiogenesis occurs in five phases: (1) endothelial cell activation, (2) base membrane degradation, (3) endothelial cell migration, (4) vessel formation, and (5) angiogenic remodeling. Hypoxia induces upregulation of proteins like vascular endothelial growth factor (VEGF), platelet-derived growth factor (PDGF), or tumor necrosis factor-α (TNF-α) [22]. The persistent stimulation from abnormally activated growth signals and incomplete remodeling during the final phase leads to irregular shaped, dilated, and tortuous tumor blood vessels [4,19,23]. As a consequence, the newly formed tumor blood vessels are "leaky" with large interendothelial junctions (10–1000 nm) [23–25].

Also, the lymphatic vessels are impaired and rather absent toward the center of the tumor and do not drain lymphatic fluid efficiently. This leads to an increased interstitial fluid pressure (IFP) in tumors. Many cancer drugs that traverse the interstitial space by convection rather than diffusion encounter IFP and thus have diminished access to the tumor tissue [26].

The effect of IFP on nano-sized drug carriers would be minimum due to their size. Instead, it facilitates the extravasation of nanoparticles into the tumor tissue. Coupled with the "leaky" vasculature and erratic lymphatic drainage, it leads to what has been termed as the "enhanced permeability and retention (EPR) effect" for nanoparticles in tumor tissue [27,28]. Thus, many types of nanoparticles are selectively able to accumulate in the tumor tissue over time allowing passive targeting of tumors with cytotoxic drugs loaded in or on nano-sized carriers [29]. Unfortunately, tumor vasculature is not uniformly permeable. The blood flow in tortuous blood vessels is slower than normal leading to poorly perfused areas in the tumor [30,31]. In some cases, highly proliferating cancer cells and stromal cells compress the tumor blood vessels causing them to collapse [32]. The nonuniform vasculature and dense ECM are existing challenges in delivering nanotherapeutics uniformly in the tumor tissue.

13.3 NANOTECHNOLOGY FOR CANCER THERAPY

The complete elimination of cancer requires the need for effective chemotherapeutic protocols that will selectively destroy the abnormal cells yet leave the healthy cells unharmed. Nanotechnology has great potential to enhance the efficacy of current therapies and provide the basis for developing new strategies for clinical treatment of cancer. However, it must be appreciated that new technologies require time and innovative technologies are usually accompanied by high failure rate [2]. The concept of nanomedicine is very new, but it has recently started to make inroads into both therapeutic and diagnostic clinical medicine. For example, the US Food and Drug Administration (USFDA) recently approved the use of a gold nanoparticle-based nucleic acid test [8]. This platform is used to identify bacteria and detect genes that provide antibiotic resistance in bloodstream pathogens [33]. Several other nanomedicines have been approved by the USFDA and European Medicines Agency (EMA) for clinical treatment of different types of cancer, and some are listed in Table 13.1 (reproduced from Patra and Turner [34]). However, it should be noted that so far most nanomedicines have only been able to improve quality of life by increasing tumor response but come short of improving overall survival of treated patients. Some of the early nanomedicines that have reached the clinic, such as Doxil (doxorubicin) or Abraxane (albumin-bound taxane), were designed with the specific goal to reduce the nonspecific cytotoxicity of their payload. They are successful at reducing some adverse effects; however, this small benefit is insufficient in the current economic and regulatory climate [2,35]. Nevertheless, it is important to keep in perspective that the nanomedicine "revolution" is barely a decade old, which is an extremely short period to have realized any clinical success, and suffice it to say, a lot is yet to be learned.

TABLE 13.1 Nanoscale Delivery Systems Used in the Clinic[a]

Name	Target	Active Component	Nanosystem
Myocet™	Breast cancer	Doxorubicin	Liposomal
Abraxane	Metastatic breast cancer	Paclitaxel	Albumin bound
Doxil	Ovarian cancer	Doxorubicin	PEGylated liposomal
Eligard	Prostate cancer	Leuprolide acetate	Biodegradable polymer
Caelyx	Kaposi's sarcoma	Doxorubicin	PEGylated liposomal
DaunoXome	Kaposi's sarcoma	Daunorubicin	PEGylated liposomal
Genexol-PM	Breast and lung cancer	Paclitaxel	Polymeric micelle
Oncaspar	Lymphoblastic leukemia	L-Asparaginase	Polymeric micelle

[a]As of 2013. Taken from Table 3, Ref. 34.

Current research in nano-sized cancer therapeutics is mainly focused toward developing safe and targeted delivery of antitumor therapeutic drugs or nucleic acids. Recent reports of phase I clinical trial results for the delivery of siRNA and docetaxel using nanocarriers have been promising [35–37]. Nanoparticles exploit the EPR effect to accumulate in tumors. As mentioned earlier, EPR arises from fenestrated blood vessels and limited lymphatic drainage. This is termed as passive targeting. Moreover, nanocarriers can be loaded with several molecules of a single drug or multiple drugs that would potentially be released in the tumor tissue. This allows a concentration effect, as opposed to traditional therapy where the drug molecules enter circulation and are distributed over the entire body. Use of multiple drugs for therapy and the concentration effect may be key to prevent the development of drug resistance in cancer cells as the abnormal cells can be more effectively eliminated. Alternatively, active targeting can be employed where conjugation of a "sensor" molecule (e.g., antibodies, growth factors, or specific ligands) that binds to a particular receptor that is selectively overexpressed on tumor cells allows homing directly to the neoplastic site and subsequent delivery of a therapeutic agent [38].

13.4 SUGARS, CANCER PROGRESSION, AND METASTASIS

Carbohydrates are ubiquitous in nature. Most eukaryotic cells make proteins that are decorated with one or more glycans that modulate the structure and function of both cell surface and secreted proteins. Some of the processes that glycans mediate are protein folding, cell signaling, fertilization, embryogenesis, pathogen recognition, inflammation, immune responses, tissue organization, and cell proliferation [39–42]. Many of these processes are mediated by the glycan interacting directly with a carbohydrate-binding protein called a lectin. Lectins come in a variety of flavors, as they can be expressed in almost any cellular context depending on the tissue and stage of development and they bind selectively (or many times, promiscuously) to specific sugar structures [43]. Monovalent glycan–lectin dissociation constants are usually quite high, that is, very weak binding, that can be lowered by multivalent interactions (the so-called clustered glycoside effect) [44–46]. A special lectin subset are the galectins, β-galactoside-binding proteins that have been implicated in a number of disease states, including cancer [47–50].

Oncogenesis is a multistep process consisting of diverse genetic and epigenetic changes in malignant cells. Some of the hallmarks of cancer are persistent growth, insensitivity to anti-growth signals, evasion of apoptosis, unlimited replication potential, angiogenesis, invasion, and metastatic potential [51]. It is now well established that another constant feature of tumors is the expression of modified glycan chains due to aberrations in cellular glycoprocessing machinery. The presentation of these aberrant sugar chains is directly associated with cancer progression and metastasis [52–54]. These atypical glycan chains on proteins can be used as biomarkers for detecting cancerous tissue and its progression toward metastasis [53,55]. Malignant transformation of cells greatly resembles early development and cellular activation in vertebrates, since many tumor-associated carbohydrates closely resemble those "immature" glycans displayed on embryonic tissue [56].

Naturally occurring glycans can be classified into several types including N- and O-linked glycans, primarily displayed on cell surface proteins along with glycolipids and glycosaminoglycans (GAG) (Fig. 13.1). All of these families have a unique pathway of biosynthesis, but as mentioned earlier, each glycan type is modified or aberrant in some way when a tumor phenotype develops. N-linked glycans become extended and/or more branched, due to increased expression of Golgi enzymes that lead to highly branched structures involving the bisecting GlcNAc, β–(1-6)-GlcNAc, and the core fucose (α–(1-3)-Fuc) [57]. These structures are associated with several molecules of cell adhesion and cancer metastasis. Overexpression of *N*-acetylglucosaminyltransferase-V (also called Mgat-5) is responsible for β–(1-6)-GlcNAc branching that leads to tetra-antennary N-glycans and increased metastatic potential in many tumors [58,59].

O-glycans play an important role in the process of cellular differentiation and regulation of cell growth through apoptosis and proliferation pathways. It is also well established that aberrant O-glycans play an important role in the process of tumor metastasis; the detailed molecular mechanisms are not yet clear [60]. O-linked glycans can either promote or suppress malignancy depending on their structure. Most of these biological effects are mediated by truncated O-linked glycans attached to cell surface mucins, very large glycoproteins expressed on the surface of, or secreted by epithelial cells. They are decorated with large number of clustered O-glycans attached to repeating peptide motifs containing many serine and threonine amino acids. Some examples of these aberrant glycans are the Tn (GalNAc-α-O-Ser/Thr), TF (Galβ–(1-3)-GalNAc-α-O-Ser/Thr), sialyl-Tn (GalNAc(α–(2-6)-Neu5Ac)-α-O-Ser/Thr) and sialyl-TF (Neu5Acα–(2-3)-Galβ–(1-3)-GalNAc-α-O-Ser/Thr or Galβ–(1-3)-GalNAc(α–(2-6)-Neu5Ac)-α-O-Ser/Thr) structures, along with the Lewis antigens, sialyl LewisX (sLe^X, (Neu5Acα-(2-3)-Galβ–(1-4)-[Fucα-(1-3)]GlcNAc) and sialyl LewisA (sLeA, Neu5Acα-(2-3)-Galβ–(1-3)-[Fucα(1-4)]GlcNAc). O-glycans have been shown to influence cancer metastasis. In addition, these glycans, along with aberrant glycolipid saccharides, are considered tumor-associated carbohydrate antigens (TACAs), as they are

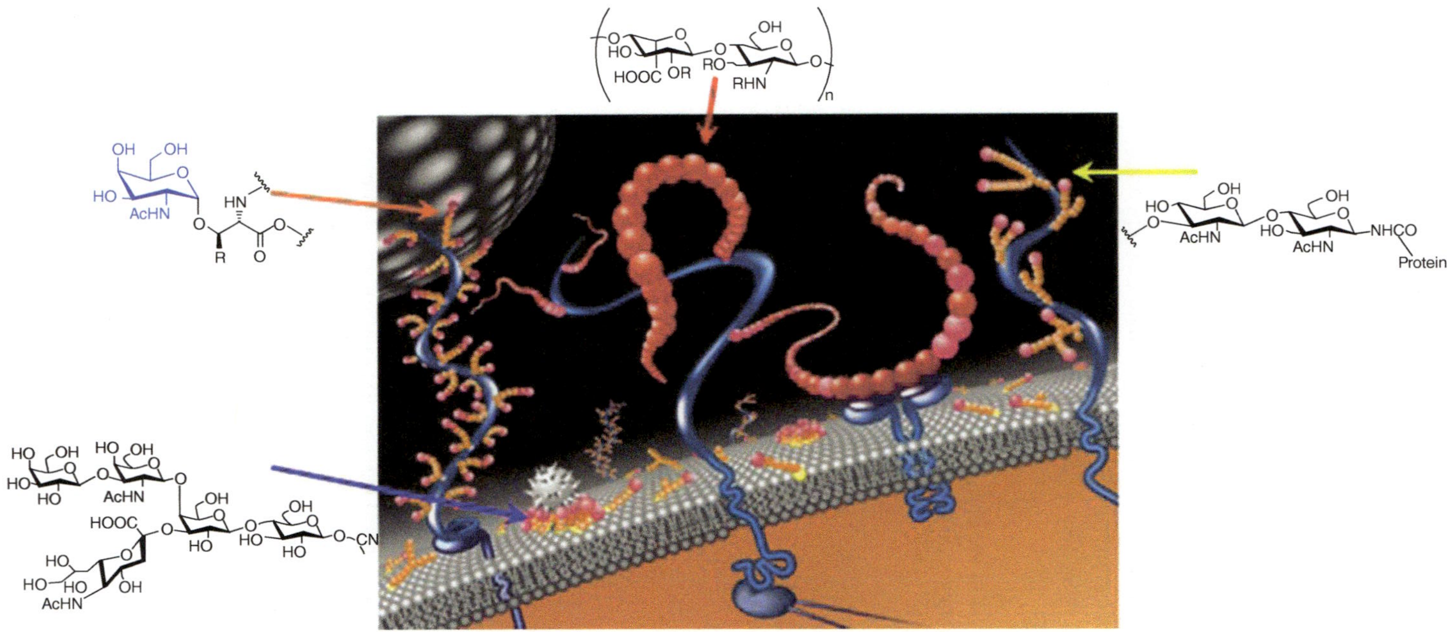

FIGURE 13.1 Various cell surface forms of glycans with the associated reducing end structures. *Source*: Reproduced from Glycoforum: http://www.glycoforum.gr.jp.

recognized by the mammalian immune system as "foreign." Hence, a variety of approaches have been developed to raise immune responses to these sugar chains with synthetic vaccines and immunotherapies [51,61].

13.5 GLYCONANOTECHNOLOGY AND CANCER

From the previous discussion, it seems clear that intervention at various stages of tumor progression and metastasis that involve aberrations in cellular glycoprocessing machinery could benefit from nanomedicine approaches. In the past 15 years, there have been increasing numbers of reports of coating various nanoparticles with carbohydrate molecules and evaluation of them in varied assays [62–66]. However, the number of reports where a glyconanomaterial was evaluated as tool in anticancer discovery is more modest. However, this field is growing dramatically, as evidenced by the number of reviews that have been published or are in press regarding cancer nanotechnology [67–74]; subsequent years will see an equally dramatic rise in cancer glyconanotechnology research. The following sections will be devoted to work performed in the most recent literature. We last briefly reviewed glyconanotechnology in cancer in 2011 [75], and the rest of this chapter will reiterate some of those more relevant studies and list others that were not covered, but focus primarily on the latest developments in the field. The remainder of the chapter will be divided into sections that describe (i) the type of platform used in glyconanotechnology, (ii) whether the attached glycan on the platform is used for *targeting* tumors or as a *therapeutic functionality*, and (iii) the glyconanomaterials that are actually *constructed* of glycans/polysaccharides.

13.5.1 Glyconanomaterial Platforms for Cancer Nanomedicine

The platforms used to create glyconanomaterials are primarily those that have been used for attaching other bio- or small-molecule entities. Liposomes, gold nanoparticles, quantum dots (Qdots), magnetic nanoparticles (MNPs) (superparamagnetic iron oxide nanoparticles (SPIONs)), various polymers, and carbon nanotubes (Fig. 13.2) have all been coated with various carbohydrates for variety of applications. However, not all have been used in specific anticancer applications; the following will concentrate only on those that have reference to anticancer therapeutic design. Glycodendrimers, another very well-studied platform for the production of useful biomedical tools and therapies, will not be covered here; the reader is referred to some recent excellent reviews in this area [65,76].

13.5.2 Platforms Where Glycan Is Used for Targeting

13.5.2.1 ***Liposomes*** These are circular artificial lipid bilayer vesicles formed from different phospholipids similar to ones found in cellular membranes. Although they are sized in the 100–500 nm range, they are still considered "nanoparticles," and hundreds have been prepared for various medical applications. Liposome-encapsulated drugs

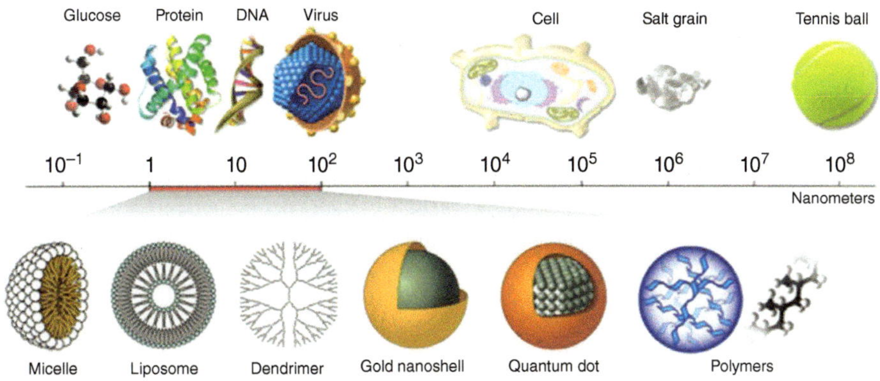

FIGURE 13.2 Various nanoparticles used in medicine and their relative sizes. From Wich Lab home page, Permission by personal communication, http://wichlab.com/research.

were the first examples of "nanomedicine" approved by the FDA (Table 13.1). Hence, the inner core of a liposome can trap water-soluble molecules (drugs, nucleic acids), while the lipid membrane can also embed hydrophobic drugs or other entities. In addition, the phospholipid scaffold that makes up the liposome can be embedded with a variety of molecules that can be used to target the liposome to various tissues or organs. This is the primary application of using glycan-coated liposomes: the glycan is used as a *targeting agent* to direct the liposome to cells or tissues where carbohydrate-binding proteins are upregulated. Two recent reviews outline much of this research, one on the use of glycosylated carriers to deliver nucleic acids selectively to cells [77] and the other on more recent compilation of glycolipid-based systems for the delivery of drugs and as vaccine adjuvants [78].

Several liposome preparations have been reported that display various carbohydrates on the outer membrane surface [79]. The synthesis usually employs a carbohydrate conjugated to a lipid that is an integral building block in the liposome bilayer [80]. Alternatively, the sugar can be attached via an electrophilic functional group (e.g., PEG-maleimide) displayed on the liposome surface [81]. There have been a host of sugars that have been attached to liposomes to target antigen-presenting cells (APCs) via their type-specific lectin expression. For example, liposomes that carry carbohydrates on their surface have been used in this manner for more than two decades to target the mannose receptor on various APCs such as dendritic cells (DC's) [82–85]. DC-specific intercellular adhesion molecule-3-grabbing nonintegrin (DC-SIGN) is another lectin receptor expressed on macrophages and dendritic cells that binds mannose glycans as well as those with fucose (e.g., Lewis antigen structures) and has been less used as a targeting receptor (vide infra) [86]. Liposomes can be prepared by standard methods for encapsulating antigen, while mannosylation is performed by one of the methods described earlier. Delivery in this manner has been used in both vaccine constructs and to delivery chemotherapeutic drugs.

The number of reports for the use of mannosylated and sugar-encapsulated liposomes for targeting APCs are too numerous to outline individually, but several reviews have been compiled in the past decade to cover most of the literature (vide supra). Various recent reports have shown some novel approaches where many different surface ligands and cargo have been used for delivery of liposomes. In 2013, Thomann *et al.* prepared liposomes displaying both T-helper (Th) and cytotoxic T-lymphocyte (CTL) epitope peptides from the ErbB2 protein and injected these into mice with renal cell tumors that overexpress ErbB2 [87]. The innovative strategy employed Toll-like receptor (TLR) agonist to either TLR1/2 or TLR2/6 where the peptides were not conjugated directly to these adjuvants but embedded into the liposome membrane separately, unlike prior studies. Mannosylation was also incorporated for targeting purposes (Fig. 13.3). The authors found that the TLR2/6 ligands

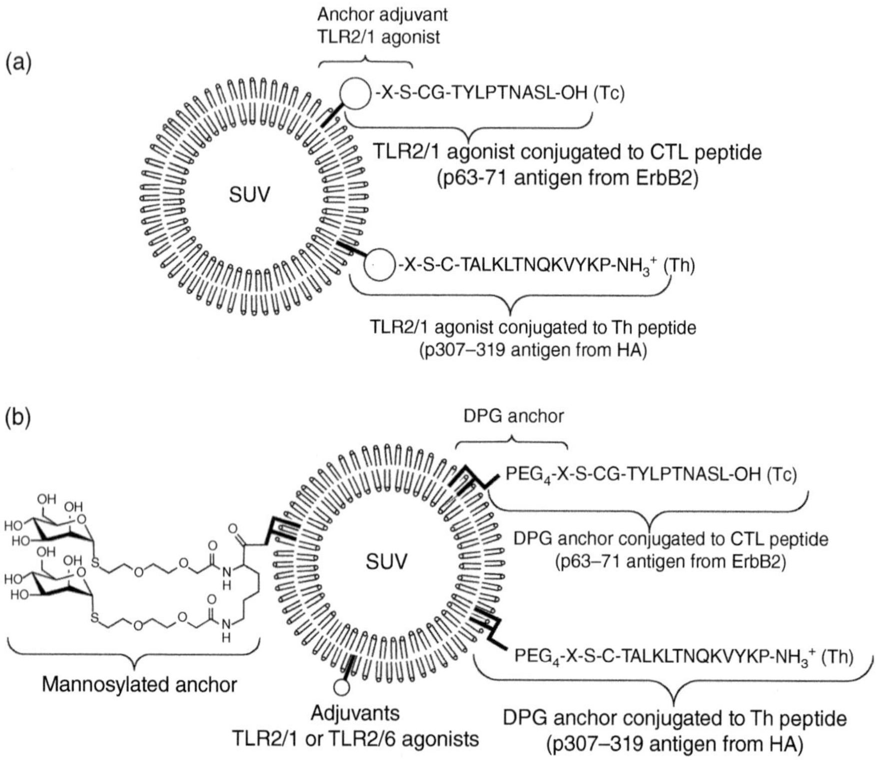

FIGURE 13.3 (a) First generation of diepitope liposomal constructs. CTL epitope ErbB2 (p63–71) and Th(HA307-319) epitope peptides, derivatized with a CG- or C-linker, respectively, were conjugated to liposomes (PC/PG/Chol, 75/20/50 molar ratio) by reaction with performed vesicles (SUV) containing 5 mol% thiol-reactive functionalized lipopeptide anchors Pam_3CAGBr or $Pam_3CAGMal$ [13]. (b) Second generation of diepitope liposomal constructs. The CTL and Th epitope peptides were conjugated to performed liposomes (PC/PG/Chol, 75/20/50 molar ratio, SUV) containing 5 mol% of thiol-reactive anchors DPGMal or DPGBr. The TLR ligands—Pam_3CAG, Pam_2CAG, or Pam_2CGD—were incorporated in the vesicles, at given concentrations, during their preparation. X = maleimide or bromoacetyl functions. *Source*: Reprinted with permission from Ref. 87.

were superior for eradicating the ErbB2-expressing tumors and active targeting through the mannose receptor enhanced the therapeutic efficacy of the constructs so lower amounts of peptide could be incorporated into the liposomes. Another innovative strategy used encapsulated mRNA of a tumor-associated antigen (TAA) protein into specialized liposomes in the relatively new field of mRNA vaccine development [88]. Liposomes were prepared after complexation of MART-1 mRNA with a cationic polymer. The mRNA–polymer complex, called a polyplex, is encapsulated into liposomes to form a lipopolyplex. Further functionalization of the outer surface with histidinylated polylysine was employed to allow endosomal escape after protonation of histidine that destabilizes the endosomal membrane. This allows the mRNA to escape into the cytosol. Targeting with mannose incorporated into the lipid membrane of the lipopolyplex completes the "glyconanomaterial" as an effective anticancer vaccine construct. Finally, a dual-purpose strategy was employed by the Huang group that took advantage of two separate carbohydrate-bearing nanoparticles [89]. First, a lipid calcium phosphate (LCP) nanoparticle was prepared for efficient delivery of a small TAA peptide with adjuvant (CpG oligonucleotide) for selective generation of CTL against B16F10 melanoma. This construct also contained a mannose coating for delivery to dendritic cells. However, this vaccine was not effective in the later stages of the disease due to the expression of immunosuppressive cytokines (TGF-β) after treatment with the LCP particles. Another construct was prepared consisting of protamine and hyaluronic acid complexed with an siRNA to TGF-β [90]. This complex was encapsulated with cationic liposomes and used to deliver siRNA to knock down gene expression of TGF-β after initial vaccination. This treatment caused a 50% knockdown of protein and resulted in 53% boost in tumor cell, killing over the initial treatment with LCP particles alone.

In an elegant study by the van Kooyk group, a DC-SIGN ligand was exploited as a targeting agent in a liposome preparation [81]. Glycoliposomes bearing either sialyl Lewis^X or sialyl Lewis^A tetrasaccharides showed a 100-fold enhanced uptake into bone marrow-derived dendritic cells and presentation of encapsulated antigen to CD4$^+$ and CD8$^+$ T cells than unmodified vesicles (schematic representation in Fig. 13.4). When the melanoma antigen MART-1 was encapsulated into these carbohydrate-coated liposomes, similarly enhanced antigen presentation to MART-1-specific CD8$^+$ T-cell clones was observed. This particular vaccine platform was able to affect primary CD8$^+$ T cells to differentiate into tetramer-specific IFN-γ-producing effector T cells. This relatively simple system seems to be highly effective for antigen uptake into APCs, and future work should prove whether this platform could be clinically viable.

13.5.2.2 MNPs SPIONs have seen increased growth in the past several years due to their ease of preparation and broad biofunctional utility for applications such as imaging, photothermal therapy, and magnetic resonance imaging (MRI) contrast agents [91]. Many efforts have employed a variety of targeting agents, such as folic acid, transferrin, or integrins through RGD-based peptide sequences [92]. Recent reports have shown that carbohydrate-ligated MNPs are also unique systems for targeting and imaging of cancer tissue [64]. It is interesting to note that several

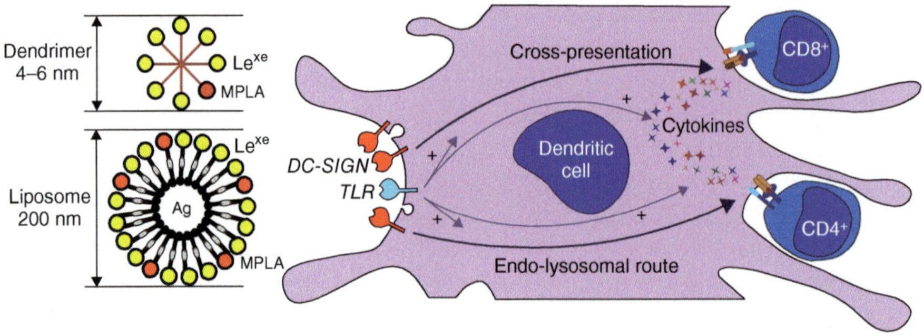

FIGURE 13.4 Glycan-modified dendrimers and liposomes enhance antigen presentation. Although dendrimers and liposomes greatly differ in size and molecular properties, both systems allow for the display of DC-SIGN ligands, such as Le^x or Le^b, in a multivalent form. *Source*: Reprinted with permission from Ref. 81.

syntheses of MNPs use a stabilizing coating on the ferric core and often these are polysaccharide in nature (dextran, starch, and alginate) [74].

Very little work on carbohydrate-coated MNPs was performed prior to 2007. In 2010, El Boubbou *et al.* published on a series of MNPs coated with a variety of monosaccharides and their binding to various tumor cell lines [93]. The work established a coded system that made it capable to differentiate among the various lines based on the binding characteristics and the changes associated with the MRI relaxivity measurements. It is well known that MNPs will decrease the weighted T_2 signals of specific tissue in MRI since the relaxation rate will be increased by proximity of the magnetic entity. Well-known binding profiles were recapitulated and others were newly found. In addition, B16F10 tumor cell adhesion was reduced by 50% by the galactose-bearing MNPs. The same group developed a new method for preparing hyaluronan (HA)-coated MNPs for targeting CD44-bearing cells such as activated macrophages [94]. The synthesis involved the preparation of stabilized dextran-coated iron oxide particles that were functionalized with amines; peptide coupling technology allowed conjugation with hyaluronic acid. These MNPs do target macrophages expressing CD44 and are also taken up by these cells in a CD44-dependent manner. Although not directly related to targeting tumor cells, these constructs would be valuable tools in antiadhesive therapy since CD44–HA interactions are critical to metastasis of many solid tumors [95, 96]. A follow-up study showed that similar particles only extended with a functional linker that allowed facile attachment of a drug molecule (doxorubicin) were taken up by tumor cells and were highly cytotoxic when conjugated to the antitumor drug [97]. Finally, these authors evaluated the *in vivo* efficacy of these SPION–DOX conjugates in an ovarian cancer tumor model in mice [98]. The constructs were injected into mice with SKOV-5 tumors, and the particles accumulated selectively in the tumor, reduced tumor growth, delayed tumor development, and extended the survival of mice. In addition, there are the added diagnostic properties of these particles that allow them to be truly considered as antitumor "theranostics" tools. The continued development of these novel agents may prove clinically useful.

The Penãdes group has published extensively on the use of gold glyconanoparticles (GGNPs) for biomedical applications, and has pioneered the development of gadolinium-containing GGNPs as MRI contrast agents [99–101]. They had shown that different sugars (mono- and disaccharides) could be directed to different tumors, depending on their cell surface lectin profile, akin to the studies outlined by Huang earlier [100]. Thus, contrast agents can potentially be directed to different sites for imaging purposes. A recent report by this group outlines the synthesis of intriguing carbohydrate-functionalized "gold-coated" magnetite nanoparticles for uptake into C33 tumor cells [102]. A study by Lee *et al.* was reported where HA-coated SPION particles containing Cy5.5 dye were able to separately detect areas of high hyaluronidase activity where quenched fluorescence is restored on polysaccharide cleavage and they could also be used as MRI contrast agents [103]. It appears obvious that the use of carbohydrates as NP coatings in these systems is a vital strategy to allow specific active targeting.

13.5.2.3 Qdots Selectively imaging tumor and the ability to tag residual circulating tumor cells have become a highly important area of cancer research. New imaging modalities and novel agents that are stable in an *in vivo* environment are essential to the successful development of clinically useful bioimaging methods [104]. Qdots have been a mainstay in this area due to their highly unique properties of size-dependent, extremely narrow emission spectra and strong resistance to photobleaching. Many useful Qdot entities have been synthesized and used for a variety of important imaging applications. Stable polymer-coated Qdots are now commercially available with a host of organic functional groups for facile bioconjugation to almost all molecular families. Glycofunctionalized Qdots have seen a variety of applications in recent years, but few have dealt with tumor targeting or therapy [105]. The following discussion will highlight a handful of recent reports where carbohydrate molecules were an integral part of the Qdot coating or assembly.

A host of reviews have been published on the usefulness of Qdots in biological imaging [106–113]. Although Qdots are incredibly useful tools in many oncology-related research areas, there is concern about their use in humans due to their elemental composition (cadmium and other toxic transition metals) [114]. Some early reports of Qdots coated with carbohydrates came from the labs of Penãdes [115], Jovin [116], and our work [117]. In Jovin's work, streptavidin Qdots were coated with either polymeric or monomeric biotinylated sialyl Lewis^X, gp120, or a DC-SIGN antibody. These were designed to target DC-SIGN on dendritic cells in an effort to internalize the Qdots and follow the processing of antigen within the cells themselves. Although not directly related to an anticancer application, targeting APCs could theoretically allow specific TAA or TACA processing in live cells. Interestingly, only the polymeric Le^X Qdots were recognized by DC-SIGN, not the monomeric, indicative of the dependency of multivalency on binding. Targeting of the Le^X could be inhibited by polymeric mannan. In 2009, Qdots with an electrostatically adhered coating of hyaluronic acid were used to label HeLa cells and lymphatic endothelial cells (LECs) [118]. These nanoparticles targeted the lymphatic vessel endothelial receptor 1 (LYVE-1) that is highly expressed on both HeLa cells and LECs. The authors state that these may be used for cancer cell

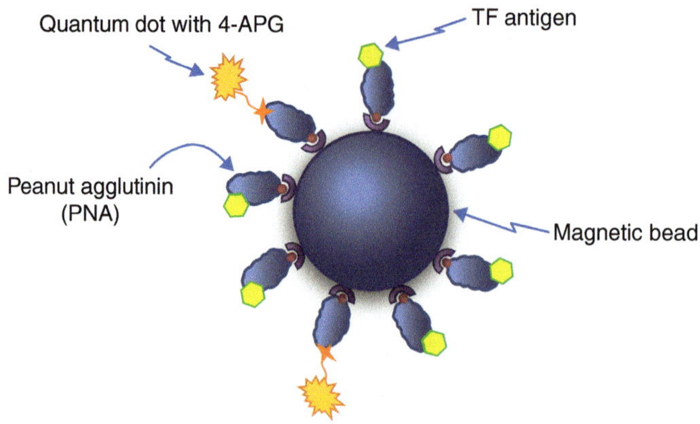

FIGURE 13.5 Schematic representation of the detection system for TF antigen using a quantum dot coated with 4-aminophenyl β-D-galactopyranoside (4-APG) that competes with TF antigen for PNA-binding sites. *Source*: Reprinted with permission from Ref. 119. Copyright (2014) American Chemical Society.

imaging, an assumption that should prove valid since CD44, another HA receptor, is found in large amounts in aggressive tumorous tissue. In 2013, Li *et al.* developed a sensitive detection system for the TF antigen by using carbohydrate-coated Qdots to compete with TF antigen binding to peanut agglutinin (PNA) attached onto magnetic beads (Fig. 13.5) [119]. The Qdots were conjugated with 4-aminophenyl β-D-galactopyranoside (4-APG), a molecule that competes with other sugars that also may bind PNA on the bead. Increasing concentrations of TF antigen will result in fluorescence signal loss. The authors were able to show that addition of potentially interfering proteins and cell lysates do not reduce with the sensitivity of the assay. This method could potentially detect TF antigen on tumor cells where the aberrant carbohydrate composition is unknown. In a slight "twist" for coating Qdots with sugars, the Kikkeri group has developed an ingenious method to display carbohydrates on the surface of Qdots by utilizing the strong host–guest interaction of cyclodextrins (CDs) with the carbon chains of trioctylphosphine oxide (TOPO) [120]. The optimal synthesis of Qdots uses a precursor TOPO-coated core-shell system followed by a place-exchange reaction with the ligand of interest to be displayed, thus removing the TOPO coat. This design makes use of the octyl chains as guests for carbohydrate-conjugated CDs where this noncovalent interaction serves to hold the system together (Fig. 13.6). Galactose-capped CD was attached to TOPO-coated Qdots, and their binding to lectins and *in vitro* cytotoxicity were measured. They were shown to bind tightly to HepG2 cells but exhibit very little cytotoxicity, thereby confirming their utility in *in vitro* and *in vivo* imaging. Finally, Penãdes' group has recently published a one-pot synthesis of CdTe-based Qdots in water [121], similar to our original synthesis [117]. The study showed that these Qdots when coated with small molecules could yield quantum yields up to 40%. Two small molecules, cysteine and mannose, were able to attenuate the known

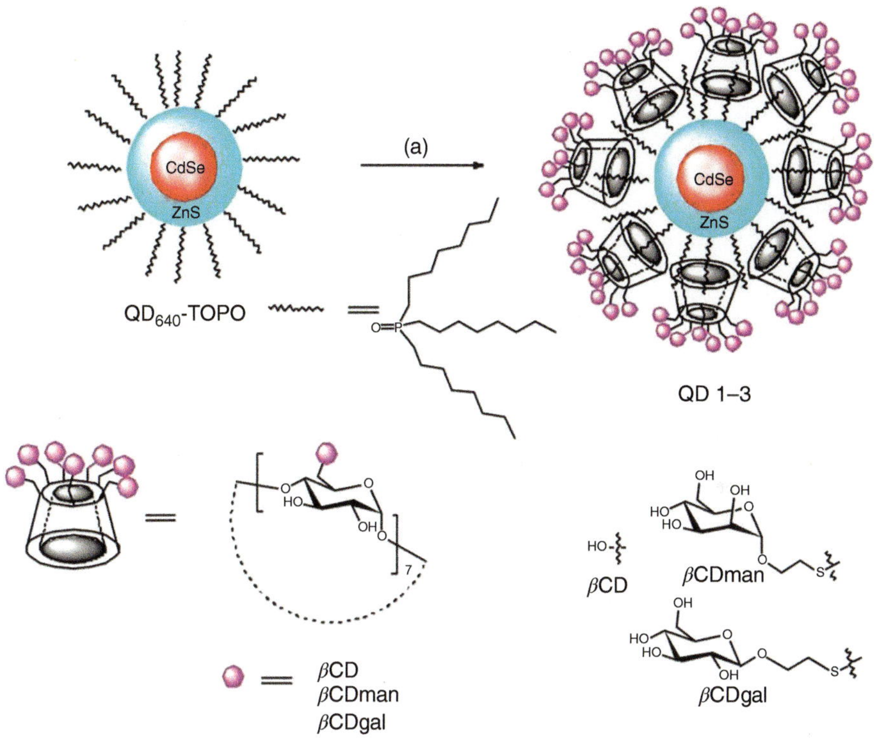

FIGURE 13.6 Preparation and self-assembly of glycol-QDs: (a) β-CD derivatives in ethanol/water (1:10). *Source*: Reprinted with permission from Ref. 120.

toxicity of CdTe core Qdots. Addition of a ZnTe shell also reduced standard cytotoxicity toward different tumor cell lines.

13.5.2.4 Carbon Nanotubes Carbon-based material science as it relates to biomedical research has seen a rapid expansion in the past decade, keeping pace with the many other nanosystems being explored. Carbon nanotubes are excellent platforms for the attachment of various molecular families due to their high surface area and wealth of compatible chemistries that can be employed for functionalization. Single-walled carbon nanotubes (SWCTs), which have unique electronic and mechanical properties, are the most popular platform for decorating with carbohydrates. More details regarding carbohydrate-functionalized nanotubes in other arenas can be found in Chapter 5. A few important applications to cancer targeting research have been reported for glycosylated SWCTs.

Since 2004, the Bertozzi group has pioneered the development of mimics of the highly glycosylated mucin proteins by decorating carbon nanotubes with a C18-terminated polymer where GalNAc residues were attached via oxime chemistry to ketone units of methyl vinyl ketone [122]. These constructs were water soluble, biocompatible, and capable of interacting with carbohydrate-binding proteins.

A follow-up study showed the interaction of these glyconanomaterials with living cells and the lack of toxicity toward those cells [123]. This will allow the use of these reagents without fear of limiting cytotoxicity. These being mucin mimics that display the Tn antigen (α-GalNAc) make them useful surrogates of mucins that are displayed on cancer cells. The Narain group in Edmonton has also pioneered a series of unique glyconanomaterials as important tools in biomedical research. They have developed glycan-coated carbon nanotubes where a diblock copolymer was prepared from three monomers, 3-aminopropyl methacrylamide (APMA), 3-gluconamidopropyl methacrylamide (GAPMA), and 2-lactobionamidoethyl methacrylamide (LAEMA) by the reversible addition–fragmentation chain transfer (RAFT) method [124]. This method avoids the use of potentially toxic metal catalysts and produces polymers with an attached glycan as well as pendant free amino groups resulting in a cationic assembly. This was "grafted" to carboxyl-bearing carbon nanotubes to yield a net positive polymeric nanotube that can associate with nucleic acids and thus be used as a gene delivery/transfection vehicle. These constructs were also biocompatible and were able to deliver genes *in vitro* with efficiencies that rival the commercial reagent Transfectamine.

13.5.2.5 Gold Nanoparticles The glyconanoparticle field has arguably seen some of the most growth in the area of spherical GGNPs, since the first publications of "glyconanotechnology" were reported [125–127]. The majority of the applications of GGNPs are in therapeutic design and not simple targeting. There are examples of many hybrid particles that include gold in one form of the assembly that have been used for delivery and targeting. For example, the Park group has published on HA-coated GGNPs as detection probes for rheumatoid arthritis and metastatic cancer [128]. Digested HA was an end thiol capped by a known method with cystamine and conjugated with a near-infrared (IR) dye to the carboxyl groups of the HA 6′-glucuronic acid position. These were attached to gold particles through the thiol group, and this process of having a local fluorophore near the gold core quenches its fluorescence. When the particles contacted tissue with high concentrations of hyaluronidase enzymes, the HA is cleaved releasing the fluorophore, thus recapitulating fluorescence. The authors were able to detect tumors *in vivo* in an ovarian cancer model, follow biodistribution, and show that the near IR signal is strongest in tumor tissue as opposed to the liver and spleen. Others have utilized HA-coated particles for targeting and detection. In another interesting design, the preparation of doxorubicin-conjugated HA-encapsulated micelles coated with a gold half shell for theranostic applications was also reported [129]. There has also been a wealth of research on preparing "CD-based multivalent glycodisplays" as pioneered by the Garcia Fernandez group in Sevilla [130]. An ingenious entry into this type of design was reported by Vargas-Berenquel and coworkers [131]. They synthesized a multifunctional GGNP bearing lactose groups for targeting β-galactose-binding lectins such as PNA or galectin-3 and CDs that could simultaneously encapsulate anticancer drugs like the folate analogue methotrexate. The authors showed that binding to the lectins and encapsulation of methotrexate could be accomplished, but direct tumor targeting was not attempted in this study.

13.5.3 Platforms Where Glycan Is Used for Therapy

Many of the reports described in Section 13.5.2 could also be listed in this section, as targeting and therapy tend to go hand in hand. The following will concentrate on only nanoparticles where delivery of a drug, antigen, or toxin showed efficacy in an *in vitro* or *in vivo* model and thus there is a clear potential therapeutic use.

13.5.3.1 Liposomes As was described earlier, liposomes are the quintessential and original "nano" platform for the incorporation of various molecular entities either embedded within the membrane structure to be displayed on the phospholipid surface or encapsulated within the hydrophilic core to be delivered as a therapeutic agent. Many recent reviews have been published on the use of a wide breadth of nanosystems as anticancer therapeutic intervention delivery agents or tools [132–135], as well as those related specifically to liposomes [136]. Again, the field is somewhat more limited with regard to glyconanosystems in this area. But many of the groups described in Section 15.5.2.1 have used their systems for therapeutic design. Other notable work on the use of glycan liposomes in anticancer therapy comes from the Paulson group who studied the therapeutic effect of liposomes coated with ligands to CD22, the so-called sialic acid-binding immunoglobulin-like lectin (Siglec), glycoproteins that bind sialic acid-containing glycoproteins and glycolipids and regulate B-cell signaling. This approach was in contrast to the use of antibodies to CD22, a standard therapeutic modality for B-cell lymphomas. A novel aromatic analogue of the natural Neu5Acα–(2-6)-Galβ–(1-4)-GlcNAc ligand was linked via a PEG spacer to a lipid tail for incorporation into liposomes (Fig. 13.7) [137–139]. These liposomes were recognized and endocytosed by CD22 on B cells. Treatment of mice implanted with a Daudi B-cell lymphoma model with a series of liposomes, either liganded or untargeted, and both with or without loaded doxorubicin showed that the liganded particles performed superior to those unliganded and the targeted particles were more effective at reducing tumor burden than even the untargeted DOX-loaded particles. In another series of studies, the authors also targeted CD169 (sialoadhesin, Siglec-1), another sialic acid-binding Siglec on macrophages, and were able to deliver antigen in liganded liposomes to APCs [140–142]. Thus, delivery of lipid antigen to macrophages induced invariant natural killer T-cell expression. Structure-based design of ligands for Siglecs seems like a new and highly worthwhile approach to anticancer therapeutic design and intervention.

Another well-studied target for tumor cell adhesion is E-selectin as described earlier. For several years, liposomal preparations with sLe^X or sLe^A incorporated into the lipid were used as competitive inhibitors of tumor cell adhesion to E-selectin on activated endothelial cells [143–146]. sLe^x incorporated into liposomes was able to inhibit tumor cell adhesion of Lewis lung and HT29 colon tumor cells to E-selectin [144]. A study by Hirai *et al.* showed that sLe^X-capped liposomes could be loaded with cisplatin and targeted to E-selectin-expressing tumors [147]. The cisplatin-loaded sLe^X liposomes showed much reduced *in vivo* acute cytotoxicity as opposed to treatment with free cisplatin and very high accumulation into tumors relative to untargeted drug-loaded liposomes, and a reduction in tumor growth and tumor volume *in vivo* was seen specifically for the sLe^X-targeted particles. Other uses of

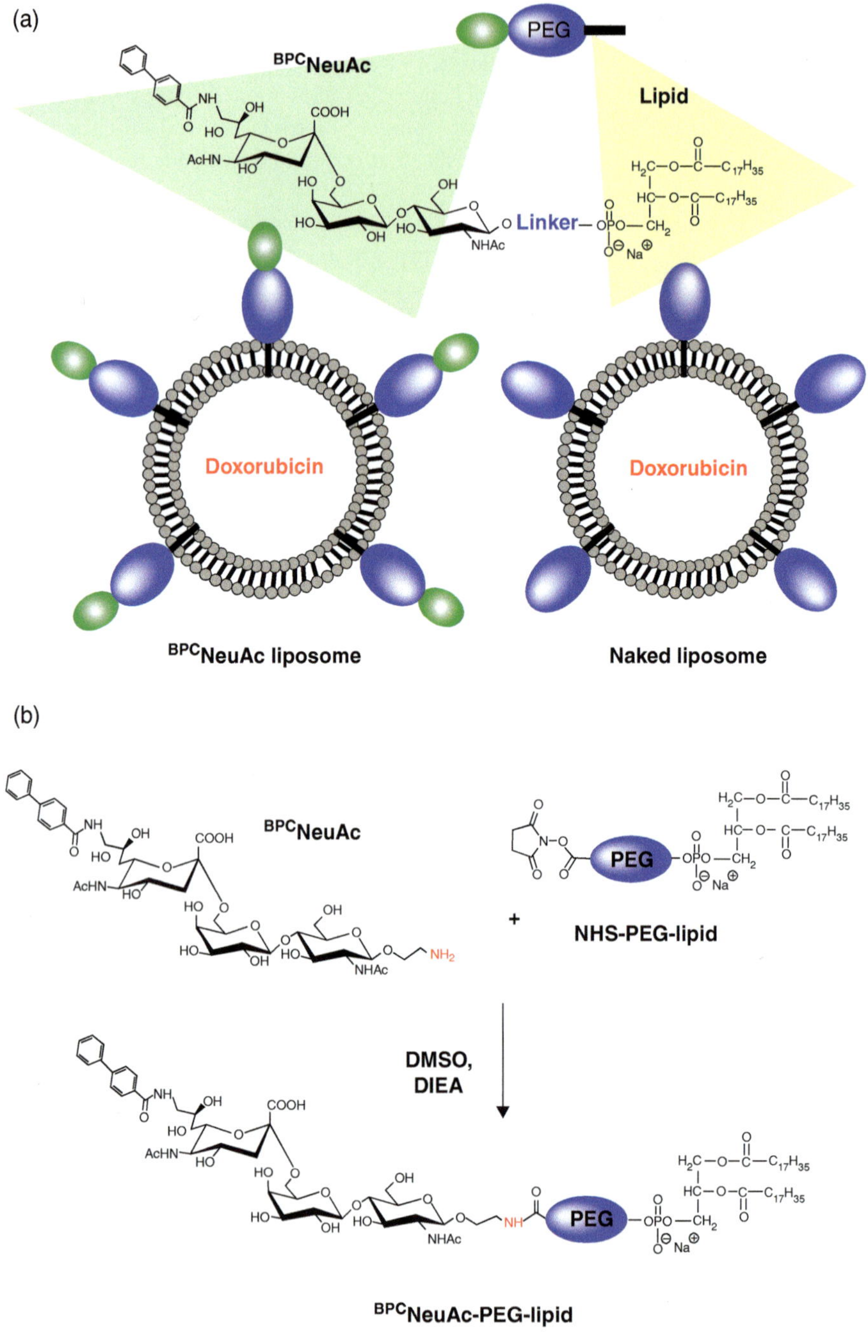

FIGURE 13.7 Liposomal nanoparticles displaying glycan ligands of CD22 for targeting and killing B-cell lymphoma. (a) Schematic illustration of a chemotherapeutic-loaded liposomal formulation comprising BPCNeuAc lipids for active targeting to CD22. (b) Synthesis of BPCNeuAc-PEGylated lipids. *Source*: Reprinted with permission from Ref. 138.

C-type lectin-targeted therapies are found for DC-SIGN and mannose receptors on APCs (vide supra) and, once again, many Siglecs that are involved in a range of disease states [148].

One novel design was developed by Assail *et al.*, with the report of what they refer to as "glyconanosomes" [149]. These are not liposomes and not nanotubes: they are initially combinations of carbon nanotubes wrapped with a hemimicellar structure of neoglycolipids via supramolecular self-assembly around the hydrophobic nanotube. The lipids consist of a PEG-linked diacetylene tails and a polar lactose head. After micelle formation around the nanotube, polymerization of the diacetylene units "zips" up the micelles in a "tube-through-a-ring" assembly (Fig. 13.8). Subsequent removal of the nanotube-threaded/templated "glyconanoring" by sonication yields the glyconanosome, capable of encapsulating hydrophobic drugs such as camptothecin. These nanostructures can bind to β-galactose-binding lectins, hence deliver drug cargo to lectin-bearing tumor cells. These materials were able to encapsulate/solubilize camptothecin, an important topoisomerase I inhibitor and a clinically used anticancer agent. Encapsulation in glyconanosomes increased cellular toxicity relative to camptothecin alone. These novel building blocks should find general use if the synthesis is streamlined for use with other glycolipids. Finally, another ingenious design comes from the Dong group in Shanghai [150]. They prepared liposomes by first preparing two types of dendrimers, one with attached lactose in various generations (0,1,2,...) and another G3 dendrimer coupled to the hydrophobic diazonaphthoquinone (DNQ). These were terminated with alkyne and azide, respectively, and "clicked" together to give a construct with opposing hydrophobic and hydrophilic faces. These spontaneously self-assembled into micelles with a DNQ core and lactose exposed. Loading of these micelles with doxorubicin gave a new nanomedicine that was shown to enter HeLa cells. Near-infrared irradiation of these particles allowed a time-dependent Wolff rearrangement of the DNQ to occur within the micelle, thus destabilizing the assembly and allowing cargo/drug release. Cytotoxicity was shown to be dependent on phototriggering DOX release.

13.5.3.2 Qdots Qdots are extremely useful particles due to their bright imaging capabilities and resistance to photobleaching. The majority of Qdot use in cancer therapy is for guided surgery where the Qdot is targeted to diseased tissue to guide surgical procedures. This is extremely important to micrometastasis removal and lymph node surgery [114]. Reports on the use of glyco-Qdots for cancer therapy are very sparse. The majority of this work is performed not with specific targeted glycans attached to the particle, but where hyaluronic acid or dextran are used as functional group-containing coatings that render the particles almost nontoxic and allow conjugation of various targeting molecules to the polysaccharide coating [151]. The fact that the known HA ligand CD44 is overexpressed in many tumor tissues renders these constructions truly vital to tumor targeting and potential delivery of therapeutic molecules [151].

13.5.3.3 Gold Nanoparticles AuNPs are very versatile platforms owing to their ease of synthesis, range of sizes available in the "nano" scale (1–100 nm), broad capabilities of conjugating all molecular families, control of surface chemistry, and

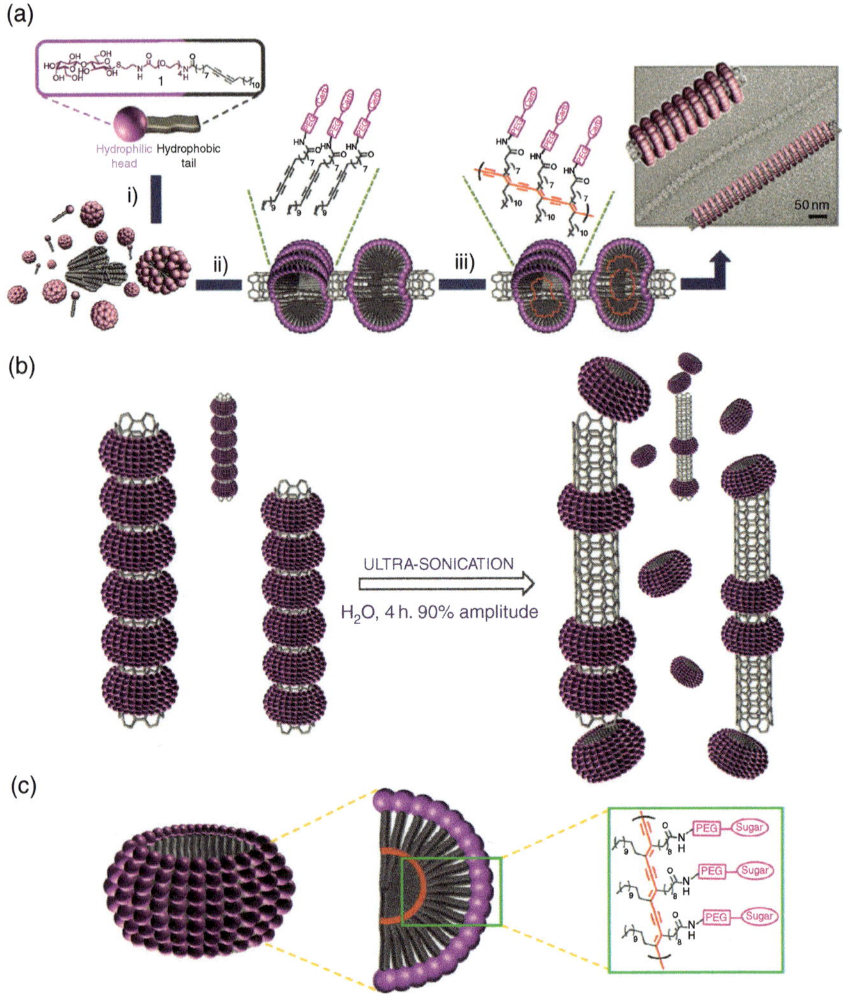

FIGURE 13.8 Procedure for SWCNT/**1** nanoassembly formation and synthesis for glyconanosome (GNS). (a) (i) Mix of neoglycolipid **1** at its critical micelle concentration in water with SWCNTs. (ii) Sonication-promoted supramolecular self-assembly of neoglycolipid **1** in concentric hemimicelles around SWCNTs. (iii) Intermolecular photopolymerization of neoglycolipid **1** hemimicelles into homogeneous glyconanorings (GNRs). Above, the TEM micrograph of SWCNT/**1** nanoassemblies and their idealized representative figures. (b) Schematic representation of the ultrasonication setup for obtaining GNS. (c) Structural and chemical composition of a glyconanosome (GNS). *Source*: Reprinted with permission from Ref. 149. Copyright (2014) American Chemical Society.

access to use of $Au^{(0)}$ in other Au hybrid materials. It is no wonder a simple "Google" search of "gold nanoparticles and cancer" yields nearly half of a million hits. Not very long ago, this field was validated by the progression of an AuNP to phase I clinical trials [152]. Aurimune© [153–156], developed by CytImmune corporation in Gaithersburg, MD., is now in phase II clinical trials and so far is performing well. This is a 12 nm AuNP-bearing TNF-α in a defined copy number with additional PEG units for biocompatibility. This formulation renders the protein's dose-limiting toxicity to dramatically higher concentrations than the free drug. Although not a glycan-coated particle, there is plenty of room for alternative formulations with sugars that may prove clinically viable in the future. Thus, GGNPs have recently found their place in the armamentarium of potential nanomedicines for cancer. The following discussion will be limited to only the latest publications in this arena.

Again, Penãdes and coworkers began in this field in 2000, and since then, research in the field of therapeutic GGNPs has reached a fever pitch in recent years. In 2004, this group published an *in vivo* study that showed that lactose-coated GGNPs were able to inhibit lung metastasis of B16F10 melanoma cells in mice [157]. However, further *in vivo* results from this group have since not been published. Near the same time, our group had synthesized TF antigen-coated GGNPs [158], and we also showed inhibition of metastasis in a 4T1 breast cancer model. However, we did not publish this work since there was difficulty in reproducing the results. We recently prepared GGNPs with the same TF antigen but with the sugar linked to either a threonine or serine amino acid and showed differential cytotoxicity to galectin-3-expressing cells [159]. We also have data that show inhibition of metastasis similar to our original study (to be published). The proper formulation and delivery vehicle seem to be the key in obtaining a therapeutically viable GGNP candidate as a true antiadhesive/metastatic agent that is consistently active *in vivo*.

Recent work on the development of novel gold hybrid constructions has proved fruitful as delivery agents and cytotoxins against tumor cells. Adokoh *et al.* synthesized a glycopolymeric gold nanoparticle that also contains a Au(I)-based antitumor agent, Au(I) triphenylphosphine [160]. This "gold-on-gold" design combines a known but not widely used therapeutic conjugated to pendant functional groups on the outside of the Au(0) particle that were formed by RAFT polymerization. Multivalent β-galactose saccharides are then displayed with adjacent linkers bearing a dithiocarbamate moiety; this group served to attach the polymers to the Au(0) nanoparticles as well as the complex Au(I) triphenylphosphine to the external polymer (Fig. 13.9). These particles were cytotoxic to MCF7 and HepG2 cells *in vitro*. Galactose-functionalized particles were more potent than both cisplatin and cytarabine against HepG2cells due to targeting of the asialoglycoprotein receptor (ASGPR) in the liver by the multivalent galactose groups.

Gold nanoparticles have recently been employed to generate immune responses to carbohydrate antigens by us [161] as well as the Cameron [162] and Penãdes [163] groups. In 2007, Penãdes and coworkers published on a series of hybrid GGNPs with the TACAs sialyl Tn and Lewis^Y, T-cell epitopes and an additional glucose ligand in a variety of densities and combinations; however, no immune responses were reported. The Cameron group reported a very novel design of using polymerized

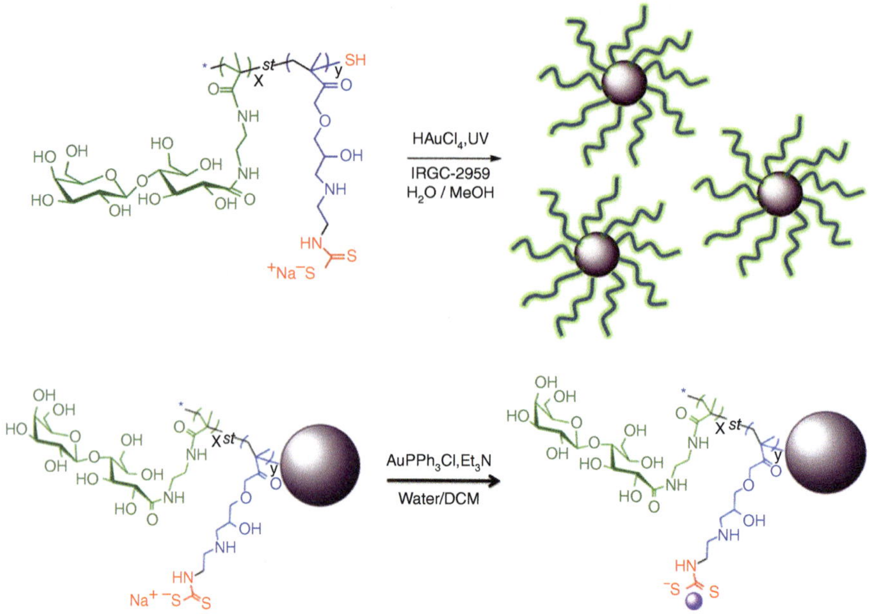

FIGURE 13.9 Synthesis of polymer-functionalized gold nanoparticles and their conjugation to the anticancer drug, Au(1)PPh_3. *Source*: Reprinted with permission from Ref. 159. Copyright (2014) American Chemical Society.

multivalent sugars bearing the Tn antigen that were then attached to the AuNP, thus so-called "multicopy-multivalent" construction. They reported strong long-lasting responses where the antisera reacted with MUC1-overexpressing tumors. Our design employed full mucin extracellular domain repeating peptide units conjugated with the TF disaccharide at various positions. We also generated statistically significant antibody titers with ternary constructions with the inclusion of a "molecular adjuvant" peptide from the d-subunit of complement-derived protein 3 (C3d) [164,165]. We have also shown that vaccination with one particular construct with the TF antigen in a defined position on the MUC4 peptide was able to better protect mice from 4T1 breast tumor growth and metastasis (Unpublished results). The work outlined here strongly suggests that the use of GGNPs in vaccine design against carbohydrate and glycopeptide antigens has a potentially very bright future [166] given the known low toxicity [167] and immunostimulating properties of gold nanoparticles [168–172].

13.5.4 Polysaccharide-Based Nanoparticle Platforms

Carbohydrates provide biocompatibility and lower the toxicity of nanoparticles at a relatively low cost. For the same reasons, long-chain polysaccharides have been used to construct nanoparticle platforms, which also provide the added advantage of being biodegradable. Polysaccharides are formed from linear or branched attachment of multiple monosaccharide units (usually >15 sugars). They vary in their composition, molecular weight, physicochemical properties, and thus biological activities. The presence of multiple functional groups provides chemical handles and allowing for simple modification to alter their properties and create tailored derivatives specific to the desired application [173–175]. A variety of natural polysaccharides have been used to make nanoparticle platforms for diagnostics and therapy. These include but are not limited to chitosan (β-(1-4)-linked D-glucosamine and *N*-acetyl-D-glucosamine), arabinogalactan (highly branched structures of galactose and arabinose), pullulan (homopolysaccharide of α-(1-6)-linked maltotriose residues), HA (repeating disaccharide units of β-(1-4)-linked D-glucuronic acid to D-glucosamine), and dextran (α-(1-6)-linked glucose) [176]. CDs, which are cyclic oligomers of α-(1-4)-linked D-glucose, make another attractive platform for nanoformulations. There are three main members of the CD family that are composed of six-membered rings (α-CD), seven-membered ring (β-CD), and eight-membered rings (γ-CD). CDs have a hydrophobic cavity and a hydrophilic exterior that allow them to form host–guest inclusion complexes with water-insoluble drugs, thus improving their water solubility and bioavailability [177,178]. The polysaccharide-based nanoparticles can be assembled by a variety of mechanisms and have been reviewed elsewhere [176]. The major mechanisms of polysaccharide–nanoparticle assembly include cross-linking with interconnecting molecules to create a three-dimensional network, self-assembly, polysaccharide–drug conjugations, and polyelectrolyte assembly. Nanoparticles made from polysaccharides have demonstrated excellent safety, and many have been FDA approved for various medical and pharmaceutical uses [176]. Like many polymeric nanoparticles, polysaccharide-based nanoparticles are classified as stealth nanoparticles, since they are able to achieve prolonged circulation times while avoiding clearance by the immune system.

13.5.4.1 Chitosan-Based Nanoparticles Chitosan is obtained by alkaline deacetylation of naturally occurring chitin. Chitosan carries several amino groups that are protonated at physiological pH. Thus, chitosan-based nanoparticles selectively adsorb to and neutralize the negatively charged tumor cell surface allowing them to target drugs to the liver, spleen, lung, and colon [179]. However, several modifications have been made to chitosan to change the chemical and/or physical properties of the nanoparticles formed and thus improve the EPR effect. PEGylation appears to be the most common modification to chitosan and is primarily to reduce the adsorption of serum proteins and clearance by the reticuloendothelial system (RES) [180]. The EPR effect alone does not provide successful therapeutic delivery to tumor tissue, and hence, like most nanoparticles, chitosan nanoparticles are also conjugated to a variety of ligands for active targeting such as biotin [181], folic acid [182], peptides [183], and sugars [184,185]. The therapeutic drug or imaging agent is loaded into the nanoparticles either by direct conjugation to the chitosan molecules or by encapsulation. Alternatively, drug conjugates with other negatively charged biopolymers such as HA have been employed to form nanoparticles with chitosan utilizing electrostatic interactions [186].

The folate receptor (FR) is overexpressed among human tumors, which makes it a good candidate for targeted drug delivery. Particles conjugated to folic acid undergo receptor-mediated endocytosis via the FR providing entry for cytotoxic drugs into tumor cells. Hence, FR is extensively used for nanoparticle targeting. Folic acid can be grafted onto the chitosan backbone by direct amide bond formation with the chitosan amino group via its γ-carboxylic acid using a DCC- or NHS-mediated coupling reaction [187,188]. Small linkers like 2,2′-(ethylenedioxy)bis(ethylamine) (EDBE) have also been employed to conjugate folic acid to chitosan or its derivatives [189]. Chitosan nanoparticles have also been designed to target specific organs such as the liver by modifying them with targeting ligands like galactose. Hepatoma cells overexpress the ASGPR on their surfaces that binds selectively to galactose [190]. Zhong and coworkers synthesized galactosylated nanoparticles from lactobionic acid-functionalized chitosan. They demonstrated the encapsulation of oridonin, a traditional Chinese drug with anticancer properties, and increased residence time of drug-encapsulated nanoparticles leading to enhanced accumulation in the liver of rats compared to the naked drug. They observed a two-stage exponential release of the encapsulated drug. The first fast release was from the nonencapsulated drug molecules and drugs located at the surface of the particles, which was followed by a slow release of the encapsulated drug [185]. A variety of chemotherapy drugs have been tested for delivery into tumor tissue using chitosan-based nanoparticles, such as paclitaxel, methotrexate, 5-fluorouracil, camptothecin, 10-hydroxycamptothecin, and doxorubicin [182]. FR-targeting nanoparticles are suggested to improve the efficiency of carrier internalization while bypassing the efflux pumps that provide cancer cells with resistance to multiple drugs [191].

Chitosan has found wide applicability in the formulation of nanocarriers for the delivery of siRNA and DNA molecules. The negatively charged nucleic acids efficiently form ionic complexes with chitosan due to its positively charged nature, and this interaction has been vastly exploited to form nanoparticles [192]. The molecular weight of chitosan used in the formulations has a critical impact on gene silencing using siRNA. The molecular weight range of 64.8–170 kDa has been recommended as optimal size [193]. Low molecular weight chitosan molecules allow the formation of smaller-sized particles but do not offer sufficient protection against nucleases and form unstable complexes with the nucleic acids [192]. If the chitosan polymers are too long, they result in large (>300 nm) particles. Also, chitosan facilitates mucosal delivery of siRNA due to its superior mucoadhesive properties and permeation-enhancing capability [194]. Transfection efficiency of siRNA is improved with the increasing degree of deacetylation in chitosan. The reason for this effect is the higher positive charge density that enables better complexation of topologically stiff nucleic acids [195]. The presence of higher number of amine groups on the nanoparticle has also been hypothesized to allow particles to escape from endosomes for the delivery of cargo into the cytoplasm. This step is critical for the survival of the cargo, since endosomes usually fuse into highly acidic lysosomes where the integrity of the nanoparticle is compromised under harsh conditions [196]. Though it is argued that chitosan alone is not sufficient to cause endosomal rupture, it has been proposed that the amine groups on chitosan act as a proton sponge during the acidification of endosomes leading to an influx of chloride ions and rupture under osmotic pressure [196,197]. Chitosan with all these advantages is still not perfect as a delivery vehicle and has its limitations, particularly low water solubility at physiological pH and relatively low transfection efficiency. Toward the goal of improving these imperfections, chitosan has been modified with a variety of molecules that tailor its properties for nucleic acid delivery. These modifications include conjugation of amine-rich polymers such as polyethylene amine for better endosomal escape and higher transfection efficiency, targeting molecules like cell-penetrating peptides and CDs and PEG for improving viability and solubility [192]. In a recent study, mannosylated chitosan (MCS) was used for the delivery of a preventative DNA vaccine against prostate carcinoma via the nasal mucosal route in mice. The MCS nanoformulated vaccine showed significantly higher titers of anti-gastrin-releasing peptide (anti-GRP) IgG that lasted for 11 weeks relative to the peptide administered alone or as a nanoformulation of unmannosylated chitosan [185].

Photodynamic therapy (PDT) is a medical treatment in which cytotoxic singlet oxygen species are generated by chemical photosensitizers upon irradiation. Since the process is light triggered, it can be spatially controlled, and only those drug molecules in the tumor tissue that are exposed to light are activated [198]. Several reports have shown the efficacy of chitosan nanoparticles in targeting photosensitizers to tumor tissue over inorganic and polymeric nanoparticles [199–201]. Moreover, polysaccharides like chitosan provide a biocompatible, solubilizing platform for the hydrophobic photosensitizers. Choi and coworkers developed a chitosan-based "on/off" system, which allows specific recovery of the photosensitizer's activity in tumor tissue [202]. They conjugated the photosensitizer protoporphyrin IX (PpIX) directly to glycol-modified chitosan, which self-assembled to form stable nanoparticles in aqueous medium due to the hydrophobic effect and π–π stacking of the phenanthrene rings of about 280 nm in size. The dense cores represented the "off" state via self-quenching with no observed fluorescence or singlet oxygen generation. Upon cellular uptake, the integrity of the particles decreased and resulted in the turning "on" of the fluorescence and therapeutic activity of particles. These particles also showed prolonged blood circulation in tumor-bearing mice, thus improving the *in vivo* therapeutic efficiency through enhanced tumor accumulation.

Incorporation of photosensitizer molecules in nanoparticles also allows the modulation of its photophysical properties by coencapsulating functional molecules. These molecules impose intraparticle intermolecular interaction on the photosensitizer and thus improve overall singlet oxygen generation [203,204]. Kim and coworkers synthesized an iodinated photosensitizing glycol chitosan nanoparticle to enhance the singlet oxygen generating efficiency of chlorin e6 (Ce6). The incorporation of iodine leads to the intraparticle heavy atom effect that facilitates intersystem crossing of the photoexcited Ce6 molecule from the singlet state to the long-lived triplet state where it sensitizes the surrounding molecular oxygen to generate $^1O_2$ species.

13.5.4.2 Other Polysaccharide-Based Nanoparticle Platforms HA, dextran, alginates, pullulan, and CDs are other widely employed polysaccharide platforms for nanoformulation of chemotherapeutic drugs. HA is a linear high molecular weight GAG composed of disaccharide repeating units of D-glucuronic acid β-(1-4)-linked to *N*-acetyl-D-glucosamine. It is a hydrophilic biopolymer with remarkable ability to retain water apart from being biodegradable and biocompatible that makes it a good candidate for drug delivery platforms. It is known to bind to CD44, a receptor overexpressed in several malignant tumor cells such as ovarian, colon, stomach, glioma, leukemia, and others [205]. Thus, nanoparticles made from HA allow direct targeting of these tumors and deliver their payload into the cell via receptor-mediated endocytosis. However, a major drawback of HA-based nanoparticles is that they preferentially accumulate in the liver after systemic administration. Young and coworkers showed that PEGylation of HA leads to the formation of self-assembled nanoparticles that showed reduced uptake by the liver and extended circulation time in the blood leading to enhanced accumulation in tumor tissue [206]. These nanoparticles had reduced cellular uptake *in vitro*; however, cancer cells overexpressing CD44 showed extensive uptake of the particles *in vivo*.

Dextrans are high molecular weight polysaccharides composed of α-(1-6)-linked glucose. Nanoparticles formed by conjugation of dextran to poorly water-soluble drugs were shown to be stable in the pH range of 4 and 11 with high loading efficiency [207]. Park and coworkers have reported hypoxia-responsive nanoparticles based on carboxymethyl-dextran. Hypoxia is a condition where the tissue is deprived of oxygen. Hypoxic conditions are often present at the interior of tumor tissue. The authors conjugated 2-nitroimidazole, which is transformed into hydrophilic 2-aminoimidazoles via a series of selective bioreductions [208]. This transformation allowed the release of encapsulated doxorubicin in the tumor tissue. These particles showed higher toxicity to hypoxic cells than normoxic cells and selectively accumulated in the hypoxic tumor tissue. There are several more examples of these systems, but they are outside the scope of this chapter. The reader is referred to several excellent reviews on these platforms for additional information [176,209].

13.6 CONCLUSIONS AND OUTLOOK

The importance of sugars in cancer biology is unmistakable, and more and more researchers are integrating some form of "glycoscience" into their programs. As is evident from the previous information, the combination of nanotechnology, glycoscience, and anticancer therapeutic design has become inextricably linked and, from the results reported to date, is poised to bring an entire set of novel therapeutic agents to the clinic in the next decade. As regulatory agencies speed up the process of toxicological evaluation of nanosystems for human, there should be an oncoming set of new materials where the carbohydrates/glycans play a central role in not only tumor targeting but also therapeutic efficacy. Future treatises on this topic will hopefully be able to report on clinical success of these unique entities.

REFERENCES

1. Tong, R.; Tang, L.; Cheng, J. Nanostructure Science and Technology, "*Multifunctional Nanoparticles for Drug Delivery Applications: Imaging, Targeting and Delivery*", Ed. by Svenson, S and Prudhomme, RK., Springer, New York **2012**, Chapter *3*, 31–46.
2. Sengupta, S.; Kulkarni, A. *ACS Nano* **2013**, *7*, 2878–2882.
3. Sandhiya, S.; Dkhar, S. A.; Surendiran, A. *Fundamental and Clinical Pharmacology* **2009**, *23*, 263–269.

4. Jain, R. K.; Stylianopoulos, T. *Nature Reviews. Clinical Oncology* **2010**, *7*, 653–664.
5. Moyret-Lalle, C.; Falette, N.; Grelier, G.; Puisieux, A. *Bull. Cancer* **2008**, *95*, 923–930.
6. Shackleton, M.; Quintana, E.; Fearon, E. R.; Morrison, S. J. *Cell* **2009**, *138*, 822–829.
7. Dick, J. E. *Nature Biotechnology* **2009**, *27*, 44–46.
8. Furth, J.; Kahn, M. C.; Breedis, C. *The American Journal of Cancer* **1937**, *31*, 276–282.
9. Bonnet, D.; Dick, J. E. *Nature Medicine* **1997**, *3*, 730–737.
10. Al-Hajj, M.; Wicha, M. S.; Benito-Hernandez, A.; Morrison, S. J.; Clarke, M. F. *Proceedings of the National Academy of Sciences of the United States of America* **2003**, *100*, 3983–3988.
11. Tang, C.; Ang, B. T.; Pervaiz, S. *FASEB Journal* **2007**, *21*, 3777–3785.
12. Dean, M.; Fojo, T.; Bates, S. *Nature Reviews Cancer* **2005**, *5*, 275–284.
13. Oravecz-Wilson, K. I.; Philips, S. T.; Yilmaz, O. H.; Ames, H. M.; Li, L.; Crawford, B. D.; Gauvin, A. M.; Lucas, P. C.; Sitwala, K.; Downing, J. R.; Morrison, S. J.; Ross, T. S. *Cancer Cell* **2009**, *16*, 137–148.
14. Hirschmann-Jax, C.; Foster, A. E.; Wulf, G. G.; Nuchtern, J. G.; Jax, T. W.; Gobel, U.; Goodell, M. A.; Brenner, M. K. *Proceedings of the National Academy of Sciences of the United States of America* **2004**, *101*, 14228–14233.
15. Bao, S.; Wu, Q.; McLendon, R. E.; Hao, Y.; Shi, Q.; Hjelmeland, A. B.; Dewhirst, M. W.; Bigner, D. D.; Rich, J. N. *Nature* **2006**, *444*, 756–760.
16. Zhou, Y.; Kopecek, J. *Journal of Drug Targeting* **2013**, *21*, 1–26.
17. Danhier, F.; Feron, O.; Preat, V. *Journal of Controlled Release: Official Journal of the Controlled Release Society* **2010**, *148*, 135–146.
18. Friedl, P.; Alexander, S. *Cell* **2011**, *147*, 992–1009.
19. Hanahan, D.; Coussens, L. M. *Cancer Cell* **2012**, *21*, 309–322.
20. Ricci-Vitiani, L.; Pallini, R.; Biffoni, M.; Todaro, M.; Invernici, G.; Cenci, T.; Maira, G.; Parati, E. A.; Stassi, G.; Larocca, L. M.; De Maria, R. *Nature* **2010**, *468*, 824–U121.
21. Wang, Y. D.; Oliver, G. *Genes & Development* **2010**, *24*, 2115–2126.
22. Carmeliet, P. *Nature Medicine* **2000**, *6*, 389–395.
23. Roberts, W. G.; Palade, G. E. *Cancer Research* **1997**, *57*, 765–772.
24. Hashizume, H.; Baluk, P.; Morikawa, S.; McLean, J. W.; Thurston, G.; Roberge, S.; Jain, R. K.; McDonald, D. M. *American Journal of Pathology* **2000**, *156*, 1363–1380.
25. Jain, R. K.; Munn, L. L.; Fukumura, D. *Nature Reviews Cancer* **2002**, *2*, 266–276.
26. Jain, R. K. *Cancer and Metastasis Reviews* **1987**, *6*, 559–593.
27. Maeda, H.; Sawa, T.; Konno, T. *Journal of Controlled Release* **2001**, *74*, 47–61.
28. Matsumura, Y.; Maeda, H. *Cancer Research* **1986**, *46*, 6387–6392.
29. Bouzin, C.; Feron, O. *Drug Resistance Updates* **2007**, *10*, 109–120.
30. Leunig, M.; Yuan, F.; Menger, M. D.; Boucher, Y.; Goetz, A. E.; Messmer, K.; Jain, R. K. *Cancer Research* **1992**, *52*, 6553–6560.
31. Yuan, F.; Salehi, H. A.; Boucher, Y.; Vasthare, U. S.; Tuma, R. F.; Jain, R. K. *Cancer Research* **1994**, *54*, 4564–4568.
32. Padera, T. P.; Stoll, B. R.; Tooredman, J. B.; Capen, D.; di Tomaso, E.; Jain, R. K. *Nature* **2004**, *427*, 695.
33. Taton, T. A.; Mirkin, C. A.; Letsinger, R. L. *Science* **2000**, *289*, 1757–1760.
34. Patra, H. K.; Turner, A. P. F. *Trends in Biotechnology* **2014**, *32*, 21–31.

35. Dorothy, F.; George, H.; Piotr, G. In *Nanotoxicology*; CRC Press, Boca Raton: **2014**, p 3–20.
36. Davis, M. E.; Zuckerman, J. E.; Choi, C.–H. J.; Seligson, D.; Tolcher, A.; Alabi, C. A.; Yen, Y.; Heidel, J. D.; Ribas, A. *Nature* **2010**, *464*, 1067–1070.
37. Tabernero, J.; Shapiro, G. I.; LoRusso, P. M.; Cervantes, A.; Schwartz, G. K.; Weiss, G. J.; Paz-Ares, L.; Cho, D. C.; Infante, J. R.; Alsina, M.; Gounder, M. M.; Falzone, R.; Harrop, J.; White, A. C. S.; Toudjarska, I.; Bumcrot, D.; Meyers, R. E.; Hinkle, G.; Svrzikapa, N.; Hutabarat, R. M.; Clausen, V. A.; Cehelsky, J.; Nochur, S. V.; Gamba-Vitalo, C.; Vaishnaw, A. K.; Sah, D. W. Y.; Gollob, J. A.; Burris, H. A., III, *Cancer Discovery* **2013**, *3*, 406–417.
38. Prados, J.; Melguizo, C.; Ortiz, R.; Perazzoli, G.; Cabeza, L.; Alvarez, P. J.; Rodriguez-Serrano, F.; Aranega, A. *Anti-Cancer Agent Medicinal Chemistry* **2013**, *13*, 1204–1216.
39. Varki, A. *Glycobiology* **1993**, *3*, 97–130.
40. Roth, J. *Chemical Reviews* **2002**, *102*, 285–303.
41. Ohtsubo, K.; Marth, J. D. *Cell* **2006**, *126*, 855–867.
42. Hart, G. W.; Copeland, R. J. *Cell* **2010**, *143*, 672–676.
43. Leffler, H.; Masiarz, F. R.; Barondes, S. H. *Biochemistry-USA* **1989**, *28*, 9222–9229.
44. Dimick, S. M.; Powell, S. C.; McMahon, S. A.; Moothoo, D. N.; Naismith, J. H.; Toone, E. J. *Journal of the American Chemical Society* **1999**, *121*, 10286–10296.
45. Mammen, M.; Choi, S. K.; Whitesides, G. M. *Angewandte Chemie International Edition* **1998**, *37*, 2755–2794.
46. Lundquist, J. J.; Toone, E. J. *Chemical Reviews* **2002**, *102*, 555–578.
47. Leffler, H. *Trends in Glycoscience Glycotechnology* **1997**, *9*, 9–+.
48. Perillo, N. L.; Marcus, M. E.; Baum, L. G. *Journal of Molecular Medicine* **1998**, *76*, 402–412.
49. Danguy, A.; Camby, I.; Kiss, R. *BBA: General Subjects* **2002**, *1572*, 285–293.
50. Hasan, S. S.; Ashraf, G. M.; Banu, N. *Cancer Letters* **2007**, *253*, 25–33.
51. Li, M.; Song, L. J.; Qin, X. Y. *Journal of Biosciences* **2010**, *35*, 665–673.
52. Brockhausen, I. *EMBO Reports* **2006**, *7*, 599–604.
53. Dube, D. H.; Bertozzi, C. R. *Nature Reviews Drug Discovery* **2005**, *4*, 477–488.
54. Fuster, M. M.; Esko, J. D. *Nature Reviews Cancer* **2005**, *5*, 526–542.
55. Cazet, A.; Julien, S.; Bobowski, M.; Krzewinski-Recchi, M.-A.; Harduin-Lepers, A.; Groux-Degroote, S.; Delannoy, P. *Carbohydrate Research* **2010**, *345*, 1377–1383.
56. Varki, A.; Kannagi, R.; Toole, B. P.; Cold Spring Harbor Laboratory Press: New York, **2009**, p 617–632.
57. Kobata, A.; Amano, J. *Immunology and Cell Biology* **2005**, *83*, 429–439.
58. Takenaka, Y.; Fukumori, T.; Raz, A. Proceedings of the 2nd International Conference on Tumor Microenvironment Progression, Therapy & Prevention, Baden, Austria, **2002**, pp. 67–82.
59. Pierce, J. M. *Handbook of Glycomics* Academic Press: Amsterdam **2009**, 399–429.
60. Brockhausen, I. *Biochimica et Biophysica Acta (BBA): General Subjects* **1999**, *1473*, 67–95.
61. Heimburg-Molinaro, J.; Lum, M.; Vijay, G.; Jain, M. T.; Almogren, A.; Rittenhouse-Olson, K. *Vaccine* **2011**, *29*, 8802–8826.

62. De la Fuente, J. M.; Penades, S. *BBA: General Subjects* **2006**, *1760*, 636–651.
63. Marradi, M.; Garcia, I.; Penades, S. *Progress in Molecular Biology and Translational Science* **2011**, *104*, 141–173.
64. El-Boubbou, K.; Huang, X. F. *Current Medicinal Chemistry* **2011**, *18*, 2060–2078.
65. Roy, R.; Shiao, T. C.; Rittenhouse-Olson, K. *Brazilian Journal of Pharmaceutical Sciences* **2013**, *49*, 85–108.
66. Marradi, M.; Chiodo, F.; Garcia, I.; Penades, S. *Chemical Society Reviews* **2013**, *42*, 4728–4745.
67. Patra, C. R.; Bhattacharya, R.; Mukherjee, P. *Journal of Materials Chemistry* **2010**, *20*, 547–554.
68. Allegra, A.; Penna, G.; Alonci, A.; Rizzo, V.; Russo, S.; Musolino, C. *Anti-Cancer Agents in Medicinal Chemistry* **2011**, *11*, 669–686.
69. Al-Jamal, W. T.; Kostarelos, K. *Accounts of Chemical Research* **2011**, *44*, 1094–1104.
70. Kievit, F. M.; Zhang, M. Q. *Accounts of Chemical Research* **2011**, *44*, 853–862.
71. Melancon, M. P.; Stafford, R. J.; Li, C. *Journal of Controlled Release* **2012**, *164*, 177–182.
72. Babu, A.; Templeton, A. K.; Munshi, A.; Ramesh, R. *Journal of Nanomaterials* **2013**, 2013, 11 pp.
73. Chen, H. M.; Zhen, Z. P.; Todd, T.; Chu, P. K.; Xie, J. *Materials Science and Engineering: Reports* **2013**, *74*, 35–69.
74. Bogart, L. K.; Pourroy, G.; Murphy, C. J.; Puntes, V.; Pellegrino, T.; Rosenblum, D.; Peer, D.; Levy, R. *ACS Nano* **2014**, *8*, 3107–3122.
75. Barchi, J. J., Jr In *ACS Symposium Series*; Huang, X., Barchi, J. J. Jr., Ed.; Oxford University Press: Washington DC, **2011**; Vol. *1091*, p 213.
76. Roy, R. *Trends in Glycoscience and Glycotechnology* **2003**, *15*, 291–310.
77. Wijagkanalan, W.; Kawakami, S.; Hashida, M. *Frontiers in Bioscience: Landmark* **2011**, *16*, 2970–2987.
78. Benvegnu, T. L., L.; Ballet, C.; Portier, Y.; Plusquellec, D. *Carbohydrate Chemistry* **2014**, *40*, 341–377.
79. Jain, K.; Kesharwani, P.; Gupta, U.; Jain, N. K. *Biomaterials* **2012**, *33*, 4166–4186.
80. Sarkar, S.; Salyer, A. C. D.; Wall, K. A.; Sucheck, S. J. *Bioconjugate Chemistry* **2013**, *24*, 363–375.
81. Unger, W. W. J.; van Beelen, A. J.; Bruijns, S. C.; Joshi, M.; Fehres, C. M.; van Bloois, L.; Verstege, M. I.; Ambrosini, M.; Kalay, H.; Nazmi, K.; Bolscher, J. G.; Hooijberg, E.; de Gruijl, T. D.; Storm, G.; van Kooyk, Y. *Journal of Controlled Release* **2012**, *160*, 88–95.
82. Copland, M. J.; Baird, M. A.; Rades, T.; McKenzie, J. L.; Becker, B.; Reck, F.; Tyler, P. C.; Davies, N. M. *Vaccine* **2003**, *21*, 883–890.
83. White, K. L.; Rades, T.; Furneaux, R. H.; Tyler, P. C.; Hook, S. J. *Pharmacy and Pharmacology* **2006**, *58*, 729–737.
84. Espuelas, S.; Thumann, C.; Heurtault, B.; Schuber, F.; Frisch, B. *Bioconjugate Chemistry* **2008**, *19*, 2385–2393.
85. Irache, J. M.; Salman, H. H.; Gamazo, C.; Espuelas, S. *Expert Opinion on Drug Delivery* **2008**, *5*, 703–724.
86. Lepenies, B.; Lee, J.; Sonkaria, S. *Advanced Drug Delivery Reviews* **2013**, *65*, 1271–1281.

87. Thomann, J.-S.; Heurtault, B.; Weidner, S.; Braye, M.; Beyrath, J.; Fournel, S.; Schuber, F.; Frisch, B. *Biomaterials* **2011**, *32*, 4574–4583.
88. Kramps, T.; Probst, J. *Wires RNA* **2013**, *4*, 737–749.
89. Xu, Z.; Wang, Y.; Zhang, L.; Huang, L. *ACS Nano* **2014**, *8*, 3636–3645.
90. Chono, S.; Li, S.-D.; Conwell, C. C.; Huang, L. *Journal of Controlled Release* **2008**, *131*, 64–69.
91. He, W. J.; Hosseinkhani, H.; Hong, P. D.; Chiang, C. H.; Yu, D. S. *International Journal of Nanotechnology* **2013**, *10*, 930–944.
92. Rosen, J. E.; Chan, L.; Shieh, D.-B.; Gu, F. X. *Nanomedicine: Nanotechnology, Biology and Medicine* **2012**, *8*, 275–290.
93. El-Boubbou, K.; Zhu, D. C.; Vasileiou, C.; Borhan, B.; Prosperi, D.; Li, W.; Huang, X. F. *Journal of the American Chemical Society* **2010**, *132*, 4490–4499.
94. Kamat, M.; El-Boubbou, K.; Zhu, D. C.; Lansdell, T.; Lu, X. W.; Li, W.; Huang, X. F. *Bioconjugate Chemistry* **2010**, *21*, 2128–2135.
95. Rudzki, Z.; Jothy, S. *Journal of Clinical Pathology–Molecular Pathology* **1997**, *50*, 57–71.
96. Jothy, S. *Clinical and Experimental Metastasis* **2003**, *20*, 195–201.
97. El-Dakdouki, M. H.; Zhu, D. C.; El-Boubbou, K.; Kamat, M.; Chen, J.; Li, W.; Huang, X. *Biomacromolecules* **2012**, *13*, 1144–1151.
98. El-Dakdouki, M. H.; Xia, J.; Zhu, D. C.; Kavunja, H.; Grieshaber, J.; O'Reilly, S.; McCormick, J. J.; Huang, X. *ACS Applied Materials & Interfaces* **2013**, *6*, 697–705.
99. Marradi, M.; Alcantara, D.; de la Fuente, J. M.; Garcia-Martin, M. L.; Cerdan, S.; Penades, S. Chemical Communications **2009**, 3922–3924.
100. Irure, A.; Marradi, M.; Arnaiz, B.; Genicio, N.; Padro, D.; Penades, S. *Biomaterials Science: UK* **2013**, *1*, 658–668.
101. Candiota, A. P.; Acosta, M.; Simoes, R. V.; Delgado-Goni, T.; Lope-Piedrafita, S.; Irure, A.; Marradi, M.; Bomati-Miguel, O.; Miguel-Sancho, N.; Abasolo, I.; Schwartz, S.; Santamaria, J.; Penades, S.; Arus, C. *Journal of Nanobiotechnology* **2014**, *12*, 12.
102. Gallo, J.; Genicio, N.; Penades, S. *Advanced Healthcare Materials* **2012**, *1*, 302–307.
103. Lee, D. E.; Kim, A. Y.; Saravanakumar, G.; Koo, H.; Kwon, I. C.; Choi, K.; Park, J. H.; Kim, K. *Macromolecular Research* **2011**, *19*, 861–867.
104. Zhang, F.; Yi, D.; Sun, H.; Zhang, H. *Journal of Nanoscience and Nanotechnology* **2014**, *14*, 1409–1424.
105. Weingart, J. J.; Sun, X. L. *Petite and Sweet: Glyco-Nanotechnology as a Bridge to New Medicines* **2011**, Oxford University Press, New York, vol. *1091*, pp. 105–121.
106. Shao, L. J.; Gao, Y. F.; Yan, F. *Sensors (Basel)* **2011**, *11*, 11736–11751.
107. Hutter, E.; Maysinger, D. *Microscopy Research and Technique* **2011**, *74*, 592–604.
108. Jin, S.; Hu, Y. X.; Gu, Z. J.; Liu, L.; Wu, H. C. Journal of Nanomaterials **2011**.
109. Huang, H.-C.; Barua, S.; Sharma, G.; Dey, S. K.; Rege, K. J. *Controlled Release* **2011**, *155*, 344–357.
110. Tomczak, N.; Janczewski, D.; Dorokhin, D.; Han, M. Y.; Vancso, G. J. *Nanotechnology in Regenerative Medicine: Methods and Protocols* **2012**, *811*, 245–265.
111. Zhu, Y.; Hong, H.; Xu, Z. P.; Li, Z.; Cai, W. *Current Molecular Medicine* **2013**, *13*, 1549–1567.

112. Probst, C. E.; Zrazhevskiy, P.; Bagalkot, V.; Gao, X. H. *Advanced Drug Delivery Reviews* **2013**, *65*, 703–718.

113. Chinnathambi, S.; Chen, S.; Ganesan, S.; Hanagata, N. *Advanced Healthcare Materials* **2014**, *3*, 10–29.

114. Luo, G. P.; Long, J.; Zhang, B.; Liu, C.; Ji, S. R.; Xu, J.; Yu, X. J.; Ni, Q. X. *Expert Opinion on Drug Delivery* **2012**, *9*, 47–58.

115. de la Fuente, J. D. M.; Penades, S. *Tetrahedron: Asymmetry* **2005**, *16*, 387–391.

116. Cambi, A.; Lidke, D. S.; Arndt-Jovin, D. J.; Figdor, C. G.; Jovin, T. M. *Nano Letters* **2007**, *7*, 970–977.

117. Svarovsky, S. A.; Barchi, J. J. *ACS Symposium Series* **2007**, *960*, 375–392.

118. Bhang, S. H.; Won, N.; Lee, T.-J.; Jin, H.; Nam, J.; Park, J.; Chung, H.; Park, H.-S.; Sung, Y.-E.; Hahn, S. K.; Kim, B.-S.; Kim, S. *ACS Nano* **2009**, *3*, 1389–1398.

119. Li, N.; Chow, A. M.; Ganesh, H. V. S.; Brown, I. R.; Kerman, K. *Analytical Chemistry* **2013**, *85*, 9699–9704.

120. Bavireddi, H.; Kikkeri, R. *Analyst* **2012**, *137*, 5123–5127.

121. Gallo, J.; Garcia, I.; Genicio, N.; Penades, S. *Particle and Particle Systems Characterization* **2014**, *31*, 126–133.

122. Chen, X.; Lee, G. S.; Zettl, A.; Bertozzi, C. R. *Angewandte Chemie International Edition* **2004**, *43*, 6111–6116.

123. Chen, X.; Tam, U. C.; Czlapinski, J. L.; Lee, G. S.; Rabuka, D.; Zettl, A.; Bertozzi, C. R. *Journal of the American Chemical Society* **2006**, *128*, 6292–6293.

124. Ahmed, M.; Jiang, X. Z.; Deng, Z. C.; Narain, R. *Bioconjugate Chemistry* **2009**, *20*, 2017–2022.

125. Bertozzi, C. R.; Bednarski, M. D. *Journal of the American Chemical Society* **1992**, *114*, 2242–2245.

126. de la Fuente, J. M.; Barrientos, A. G.; Rojas, T. C.; Rojo, J.; Canada, J.; Fernandez, A.; Penades, S. *Angewandte Chemie International Edition* **2001**, *40*, 2258–2261.

127. Rojas, T. C.; de la Fuente, J. M.; Barrientos, A. G.; Penades, S.; Ponsonnet, L.; Fernandez, A. *Advanced Materials* **2002**, *14*, 585–588.

128. Lee, H.; Lee, K.; Kim, I. K.; Park, T. G. *Biomaterials* **2008**, *29*, 4709–4718.

129. Kim, K. S.; Park, S. J.; Lee, M. Y.; Lim, K. G.; Hahn, S. K. *Macromolecular Research* **2012**, *20*, 277–282.

130. Martinez, A.; Ortiz Mellet, C.; Garcia Fernandez, J. M. *Chemical Society Reviews* **2013**, *42*, 4746–4773.

131. Aykac, A.; Martos-Maldonado, M. C.; Casas-Solvas, J. M.; Quesada-Soriano, I.; Garcia-Maroto, F.; Garcia-Fuentes, L.; Vargas-Berenguel, A. *Langmuir* **2014**, *30*, 234–242.

132. Mattheolabakis, G.; Rigas, B.; Constantinides, P. P. *Nanomedicine (UK)* **2012**, *7*, 1577–1590.

133. Goncalves, A. S.; Macedo, A. S.; Souto, E. B. *Clinical and Translational Oncology* **2012**, *14*, 883–890.

134. Ranganathan, R.; Madanmohan, S.; Kesavan, A.; Baskar, G.; Krishnamoorthy, Y. R.; Santosham, R.; Ponraju, D.; Rayala, S. K.; Venkatraman, G. *International Journal of Nanomedicine* **2012**, *7*, 1043–1060.

135. Matsumura, Y. *Japanese Journal of Clinical Oncology* **2014**, *44*, 515–525.

136. Kozako, T.; Arima, N.; Yoshimitsu, M.; Honda, S. I.; Soeda, S. *International Journal of Nanomedicine* **2012**, *7*, 4943–4951.

137. Collins, B. E.; Blixt, O.; Han, S. F.; Duong, B.; Li, H. Y.; Nathan, J. K.; Bovin, N.; Paulson, J. C. *Journal of Immunology* **2006**, *177*, 2994–3003.

138. Chen, W. C.; Completo, G. C.; Sigal, D. S.; Crocker, P. R.; Saven, A.; Paulson, J. C. *Blood* **2010**, *115*, 4778–4786.

139. Chen, W. H. C.; Sigal, D. S.; Saven, A.; Paulson, J. C. *Leukemia & Lymphoma* **2012**, *53*, 208–210.

140. Kawasaki, N.; Nycholat, C.; Rademacher, C.; Chen, W. H.; Vela, J.; Maricic, I.; Crocker, P.; Kumar, V.; Kronenberg, M.; Paulson, J. *Glycobiology* **2011**, *21*, 1457–1457.

141. Chen, W. H. C.; Kawasaki, N.; Nycholat, C. M.; Han, S. F.; Pilotte, J.; Crocker, P. R.; Paulson, J. C. PLoS One **2012**, 7.

142. Kawasaki, N.; Vela, J. L.; Nycholat, C. M.; Rademacher, C.; Khurana, A.; van Rooijen, N.; Crocker, P. R.; Kronenberg, M.; Paulson, J. C. *Proceedings of the National Academy of Sciences of the United States of America* **2013**, *110*, 7826–7831.

143. Stahn, R.; Grittner, C.; Zeisig, R.; Karsten, U.; Felix, S. B.; Wenzel, K. *Cellular and Molecular Life Sciences* **2001**, *58*, 141–147.

144. Zeisig, R.; Stahn, R.; Wenzel, K.; Behrens, D.; Fichtner, I. *BBA: Biomembranes* **2004**, *1660*, 31–40.

145. Mitchell, M. J.; Chen, C. S.; Ponmudi, V.; Hughes, A. D.; King, M. R. *Journal of Controlled Release* **2012**, *160*, 609–617.

146. Jubeli, E.; Moine, L.; Vergnaud-Gauduchon, J.; Barratt, G. *Journal of Controlled Release* **2012**, *158*, 194–206.

147. Hirai, M.; Minematsu, H.; Hiramatsu, Y.; Kitagawa, H.; Otani, T.; Iwashita, S.; Kudoh, T.; Chen, L.; Li, Y.; Okada, M.; Salomon, D. S.; Igarashi, K.; Chikuma, M.; Seno, M. *International Journal of Pharmaceutics* **2010**, *391*, 274–283.

148. O'reilly, M. K.; Paulson, J. C. *Trends in Pharmacological Sciences* **2009**, *30*, 240–248.

149. Assali, M.; Cid, J. J.; Pernia-Leal, M.; Munoz-Bravo, M.; Fernandez, I.; Wellinger, R. E.; Khiar, N. *ACS Nano* **2013**, *7*, 2145–2153.

150. Sun, L.; Ma, X.; Dong, C.-M.; Zhu, B.; Zhu, X. *Biomacromolecules* **2012**, *13*, 3581–3591.

151. Wang, H.; Sun, H.; Wei, H.; Xi, P.; Nie, S.; Ren, Q. *Journal of Nanoparticle Research* **2014**, *16*, 1–13.

152. Libutti, S. K.; Paciotti, G. F.; Myer, L.; Haynes, R.; Gannon, W.; Walker, M.; Seidel, G.; Byrnes, A.; Yuldasheva, N.; Tamarkin, L. *Journal of Clinical Oncology* **2009**, *27*, 3586.

153. Weinreich, D. M.; Puhlmann, M.; Turner, E. M.; Paciotti, G.; Tamarkin, L.; Myer, L.; Alexander, H. R. *Annals of Surgical Oncology* **2002**, *9*, S44–S44.

154. Paciotti, G. F.; Myer, L.; Weinreich, D.; Goia, D.; Pavel, N.; McLaughlin, R. E.; Tamarkin, L. *Drug Delivery* **2004**, *11*, 169–183.

155. Paciotti, G. F.; Kingston, D. G. I.; Tamarkin, L. *Drug Development Research* **2006**, *67*, 47–54.

156. Farma, J. M.; Puhlmann, M.; Soriano, P. A.; Cox, D.; Paciotti, G. F.; Tamarkin, L.; Alexander, H. R. *International Journal of Cancer* **2007**, *120*, 2474–2480.

157. Rojo, J.; Diaz, V.; de la Fuente, J. M.; Segura, I.; Barrientos, A. G.; Riese, H. H.; Bernade, A.; Penades, S. *Chembiochem* **2004**, *5*, 291–297.

158. Svarovsky, S. A.; Szekely, Z.; Barchi, J. J. *Tetrahedron: Asymmetry* **2005**, *16*, 587–598.

159. Biswas, S.; Medina, S. H.; Barchi, J. J. *Jr. Carbo. Res.* **2015**, *405*, 93–101.

160. Adokoh, C. K.; Quan, S.; Hitt, M.; Darkwa, J.; Kumar, P.; Narain, R. *Biomacromolecules* **2014**, *15*, 3802–3810.

161. Brinas, R. P.; Sundgren, A.; Sahoo, P.; Morey, S.; Rittenhouse-Olson, K.; Wilding, G. E.; Deng, W.; Barchi, J. J. *Bioconjugate Chemistry* **2012**, *23*, 1513–1523.

162. Parry, A. L.; Clemson, N. A.; Ellis, J.; Bernhard, S. S. R.; Davis, B. G.; Cameron, N. R. *Journal of the American Chemical Society* **2013**, *135*, 9362–9365.

163. Ojeda, R.; de Paz, J. L.; Barrientos, A. G.; Martin-Lomas, M.; Penades, S. *Carbohydrate Research* **2007**, *342*, 448–459.

164. Thornton, B. P.; Vetvicka, V.; Ross, G. D. *Clinical and Experimental Immunology* **1996**, *104*, 531–537.

165. Nielsen, C. H.; Fischer, E. M.; Leslie, R. G. Q. *Immunology* **2000**, *100*, 4–12.

166. Niikura, K.; Matsunaga, T.; Suzuki, T.; Kobayashi, S.; Yamaguchi, H.; Orba, Y.; Kawaguchi, A.; Hasegawa, H.; Kajino, K.; Ninomiya, T.; Ijiro, K.; Sawa, H. *ACS Nano* **2013**, *7*, 3926–3938.

167. Thakor, A. S.; Jokerst, J.; Zavaleta, C.; Massoud, T. F.; Gambhir, S. S. *Nano Letters* **2011**, *11*, 4029–4036.

168. Dykman, L. A.; Sumaroka, M. V.; Staroverov, S. A.; Zaitseva, I. S.; Bogatyrev, V. A. *Biological Bulletin* **2004**, *31*, 75–79.

169. Dykman, L. A.; BogatyreV, V. A.; Staroverov, S. A.; Pristensky, D. V.; Shchyogolev, S. Y.; Khlebtsov, N. G. *Proceedings of the Photo-Optical Instrumentation Engineers* **2006**, *6164*, 16404–16404.

170. Staroverov, S. A.; Vasilenko, O. A.; Gabalov, K. P.; Pristensky, D. V.; Yermilov, D. N.; Aksinenko, N. M.; Shchyogolev, S. Y.; Dykman, L. A. *International Immunopharmacology* **2008**, *8*, 1418–1422.

171. Staroverov, S. A.; Aksinenko, N. M.; Gabalov, K. P.; Vasilenko, O. A.; Vidyasheva, I. V.; Shchyogolev, S. Y.; Dykman, L. A. *Gold Bulletin* **2009**, *42*, 153–156.

172. Dykman, L. A.; Staroverov, S. A.; Bogatyrev, V. A.; Shchyogolev, S. Y. Nanotechnologies in Russia **2012**, *5*, 748–761.

173. Yang, J. S.; Xie, Y. J.; He, W. *Carbohydrate Polymers* **2011**, *84*, 33–39.

174. Sinha, V. R.; Kumria, R. *International Journal of Pharmaceutics* **2001**, *224*, 19–38.

175. Baldwin, A. D.; Kiick, K. L. *Biopolymers* **2010**, *94*, 128–140.

176. Mizrahy, S.; Peer, D. *Chemical Society Reviews* **2012**, *41*, 2623–2640.

177. Davis, M. E.; Brewster, M. E. *Nature Reviews Drug Discovery* **2004**, *3*, 1023–1035.

178. Ortiz Mellet, C.; Garcia Fernandez, J. M.; Benito, J. M. *Chemical Society Reviews* **2011**, *40*, 1586–1608.

179. Park, J. H.; Saravanakumar, G.; Kim, K.; Kwon, I. C. *Advanced Drug Delivery Reviews* **2010**, *62*, 28–41.

180. Hu, F.-Q.; Meng, P.; Dai, Y.-Q.; Du, Y.-Z.; You, J.; Wei, X.-H.; Yuan, H. *European Journal of Pharmaceutics and Biopharmaceutics* **2008**, *70*, 749–757.

181. Tian, X.; Yin, H.; Zhang, S.; Luo, Y.; Xu, K.; Ma, P.; Sui, C.; Meng, F.; Liu, Y.; Jiang, Y.; Fang, J. *European Journal of Pharmaceutics and Biopharmaceutics* **2014**, *87*, 445–453.

182. Vllasaliu, D.; Casettari, L.; Bonacucina, G.; Cespi, M.; Palmieri, G. F.; Illum, L. *Pharmaceutical Nanotechnology* **2013**, *1*, 184–203.

183. Jie, L.-Y.; Cai, L.-L.; Wang, L.-J.; Ying, X.-Y.; Yu, R.-S.; Zhang, M.-M.; Du, Y.-Z. *International Journal of Nanomedicine* **2012**, *7*, 3981–3989.

184. Yao, W.; Peng, Y.; Du, M.; Luo, J.; Zong, L. *Molecular Pharmaceutics* **2013**, *10*, 2904–2914.

185. Zheng, D.; Duan, C.; Zhang, D.; Jia, L.; Liu, G.; Liu, Y.; Wang, F.; Li, C.; Guo, H.; Zhang, Q. *International Journal of Pharmaceutics* **2012**, *436*, 379–386.
186. Jain, A.; Jain, S. K.; Ganesh, N.; Barve, J.; Beg, A. M. *Nanomedicine* **2010**, *6*, 179–190.
187. Li, P. W.; Wang, Y. C.; Zeng, F. B.; Chen, L. J.; Peng, Z.; Kong, L. X. *Carbohydrate Research* **2011**, *346*, 801–806.
188. Yang, S. J.; Lin, F. H.; Tsai, K. C.; Wei, M. F.; Tsai, H. M.; Wong, J. M.; Shieh, M. J. *Bioconjugate Chemistry* **2010**, *21*, 679–689.
189. Sahu, S. K.; Maiti, S.; Maiti, T. K.; Ghosh, S. K.; Pramanik, P. *Macromolecular Bioscience* **2011**, *11*, 285–295.
190. Fallon, R. J.; Schwartz, A. L. *Journal of Biological Chemistry* **1988**, *263*, 13159–13166.
191. Goren, D.; Horowitz, A. T.; Tzemach, D.; Tarshish, M.; Zalipsky, S.; Gabizon, A. *Clinical Cancer Research* **2000**, *6*, 1949–1957.
192. Ragelle, H.; Vandermeulen, G.; Préat, V. *Journal of Controlled Release* **2013**, *172*, 207–218.
193. Liu, X.; Howard, K. A.; Dong, M.; Andersen, M. Ø.; Rahbek, U. L.; Johnsen, M. G.; Hansen, O. C.; Besenbacher, F.; Kjems, J. *Biomaterials* **2007**, *28*, 1280–1288.
194. Sezer, A. D.; Cevher, E. *Expert Opinion on Drug Delivery* **2012**, *9*, 1129–1146.
195. Mao, S.; Sun, W.; Kissel, T. *Advanced Drug Delivery Reviews* **2010**, *62*, 12–27.
196. Nguyen, J.; Szoka, F. C. *Accounts of Chemical Research* **2012**, *45*, 1153–1162.
197. Lai, W. F.; Lin, M. C. M. *Journal of Controlled Release* **2009**, *134*, 158–168.
198. Dolmans, D. E. J. G. J.; Fukumura, D.; Jain, R. K. *Nature Reviews Cancer* **2003**, *3*, 380–387.
199. Lee, S. J.; Park, K.; Oh, Y. K.; Kwon, S. H.; Her, S.; Kim, I. S.; Choi, K.; Lee, S. J.; Kim, H.; Lee, S. G.; Kim, K.; Kwon, I. C. *Biomaterials* **2009**, *30*, 2929–2939.
200. Simon, V.; Devaux, C.; Darmon, A.; Donnet, T.; Thiénot, E.; Germain, M.; Honnorat, J.; Duval, A.; Pottier, A.; Borghi, E.; Levy, L.; Marill, J. *Photochemistry and Photobiology* **2010**, *86*, 213–222.
201. Bae, Y. H.; Yin, H. *Journal of Controlled Release* **2008**, *131*, 2–4.
202. Lee, S. J.; Koo, H.; Lee, D.-E.; Min, S.; Lee, S.; Chen, X.; Choi, Y.; Leary, J. F.; Park, K.; Jeong, S. Y.; Kwon, I. C.; Kim, K.; Choi, K. *Biomaterials* **2011**, *32*, 4021–4029.
203. Kim, S.; Ohulchanskyy, T. Y.; Bharali, D.; Chen, Y.; Pandey, R. K.; Prasad, P. N. *The Journal of Physical Chemistry C* **2009**, *113*, 12641–12644.
204. Lim, C.-K.; Shin, J.; Lee, Y.-D.; Kim, J.; Park, H.; Kwon, I. C.; Kim, S. *Small* **2011**, *7*, 112–118.
205. Platt, V. M.; Szoka, F. C. *Molecular Pharmaceutics* **2008**, *5*, 474–486.
206. Choi, K. Y.; Min, K. H.; Yoon, H. Y.; Kim, K.; Park, J. H.; Kwon, I. C.; Choi, K.; Jeong, S. Y. *Biomaterials* **2011**, *32*, 1880–1889.
207. Hornig, S.; Bunjes, H.; Heinze, T. *Journal of Colloid and Interface Science* **2009**, *338*, 56–62.
208. Edwards, D. I. *Journal of Antimicrobial Chemotherapy* **1993**, *31*, 9–20.
209. El-Dakdouki, M.; Huang, X. F. *Petite and Sweet: Glyco-Nanotechnology as a Bridge to New Medicines* **2011**, Oxford University Press, New York, vol. *1091*, pp. 181–213.

14

CARBOHYDRATE NANOTECHNOLOGY APPLIED TO VACCINE DEVELOPMENT

RAJESH SUNASEE[1] AND RAVIN NARAIN[2]

[1] *Department of Chemistry, State University of New York at Plattsburgh, Plattsburgh, NY, USA*
[2] *Chemical and Materials Engineering, University of Alberta, Edmonton, Alberta, Canada*

14.1 INTRODUCTION

Vaccination is perhaps one of the most crucial parts of modern medicine, and recent years have witnessed an increase in the development of vaccination technology toward the search for the ideal vaccine. The extraordinary advancement in the fields of glycochemistry, glycobiology, glycomedicine, and immunology in the last decades has led to an acceleration of the pace of vaccine development [1–3]. Many infectious diseases, which were previously considered fatal, are now easily prevented or controlled, and many have been eradicated altogether, for examples, smallpox, poliomyelitis, diphtheria, and neonatal tetanus. Vaccines offer protection by inducing humoral and/or cellular immunity to disease-causing pathogens. Typically, a vaccine must satisfy the following criteria: it should be safe, stable, and reproducible and must be able to elicit the appropriate immune response. Other issues such as cost, immunization route, or number of administrations must also be taken into consideration [4]. Conventional vaccines have been developed using live attenuated organisms (bacillus Calmette–Guérin (BCG), measles, mumps, rubella, and varicella) [5], killed or inactivated whole organisms (e.g., influenza) [6], or inactivated toxins (including diphtheria and tetanus) [7]. However, these vaccines

Carbohydrate Nanotechnology, First Edition. Edited by Keith J. Stine.

have several limitations such as intrinsic instability, serious risk of reverting to more active forms, weaker immune responses, and the need for multiple administrations. An adjuvant is usually added to render them effective [8]. Consequently, subunit vaccines and DNA vaccines were developed and were found to have higher safety profile over traditional vaccines [9, 10]. However, their relatively low immunogenicity requires the use of adjuvants and/or vaccine delivery systems [11]. Thus, adjuvants and novel delivery systems that boost immunogenicity are in great need, and as such, nanotechnology is nowadays playing a significant role in vaccine development [12–16]. While nanotechnology is likely to have a huge implication in medicine (the so-called field of nanomedicine), vaccines are no exception. Nanotechnology has gained significant momentum in recent years, and nanotechnology platforms offer the opportunity to medical researchers to design vaccines that can work without the use of any real viral material. The ability to engineer nanoparticles varying in size, composition, shape, and surface properties has led to an increase use of nanotechnology in the field of vaccinology, which eventually gives birth to the term *nanovaccinology* [17]. A number of vaccine nanocarriers were developed for the delivery of antigens and adjuvants to immune cells in order to activate or enhance immunity. The nanoscale particle size allows efficient cellular uptake and hence a better antigen recognition and presentation. Furthermore, the surface of nanocarriers can be easily modified with targeting moieties for the delivery of antigens to specific cell surface receptors. Typical nanocarrier vaccine systems that have been investigated include liposomes, emulsions, polymer-based micelles, metallic nanoparticles, biodegradable nanoparticles, viruslike particles, immunostimulating complex (ISCOM), proteasomes, dendrimers, and cholesterol-bearing hydrophobized pullulan nanoparticles [12,13,16]. Carbohydrate-based vaccine design has also received significant attention in recent years [18–22] due to advances in the chemical synthesis of neoglycoconjugates [23,24] and oligo- or polysaccharides [25,26], one-pot protection of monosaccharides [27], and de novo strategies [28–30]. However, the nonimmunogenicity of carbohydrates remains one of the main challenges for the development of carbohydrate-based vaccines (natural or synthetic). This chapter will provide an overview of recent advances in carbohydrate-based polymeric and nanoparticulate vaccine systems with the main focus on the emerging roles of glycopolymers and glyconanoparticles (GNPs) in vaccine development.

14.2 GLYCOPOLYMERS IN VACCINE DEVELOPMENT

Carbohydrates are highly involved in a myriad of biological processes including adhesion, cell growth regulation, cellular trafficking, cancer cell metastasis, fertilization, and inflammation, which led to the development of carbohydrate-based systems for therapeutic needs [31,32]. Carbohydrates continue to be attractive targets for vaccine development given the dense surface distribution of unique glycan structures on various pathogens and malignant cells. While the field of carbohydrate vaccine design has evolved tremendously in the past few years due to technological advances, a number of emerging specific challenges remain to be tackled within each area of carbohydrate vaccine design and development [19,21,33]. Modern organic and carbohydrate chemistries have allowed chemists to assemble carbohydrate-containing synthetic polymer conjugates, the so-called glycopolymers. Glycopolymers (synthetic polymer featuring pendant carbohydrate moieties) have been widely investigated as carbohydrate mimics for therapeutic potentials and can provide numerous advantages over those of natural glycoconjugates [31,32,34–36]. The saccharide units in glycopolymers help in molecular recognition and interact with specific carbohydrate receptors. Their high biocompatibility, multivalency, and target specificity render them as suitable materials for biomimetic models of glycoconjugates. With the advent of powerful polymerization techniques and recent advances in controlled polymerization reactions, diverse glycopolymer topologies can be easily accessed [35,37–39]. These robust tailored glycopolymers have been exploited as therapeutics, drug and gene delivery systems, and protein stabilization and signal transduction processes [32, 40–44]. In the burgeoning field of vaccinology, a few reports disclosed the use of synthetic glycopolymers in the design of vaccines. Other than cancer vaccinations, glycopolymers can also play significant roles in immunotherapy applications such as vaccination against bacterial pathogens, viral inhibition, macrophage targeting, transplant rejection, and complement activation [45].

14.2.1 Glycopolymers for Complement Passivation and Activation

The complement system is a group of soluble and cell surface proteins that serves as a first line of defense against pathogens. A recent study by Narain and coworkers showed that hyperbranched lactose glycopolymers did not trigger complement activation at the studied concentrations of 0.1–0.5 mg/ml (Fig. 14.1b) [46]. These hyperbranched glycopolymers of varying molecular weights (15–38 kDa) were synthesized by reversible addition–fragmentation chain transfer (RAFT) polymerization (Fig. 14.1a) and were also found not to induce blood coagulation as well as platelet activation.

While studies have been focusing on inhibiting complement activation to avoid effects such as implant rejection and clearance of systemic drug delivery carriers, it is sometimes useful to induce complement activation particularly for vaccine. Complement activation could be exploited as a potential strategy for vaccine adjuvants [47, 48]. Haddleton and coworkers reported the preparation of well-defined neoglycopolymer–protein biohybrid materials using a combination of copper-catalyzed living radical polymerization and alkyne–azide [2+3] cycloaddition ("click") techniques, followed by conjugation with a nonglycosylated model protein, bovine serum albumin (BSA) (Fig. 14.2) [49]. The resulting glycopolymer conjugates were found to bind with mannose-binding lectin (MBL) with deposition of complement protein C9. Enzyme-linked immunosorbent assay (ELISA) revealed that the glycopolymer–BSA conjugates could significantly activate complement via the lectin pathway when compared with native unmodified BSA.

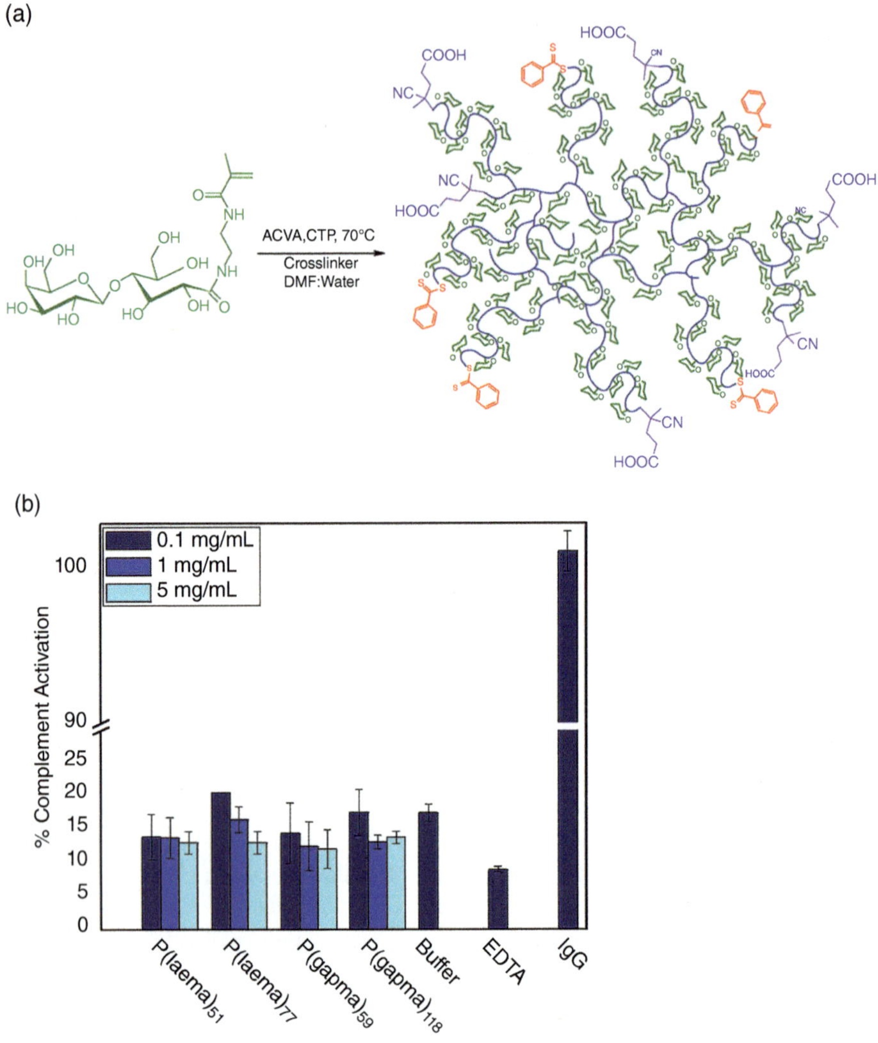

FIGURE 14.1 (a) Synthesis of hyperbranched glycopolymers by RAFT. (b) Percent complement consumption caused by hyperbranched glycopolymers at different concentrations in serum after incubation for 1 h (IgG is used as a positive control, and EDTA is used as a negative control). *Source*: Adapted with permission from Ref. 46, Copyright 2012, ACS.

14.2.2 Glycopolymers as Therapeutic Materials for Influenza, Leishmaniasis, and Human Immunodeficiency Virus

Glycopolymers have been used to competitively inhibit binding between pathogens and host cells, hence a useful strategy for disease prevention. Neuraminidase inhibition is a one such common strategy for the treatment of influenza. Each year, thousands of people worldwide contract influenza virus, which can lead to annual epidemics with significant morbidity and mortality, even in countries with the most

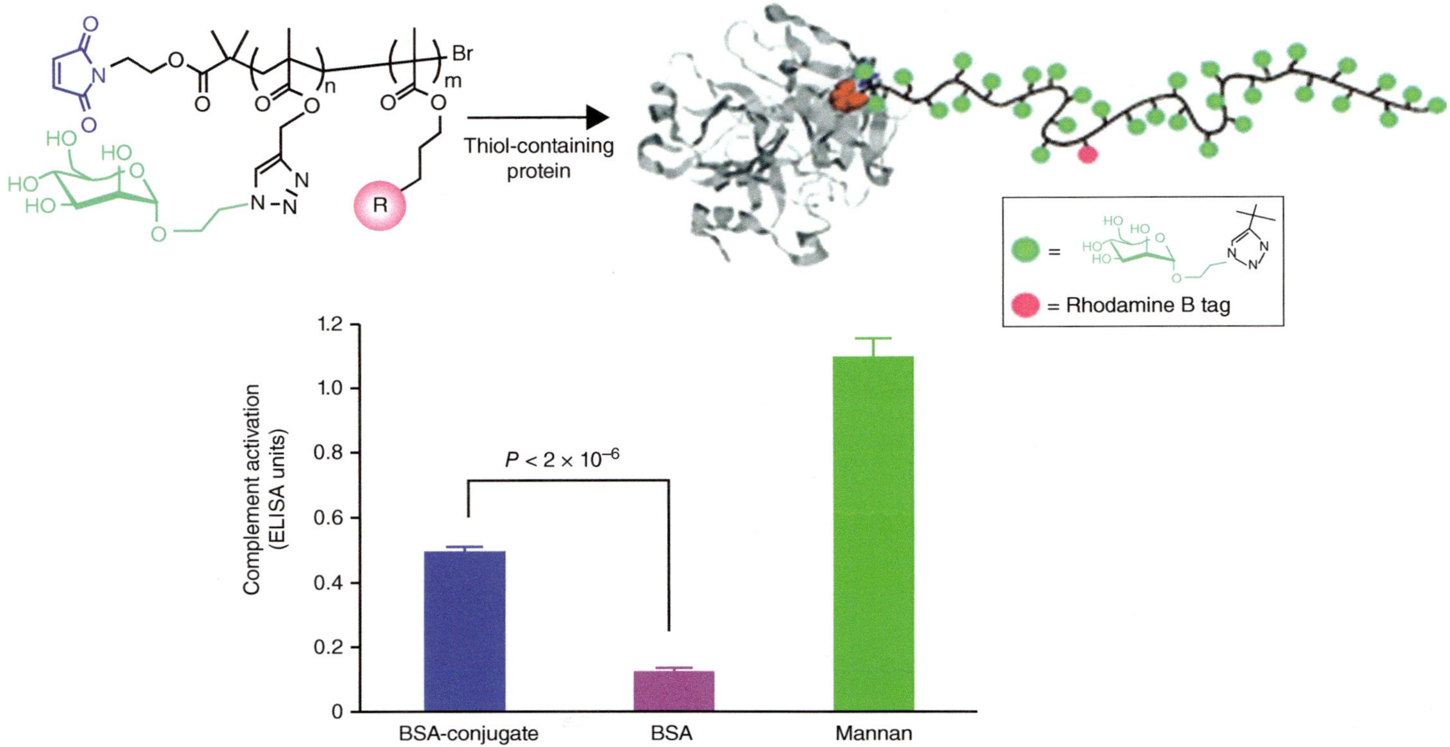

FIGURE 14.2 Synthesis of well-defined neoglycopolymer–protein biohybrid materials and the level of activated complement C9 deposition. *Source*: Adapted with permission from Ref. 49, Copyright 2007, ACS.

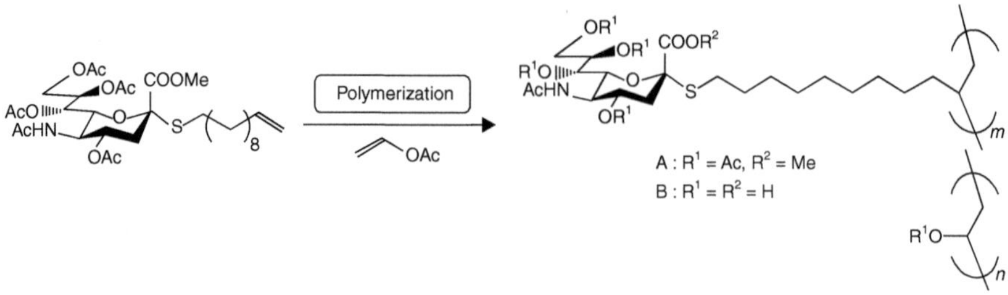

FIGURE 14.3 Linear thiosialoside glycopolymers as candidates for neuraminidase inhibitor. *Source*: Reprinted with permission from Ref. 54, Copyright 2007, Elsevier.

advanced health-care systems. Influenza vaccination remains the main method of prevention. Several sialic acid-derived glycopolymers have been designed to act as neuraminidase inhibitors with various polymeric backbones such as poly(acrylic acid), poly(acrylamide), poly(glutamic acid), and poly(styrene) [45,50–53]. Recently, a new class of linear glycopolymers containing thiosialoside residues was prepared by radical copolymerization, and preliminary studies showed that these glycopolymers have potent inhibitory activity against the neuraminidases (Fig. 14.3) [54].

Adaptive immune system, also known as the acquired immune system, eliminates or prevents pathogen growth and provides specific long-term defense against infection. One of the major functions of the adaptive immune system is the development of immunological memory whereby each pathogen is remembered by a signature antibodies or T-cell receptors. Leishmaniasis, a disease caused by protozoan parasites of the genus *Leishmania*, is easily transmitted by the bite of infected sandflies and currently affects about 12 million people worldwide [55]. Vaccination against leishmaniasis has been investigated using a number of strategies such as inoculation of virulent parasites and immunization with killed parasite preparations, live attenuated parasites, or recombinant proteins or plasmid DNA coding for defined *Leishmania* antigens [56]. In a preliminary report, Merkle and Seeberger *et al.* disclosed a promising strategy for the delivery of synthetic *Leishmania* antigen promoting T-cell-mediated specific immune responses by modifying the surface of biodegradable poly(D,L-lactic-*co*-glycolic acid) (PLGA) microparticles with a glycopolymer [57]. Poly(L-lysine)-*graft*-poly(ethylene glycol) (PLL-*g*-PEG) brush copolymers were conjugated with *Leishmania* tetrasaccharide (assembled by automated solid-phase synthesis) terminated with a mercaptopropyl linker at the reducing end, followed by adsorption onto PLGA microparticles (1–10 μm in size). The microparticles were loaded with a model protein in order to enhance immunogenicity. The choice of PLGA as the matrix for vaccine formulations is due to its biodegradability, long-term safety, and the commercial availability of a number of polymers of different molecular weights and monomer ratios [58]. These surface-modified PLGA microparticles could further be explored as antigen carriers to the immune system; however, studies in an animal model remain challenging. While several vaccines are being developed against leishmaniasis, there are none available as of 2013.

The treatment of the human immunodeficiency virus (HIV) infection remains a challenge, and current efforts are in place to look for preventive compounds that can inhibit the entry and persistence of viruses in the body [59–62]. In this regard, sulfated maltoheptaose-derived methacrylate glycopolymers (Fig. 14.4a) have been investigated, and the anti-HIV activity was found to increase with increase in degree of sulfation [63]. Furthermore, the distance between branched maltoheptaose units in the polymethacrylate main chain was vital for high anti-HIV activities. These glycopolymers also have low cytotoxicity and are promising therapeutic materials. Recently, Becer and coworkers successfully illustrated the inhibitory effect of mannosylated methacrylate-based glycopolymers (Fig. 14.4b) on the binding of gp120 to dendritic cell-specific ICAM-3-grabbing nonintegrin (DC-SIGN), a major C-type lectin receptor present on dendritic cells [64]. The glycopolymers were synthesized using a postpolymerization modification approach employing "click chemistry." As the mannose density increases, the ability to disrupt the gp120–DC-SIGN interaction also increases. These preliminary results provide useful insights into the mechanisms of HIV infection and potential new therapeutics. The same group later reported the effect of glycopolypeptides (Fig. 14.4c) in binding to DC-SIGN in comparison to poly(methacrylate) glycopolymers [65]. The results indicated a slower binding kinetics for the glycopolypeptides but much stronger and could potentially be used as effective inhibitory agents to prevent HIV-1 gp120 binding to DC-SIGN.

14.2.3 Glycopolymers for *Candida* Vaccination and as Anticancer Drugs

The Bundle group reported the synthesis of a fully functional glycopolymer conjugate vaccine construct deriving B-cell epitopes from the glycopolymer and T-cell peptides from the immunogenic carrier protein, chicken serum albumin (Fig. 14.5) [66]. The glycopolymer acts as a carrier for oligosaccharide haptens. The synthesis started with the copolymerization of acrylamide with an acrylamide-activated trisaccharide to afford a linear polyacrylamide with pendant oligosaccharides (about 33 trisaccharide haptens). The glycopolymer was further modified to introduce a reactive azide end group for subsequent conjugation to a propargylated albumin via the *azide–alkyne* Huisgen cycloaddition process. The resulting glycopolymer–protein conjugate was found to be highly immunogenic in mice and could induce a more robust immune response than a tetanus toxoid conjugate vaccine (8–12 trisaccharides). These glycoconjugates could be useful tools for the generation of high-titer sera or monoclonal antibodies specific for oligosaccharide epitopes as well as low molecular weight haptens. However, a detailed investigation still remains to elucidate whether the improved immunogenicity of the glycopolymer vaccine stems from the higher payload of sugar in the construct or better presentation of haptens onto the long polymeric backbone.

In a recent study, cationic glycopolymers with varying degree of polymerization were prepared by RAFT and subsequently conjugated with dithiocarbamates (DTC). The resulting glycopolymer-based DTC conjugates were modified with gold (I) phosphine, and the gold complexes were tested for their *in vitro* toxicity in both normal

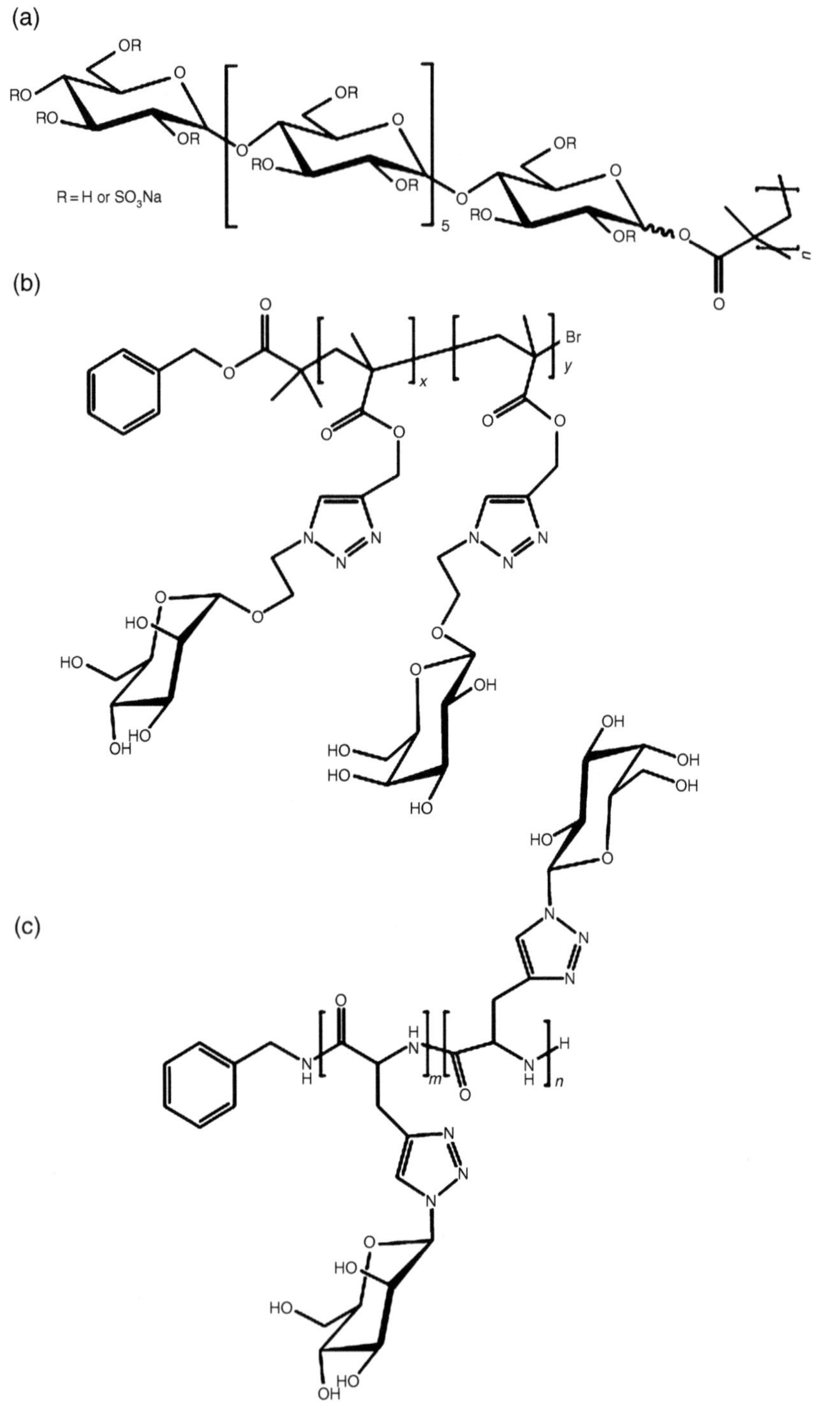

FIGURE 14.4 Glycopolymers employed as inhibitory agents in HIV studies. (a) Sulfated maltoheptaose-derived methacrylate glycopolymer. (b) Mannosylated methacrylate-based glycopolymer. (c) Glycopolypeptide.

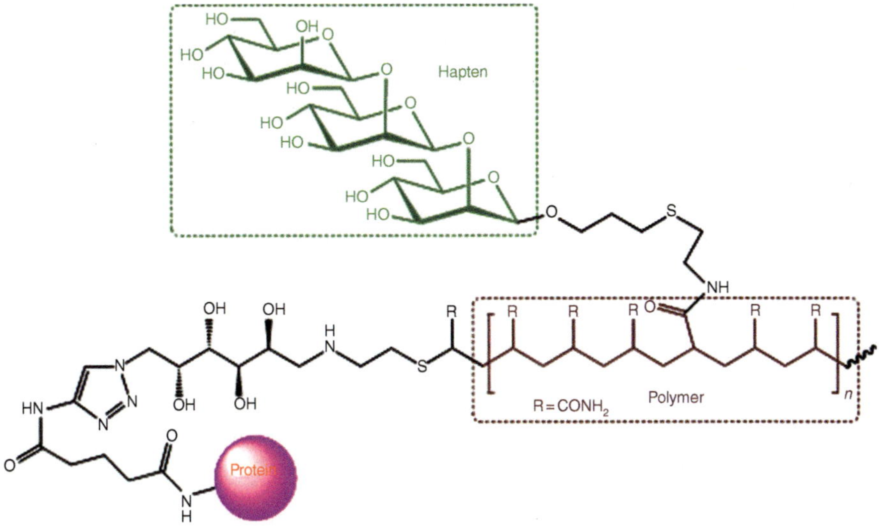

FIGURE 14.5 Bundle's glycopolymer vaccines. *Source*: Reprinted with permission from Ref. 31, Copyright 2013, Wiley-VCH.

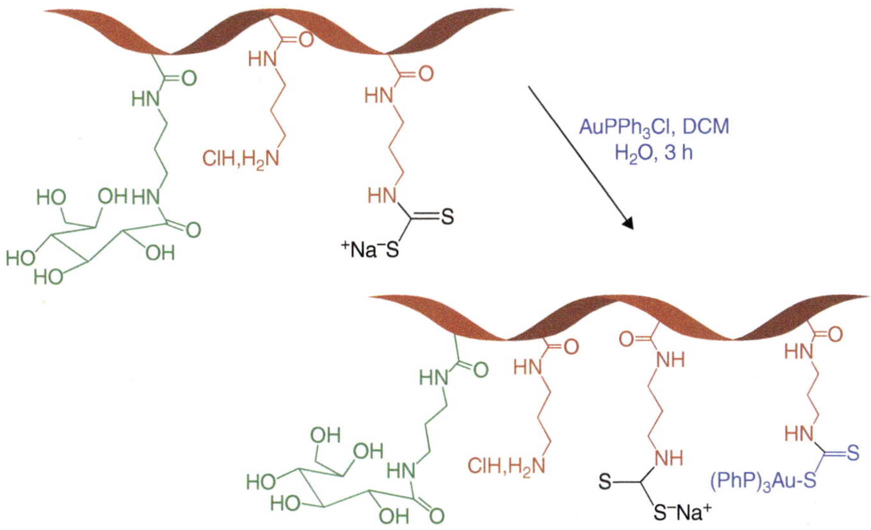

FIGURE 14.6 Gold glycoconjugates as anticancer agents. *Source*: Reprinted with permission from Ref. 67, Copyright 2013, ACS.

and cancer cell lines (Fig. 14.6) [67]. It was shown that these glycopolymer-based DTC derivatives and their gold conjugates displayed higher accumulation as well as cytotoxicity to cancer cells under hypoxic conditions in comparison to the normoxic ones. Overall, the polymeric gold glycoconjugate can serve as an anticancer agent for the treatment of breast cancer, without significantly affecting normal tissues.

14.3 GNPs IN VACCINE DEVELOPMENT

The use of carbohydrate-coated gold nanoparticles, the so-called GNPs, for the design of carbohydrate-based vaccines is still at its infancy with some initial promising results for potential antitumor therapies. GNPs can actually make a huge impact in postoperative metastasis of tumors. GNPs can offer a number of advantages in the vaccination area such as targeting of cell–cell interactions, targeted immune responses in communicable and noncommunicable diseases, antigen delivery, ability to target defined cells, and particle features that mimic cell walls, hence creating a target for the primary immune response [68].

14.3.1 GNPs as Anticancer Agents

In 2005, the Barchi group disclosed the synthesis of gold nanoparticles coated with Thomsen–Friedenreich (TF) antigen, which could potentially inhibit lung metastasis in the murine 4T1 breast cancer model [69]. However, the nonuniform size distribution of their nanoparticles in solution was suspected to be an obstacle during *in vivo* studies with nonreproducible results. Thus, attempt to refine the synthesis to improve the monodispersity of the construct [70] and further detailed investigations to improve the vaccine design are currently underway in the Barchi laboratory [71]. More recently, the same group developed a novel gold nanoparticle-based glycopeptide vaccine construct as possible carriers for tumor-associated carbohydrate antigens (TACAs) and TACA-conjugated glycopeptides [72] (Fig. 14.7). The peptide sequences are from MUC4, which is a valid starting target for vaccine development. A 28-residue peptide from the complement-derived protein C3d was incorporated in

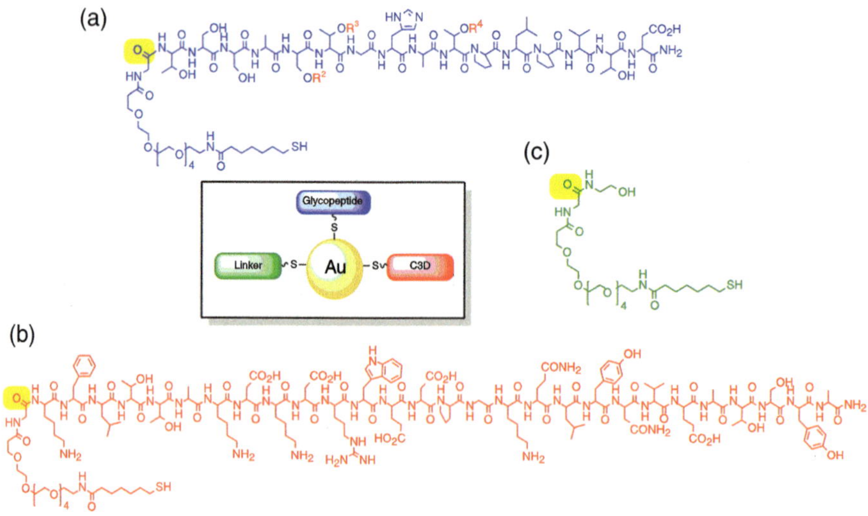

FIGURE 14.7 GNP vaccine design. *Source*: Reprinted with permission from Ref. 72, Copyright 2012, ACS.

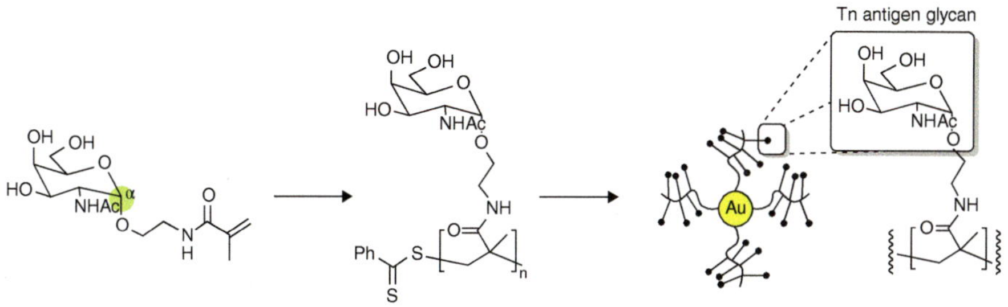

FIGURE 14.8 A fully synthetic vaccine consisting of gold nanoparticle decorated with *N*-acetylgalactosamine. *Source*: Reprinted with permission from Ref. 73, Copyright 2013, ACS.

the construct to act as a B-cell activating "molecular adjuvant." The immune responses of the nanoparticles were evaluated after mice immunization and gave measurable and statistically significant responses. While preliminary work looks promising, further studies such as uniformity and immunogenicity of these nanoparticles are required to fully exploit this novel vaccine construct.

Cameron and coworkers described the synthesis of "multicopy-multivalent" nanoparticles decorated with tumor-associated (Tn) antigen glycans (Fig. 14.8) [73]. The synthesis involves a polymerizable version of the Tn-antigen glycan, which was converted into well-defined glycopolymers by RAFT polymerization followed by conjugation to gold nanoparticles. Immunological studies indicated that the glycosylated nanoparticles were able to generate antibodies against the Tn antigen that were also cross-reactive with natural mucins displaying the antigen. Although the reported absolute titers are lower than expected, this fully synthetic protein- and peptide-free glycoconjugate system has potential as synthetic cancer vaccines.

Recently, the Narain group disclosed the first study on direct conjugation of toxic cancer drugs onto GNPs toward the design of powerful anticancer agents [74]. The hypothesis was based on the fact that by combining glycopolymer DTC gold nanoparticles and gold(I) triphenylphosphine, it might improve the biological efficacy of gold(I) conjugate and lead to a new class of gold-based anticancer drugs. Statistical glycopolymers decorated with DTC of varying molecular weights were synthesized by RAFT polymerization and was subsequently used to prepare GNPs and conjugated GNPs with the anticancer drug, gold(I) triphenylphosphine (Fig. 14.9). The GNPs and their Au(I)PPh_3 conjugates were active against MCF7 and HepG2 cells, but galactose-functionalized GNPs were found to be a fourfold more potent anticancer agent (IC_{50} = 4.13 ± 0.73 μg/ml) in HepG2 cells as compared to the corresponding glucose derivative in ASGPR-deficient HeLa cells and significantly more toxic compared to standard chemotherapeutic reagents (cisplatin and cytarabine).

Penadés and coworkers designed multifunctional GNPs as a potential anticancer vaccine platform [75]. A small library of GNPs was prepared in one step, which incorporated sialyl Tn and Lewis^y antigens, T-cell helper peptides (TT), and glucose in varying densities on the particles (Fig. 14.10). This method allows the type, the ratio, and the density of carbohydrate antigens and T-cell helper peptides on the multivalent

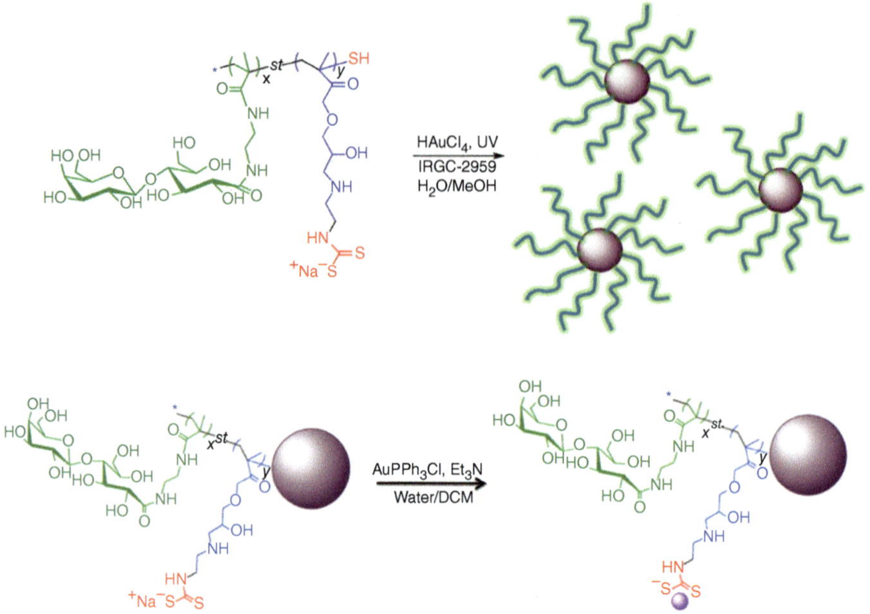

FIGURE 14.9 Glyconanoparticles and their conjugation to the anticancer drug, Au(1)PPh_3. *Source*: Reprinted with permission from Ref. 74, Copyright 2014, ACS.

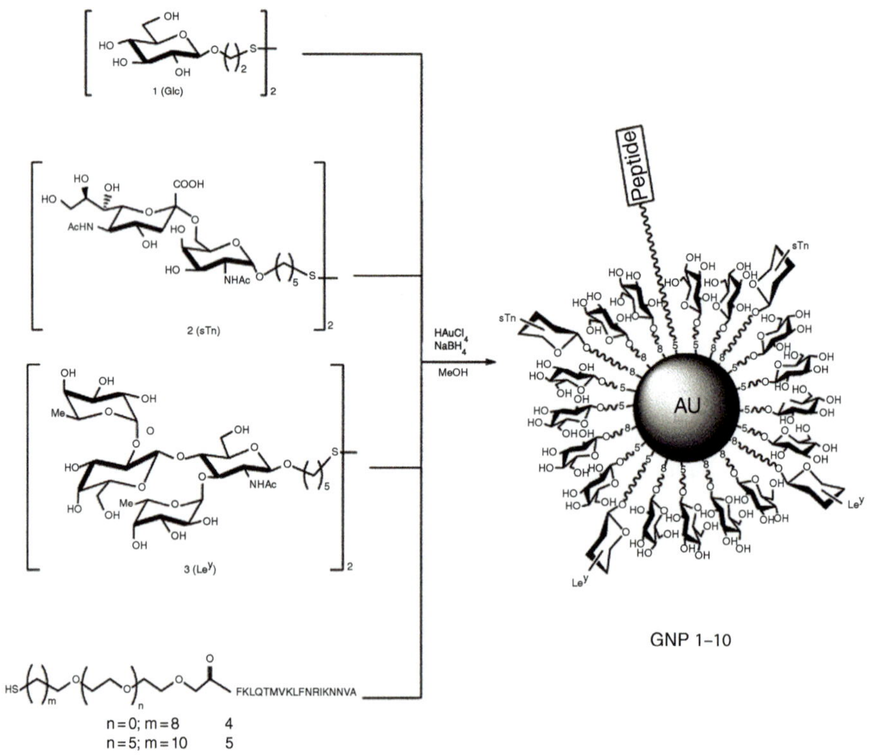

FIGURE 14.10 One-step synthesis of GNPs as anticancer vaccine platform. *Source*: Reprinted with permission from Ref. 75, Copyright 2007, Elsevier.

construct to be varied in a straightforward way during the self-assembly process. While preliminary *in vivo* studies of these multifunctional GNPs have been attempted, however concrete immunological data about these constructs has yet to be investigated.

14.3.2 GNPs as Anti-HIV Agents

In the world of drug delivery system against HIV, very few studies described the application of gold nanoparticles for HIV treatment [76–78]. The Penadés group pioneered the use of GNPs as a platform for carbohydrate-based vaccines against HIV [78–80]. GNPs coated with synthetic oligomannosides of the gp120 (manno-GNPs) were able to inhibit the DC-SIGN-mediated HIV-1 transinfection of human T cells [80], while GNPs coated with sulfated ligands were shown to interfere with the adhesion/fusion of HIV during its entry [81]. Recently, the same group further explored the use of GNPs as a scaffold to insert antiretroviral drugs to construct new multivalent anti-HIV systems [82]. The use of biocompatible GNPs as scaffolds for the antiviral drugs could lead to the improvement of the solubility in water and biological media of the drugs and better cellular uptake due to the presence of carbohydrates on the GNPs. Furthermore, antiviral activity could be improved due to a local increase of the drug concentration on the gold surface. Glucose-coated gold nanoparticles were prepared, followed by their functionalization with in clinical use antiviral drugs abacavir and lamivudine (Fig. 14.11). The presence of hydrolysable ester groups allows the release of the drug from the GNPs by pH or enzymatic mediated hydrolysis. Results indicated that the drugs were able to inhibit viral replication in cellular assays with IC_{50} values similar to the free drugs. Overall, this new strategy opens up the possibility of creating multivalent nanodelivery systems against HIV.

14.4 CONCLUSIONS AND OUTLOOK

This chapter has focused on the recent developments of carbohydrate-based polymeric (glycopolymers) and GNPs as possible scaffolds for vaccine development. The preparation of advanced glycopolymeric materials with predetermined compositions, molecular weights, and architectures will continue to be a major player in such developments. GNPs, with their nanometric size, high water solubility, and biocompatibility, display great potential in the field of nanovaccinology. While initial attempts look promising, several issues need to be addressed before the application of these materials into clinical use. A better understanding of how nanoparticles interact with biological systems is required. Toxicity issues of these materials remain the most important one that needs to be addressed. As for HIV treatment, further investigations are required to fully understand the molecular mechanisms of the described antiviral activity. A cellular tracking of the GNPs could perhaps give a molecular explanation of their behavior in the intracellular milieu. With the continued convergence of areas such as glycochemistry, glycobiology, glycomedicine, and immunology, we should expect significant developments in the application of carbohydrate nanotechnology to vaccine development.

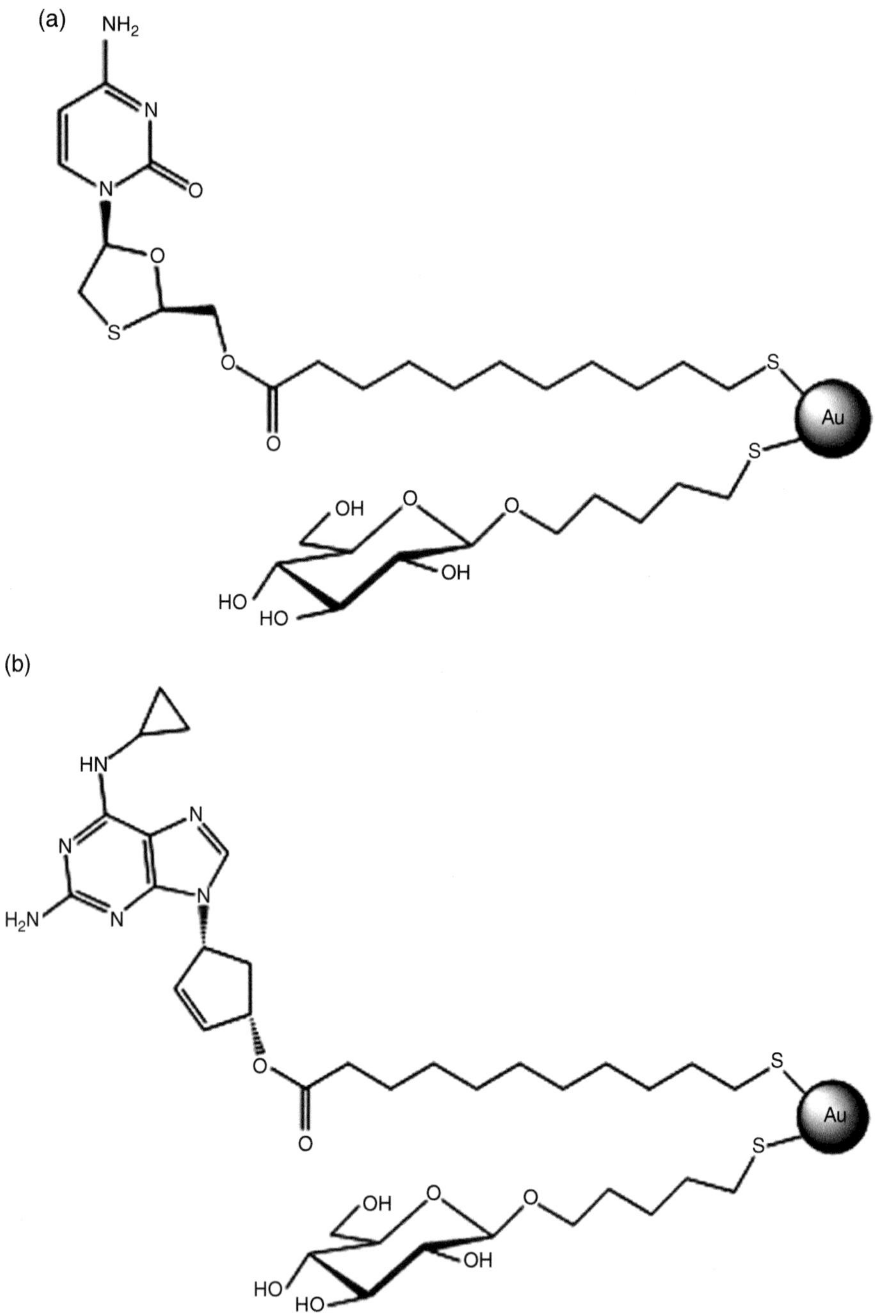

FIGURE 14.11 (a) Lamivudine GNPs. (b) Abacavir GNPs. *Source*: Adapted with permission from Ref. 82, Copyright 2014, Creative Commons Attribution License.

REFERENCES

1. Galonic, D. P.; Gin, D. Y. *Nature* **2007**, *446*, 1000–1007.
2. Ouerfelli, O.; Warren, J. D.; Wilson, R. M.; Danishefsky, S. J. *Expert Rev. Vac.* **2005**, *4*, 677–685.
3. Kaufmann, S. H. E. *Nat. Rev. Microbiol.* **2007**, *5*, 491–504.
4. Perrie, Y.; Mohammed, A. R.; Kirby, D. J.; McNeil, S. E.; Bramwell, V. W. *Int. J. Pharm.* **2008,** *364*, 272–280.
5. Naiman, B. M.; Alt, D.; Bolin, C. A. *et al. Infect. Immun.* **2001**, *69*, 7550–7558.
6. Epstein, J. E.; Tewari, K.; Lyke, K. E. *et al. Science* **2011**, *334*, 475–480.
7. Saha, A.; Chowdhury, M. I.; Nazim, M. *et al. Vaccine* **2013**, *31*, 647–652.
8. Aguilar, J. C.; Rodríguez, E. G. *Vaccine* **2007**, *25*, 3752–376.
9. Norton E. B.; Lawson, L.B.; Mahdi Z. *et al. Infect. Immun.* **2012**, *80*, 2426–2435.
10. Luke, J. M.; Simon, G. G.; Söderholm, J. *et al. J. Virol.* **2011**, *85*, 1370–1383.
11. Bramwell, V. W.; Eyles, J. E.; Alpar, H. O. *Adv. Drug Deliv. Rev.* **2005**, *57*, 1247–1265.
12. Zhao, L.; Seth, A.; Wibowo, N.; Zhao, C.-X.; Mitter, N.; Yu, C.; Middelberg, A. P. J. *Vaccines* **2014**, *32*, 327–337.
13. Kim, M.-G.; Park, J. Y.; Shon, Y.; Kim, G.; Shim, G.; Oh, Y.-K. *Asian J. Pharm. Sci.* **2014**, *9*(5), 227–235; doi:10.1016/j.ajps.2014.06.002
14. Salvador, A.; Igartua, M.; Hernández, R. M.; Luis Pedraz, J. *J. Drug Deliv.* **2011**, *18*, 181646.
15. Nandedkar, T. D. *J. Biosci.* **2009**, *34*, 995–1003.
16. Peek, L. J.; Middaugh, C. R.; Berkland, C. *Adv. Drug Deliv. Rev.* **2008**, *60*, 915–928.
17. Mamo, T.; Poland, G. A. *Vaccine* **2012**, *30*, 6609–6611.
18. Danishefsky, S. J.; Allen, J. R. *Angew. Chem. Int. Ed.* **2000**, *39*, 836–863.
19. Huang, Y. L.; Wu, C. Y. *Expert. Rev. Vac.* **2010**, *9*, 1257–1274.
20. Hecht, M. L.; Stallforth, P.; Silva, D. V.; Adibekian, A.; Seeberger, P. H. *Curr. Opin. Chem. Biol.* **2009**, *13*, 354–359.
21. Astronomo, R. D.; Burton, D. R. *Nat. Rev. Drug. Discov.* **2010**, *9*, 308–324.
22. Guo, Z.; Wang, Q. *Curr. Opin. Chem. Biol.* **2009**, *13*, 608–617.
23. Foley, T. L.; Burkart, M. D. *Curr. Opin. Chem. Biol.* **2007**, *11*, 12–19.
24. Bernardes, G. J.; Castagner, B.; Seeberger, P. H. *ACS Chem. Biol.* **2009**, *4*, 703–713.
25. Seeberger, P. H. *Chem. Soc. Rev.* **2008**, *37*,19–28.
26. Sears, P.; Wong, C. H. *Science* **2001**, *291*, 2344–2350.
27. Wang, C. C.; Lee, J. C.; Luo, S. Y.; Kulkarni, S. S.; Huang, Y. W.; Lee, C. C.; Chang, K. L.; Hung, S. C. *Nature* **2007**, *446*, 896–899.
28. Stallforth, P.; Adibekian, A.; Seeberger, P. H. *Org. Lett.* **2008**, *10*, 1573–1576.
29. Adibekian, A.; Bindschädler, P.; Timmer, M. S.; Noti, C.; Schützenmeister, N.; Seeberger, P. H. *Chemistry* **2007**, *13*, 4510–4522.
30. Guo, H.; O'Doherty, G. A. *Angew. Chem. Int. Ed.* **2007**, *46*, 5206–5208.
31. Sunasee, R.; Narain, R. *Macromol. Biosci.* **2013**, *13*, 9–27.
32. Sunasee, R.; Adokoh, C. K.; Dwarka, J.; Narain, R. *Expert Opin. Drug Deliv.* **2014**, *11*, 1–18.

33. Anish, C.; Schumann, B.; Pereira, C. L.; Seeberger, P. H. *Chem. Biol.* **2014**, *21*, 38–50.
34. Ladmiral, V.; Melia, E.; Haddleton, D. M. *Eur. Polym. J.* **2004**, *40*, 431–449.
35. Vàzquez-Dorbatt, V.; Lee, J.; Lin, E.-W.; Maynard, H. D. *ChemBioChem* **2012**, *13*, 2478–2487.
36. Miura, Y. *Polym. J.* **2012**, *44*, 679–689.
37. Matyjaszewski, K. *Science* **2011**, *333*, 1104–1105.
38. Braunecker, W. A.; Matyjaszewski, K. *Prog. Polym. Sci.* **2007**, *32*, 93–146.
39. Ghadban, A.; Albertin, L. *Polymers* **2013**, *5*, 431–526.
40. Ahmed, M.; Narain, R. *Biomaterials* **2011**, *32*, 5279–5290.
41. Ahmed, M.; Narain, R. *Biomaterials* **2012**, *33*, 3990–4001.
42. Ahmed, M.; Narain, R. *Biomaterials* **2013**, *34*, 4368–4376.
43. Mancini, R. J.; Lee, J.; Maynard, H. D. *J. Am. Chem. Soc.* **2012**, *134*, 8474–8479.
44. Kiessling, L. L.; Grim, J. C. *Chem. Soc. Rev.* **2013**, *42*, 4476–4491.
45. Lin, K.; Kasko, A. M. *ACS Macro Lett.* **2014**, *3*, 652–657.
46. Ahmed, M.; Lai, B. F. L.; Kizhakkedathu, J. N.; Narain, R. *Bioconjugate Chem.* **2012**, *23*, 1050–1058.
47. Reddy, S. T.; van der Vlies, A. J.; Simeoni, E. Angeli, V.; Randolph, G. J.; O'Neill C. P.; Lee, L. K.; Swartz, M. A.; Hubbell, J. A. *Nat. Biotechnol.* **2007**, *25*, 1159–1164.
48. Liu, Y.; Yin, Y.; Wang, L.; Zhang, W.; Chen, X.; Yang, X.; Xu, J.; Ma, G. *Biomacromolecules* **2013**, *14*, 3321–3328.
49. Geng, J.; Mantovani, G.; Tao, L.; Nicolas, J.; Chen, G.; Wallis, R.; Mitchell, D. A.; Johnson, B. R. G.; Evans, S. D. Haddleton, D. M. *J. Am. Chem. Soc.* **2007**, *129*,15156–15163.
50. Choi, S.; Mammen, M.; Whitesides, G. M. *J. Am. Chem. Soc.***1997**, *119*, 4103–4111.
51. Tsuchida, A.; Kobayashi, K.; Matsubara, N.; Muramatsu, T.; Suzuki, T.; Suzuki, Y. *Glycoconjugate J.* **1998**, *15*, 1047–1054.
52. Furuike, T.; Aiba, S.; Suzuki, T.; Takahashi, T.; Suzuki, Y.; Yamada, K.; Nishimura S.-I. *J. Chem. Soc., Perkin Trans. 1* **2000**, 3000–3005.
53. Totani, K.; Kubota, T.; Kuroda, T.; Murata, T.; Hidari, K. I.-P.J.; Suzuki, T.; Suzuki Y.; Kobayashi, K.; Ashida, H.; Yamamoto, K.; Usui, T. *Glycobiology* **2003**, *13*, 315–326.
54. Matsuoka, K.; Takita, C.; Koyama, T. *et al. Bioorg. Med. Chem. Lett.* **2007**, *17*, 3826–3830.
55. Barrett, M. P.; Croft S. L. *British Med. Bull.* **2012**, *104*, 175–196.
56. de Oliveira, C.I.; Nascimento, I. P.; Barral, A.; Soto, M.; Barral-Netto, M. *Parasitol. Int.* **2009**, *58*, 319–324.
57. Csaba, N.; Coullerez, G.; Liu, X. Textor, M.; Seeberger P. H.; Merkle, H. P. *Eur. Cells Mater.* **2007**, *14*, 79.
58. Singh, M.; Li, X. M.; Wang, H. Y.; McGee, J. P.; Zamb, T.; Koff, W.; Wang, C. Y.; O'Hagan, D. T. *Infect. Immun.* **1997**, *65*, 1716–1721.
59. McReynolds, K. D.; Gervay-Hague, J. *Chem. Rev.* **2007**, *107*, 1533–1552.
60. Borrok, M. J.; Kiessling, L. L. *J. Am. Chem. Soc.* **2007**, *129*, 12780–12785.
61. Mahalingam, A.; Geonnotti, A. R.; Balzarini, J.; Kiser, P. F. *Mol. Pharm.* **2011**, *8*, 2465–2475.
62. Danial, M.; Root, M. J.; Klok, H.-A. *Biomacromolecules* **2012**, *13*, 1438–1447.

63. Yoshida, T.; Akasaka, T.; Choi, Y. *et al. J. Polym. Sci. Part A Polym. Chem.* **1999**, *37*, 789–800.

64. Becer, C. R.; Gibson, M. I.; Geng, J.; Ilyas, R.; Wallis, R.; Mitchell, D. A.; Haddleton, D. M. *J. Am. Chem. Soc.* **2010**, *132*, 15130–15132.

65. Huang J.; Zhang, Q.; Li, G.-Z.; Haddleton, D. M.; Wallis, R.; Mitchell, D.; Heise, A.; Becer, C. R. *Macromol. Rapid Commun.* **2013**, *34*, 1542–1546.

66. Lipinski, T.; Kitov, P. I.; Szpacenko, A.; Paszkiewicz, E.; Bundle, D. R. *Bioconjugate Chem.* **2011**, *22*, 274–281.

67. Ahmed, M.; Mamba, S.; Yang, X.-H.; Darkwa, J.; Kumar, P.; Narain, R. *Bioconjugate Chem.* **2013**, *24*, 979–986.

68. McL Hill, D. *Innov Pharma Technol* **2005**, 94–97.

69. Svarovsky, S. A.; Szekely, Z.; Barchi Jr., J. J. *Tetrahedron Asymm.* **2005**, *16*, 587–598.

70. Sundgren A.; Barchi Jr., J. J. *Carbohydr. Res.* **2008**, *343*, 1594–1604.

71. Barchi Jr., J. J. *ACS Symposium Book Series "Petite and Sweet: Glyco-Nanotechnology as a Bridge to New Medicines"*. Huang X.; Barchi Jr., J. J., eds. **2011,** *1091*, 161–179.

72. Brinas, R. P.; Sundgren, A.; Sahoo, P.; Morey, S.; Rittenhouse-Olson, K.; Wilding, G. E.; Deng, W.; Barchi Jr., J. J. *Bioconjugate Chem.* **2012**, *23*, 1513–1523.

73. Parry, A. L.; Clemson, N. A.; Ellis, J.; Bernhard, S. S. R.; Davis, B. G.; Cameron, N. R. *J. Am. Chem. Soc.* **2013**, *135*, 9362–9365.

74. Adokoh, C. K.; Quan. S.; Hitt, M.; Darkwa, J.; Kumar, P.; Narain, R. *Biomacromolecules* **2014**, *15*, 3802–3810.

75. Ojeda, R.; de Pazm J. L.; Barrientos, A. G.; Martin-Lomas, M.; Penadés, S. *Carbohydr. Res.* **2007**, *342*: 448–459.

76. Bowman, M.-C.; Ballard, T. E.; Ackerson, C. J.; Feldheim, D. L.; Margolis, D. M.; Melander, C. *J. Am. Chem. Soc.* **2008**, *130*, 6896–6897.

77. Elechiguerra, J. L.; Burt, J. L.; Morones, J. R.; Camacho-Bragado, A.; Gao, X.; Lara H. H.; Yacaman, M. J. *J. NanoBiotechnol.* **2005**, *3*, 6.

78. Shiang, Y.-C.; Ou, C.-M.; Chen, S.-J.; Ou, T.-Y.; Lin, H.-J.; Huang, C.-C.; Chang, H.-T. *Nanoscale* **2013**, *5*, 2756–2764.

79. Di Gianvincenzo, P.; Chiodo, F.; Marradi, M.; Penadés, S. *Methods Enzymol.* **2012**, *509*, 21–40.

80. Martínez-Ávila, O.; Bedoya, L. M.; Marradi, M.; Clavel, C.; Alcamí, J.; Penadés, S. *ChemBioChem* **2009**, *10*,1806–1809.

81. Di Gianvincenzo, P.; Marradi, M.; Martínez-Ávila, O.; Bedoya, L. M.; Alcamí, J.; Penadés, S. *Bioorg. Med. Chem. Lett.* **2010**, *20*, 2718–2721.

82. Chiodo, F.; Marradi, M.; Calvo, J.; Yuste, E.; Penadés, S. *Beilstein J. Org. Chem.* **2014**, *10*, 1339–1346.

15

CARBOHYDRATE NANOTECHNOLOGY AND ITS APPLICATION TO BIOSENSOR DEVELOPMENT

ANDRAS HUSHEGYI, LUDMILA KLUKOVA, TOMAS BERTOK, AND JAN TKAC

Institute of Chemistry, Slovak Academy of Sciences, Bratislava, Slovakia

15.1 GLYCOMICS

Glycans, complex carbohydrates attached to proteins, lipids, or present on the cell surfaces, play an important role in many cellular processes including immune response, tumor metastasis, infection by bacteria or viruses, inflammation, molecular recognition, and cell signaling (Fig. 15.1) [1–3]. The role of glycans in cellular recognition and function was recognized and understood in recent years [4,5], and studying glycans can be useful for developing new therapeutic and diagnostic strategies for many diseases [1,6–8]. Fluorescent glycan biochips or microarrays were developed from DNA and protein arrays in 2002 [9–13] and soon became a successful tool for highly parallel analysis of interaction of glycans with proteins or cells [14–17]. Glycan microarrays are behind numerous discoveries, but requirement to have label for signal generation and quite high detection limits are the main limitation of the approach. This is why glycan biosensors overcoming such limitations have been developed in recent years. Since there are numerous reviews [5,18–21] dealing with analytical performance of glycan biosensors for detection of various glycan-binding proteins, thus the main aim of

Carbohydrate Nanotechnology, First Edition. Edited by Keith J. Stine.

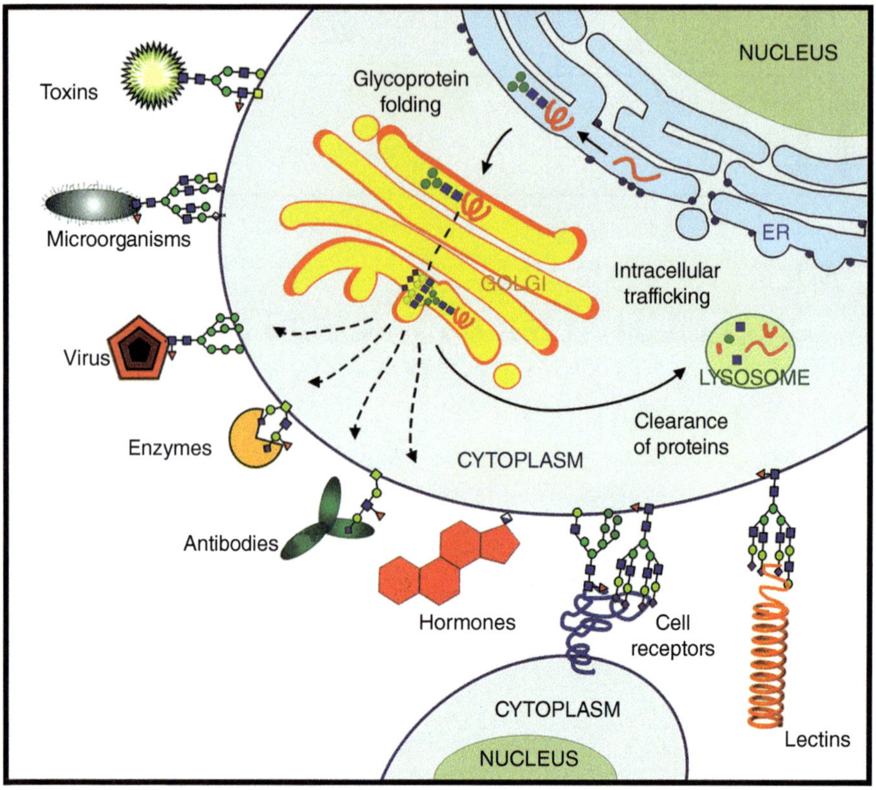

FIGURE 15.1 Involvement of glycans in the cell physiology and pathology in processes such as nascent protein folding, intracellular trafficking, cell–cell communication, interaction with antigens, or in regulation of numerous receptor–ligand interactions. *Source*: Reproduced from Ref. 213 with permission of the Royal Society of Chemistry.

this chapter is description of applicable surface chemistry for glycan immobilization, transducing label-free schemes and challenges ahead, which have to be addressed for construction of glycan biosensors.

15.2 SELF-ASSEMBLED MONOLAYERS (SAMs): MODIFICATION OF INTERFACES AT NANOSCALE

Nanotechnology is dealing with systems, devices, or materials at nanoscale (<100 nm) [22–29]. In comparison to macroscopic materials, nanosized materials show very impressive chemical and physical (thermal, electronic, optical, mechanical, or biological) properties useful in numerous applications [22–31]. Moreover, nanoscience brings a number of benefits to devices based on them such as miniaturization of bioanalytical devices (biosensors) and control over the sensing surface at nanoscale with ability to tune sensitivity and specificity of such devices as well [32–35].

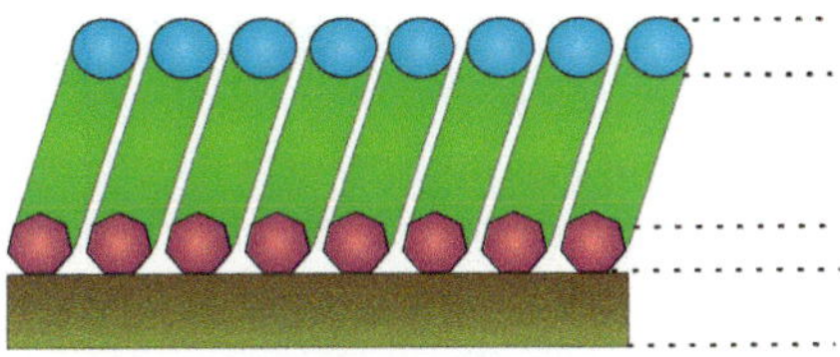

FIGURE 15.2 Scheme of SAM composed of three basic parts.

The creation of materials that use a certain molecular arrangement for a specific function in optical, electronic, or other devices is a main goal of molecular nanoscience [36]. Therefore, nanotechnology sets the current trend in material science and has also prominent position in physics, electrical engineering, biology, or medicine.

To investigate the *in vitro* (eventually *in vivo*) molecular behavior of glycans in biological systems, a well-defined and flexible model system is required. Formation of self-assembled monolayers (SAMs) is the most widely used technique to modify metal or semimetal surface in a controlled way for subsequent conjugation of biomolecules [25,28,37]. SAM is the most elementary form of nanometer-scale thick-film material with a typical thickness of 1–3 nm [28]. The term "self-assembled" describes spontaneous arrangement of atoms and molecules into a highly ordered, densely packed organic (hydrocarbon) layer by sorption on a solid surface. It was first observed as early as 1946 by Zisman's group [38]; however, the interest in the study of the SAMs rapidly increased soon after Nuzzo and Allara's breakthrough study in 1983 [39]. They reported a novel method, which allows preparations of SAM from a dilute solution of thiols on gold surfaces [39].

Since then, SAMs have become one of the most attractive tool for preparation of biosensors, and there are several reasons why SAMs are proper for studies in nanotechnology as well: (i) preparation of SAMs does not require specialized equipment; (ii) SAMs can be formed on material with different morphology including nanomaterials; (iii) SAM-modified nanomaterial (e.g., gold nanoparticles (AuNPs), graphene, carbon nanotubes (CNTs)) can form stable colloidal solutions [25,28,40]; (iv) SAMs interconnect external environment to the electronic (e.g., current–voltage responses in electrochemistry) and optical (e.g., local refractive index in surface plasmon resonance (SPR)) characteristic of metallic structures; and (v) SAMs link molecular-level structures to macroscopic interfacial phenomena such as adhesion, corrosion, friction, wetting, etc. [28,41]. Regarding SAM structure, this thin film consists of the three main parts including head group, backbone chain, and functional terminal group (Fig. 15.2) [32,36].

The head group, also called linking group, forms a strong bond with a solid surface and thus guides the self-assembly process on each type of a substrate. The main chain of variable length determines the thickness of SAMs and contributes to packing and stability of a monolayer by intermolecular forces between backbones (the longer chain, the higher stability) [42]. The terminal functional group determines surface properties (hydrophobicity, hydrophilicity) and allows chemical functionalization of SAM via incorporation of suitable active groups [40]. Among all functional groups, $-NR_2$, –OH, –CN, –COOH, and $-NH_2$ [43] are the most widely used depending on the application developed. The main advantages of SAM in preparation of glycan biosensors can be summarized as follows: (i) conjugation of glycans or nanomaterials [44]; (ii) SAMs mimic the cellular microenvironment of a lipid bilayer structure and also provide oriented attachment of glycans with good reproducibility, durability, and control [34]; and (iii) glycan density on the surface can be tuned [45].

Despite of many advantages and various applications of SAMs, such thin films prepared from one component may have some limitations, for example, undesirable properties caused by high/low surface density of functional groups [46]. It is possible to control and tune properties of SAM by using a mixture of two components (see Section 15.5.2) forming "mixed" SAMs applied in a biosensor construction, as well. The most commonly used mixed SAM is composed of a thiol carrying a functional group and a diluting thiol (e.g., thiol with terminal group $-CH_3$, $-OH$) applied to tune density of a functional one. It is beneficial for the overall interfacial design when a diluting thiol has also antifouling properties to minimize nonspecific interactions (oligoethylene glycol, sulfobetaine) besides being only a passive diluting component [5,47–50]. Glycans can be incorporated into SAM via a direct self-assembly of thiolated glycans or via conjugation on activated functional groups present in SAM [5,51,52] (see Section 15.3). There are many types of surfaces that can be used for SAM formation, but only gold, carbon-based and silicon surfaces are preferred in construction of glycan biosensors [18].

15.2.1 Formation of SAMs on Gold

The major group of SAMs is based on strong affinity of sulfur compounds (thiols, sulfides, or disulfides) to noble and coinage metals such as gold, silver, platinum, copper, mercury, or palladium [28,40,46] (Fig. 15.3). However, the most popular and studied class of self-assembly process is on gold surfaces, because gold does not form stable oxide under ambient conditions [37]. In addition, alkanethiols form a relatively strong bond (binding energy of Au–S bond is 40–45 kcal/mol [43]) with SAM resulting in well-ordered, densely packed, and stable film via the following mechanism [28]:

$$R\text{–}S\text{–}H + Au_n^{\circ} \rightarrow R\text{–}S\ Au^{+}\cdot\ Au_n^{\circ} + 1/2H_2$$

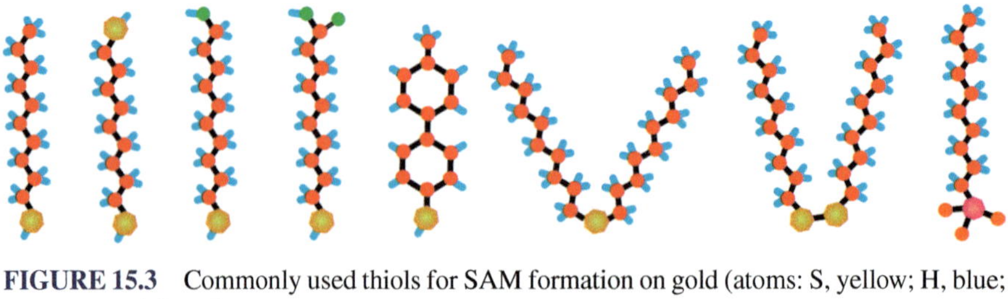

FIGURE 15.3 Commonly used thiols for SAM formation on gold (atoms: S, yellow; H, blue; O, green; and C, red).

Self-assembly of alkanethiols takes place in two distinct phases: initial fast adsorption of thiols onto gold surface and subsequent slow reorganization of alkyl chains as a consequence of interchain forces [53]. Structure and stability of thiolated SAMs are influenced and can be tuned by several critical factors such as concentration and type of thiols, purity of substrate, temperature, and reaction time [53,54]. The monolayer formed has alkyl chains in the all-*trans* extended conformation tilted from the surface normal by ~20–30° (depending on the chain length) [37]. An important advantage of thiolated SAM is that it can be prepared in a laboratory using very simple method by dipping of a clean gold surface into a solution of thiol(s). Another key element for practical application of SAM formation on gold is the fact that sulfur–gold bond is preferentially formed in the presence of other terminal functional groups present at ω-end [43]. Moreover, the bond between gold and sulfur is reversible, thus gold surface can be easily regenerated and reused. On the other hand, Au–S bond has a tendency to oxidize (in air) upon prolonged storage, low-temperature stability in solutions and low redox potential stability window [55].

Thiolate–Au bond formation is widely used not only for patterning planar gold surfaces like Au(111), Au(110), and Au(100) [56] but nowadays is becoming more and more popular to develop assemblies on gold nanoparticles (AuNPs) [57]. In comparison to planar gold, organization of alkanethiols on rough surface enables immobilizations of glycans with higher density and better availability for biorecognition.

15.2.2 Formation of SAMs on Silicon

The second popular system relying on self-assembly is silane-based SAM having substantially different mechanism of formation in comparison to thiolate–gold SAMs [58]. A typical example is reaction of trichloro- or trialkoxysilanes (containing Si atom tetrahedrally bound to three hydrolysable alkoxy groups or chlorines and having a functional group to be deposited on the interface [59]) with hydroxylated surface (mostly silicon oxide, metal oxides, but also glass or graphene oxide [18]) to form strong irreversible bond. The initial step of silanization is hydrolysis of silane molecules with subsequent condensation reaction [59,60] creating Si–O bond (Fig. 15.4). This different chemistry of silane-based SAMs depends strongly on water content in solution and degree of hydration of the surface. In addition, concentration of silanes,

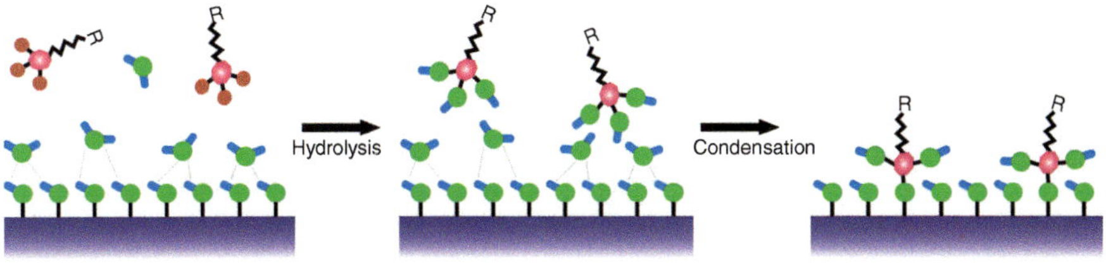

FIGURE 15.4 Simplified mechanism describing formation of silane-based SAMs (atoms: H, blue; O, green; Cl, red; and Si, pink).

reaction time and conditions [61], and pretreatment of the substrate surface (with various acid mixtures) [61] affect kinetics of monolayer formation and its quality, as well. Similar to thiolated SAMs, silane SAMs also allow introducing different functional groups [62], which can be applied for glycan immobilization. On the other hand, silane SAMs provide low surface coverage, the process of patterning is sensitive to moisture and silanes are highly reactive [63].

15.2.3 Formation of SAMs on Carbon

Economic availability, simplicity of modification, for example, by electrografting [64], and good mechanical properties and biocompatibility make carbon materials a smart choice for designing glycan biosensors [18,65]. Traditionally used carbon surfaces such as glassy carbon, graphite, or carbon black can be covalently modified by several effective methods [66], resulting in high stability of grafted layers [67]. However, the quality of formed monolayer is much lower compared to those on gold or silica [18].

Functionalization of carbon-based nanomaterials (especially nanodiamond [68], CNTs, and graphene [69]) has opened novel opportunities in glycoprofiling. CNTs are hollow graphitic nanomaterials composed of cylinders (one or more) of sp^2-hybridized carbon atoms [70,71]. They are widely used due to their remarkable physicochemical properties and simple modification via covalent or noncovalent route (π–π stacking, hydrophobic interaction) [70]. Special attention has graphene attracted since its discovery in 2004 [29]. Graphene is one-atom-thick layer formed from sp^2-hybridized carbons (as in the case of CNTs) arranged in a honeycomb crystal lattice [72]. Although relatively new, graphene has been utilized in various fields thanks to its impressive electrical, optical, and physical properties (exceptional electrical and thermal conductivity, excellent mechanical flexibility and strength, extremely large surface area, and transparency) [73–75]. Besides application of graphene, utilization of graphene oxide (GO, oxidized form of graphene containing also sp^3-hybridized atoms) is on the rise, because this graphene derivative can be produced in a cost effective way. GO production is scalable, handling in solutions is highly reproducible, it is biocompatible, and it can be dissolved in aqueous and organic solvents, as well [75–77]. GO sheet is decorated with oxygen groups (mainly hydroxyl groups and also epoxy, carboxyl, carbonyl, phenol, lactone, and quinone) [78], which are perfect for covalent attachment of glycans [79]. Thermal, chemical, or electrochemical reduction of GO [80] provides opportunities for tuning properties of graphene-based materials (size, shape, content of oxygen group, and conductivity). Graphene can be modified by covalent (free radicals or dienophiles with C=C bonds of pristine graphene, oxygen group of GO) and noncovalent functionalization (Van der Waals, hydrophobic and ionic interactions, π–π stacking) [81], but noncovalent is more preferred to preserve excellent properties of graphene. Buildup of SAMs on graphene and GO was applied in a wide range of applications [32,82,83]. Graphene and CNTs can be effectively patterned by glycans without formation of SAMs [84,85].

15.3 BIOCONJUGATION TECHNIQUES FOR GLYCAN IMMOBILIZATION

Nature provides all the necessary saccharide residues (building blocks) for oligosaccharides synthesis in biosynthetic pathways. Glycoside synthesis is very common reaction in nature, and it is characterized by much larger variation in a structure compared to other biopolymers (because of the presence of multiple stereocenters). Therefore, almost all chemical synthesis of glycoconjugates start from natural precursors (i.e., activated saccharides like UDP-Glc, UDP-Gal, UDP-GlcNAc, UDP-GalNAc, GDP-Fuc, GDP-Man, and CMP-Neu5Ac) [86].

Today, there is a great amount of different protocols for immobilization of biomolecules on surfaces using different functional groups presented on glycans. Since saccharides contain mostly hydroxyl, carbonyl, carboxyl, and amino groups, in this section, we will focus on chemistry of these groups with respect to their utilization for covalent glycan immobilization. However, due to the fact that higher monosaccharides contain several hydroxyl groups, immobilization via such groups can result in random glycan immobilization with low yield. Any reactions (modifications) of hydroxyl groups should be thus used only as a protection for the next step during a synthesis of more complex molecules and will not be discussed in this chapter. However, it is still possible to activate a hydroxyl group, most commonly using *N,N′*-disuccinimidyl carbonate (DSC), which is the smallest homobifunctional NHS (*N*-hydroxysuccinimide) ester cross-linker reagent available. In aqueous solutions, it forms two NHS molecules with release of CO_2. However, in nonaqueous solutions, it can be used for hydroxyl group activation and formation of carbamate linkage with a primary amine-containing molecules [87]. There are basically two general approaches for a glycoside bond formation with retention or exchange of anomeric oxygen (Fig. 15.5) [88].

Very often, different glycosyltransferases or even glycosidases are being involved in the synthesis and immobilization of glycans (Section 15.5.3); however, they are dependent on the presence of natural substrates [89]. Usually, glycan being immobilized on a solid surface must be chemically modified to introduce new functional groups involved in a bioconjugation process. Very often, amino group is used for immobilization [90], since aminosaccharides occur naturally, as building blocks of different polysaccharides (e.g., in hyaluronic acid or sulfated aminosaccharides in heparin). The most commonly used methods for glycan conjugation are summarized in Table 15.1 [91–94]. More comprehensive conjugation approaches are summarized in excellent review papers [1,95–97].

Aminosaccharides offer a spontaneous reaction with aldehyde groups, creating a Schiff base (Fig. 15.6), which is needed to be further reduced, so very often a homobifunctional linker glutaraldehyde is used for the surface immobilization (e.g., using SAM terminated in amine groups) [98,99]. Spontaneous reaction with epoxy group is often used in a microarray technology, where epoxy-coated slides are employed [17,100]. There are other reactions with involvement of amino groups as summarized in Figure 15.7. Other conjugation techniques involve arylating agents

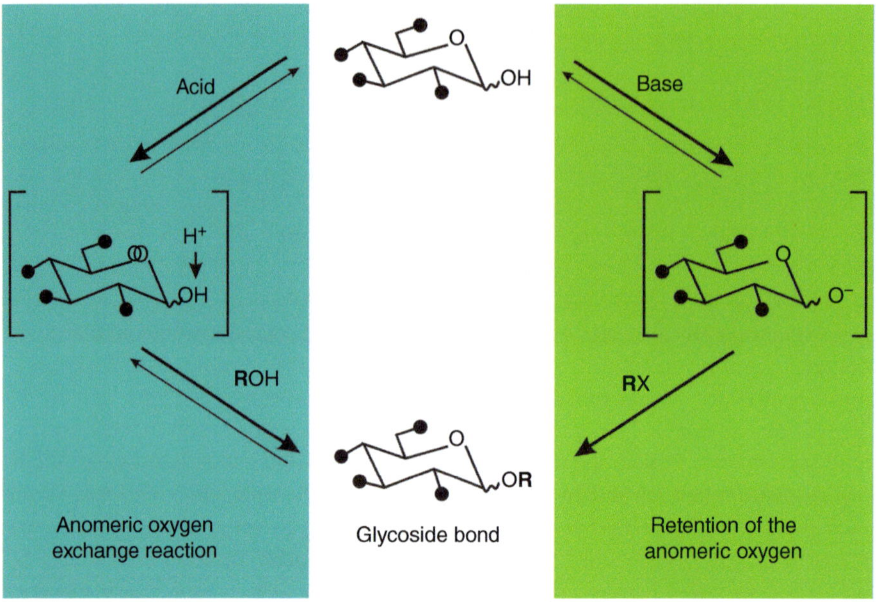

FIGURE 15.5 Two approaches for a glycoside bond formation in chemical synthesis.

TABLE 15.1 Examples of Different Functional Groups Involved in Glycan Activation/Immobilization

Glycan Functional Group	Surface Functional Group	Immobilization
Amino group	NHS ester/epoxy group	Covalent
Ene group	Diene group	Covalent
Sulfhydryl group	Maleimide	Covalent—"click" chemistry
Alkyne group	Azide group	Covalent—"click" chemistry
C_mF_{2m+1}	C_nF_{2n+1}	Noncovalent
Lipid tail	Nitrocellulose	Noncovalent
Oligonucleotide tail	Oligonucleotide	Noncovalent

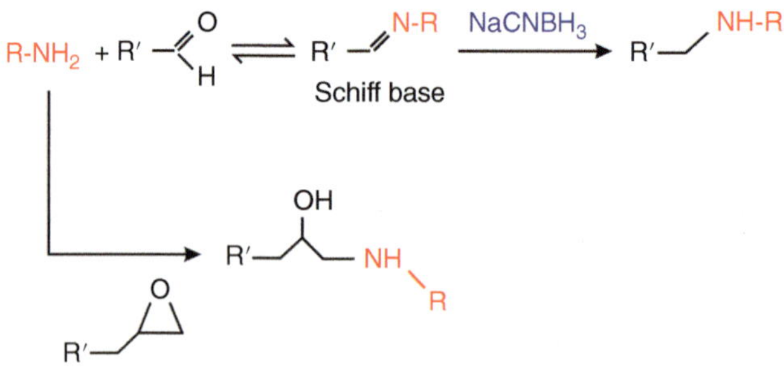

FIGURE 15.6 Reaction of primary amine with an aldose forming a Schiff base (up) and reaction with a free epoxy group (down).

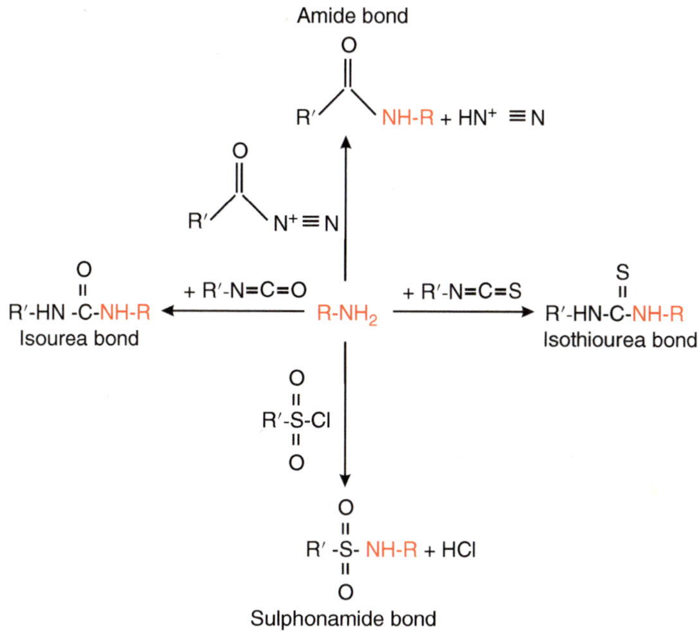

FIGURE 15.7 Overview of some commonly used bioconjugation techniques using a primary amine.

(fluorobenzene) to create arylamide bond, carbonates to create carbamate linkage, and imidoesters to create an amidine linkage [101].

Since gold nanoparticles usually form a bond with amino groups similar to that with a sulfhydryl group on planar gold, amine-containing glycans may be immobilized directly on gold nanoparticles without any activation step needed [45]. The same is true for the thiolated glycans, which can form dense and stable monolayers on planar metal surfaces, as well. These monolayers might be more stabilized using longer linkers (alkanethiol chains), since van der Waals forces stabilize neighboring alkanethiol chains [28]. Glycan monolayers on gold nanoparticles can be more accessible for biorecognition compared to glycans being immobilized on a planar surface due to curvature of the nanoparticle surface, which affects density of a monolayer [102].

In general, using different ligands, we are able to introduce any functional group into a glycan. Conjugation via reducing end [103,104] (attachment of linkers to the reducing end of an aldehyde) is shown in Figure 15.8. It is required to use longer ligand chains for glycan conjugation when a mixed SAM (having functional and diluting thiols) is applied to make glycans fully accessible for the biorecognition.

This method may also be used for the introduction of amino group to sugar residues not containing amino group for subsequent covalent immobilization. Probably, the simplest way to decorate a bioreceptive surface with glycan moieties is introduction of sulfur atom and preparation of SAMs (Fig. 15.9, left). The other approach involves preparation of SAMs using alkanethiol derivatives with different functional groups for subsequent

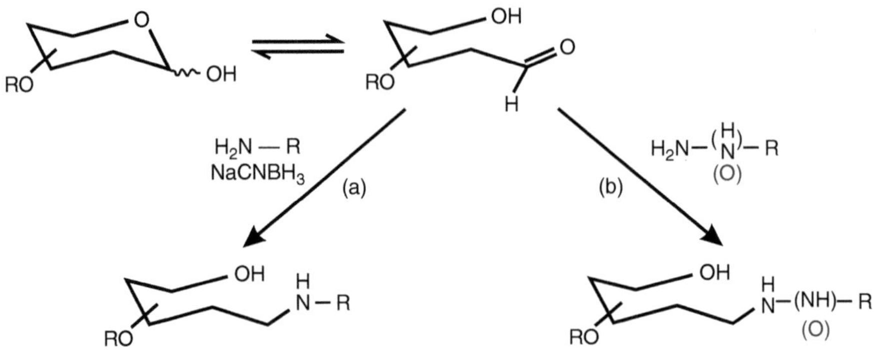

FIGURE 15.8 Standardly used (a) reductive amination and (b) hydrazide (oxyamine)-based chemistry.

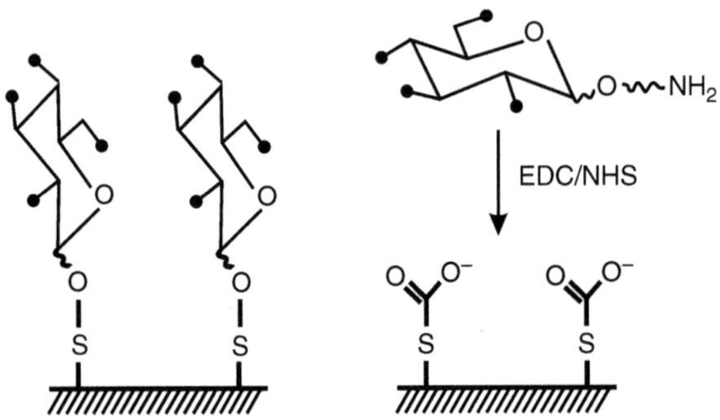

FIGURE 15.9 Immobilization of glycans directly on gold via sulfur forming a thiolate–Au bond (on left) and immobilization of amine-modified glycan on carboxyl-modified surface via EDC/NHS chemistry (on right).

conjugation reaction(s). Probably, the most common approach involves activation of carboxylic group using EDC (*N*-(3-dimethylaminopropyl)-*N′*-ethylcarbodiimide hydrochloride) and NHS (*N*-hydroxysuccinimide) with subsequent immobilization via amino group (Fig. 15.9, right) [105]. NHS is commonly used for the creation of –CO–NH– bond (even without the EDC reagent). However, reaction between carboxylate compound and EDC yields higher surface activation degree compared to just NHS [106,107].

Other routinely used reagents are, for instance, *N*-ethyl-3-phenylisoxazolium-3′-sulfonate (a zero-length cross-linker called also a Woodward's reagent K) used for amide bond formation between carboxylates and amines, or *N,N′*-carbonyldiimidazole (CDI), a highly active carbonylating agent containing two leaving acylimidazole groups (Fig. 15.10). It activates carboxylic group or hydroxyl group for conjugation with other nucleophiles, creating zero-length amide bond or one carbon atom length *N*-alkyl carbamate bond (Fig. 15.11) [101]. For the introduction of sulfur atom (commonly used for peptides), SATA (*N*-succinimidyl *S*-acetylthioacetate) or SATP (*N*-succinimidyl *S*-acetylthiopropionate) are used [108].

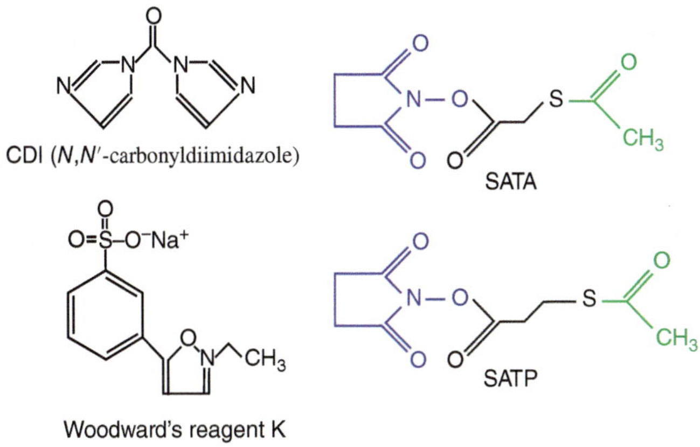

FIGURE 15.10 Structure of some chemicals applied in covalent immobilization protocols.

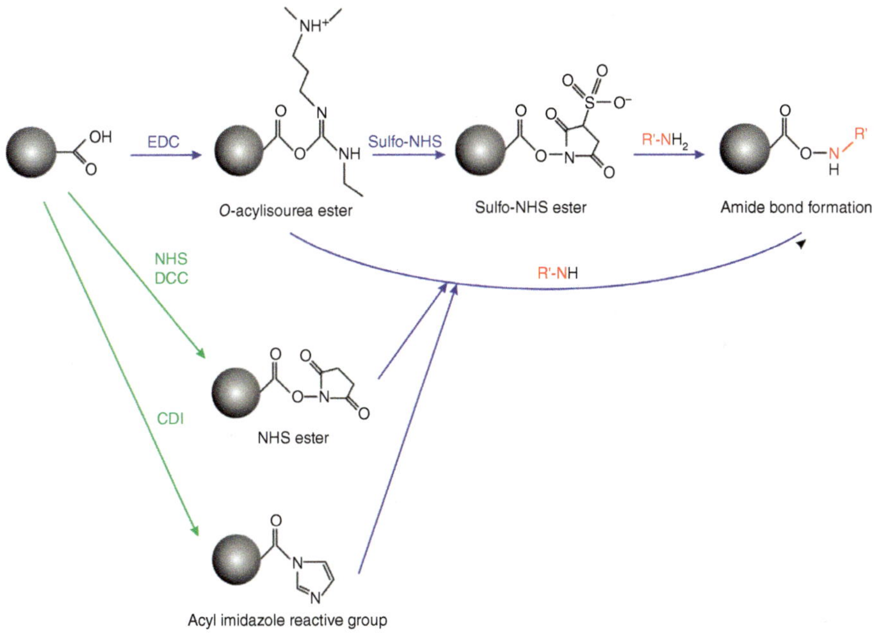

FIGURE 15.11 Possibilities for the amide bond formation with a biomolecule containing carboxylate groups (up) (blue, aqueous solution; green, nonaqueous solution) and amino group-modified surface (down).

The structure of SATA and SATP (Fig. 15.10, right), respectively, contains the *N*-hydroxysuccinimidyl residue for the introduction of protected sulfhydryl to an amine-containing molecule. The second step involving deprotection using hydroxylamine is also necessary (Fig. 15.12) [109]. Prepared molecules may be directly used (after a cleaning procedure using desalting columns) for modification of SAMs.

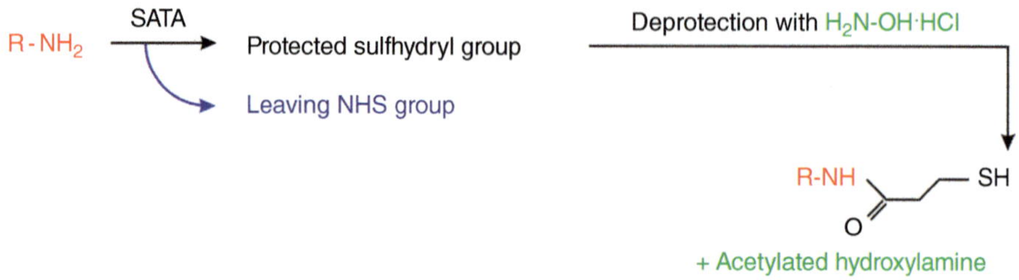

FIGURE 15.12 Introduction of a sulfhydryl group into an amine-containing molecule using SATA reagent.

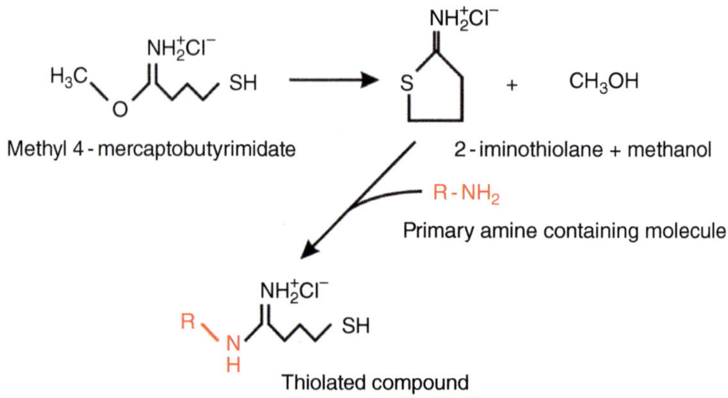

FIGURE 15.13 Mechanism of action of Traut's reagent with a primary amine to form a thiolated compound.

Other approach involves the so-called Traut's reagent (2-iminothiolane). Schematic presentation of this conjugation method is depicted in Figure 15.13 [110]. The sulfhydryl group is also quite reactive and offers a whole palette of reactions in order to create a thioether bond (with iodoacetyl, maleimide, or aziridine derivatives) or aryl thioether bond (using arylating agent, e.g., fluorobenzene) [101].

15.3.1 Cross-linkers

Very often, cross-linker reagents are used to create a bond between two different molecules. The smallest available reagents are so-called zero-length cross-linkers. This type of cross-linkers creates a bond containing no additional atoms (with no intervening linker or spacer) [111]. Commonly used group are carbodiimides such as EDC [112] or 1-cyclohexyl-3-(2-morpholinoethyl)carbodiimide (CMC) (Fig. 15.14) [113], which are the only water-soluble carbodiimides available. For the organic synthesis, diisopropyl carbodiimide (DIC) and dicyclohexyl carbodiimide (DCC) are used. Other examples involve previously mentioned Woodward's reagent or CDI [109,114].

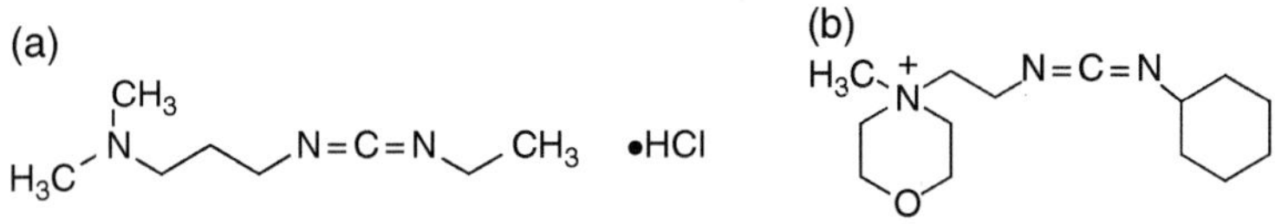

FIGURE 15.14 Two water-soluble zero-length carbodiimide cross-linkers: EDC (left) and CMC (right).

The most popular cross-linking reagent is still glutaraldehyde [115,116], forming polymerizing cyclic hemiacetal at acidic pH in aqueous solutions or α,β-unsaturated aldehyde polymer at basic pH. The formation of the linkage is due to formation of a Schiff base, which needs to be reduced using $NaCNBH_3$ [101,109].

15.3.2 Biotin–(Strept)avidin

A very popular system for the bioconjugation of different molecules and for the surface immobilization is the (strept)avidin–biotin interaction. Streptavidin (60 kDa) is of bacterial origin (*Streptomyces avidinii*) and avidin (66 kDa) is an egg white glycoprotein [101,117]. The dissociation constant of this bond is about 1.3×10^{-15} M. Each molecule of avidin can bind up to four molecules of biotin. Biotinylation of glycan structures may be a useful tool for the glycan immobilization on different surfaces, with previously involved (strept)avidin immobilization (using free amino groups of the protein molecule).

Because biotin is a relatively small molecule (244.31 Da), therefore coupling with other macromolecules is usually done without changing the activity of target molecule. As well as biotinylation, modification with (strept)avidin is also crucial for the preparation of bioconjugates. One typical example is the sodium periodate oxidation of glycan (even present on the glycoprotein molecule), creating reactive aldehydes from the hydroxyl groups, and subsequent labeling with hydrazide-activated (strept) avidin molecule [118–120].

15.3.3 Click Chemistry

Click chemistry via Cu(I)-catalyzed azide–alkyne cycloaddition (CuAAC) is a reaction between azide and alkyne group with formation of 1,4-disubstituted 1,2,3-triazole product (Fig. 15.15). It provides an elegant strategy for glycan immobilization [121,122]. The concept of CuAAC was first time defined in 2002 by independent groups of Fokin and Sharpless [123] and of Meldal et al. [124]. The Cu(I) catalyst decreases activation energy of this reaction. CuAAC became very successful, because it is easy to carry out and it is widely applicable. Moreover, the reaction can be performed at room temperature, in aqueous and also in organic solutions (toluene, chloroform, dichloromethane, DMSO, alcohols, acetone, etc.), in solutions with a wide range of pH (from 4 to 12), and with a large variety of copper catalysts. $CuSO_4$ is the most usually used source of copper atoms [125]. Salts of Cu^{II} are widely used because they are purer and cheaper, compared to Cu^{I} salts such as CuBr or CuOAc. Copper atoms

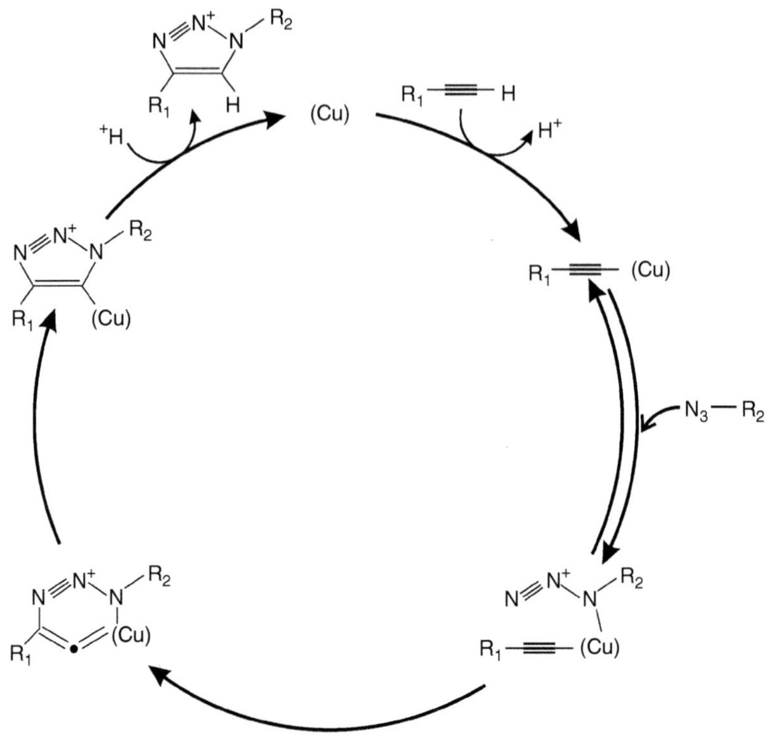

FIGURE 15.15 Mechanism of copper-catalyzed azide–alkyne cycloaddition. *Source*: Reprinted with permission from Ref. 123, Copyright 2002, John Wiley & Sons.

with oxidation number II can be easily reduced with ascorbic acid, sodium ascorbate [123], or through application of a negative potential of −300 mV [126].

When the way of reducing Cu(II) to Cu(I) is considered, there are two types of this reaction either chemically or electrochemically driven, both applicable for glycan immobilization. The first type uses ascorbic acid or sodium ascorbate as a reducing agent, propargyl-terminated reagent, and source of copper usually in mM amount, and the reaction time is usually 24 h [127,128]. The second type reduces Cu(II) with application of a negative potential of −300 mV to the electrode surface with either azide or alkyne being immobilized on this electrode. The reaction time is shorter; it takes from few minutes to few hours. Electrochemical click chemistry uses copper salt and propargyl-terminated reagent in μM concentration [55,126].

The reduced form of copper can react with oxygen and creates reactive oxygen species. In aqueous solution, Cu(I) can react with another copper atoms and form Cu(II) and Cu(0), which can contaminate the surface of the biosensor [129]. Tris(triazolyl)methyl amine ligands as tris[(1-benzyl-1*H*-1,2,3-triazol-4-yl) methyl]amine (TBTA) [125] and tris(3-hydroxypropyltriazolyl-methyl)amine (THPTA—soluble in water) stabilize Cu(I) in the solution, avoiding unwanted side effects [130].

15.4 LABEL-FREE TRANSDUCING PLATFORMS

In order to mimic natural biorecognition of glycans with its binding partners, label-free mode of detection should be preferred since presence of artificial (fluorescent or redox) labels can negatively affect a binding event [131,132]. Thus, in the following section, only label-free methods applicable for preparation of glycan biosensors are discussed.

15.4.1 Electrochemical Impedance Spectroscopy

Electrochemical impedance spectroscopy (EIS) characterizes changes in impedance, resistance, and capacitance on the biosensor surface as a result of surface patterning (SAM formation) or biorecognition [106]. Impedance is a complete resistance of a circuit consisting of resistors, capacitors, or inductors. Electrochemical processes that occur at the electrode–solution interface can be defined by an equivalent circuit. Randles and Ershler circuit model is one of the most often applied one to obtain information about electrolyte solution resistance R_s, electron transfer resistance R_{et}, double-layer capacitance C_d, and Warburg coefficient (Fig. 15.16) [133].

EIS is measured by applying a small AC sinusoidal potential [134] in presence of a redox mediator [106]. Changes in impedance are often presented in a Nyquist plot, where diameter of a semicircle (at higher frequencies) shows R_{et} and linear part (at lower frequencies) describes a diffusion-limited process (Fig. 15.16) [135]. The x-axis in a Nyquist plot represents a real component of impedance (Z') (in-phase current response) and on y-axis is an imaginary part of impedance ($-Z''$) (out-of-phase

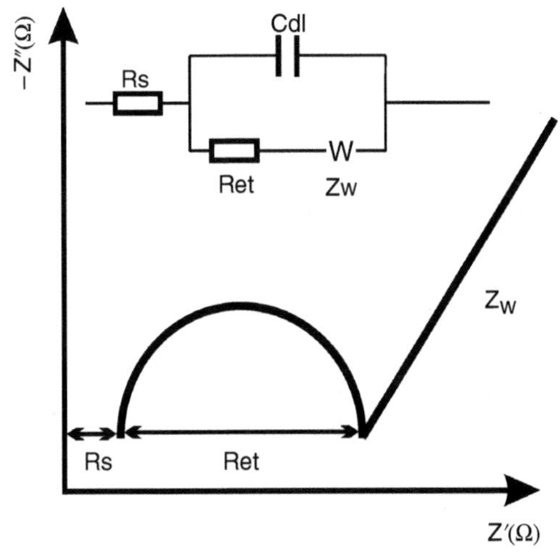

FIGURE 15.16 A Nyquist plot representation (Z' vs. Z'', where Z' is real part of impedance and Z'' is imaginary part of impedance) applied to get interfacial characteristics. In the figure inset, there is Randles circuit model shown (C_{dl}, double-layer capacitance; R_s, resistance of electrolyte solution; R_{et}, electron transfer resistance; Z_W, Warburg impedance).

current response). The applied sinusoidal excitation potential E_t and the corresponding current signal I_t are expressed as

$$E_t = E_0\left(\sin 2\pi ft\right)$$

$$I_t = I_0 \sin\left(2\pi ft + \phi\right)$$

The impedance of the system is defined as

$$Z = \frac{E_t}{I_t} = \frac{E_0 \sin\left(2\pi ft\right)}{I_0\left(\sin 2\pi ft + \phi\right)} = Z_0 \frac{\sin\left(2\pi ft\right)}{\left(\sin 2\pi ft + \phi\right)}$$

where E_0 is the voltage signal amplitude, f is the frequency (Hz), I_0 is the current signal amplitude, ϕ is the phase angle, and t is the time [135].

Impedance of the biosensor is expressed as the summary of real and imaginary part of impedance:

$$Z = \frac{R_s + R_{et}}{\left(1 + \omega^2 R_{et}^2 C_{dl}^2\right)} - \frac{j\omega R_{et}^2 C_{dl}}{\left(1 + \omega^2 R_{et}^2 C_{dl}^2\right)} = Z' + jZ''$$

$$j = \sqrt{-1}$$

Besides traditional application of a Nyquist graphical data representation plotting $-Z''$ versus Z', recent studies suggest that plotting $-C''$ versus C' can provide useful interfacial characteristics, which are changed as a result of a biorecognition event and thus can be applied in biosensing [49,136]. Typical limit of detection (LOD) of an EIS-based glycan biosensor is in low nM or sub nM concentration range [18].

15.4.2 SPR

The most widespread optical detection method applied in biosensing is SPR, measuring changes in resonance angle as a result of a biorecognition event. A thin metaling surface is illuminated by a beam of light with a certain angle of incidence. The angle of reflected light will change with changing refractive index close to the metallic surface due to interaction of an evanescent field with sample [137].

Ligand is immobilized on the SPR surface and then analyte is injected, which after binding to the ligands changes the refractive index in the vicinity of the surface resulting in change of SPR response. The SPR assay procedure consists of three phases: (1) association, introducing of analyte with constant concentration; (2) dissociation, analyte replaced by a plain buffer to follow dissociation of the analyte from the immobilized ligand; and (3) regeneration, preparation of the surface for the next measurement, establishing original baseline (Fig. 15.17) [135].

The SPR detection of molecules large as cells and bacteria is limited, because the evanescent field can penetrate only about 100 nm into the surrounding medium [138].

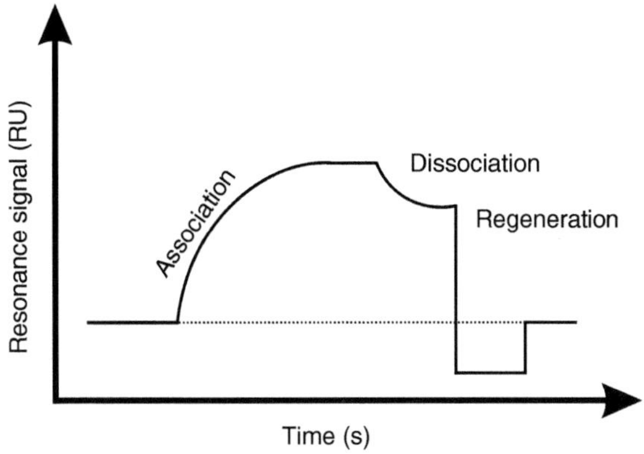

FIGURE 15.17 A typical SPR cycle with association phase during interaction with an analytes, followed by a dissociation phase with a buffer injected into the system, followed by regeneration restoring the surface before injection of an analyte.

The performance of SPR is also limited by molecular weight of an analyte, since a significant change in refractive index (mass) is needed for detection [135]. Typical LOD for glycan biosensors based on SPR is down to nM level [18].

SPR imaging is a special case of SPR system able to work in an array format of analysis, when multiplexed analysis with up to 400 sensing elements was performed [139]. SPR imaging allows to detect analyte down to nM level [18,140].

15.4.3 Localized SPR

A novel technique of SPR is called localized SPR (LSPR) working on metallic nanoparticles (MNPs) either in solution or in immobilized state on glass, gold, or an optical fiber. LSPR compared to SPR has more flexible design and lower cost [141,142]. Most commonly, MNPs are made from gold and silver. For biological applications, the use of Au is more preferred because of their unique optical properties [143]. LSPR is highly dependent on the shape and size of nanoparticles; with these parameters, the wavelength of LSPR can be tuned from visible, near-infrared to infrared region. The phenomenon called LSPR is made by an interaction between the electric field of an incident light and free electrons in the MNPs, and it results in a strong absorption of light, which can be expressed by Mie solution to a Maxwell's equation for extinction of a cross section C_{ext}:

$$C_{ext} = \frac{24\pi^2 R^3 N \varepsilon_m^{3/2}}{\lambda} \frac{\varepsilon_i}{(\varepsilon_r + 2\varepsilon_m)^2 + \varepsilon_i^2}$$

where ε_m is the dielectric constant of surrounding medium, $\varepsilon = \varepsilon_r + i\varepsilon_i$ (ε_r is the real and ε_i is the imaginary part of the dielectric function of MNPs) is the complex dielectric constant of the bulk metal, λ is the wavelength of the incident light, R

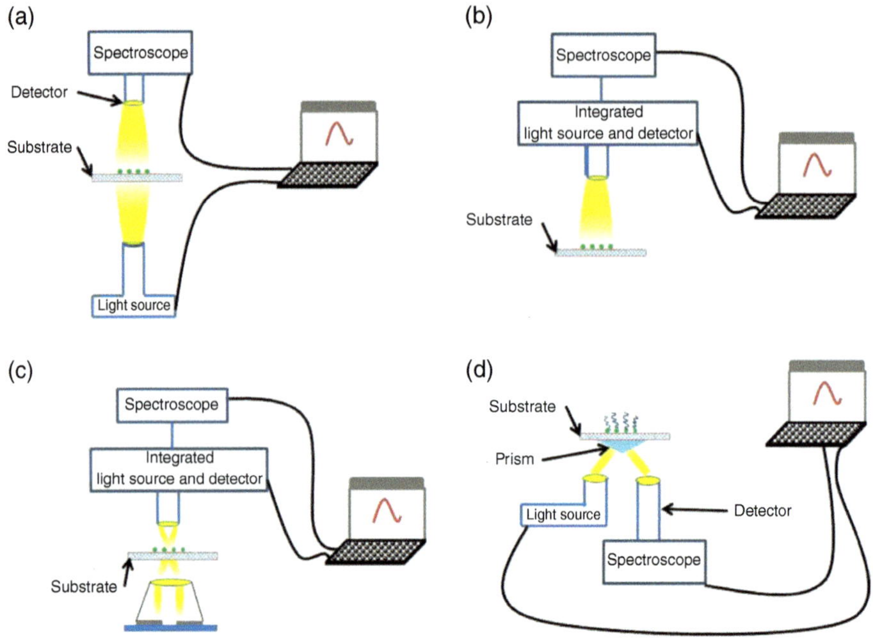

FIGURE 15.18 Diagrams of transmission (a), reflection (b), dark-field scattering (c), and total internal reflection (d) geometries of an LSPR device. *Source*: Reproduced from Ref. 142, an open-access article distributed under the Creative Commons Attribution License.

(smaller than the wavelength of light, $R/\lambda < 0.1$) is the radius of the MNPs, and N is the electron density. The plasmon absorption band appears when $\varepsilon_r = -2\varepsilon_m$ [144]. This interaction causes that the electromagnetic field of the light induces a collective coherent oscillation of the nanoparticle's surface electrons [145]. The resonant frequency of this oscillation highly depends on the size, geometry, composition, dielectric environment, and separation distance between MNPs [143]. Nanoparticles are very sensitive to changes in a local refractive index, and these changes are shown in spectral shift of extinction and scattering spectra. There are few different LSPR configurations applicable in biosensing (Fig. 15.18).

LSPR, a special case of SPR occurring on the surface of nanoparticles or nanostructured surfaces, can also be applied for studying the interaction between immobilized glycans and glycan-binding proteins with LOD of 5 nM [146,147].

15.4.4 Field-Effect Transistor-Based Sensing

Field-effect transistor (FET) is a device that measures changes in conductivity, through changes in electric field caused by small changes on a transistor dielectric interface [148]. There are two major types of FET, metal–oxide–semiconductor-based FET (MOSFET) and junction-based FET. Every FET can be prepared as n-channel (n-p-n) or p-channel (p-n-p) device. For the biosensor fabrication, n-channel-type MOSFETs

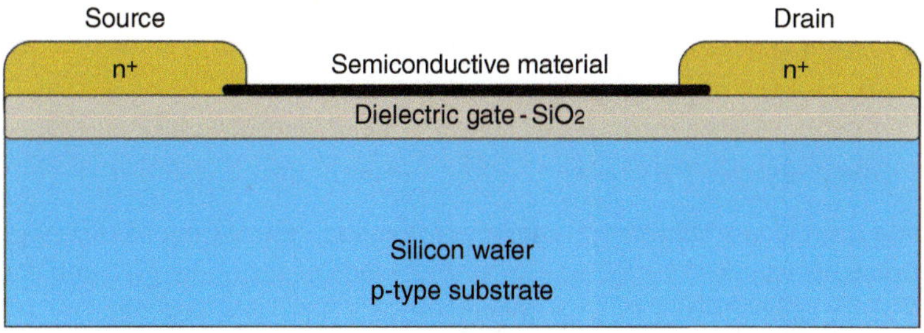

FIGURE 15.19 Structure of n-channel-type field-effect transistor.

are mostly used. In this case, the FET is prepared on p-type substrate (or body), with heavily doped n-type regions (source and drain). Between source and drain, an electrical insulator is deposited, usually a thin layer of SiO_2. This thin layer of SiO_2 is covered with a material (metal, silicon nanowires, CNTs, graphene, etc.), to form the gate. A typical FET sensor consists of a three electrode system, drain, source, and gate electrode. The current flows from source to drain, and the gate electrode modulates the channel conductance (Fig. 15.19) [149].

The configuration of FET can be divided in three types: (1) back gate (the gate electrode is placed underneath the channel region), (2) front gate (the gate electrode is placed in a contact with the surrounding analyte solution), and (3) double gate (it is a combination of both types and consists of primary (driving) and secondary (supporting) gates). The sensitivity of FET biosensors highly depends on its configuration. The double gate type allows a precise control of the channel potential, because the voltage at each of two gates can be controlled independently [150].

As it was mentioned, the thin layer of SiO_2 (electrical insulator, gate) is covered with a semiconductive material. This semiconductive material creates an active part of biosensor, because the biorecognition elements are immobilized onto these semiconductive surfaces. FET biosensors can be prepared with various surfaces as gold [151], silicon nanowires [149], CNTs [152], or graphene [153]. The most widespread FET biosensors are silicon nanowire-based ones (SiNW-FET). There are two techniques for SiNW-FET fabrication, top-down method and bottom-up method. Top-down technique uses silicon-on-insulator wafer, lithographic processes, electron beam etching, and doping technique. In the case of bottom-up technique, the first step is growing of SiNW-FETs from SiNWs in a chemical vapor deposition system [154].

Carbon-based materials have also found applications in preparation of FETs. Single-walled carbon nanotubes (SWCNTs) are rolled up graphite sheets with semiconductive properties. SWCNT-FETs are prepared by standard photolithography and a lift-off process [152]. Another very interesting and promising material for construction of FET is graphene. Compared to CNT or SiNW, which are one-dimensional materials, graphene is one-atom-thick two-dimensional material with low electronic band gap [155]. FET can be modified with graphene through chemical vapor deposition, by epitaxial growth on SiC substrate, or by mechanical exfoliation of graphene on a SiO_2 substrate [153]. FET biosensors provide a label-free detection with real-time monitoring [156]. These sensors are sensitive with LOD down to fM level [149]. It is good to point out the fact that there is also some skepticism about FET-based sensing [157].

15.4.5 Other Sensing Platforms

Ion gating transducing platform is quite interesting way how the interaction of glycan-binding proteins can be detected in a label-free mode of action by glycan biosensors [158]. A single nanochannel with tip opening of 15 ± 3 nm and base opening of 750 ± 10 nm can be used in detection of Con A (Fig. 15.20), when mannose is immobilized inside the nanochannel by measuring changes in the nanochannel conductance. The method is applicable for detection of Con A in the concentration window 10–1000 nM, and the surface can be regenerated after binding by mannose [158].

Atomic force microscopy (AFM) is based on scanning of surfaces by monitoring interaction between a sharp tip and the surface. The AFM consists of 5 basic components: a piezoelectric scanner, a flexible cantilever containing a sharp tip, laser, a photodiode detector, and feedback electronics. The principle of AFM is really simple including movements of a flexible cantilever with a sharp tip on analyzed surface, which is monitored by changes in laser deflection [159]. Photodiode releases a laser beam, which is reflected from the back of the cantilever. Small

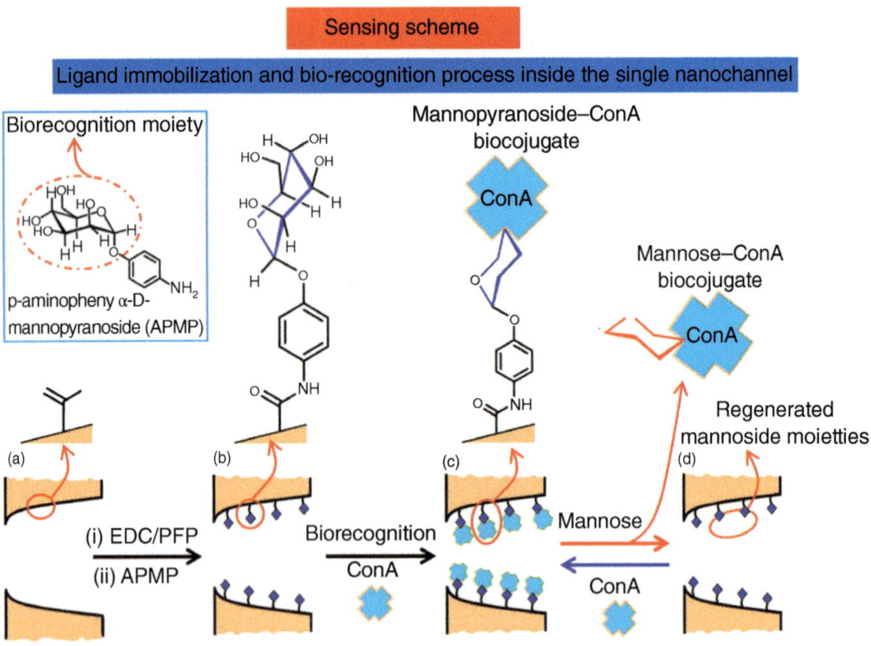

FIGURE 15.20 A scheme showing (a) asymmetric nanochannel naturally covered by –COOH groups, (b) covalent immobilization of glycan, (c) biorecognition process with Con A, and (d) regeneration of the surface by mannopyranoside. *Source*: Reprinted with permission from Ref. 158, Copyright 2013, American Chemical Society.

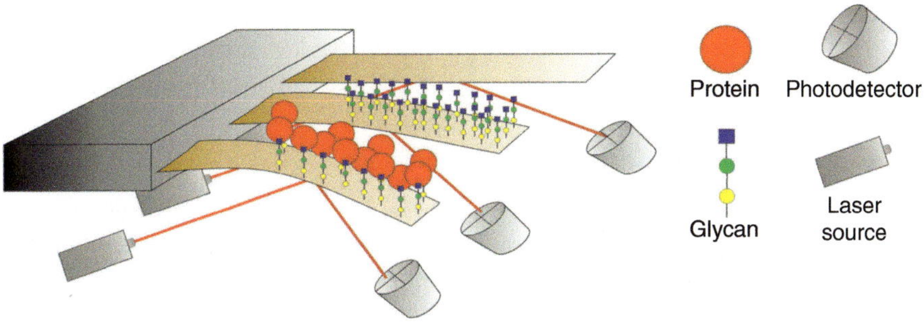

FIGURE 15.21 Detection of carbohydrate–protein interactions with cantilever microarray. *Source*: Reproduced from Ref. 18 with permission from the Royal Society of Chemistry.

changes on the deflection of the cantilever are monitored as the change in the position of the reflected laser beam on the surface of a photodiode detector. AFM allows to monitor protein–glycan interactions with high resolution [160]. There are three modes applicable for AFM imaging: (1) a contact mode, (2) a tapping mode [161], and (3) a noncontact mode [162]. In the contact mode, the cantilever is in a constant contact with the surface. In the tapping mode, the cantilever oscillates with a constant resonance frequency, resulting in intermittent contact between the tip and the surface. The changes in heights in the samples change the amplitude of the cantilever's oscillations [161]. In the case of noncontact mode, a weak force is applied between the tip and the samples on the surface. The contact and the tapping mode can destruct the samples on surface and also the tip of the cantilever. The noncontact mode is nondestructive, because the tip is not in real contact with the samples; it just monitors the interface of the surface [162].

Cantilever microarray (Fig. 15.21) is a system of highly sensitive micrometer-sized cantilevers offering sensitive, portable, cheap, highly parallel, and fast detection without labeling. It can measure changes in surface stress, temperature, and mass via a cantilever deflection. For monitoring of deflection, there are many techniques, but the most common is the one based on a laser beam focusing on the back of the cantilever (like in case of AFM). This beam is reflected from a cantilever onto a photodetector surface with ability to detect just 1 nm deflection. Piezoelectric readout system can be used as well. In this case, cantilevers are in contact with a piezoelectric sensor. Mass change generates a mechanical stress resulting in electrical potential generation. Capacitive readout is based on close placement of cantilevers to a parallel electrode, sensing changes in capacitance as a result of deflection [163].

15.5 CONSTRUCTION OF GLYCAN BIOSENSORS

There are few important aspects, which have to be considered in order to prepare highly robust and sensitive glycan biosensors. Lessons about critical aspects of glycan biosensor construction can be learnt especially from massively utilized glycan microarrays even though few glycan biosensors so far constructed shed light on some important construction features, as well. In the following section, the effect of topology, glycan density control, effect of linker attached to the glycan moiety, immobilization process itself, and some construction features will be discussed.

15.5.1 Topology of the Surface at Nanoscale

Pandey *et al.* studied construction of glycan biosensors either on planar or nanoporous gold [134,164]. Binding of Con A lectin to mannose present on gold in a form of thiolated SAM showed differences in its binding depending on the surface topology as investigated by EIS [164]. Similarly, dependence for binding of soybean agglutinin on globotriose glycan immobilized either on planar or nanoporous gold surface was studied, as well [134]. Interestingly, EIS method can be applied for calculation of association rates [134]. Nanoporous gold monolith with immobilized glycans can be applied for purification of glycan-binding proteins [165].

15.5.2 Glycan Density Control

Density of mannose-terminated glycans immobilized on the silicon surface via click chemistry was determined by FTIR-ATR spectroscopy, a technique able to provide surface density of a *Lens culinaris* lectin incubated on such surface [166]. The study confirmed that the optimal separation between mannose moieties is 1.4 nm for effective interaction with lectin having size of 5.9 nm (Fig. 15.22) [166]. For binding

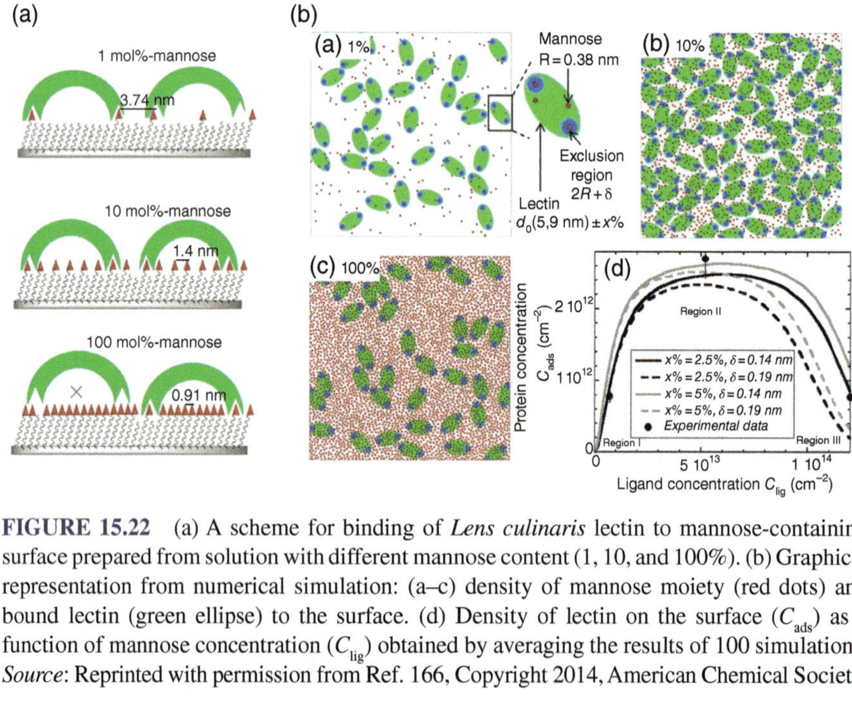

FIGURE 15.22 (a) A scheme for binding of *Lens culinaris* lectin to mannose-containing surface prepared from solution with different mannose content (1, 10, and 100%). (b) Graphical representation from numerical simulation: (a–c) density of mannose moiety (red dots) and bound lectin (green ellipse) to the surface. (d) Density of lectin on the surface (C_{ads}) as a function of mannose concentration (C_{lig}) obtained by averaging the results of 100 simulations. *Source*: Reprinted with permission from Ref. 166, Copyright 2014, American Chemical Society.

of antiglycan antibody with larger size compared to lectin with width of an antigen-binding Fab fragment of 7 nm, the optimal glycan separation is 5 nm [167]. In another study, it was confirmed that the decrease of distance between glycopolymer ligands from 35 to 10 nm printed on glycan arrays resulted in an increased affinity of lectin binding [168]. The distance between two binding sites of a glycan-binding protein can span distances from 0.9 nm (Shiga-like toxin) to 10.0 nm (*Ricinus communis* agglutinin), a feature that should be taken into consideration for designing highly sensitive glycan biosensors [169]. The optimal glycan density within a mixed SAM for binding of Con A at density of 1.8×10^{12} molecules/cm^2 was found out to be 10% [170]. Formation of a single-component glycan SAM resulted in 42-fold decreased amount of bound Con A [170]. The study showed that difference in length of 6 carbons, that is, $-(CH_2)_6-$ moiety, between glycan-containing and a diluting thiol was important for efficient Con A binding showing 5.9-fold lower K_D value compared to K_D value obtained for Con A bound to a mixed SAM layer composed of both components of the same length [170]. Multivalent topological glycan presentation in special cases can result in up to three orders of magnitude lower affinity constant K_D (i.e., higher affinity) of lectin binding [171–173].

A detailed study aimed at investigation of the effect of small change in the construction of two glycan arrays prepared on either epoxy- or hydroxysuccinimide-containing slides with only small change in the immobilization process showed that such glycan arrays might exhibit different binding profile to various lectins, monoclonal/polyclonal anticarbohydrate antibodies, viruses, and human Siglecs [174]. The result of the study is that glycan arrays with similar glycan structures but differing in glycan immobilization/presentation might provide different binding pattern [174]. Thus, glycan biosensors have to be prepared with caution, and their practical application has to be extensively validated. A linker putting together a glycan moiety with functional groups applied for surface binding has a strong influence on binding of antiglycan antibody to surface-confined glycans, when small change in the chemistry of linker can completely prevent binding of antibody to glycan, what can be seen in Figure 15.23 [175]. AFM is an additional technique applied for better understanding of the effect of glycan presentation on the affinity of glycan–protein binding [176].

Very important study describing formation of glycan clusters over time, when a mixed SAM composed of glycan thiol and a diluting thiol deposited on the gold surface, was published recently [177]. Glycans have tendency to form glycan cluster both at low and high initial density, as can be seen in Figures 15.24 and 15.25. Such observations have important implication for construction of various biosensing devices with glycans immobilized on gold surfaces in the form of mixed SAM.

Alternative to formation of SAM glycan layers on gold is the development of surfaces with glycolipids on hydrophobic surfaces as introduced by Feizi [10,178,179] to better mimic natural glycan–protein biorecognition. Glycolipids as a significant part of the plasma membrane are interacting with proteins and other biomolecules to form an array of heterogeneous clusters with diverse biological properties [180]. It was really observed that incorporation of glycans into the lipid layer deposited on indium tin oxide glass slide can be applied for studying interaction between glycan

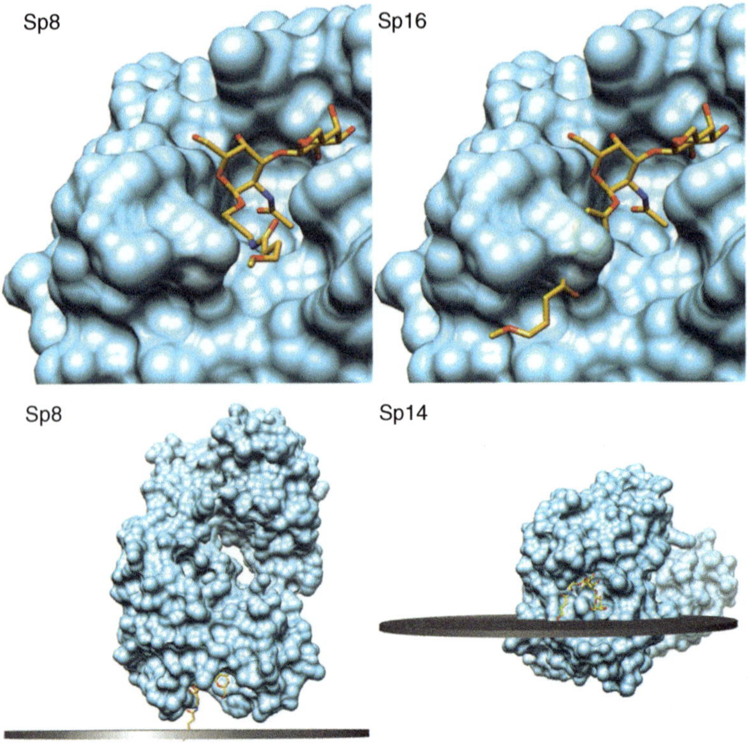

FIGURE 15.23 The effect of linker present in the glycan structure on interaction of glycan-binding protein to glycan immobilized on the surface. While linker present in the Sp8 glycan is aliphatic with a high degree of flexibility allowing efficient interaction between the protein and glycan (upper left), a more rigid, aromatic linker present in Sp16 does not allow efficient binding (upper right). A proper orientation of Sp8 glycan allows interaction with protein (lower left), while position of glycan Sp14 in the space due to presence of unsuitable linker does not expose glycan for the interaction with the protein (lower right). *Source*: With permission from Ref. 175, Oxford University Press.

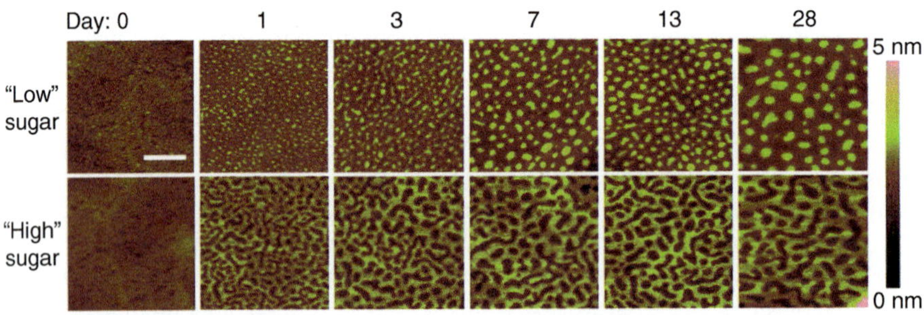

FIGURE 15.24 Formation of glycan clusters on ultraflat gold surface imaged by AFM. Mixed SAM was formed using ratio between thiolated glycan and a diluting thiol of 2:3 (upper row) or 4:1 (lower row). Scale bar = 100 nm. *Source*: Reprinted with permission from Ref. 177, Copyright 2012, American Chemical Society.

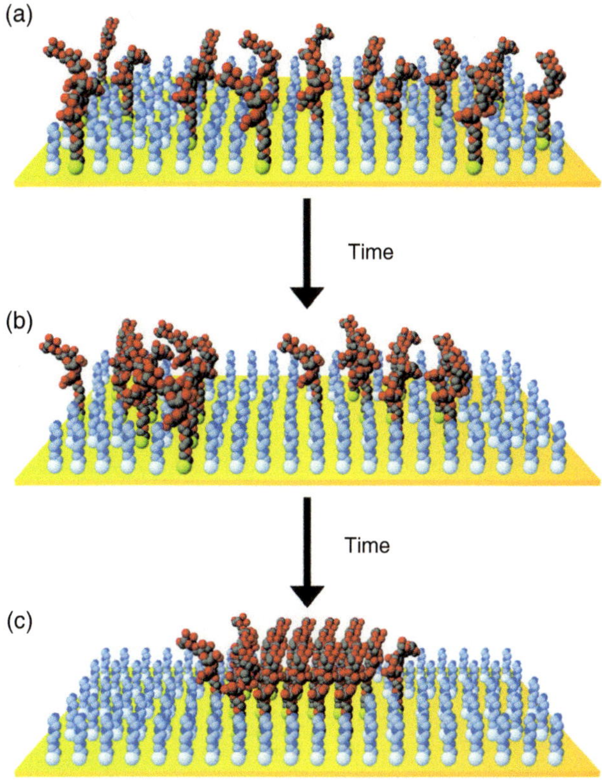

FIGURE 15.25 Proposed scheme of time-dependent formation of glycan clusters. *Source*: Reprinted with permission from Ref. 177, Copyright 2012, American Chemical Society.

and galactosyltransferase [181]. The study showed significant ninefold enhancement of enzymatic activity upon formation of a glycan cluster [181]. Moreover, different activity enhancement was observed in case mixed lipid microdomain contained *N*-acetylglucolipids and galactolipids [181]. Liposomal glyco-microarray can be applied for studying interaction between glycans and lectins or other glycan-binding proteins including lectins, bacterial toxins, and antibodies, as well [180,182].

Probably, the best method how to adjust spatial separation between neighboring glycans on the surface is to apply oligonucleotide-modified glycan sequences to the surface with bound ssDNA sequences [183,184]. This method not only allows precise separation between neighboring glycans (Fig. 15.26) but allows to attach several different glycan moieties along ssDNA chain with defined separation.

15.5.3 *In Situ* Glycan Synthesis by Enzymes

Synthesis of glycans by the action of enzymes is a vital alternative to chemical glycan synthesis in order to minimize synthetic steps and simplify overall synthetic protocols [185,186]. This approach can be effectively applied for direct *in situ* synthesis of glycans on surfaces [187,188]. It is important to take into account that density of glycan

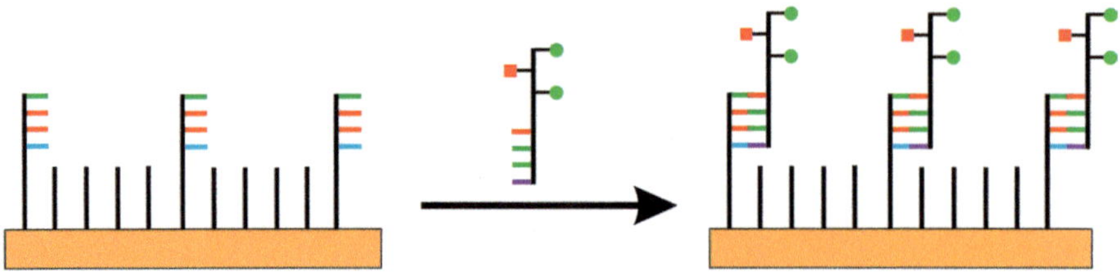

FIGURE 15.26 DNA-directed immobilization of glycans.

ligands and/or presence of glycan clusters on the surface can significantly influence the rate and affinity of glycosyltransferases/glycosidases for enzymatic action on glycans chains [181,185,189]. Incorporation of galactose by galactosyltransferase was the highest on the surface containing 70% of GlcNAc within mixed SAM [190], that is, at glycan density significantly larger than optimal glycan density for lectin recognition (10%) [170]. Interestingly, inert PEG brushes can be quite effective modified by glycans via transglycosylation reaction, when using lactose galactose can be conjugated to PEG moieties by the action of β-galactosidase [191].

15.5.4 Immobilization of Glycopeptides *versus* Glycans

Glycan recognition in nature can also involve interaction of a glycan-binding partner with a peptide sequence beneath glycan moiety as in the case of HIV-neutralizing antibodies [192]. Thus, it is not surprising that immobilized glycopeptides are studied for the ability to bind to lectins [193]. Moreover, glycopeptides can produce conformationally constrained SAM layers [194]. One recent study pointed out to the fact that immobilized glycopeptides can interact with disease-related autoantibodies, which are not bound to surfaces containing isolated peptide or glycan moieties, and such antiglycopeptide antibodies can be cancer biomarkers [195].

15.5.5 Other Construction Aspects

Glycan biosensors should be integrated also into flow-through system. Such integration has potential to significantly reduce required time (17 h vs. 3 h) of interaction between immobilized glycan and its analyte and can provide kinetic constants of interaction [196]. Integration of glycan biosensors with a detection platform at least to some degree offering multiplexing of analysis is important in order to achieve cost- and time-effective biosensing.

15.6 APPLICATION OF GLYCAN BIOSENSORS

Lectins of plant origin are first choice to show usefulness of glycan biosensors [166,168], but glycan biosensors should be applied in detection of other lectins of human origin such as galectins, which could be, for example, cancer biomarkers [197,198].

Antibodies against various types of glycans can indicate allergy [199] or various types of cancer, and thus, antiglycan antibodies should be more intensively detected by glycan biosensors as potential disease biomarkers. Two different detection platforms with immobilized glycans including suspension arrays and ELISA-like approach were compared to printed microarrays, with a focus on their analytical and diagnostic potential [200]. The study showed that the presence of antibodies in human blood against P1 glycan antigen (Galα1-4Galβ1-4GlcNAcβ) can be applied for discrimination between healthy individuals and those suffering from cancer. While printed microarrays is ideal for a discovery approach, a suspension array is suitable for diagnostic purposes [200]. Autoantibodies against P1 glycan proved to become a perspective cancer biomarker, since this biomarker provides higher selectivity and sensitivity of cancer diagnosis over routinely used CA125 biomarker [201]. Varki's group is intensively working to prove that the presence of *N*-glycolylneuraminic acid (Neu5Gc) in the human body from a dietary intake (red meat) after incorporation into tissues can elicit inflammatory response [202]. Thus, it is suggested that determination of antibodies against Neu5Gc can be used as a potential method to evaluate risk of getting inflammation-related diseases such as cancer or cardiovascular diseases [203].

Other proteins such amyloid β-protein (Alzheimer disease) [204], viral proteins hemagglutinins [205], toxic proteins such as ricin [206] or cholera toxin [207], and various enzyme activities [188,208] have been detected by glycan biosensors, but repertoire of proteins interacting with glycan biosensors should be extended to find novel perspective disease biomarkers.

Other analytes such as intact bacterial cells [209,210], viral particles [211], and tumor cells [212] have been detected by glycan biosensors, but more frequent application of this analytical platform in analysis of these large particles is still awaiting.

15.7 CONCLUSIONS

Better understanding of key attributes of interactions between glycans and glycan-processing enzymes or glycan-recognizing proteins will be without any doubt beneficial for the society. Recent studies suggest that potent vaccine against HIV with broad specificity can be produced [192]. Some of glycan-binding proteins have potential to become cancer biomarkers or biomarkers of other diseases [203].

Glycan biosensors could be a vital alternative to glycan biochips in the future, but a lot of work is still needed to fulfill this task with effort focused on improvement of sensitivity of some transducing platforms. The main limitation of glycan biosensors compared to glycan biochips is still extremely low throughput of detection. There are some detection platforms, which has potential to offer multiplexed format of analysis (LSPR and SPR imaging), but such technologies lack sensitivity of analysis. From label-free formats of detection especially EIS can offer some degree of multiplexing, while allowing detection limits down to aM levels [50].

There are still plenty of factors that have to be taken into account such as control of surface topology at nanoscale, glycan surface density, immobilization of glycopeptides along glycans, wider utilization of enzymes for glycan surface patterning, and integration of glycan biosensors with flow-through system for robust biosensing of glycan-binding biomolecules.

ACKNOWLEDGMENT

The financial support from the Slovak Scientific Grant Agency VEGA 2/0162/14 and from the Slovak Research and Development Agency APVV 0282-11 is acknowledged. The research leading to these results has received funding from the European Research Council under the European Union's Seventh Framework Program (FP/2007-2013)/ERC Grant Agreement No. 311532, and this work has received funding from the European Union's Seventh Framework Program for Research, Technological Development and Demonstration under grant agreement No. 317420. This publication was made possible by NPRP grant 6-381-1-078 from the Qatar National Research Fund (a member of the Qatar Foundation). The statements contained are entirely the responsibility of the authors.

REFERENCES

1. Park, S.; Gildersleeve, J. C.; Blixt, O.; Shin, I. *Chem. Soc. Rev.* **2013**, *42*, 4310–4326.
2. Reichardt, N. C.; Martín-Lomas, M.; Penadés, S. *Chem. Soc. Rev.* **2013**, *42*, 4358–4376.
3. Varki, A. *Essentials of glycobiology*; Cold Spring Harbor Laboratory Press: Cold Spring Harbor, NY, **2009**.
4. Kluková, Ľ.; Bertók, T.; Kasák, P.; Tkac, J. *Anal. Methods* **2014**, *6*, 4922–4931.
5. Zeng, X.; Andrade, C. S.; Oliveira, M. L.; Sun, X.-L. *Anal. Bioanal. Chem.* **2012**, *402*, 3161–3176.
6. Dalziel, M.; Crispin, M.; Scanlan, C. N.; Zitzmann, N.; Dwek, R. A. *Science* **2014**, *343*, 1235681.
7. Alley, W. R.; Mann, B. F.; Novotny, M. V. *Chem. Rev.* **2013**, *113*, 2668–2732.
8. Burton, D. R.; Poignard, P.; Stanfield, R. L.; Wilson, I. A. *Science* **2012**, *337*, 183–186.
9. Drickamer, K.; Taylor, M. E. *Genome Biol.* **2002**, *3*, 1034.
10. Fukui, S.; Feizi, T.; Galustian, C.; Lawson, A. M.; Chai, W. *Nat. Biotechnol.* **2002**, *20*, 1011–1017.
11. Houseman, B. T.; Mrksich, M. *Chem. Biol.* **2002**, *9*, 443–454.
12. Park, S.; Shin, I. *Angew. Chem. Int. Ed.* **2002**, *41*, 3180–3182.
13. Wang, D.; Liu, S.; Trummer, B. J.; Deng, C.; Wang, A. *Nat. Biotechnol.* **2002**, *20*, 275–281.
14. de Graaf, M.; Fouchier, R. A. M. *EMBO J.* **2014**, *33*, 823–841.
15. Stencel-Baerenwald, J. E.; Reiss, K.; Reiter, D. M.; Stehle, T.; Dermody, T. S. *Nat. Rev. Microbiol.* **2014**, *12*, 739–749.
16. Cohen, M.; Varki, A. In *International Review of Cell and Molecular Biology*; Kwang, W. J., Ed.; Academic Press, **2014**; Vol. *308*.
17. Donczo, B.; Kerekgyarto, J.; Szurmai, Z.; Guttman, A. *Analyst* **2014**, *139*, 2650–2657.

18. Hushegyi, A.; Tkac, J. *Anal. Methods* **2014**, *6*, 6610–6620.

19. Reuel, N. F.; Mu, B.; Zhang, J.; Hinckley, A.; Strano, M. S. *Chemical Society Reviews* **2012**, *41*, 5744–5779.

20. Reuel, N. F.; Grassbaugh, B.; Kruss, S.; Mundy, J. Z.; Opel, C.; Ogunniyi, A. O.; Egodage, K.; Wahl, R.; Helk, B.; Zhang, J.; Kalcioglu, Z. I.; Tvrdy, K.; Bellisario, D. O.; Mu, B.; Blake, S. S.; Van Vliet, K. J.; Love, J. C.; Wittrup, K. D.; Strano, M. S. *ACS Nano* **2013**, *7*, 7472–7482.

21. Mu, B.; Zhang, J.; McNicholas, T. P.; Reuel, N. F.; Kruss, S.; Strano, M. S. *Acc. Chem. Res.* **2014**, *47*, 979–988.

22. Alivisatos, A. P. *Science* **1996**, *271*, 933–937.

23. Brus, L. E. *J. Chem. Phys.* **1984**, *80*, 4403–4409.

24. Brust, M.; Walker, M.; Bethell, D.; Schiffrin, D. J.; Whyman, R. *J. Chem. Soc. Chem. Commun.* **1994**, *7*, 801–802.

25. Daniel, M.-C.; Astruc, D. *Chem. Rev.* **2003**, *104*, 293–346.

26. Iijima, S. *Nature* **1991**, *354*, 56–58.

27. Iijima, S.; Ichihashi, T. *Nature* **1993**, *363*, 603–605.

28. Love, J. C.; Estroff, L. A.; Kriebel, J. K.; Nuzzo, R. G.; Whitesides, G. M. *Chem. Rev.* **2005**, *105*, 1103–1170.

29. Novoselov, K. S.; Geim, A. K.; Morozov, S. V.; Jiang, D.; Zhang, Y.; Dubonos, S. V.; Grigorieva, I. V.; Firsov, A. A. *Science* **2004**, *306*, 666–669.

30. Pyrzynska, K. *Trends Anal. Chem.* **2013**, *43*, 100–108.

31. Yin, Y.; Talapin, D. *Chem. Soc. Rev.* **2013**, *42*, 2484–2487.

32. Newton, L.; Slater, T.; Clark, N.; Vijayaraghavan, A. *J. Mater. Chem. C* **2013**, *1*, 376–393.

33. Halik, M.; Hirsch, A. *Adv. Mater.* **2011**, *23*, 2689–2695.

34. Park, B.-W.; Yoon, D.-Y.; Kim, D.-S. *J. Electroanal. Chem.* **2011**, *661*, 329–335.

35. Shen, J.; Li, Y.; Gu, H.; Xia, F.; Zuo, X. *Chem. Rev.* **2014**, *114*, 7631–7677.

36. Turchanin, A.; Gölzhäuser, A. *Prog. Surf. Sci.* **2012**, *87*, 108–162.

37. Gooding, J. J.; Ciampi, S. *Chem. Soc. Rev.* **2011**, *40*, 2704–2718.

38. Bigelow, W. C.; Pickett, D. L.; Zisman, W. A. *J. Colloid Sci.* **1946**, *1*, 513–538.

39. Nuzzo, R. G.; Allara, D. L. *J. Am. Chem. Soc.* **1983**, *105*, 4481–4483.

40. Vericat, C.; Vela, M. E.; Benitez, G.; Carro, P.; Salvarezza, R. C. *Chem. Soc. Rev.* **2010**, *39*, 1805–1834.

41. Romashov, L. V.; Ananikov, V. P. *Chem. Eur. J.* **2013**, *19*, 17640–17660.

42. Dyne, J.; Lin, Y.-S.; Lai, L. M. H.; Ginges, J. Z.; Luais, E.; Peterson, J. R.; Goon, I. Y.; Amal, R.; Gooding, J. J. *ChemPhysChem* **2010**, *11*, 2807–2813.

43. Srisombat, L.; Jamison, A. C.; Lee, T. R. *Colloids Surf., A* **2011**, *390*, 1–19.

44. Samanta, D.; Sarkar, A. *Chem. Soc. Rev.* **2011**, *40*, 2567–2592.

45. Bertok, T.; Sediva, A.; Katrlik, J.; Gemeiner, P.; Mikula, M.; Nosko, M.; Tkac, J. *Talanta* **2013**, *108*, 11–18.

46. Shen, C.-H.; Lin, J.-C. *Colloids Surf., B* **2010**, *79*, 156–163.

47. Xu, Q.; Davis, J. J. *Electroanalysis* **2014**, *26*, 1249–1258.

48. Bertok, T.; Klukova, L.; Sediva, A.; Kasák, P.; Semak, V.; Micusik, M.; Omastova, M.; Chovanová, L.; Vlček, M.; Imrich, R.; Vikartovska, A.; Tkac, J. *Anal. Chem.* **2013**, *85*, 7324–7332.

49. Jolly, P.; Formisano, N.; Tkáč, J.; Kasák, P.; Frost, C. G.; Estrela, P. *Sens. Actuat. B: Chem.* **2014**, doi:10.1016/j.snb.2014.1011.1083
50. Klukova, L.; Bertok, T.; Petrikova, M.; Sediva, A.; Mislovicova, D.; Katrlik, J.; Vikartovska, A.; Filip, J.; Kasak, P.; Andicsová-Eckstein, A.; Mosnáček, J.; Lukáč, J.; Rovenský, J.; Imrich, R.; Tkac, J. *Anal. Chim. Acta* **2015**, *853*, 555–562.
51. Zhang, Y.; Luo, S.; Tang, Y.; Yu, L.; Hou, K. Y.; Cheng, J. P.; Zeng, X.; Wang, P. G. *Anal. Chem.* **2006**, *78*, 2001–2008.
52. Cheng, F.; Ratner, D. In *Carbohydrate Microarrays*; Chevolot, Y., Ed.; Humana Press, New York, **2012**; Vol. *808*.
53. Mandler, D.; Kraus-Ophir, S. *J. Solid State Electrochem.* **2011**, *15*, 1535–1558.
54. Penon, O.; Siapkas, D.; Novo, S.; Durán, S.; Oncins, G.; Errachid, A.; Barrios, L.; Nogués, C.; Duch, M.; Plaza, J. A.; Pérez-García, L. *Colloids Surf., B* **2014**, *116*, 104–113.
55. Ripert, M.; Farre, C.; Chaix, C. *Electrochim. Acta* **2013**, *91*, 82–89.
56. Hakkinen, H. *Nat. Chem.* **2012**, *4*, 443–455.
57. Shaw, C. P.; Fernig, D. G.; Levy, R. *J. Mater. Chem.* **2011**, *21*, 12181–12187.
58. Santoyo-Gonzalez, F.; Hernandez-Mateo, F. *Chem. Soc. Rev.* **2009**, *38*, 3449–3462.
59. Vistas, C. R.; Águas, A. C. P.; Ferreira, G. N. M. *Appl. Surf. Sci.* **2013**, *286*, 314–318.
60. Cui, H.; Kessler, M. R. *Compos. Part A-Appl. Sci.* **2014**, *65*, 83–90.
61. Guha Thakurta, S.; Subramanian, A. *Colloids Surf., A* **2012**, *414*, 384–392.
62. Haensch, C.; Hoeppener, S.; Schubert, U. S. *Chem. Soc. Rev.* **2010**, *39*, 2323–2334.
63. Gouget-Laemmel, A. C.; Yang, J.; Lodhi, M. A.; Siriwardena, A.; Aureau, D.; Boukherroub, R.; Chazalviel, J. N.; Ozanam, F.; Szunerits, S. *J. Phys. Chem. C* **2012**, *117*, 368–375.
64. Belanger, D.; Pinson, J. *Chem. Soc. Rev.* **2011**, *40*, 3995–4048.
65. Vardharajula, S.; Ali, S. Z.; Tiwari, P. M.; Eroglu, E.; Vig, K.; Dennis, V. A.; Singh, S. R. *Int. J. Nanomed.* **2012**, *7*, 5361–5374.
66. Leroux, Y. R.; Fei, H.; Noel, J. M.; Roux, C.; Hapiot, P. *J. Am. Chem. Soc.* **2010**, *132*, 14039–14041.
67. Hernández-Muñoz, L. S.; González, F. J. *Electrochem. Commun.* **2011**, *13*, 701–703.
68. Krueger, A.; Lang, D. *Adv. Funct. Mater.* **2012**, *22*, 890–906.
69. Chen, Y.; Star, A.; Vidal, S. *Chem. Soc. Rev.* **2013**, *42*, 4532–4542.
70. Kruss, S.; Hilmer, A. J.; Zhang, J.; Reuel, N. F.; Mu, B.; Strano, M. S. *Adv. Drug Deliv. Rev.* **2013**, *65*, 1933–1950.
71. Liu, W.-W.; Chai, S.-P.; Mohamed, A. R.; Hashim, U. *J. Ind. Eng. Chem.* **2014**, *20*, 1171–1185.
72. Zhang, B.-T.; Zheng, X.; Li, H.-F.; Lin, J.-M. *Anal. Chim. Acta* **2013**, *784*, 1–17.
73. Liu, J.; Liu, Z.; Barrow, C. J.; Yang, W. *Anal. Chim. Acta*, **2015**, *859*, 1–19. doi:10.1016/j.aca.2014.07.031.
74. Wu, C.; Cheng, Q.; Sun, S.; Han, B. *Carbon* **2012**, *50*, 1083–1089.
75. Filip, J.; Tkac, J. *Electrochim. Acta* **2014**, *136*, 340–354.
76. Cittadini, M.; Bersani, M.; Perrozzi, F.; Ottaviano, L.; Wlodarski, W.; Martucci, A. *Carbon* **2014**, *69*, 452–459.
77. Filip, J.; Tkac, J. *Electrochem. Commun.* **2014**, *49*, 70–74.
78. Chen, D.; Feng, H.; Li, J. *Chem. Rev.* **2012**, *112*, 6027–6053.
79. Geim, A. K. *Science* **2009**, *324*, 1530–1534.

80. Toh, S. Y.; Loh, K. S.; Kamarudin, S. K.; Daud, W. R. W. *Chem. Eng. J.* **2014**, *251*, 422–434.

81. Georgakilas, V.; Otyepka, M.; Bourlinos, A. B.; Chandra, V.; Kim, N.; Kemp, K. C.; Hobza, P.; Zboril, R.; Kim, K. S. *Chem. Rev.* **2012**, *112*, 6156–6214.

82. Wang, Z.; Mohammadzadeh, S.; Schmaltz, T.; Kirschner, J.; Khassanov, A.; Eigler, S.; Mundloch, U.; Backes, C.; Steinrück, H.-G.; Magerl, A.; Hauke, F.; Hirsch, A.; Halik, M. *ACS Nano* **2013**, *7*, 11427–11434.

83. Zhang, W.; Han, H.; Bai, H.; Tong, W.; Zhang, Y.; Ying, W.; Qin, W.; Qian, X. *Anal. Chem.* **2013**, *85*, 2703–2709.

84. Ragoussi, M.-E.; Casado, S.; Ribeiro-Viana, R.; Torre, G. d. l.; Rojo, J.; Torres, T. *Chem. Sci.* **2013**, *4*, 4035–4041.

85. Wang, H.; Gu, L.; Lin, Y.; Lu, F.; Meziani, M. J.; Luo, P. G.; Wang, W.; Cao, L.; Sun, Y.-P. *J. Am. Chem. Soc.* **2006**, *128*, 13364–13365.

86. Seeberger, P. H.; Werz, D. B. *Nature* **2007**, *446*, 1046–1051.

87. Huang, K.; Lee, B. P.; Ingram, D. R.; Messersmith, P. B. *Biomacromolecules* **2002**, *3*, 397–406.

88. Schmidt, R. R.; Castro-Palomino, J. C.; Retz, O. *Pure Appl. Chem.* **1999**, *71*, 729–744.

89. Ohtsubo, K.; Marth, J. D. *Cell* **2006**, *126*, 855–867.

90. Pattabiraman, V. R.; Bode, J. W. *Nature* **2011**, *480*, 471–479.

91. Tiefenbrunn, T. K.; Dawson, P. E. *Peptide Sci.* **2010**, *94*, 95–106.

92. Takaoka, Y.; Ojida, A.; Hamachi, I. *Angew. Chem. Int. Ed.* **2013**, *52*, 4088–4106.

93. Stephanopoulos, N.; Francis, M. B. *Nat. Chem. Biol.* **2011**, *7*, 876–884.

94. Lutz, J.-F.; Zarafshani, Z. *Adv. Drug Deliv. Rev.* **2008**, *60*, 958–970.

95. Rillahan, C. D.; Paulson, J. C. *Annu. Rev. Biochem* **2011**, *80*, 797–823.

96. Laurent, N.; Voglmeir, J.; Flitsch, S. L. *Chem. Commun.* **2008**, *37*, 4400–4412.

97. Song, X.; Heimburg-Molinaro, J.; Cummings, R. D.; Smith, D. F. *Curr. Opin. Chem. Biol.* **2014**, *18*, 70–77.

98. Györgydeák, Z.; Thiem, J. In *Advances in Carbohydrate Chemistry and Biochemistry*; Derek, H., Ed.; Academic Press, New York, **2006**; Vol. *60*.

99. Harvey, D. J. *J. Chromatogr. B* **2011**, *879*, 1196–1225.

100. Hirabayashi, J.; Yamada, M.; Kuno, A.; Tateno, H. *Chem. Soc. Rev.* **2013**, *42*, 4443–4458.

101. Hermanson, G. T. *Bioconjugate Techniques*. 3rd Edition, Academic Press, New York, **2013**.

102. Bertok, T.; Sediva, A.; Vikartovska, A.; Tkac, J. *Int. J. Electrochem. Sci.* **2014**, *9*, 890–900.

103. Von Rybinski, W.; Hill, K. *Angew. Chem. Int. Ed.* **1998**, *37*, 1328–1345.

104. Honda, S. *J. Chromatogr. A* **1996**, *720*, 183–199.

105. Grabarek, Z.; Gergely, J. *Anal. Biochem.* **1990**, *185*, 131–135.

106. Bertok, T.; Katrlik, J.; Gemeiner, P.; Tkac, J. *Microchim. Acta* **2013**, *180*, 1–13.

107. Fernández-Fernández, M.; Sanromán, M. Á.; Moldes, D. *Biotechnol. Adv.* **2013**, *31*, 1808–1825.

108. Jones, M. N. In *Methods in Enzymology.*; Nejat, D., Ed.; Academic Press, New York, **2005**; Vol. *391*.

109. www.sigmaaldrich.com 9/2014.

110. Miyata, K.; Kakizawa, Y.; Nishiyama, N.; Harada, A.; Yamasaki, Y.; Koyama, H.; Kataoka, K. *J. Am. Chem. Soc.* **2004**, *126*, 2355–2361.

111. Ghosh, S. S.; Kao, P. M.; McCue, A. W.; Chappelle, H. L. *Bioconjugate Chem.* **1990**, *1*, 71–76.

112. Nakajima, N.; Ikada, Y. *Bioconjugate Chem.* **1995**, *6*, 123–130.

113. Sheehan, J. C.; Hlavka, J. J. *J. Org. Chem.* **1956**, *21*, 439–441.

114. Nicolaou, K. C.; Vourloumis, D.; Winssinger, N.; Baran, P. S. *Angew. Chem. Int. Ed.* **2000**, *39*, 45–122.

115. Olde Damink, L. H. H.; Dijkstra, P. J.; Van Luyn, M. J. A.; Van Wachem, P. B.; Nieuwenhuis, P.; Feijen, J. *J. Mater. Sci. Mater. Med.* **1995**, *6*, 460–472.

116. Monteiro Jr, O. A. C.; Airoldi, C. *Int. J. Biol. Macromol.* **1999**, *26*, 119–128.

117. Green, N. M. *Adv. Protein Chem.* **1975**, *29*, 85–133.

118. Diamandis, E. P.; Christopoulos, T. K. *Clin. Chem.* **1991**, *37*, 625–636.

119. Michael Green, N. In *Methods in Enzymology*; Meir, W., Edward, A. B., Eds.; Academic Press, New York, **1990**; Vol. *184*.

120. Wilchek, M.; Bayer, E. A. *Anal. Biochem.* **1988**, *171*, 1–32.

121. Mahouche, S.; Mekni, N.; Abbassi, L.; Lang, P.; Perruchot, C.; Jouini, M.; Mammeri, F.; Turmine, M.; Ben Romdhane, H.; Chehimi, M. M. *Surf. Sci.* **2009**, *603*, 3205–3211.

122. Gerland, B.; Goudot, A.; Pourceau, G.; Meyer, A.; Dugas, V.; Cecioni, S.; Vidal, S.; Souteyrand, E.; Vasseur, J.-J.; Chevolot, Y.; Morvan, F. *Bioconjugate Chem.* **2012**, *23*, 1534–1547.

123. Rostovtsev, V. V.; Green, L. G.; Fokin, V. V.; Sharpless, K. B. *Angew. Chem. Int. Ed.* **2002**, *41*, 2596–2599.

124. Tornoe, C. W.; Christensen, C.; Meldal, M.. *J. Org. Chem.* **2002**, *67*, 3057–3064.

125. Liang, L. Y.; Astruc, D. *Coord. Chem. Rev.* **2011**, *255*, 2933–2945.

126. Devaraj, N. K.; Dinolfo, P. H.; Chidsey, C. E. D.; Collman, J. P. *J. Am. Chem. Soc.* **2006**, *128*, 1794–1795.

127. Pang, Y. H.; Ge, Z. Q.; Liu, Y.; Shen, X. F. *Electrochem. Commun.* **2012**, *23*, 98–101.

128. Maalouli, N.; Barras, A.; Siriwardena, A.; Bouazaoui, M.; Boukherroub, R.; Szunerits, S. *Analyst* **2013**, *138*, 805–812.

129. Devaraj, N. K.; Collman, J. P. *QSAR Comb. Sci.* **2007**, *26*, 1253–1260.

130. Hong, V.; Presolski, S. I.; Ma, C.; Finn, M. G. *Angew. Chem. Int. Ed.* **2009**, *48*, 9879–9883.

131. Gemeiner, P.; Mislovičová, D.; Tkáč, J.; Švitel, J.; Pätoprstý, V.; Hrabárová, E.; Kogan, G.; Kožár, T. *Biotechnol. Adv.* **2009**, *27*, 1–15.

132. Tkac, J.; Bertok, T.; Nahalka, J.; Gemeiner, P. In *Lectins*; Hirabayashi, J., Ed.; Springer New York, **2014**; Vol. *1200*.

133. Wang, J. In *Analytical Electrochemistry*; John Wiley & Sons, Inc., Hoboken, **2006**.

134. Pandey, B.; Tan, Y. H.; Parameswar, A. R.; Pornsuriyasak, P.; Demchenko, A. V.; Stine, K. J. *Carbohydr. Res.* **2013**, *373*, 9–17.

135. Reuel, N. F.; Mu, B.; Zhang, J. Q.; Hinckley, A.; Strano, M. S. *Chem. Soc. Rev.* **2012**, *41*, 5744–5779.

136. Lehr, J.; Fernandes, F. C. B.; Bueno, P. R.; Davis, J. J. *Anal. Chem.* **2014**, *86*, 2559–2564.

137. Safina, G. *Anal. Chim. Acta* **2012**, *712*, 9–29.

138. Ciminelli, C.; Campanella, C. M.; Dell'Olio, F.; Campanella, C. E.; Armenise, M. N. *Prog. Quant. Electron.* **2013**, *37*, 51–107.
139. Linman, M. J.; Yu, H.; Chen, X.; Cheng, Q. *ACS Appl. Mater. Interf.* **2009**, *1*, 1755–1762.
140. Smith, E. A.; Thomas, W. D.; Kiessling, L. L.; Corn, R. M. *J. Am. Chem. Soc.* **2003**, *125*, 6140–6148.
141. Cao, J.; Sun, T.; Grattan, K. T. V. *Sens. Actuat. B: Chem.* **2014**, *195*, 332–351.
142. Hammond, J.; Bhalla, N.; Rafiee, S.; Estrela, P. *Biosensors* **2014**, *4*, 172–188.
143. Petryayeva, E.; Krull, U. J. *Anal. Chim. Acta* **2011**, *706*, 8–24.
144. Mayer, K. M.; Hafner, J. H. *Chem. Rev.* **2011**, *111*, 3828–3857.
145. Willets, K. A.; Van Duyne, R. P. *Annu. Rev. Phys. Chem.* **2007**, *58*, 267–297.
146 Bhattarai, J. K.; Sharma, A.; Fujikawa, K.; Demchenko, A. V.; Stine, K. J. *Carbohydr. Res.*, **2015**, *405*, 55–65. doi:10.1016/j.carres.2014.08.019
147. Bellapadrona, G.; Tesler, A. B.; Grünstein, D.; Hossain, L. H.; Kikkeri, R.; Seeberger, P. H.; Vaskevich, A.; Rubinstein, I. *Anal. Chem.* **2011**, *84*, 232–240.
148. Hideshima, S.; Kuroiwa, S.; Kimura, M.; Cheng, S. S.; Osaka, T. *Electrochim. Acta* **2013**, *110*, 146–151.
149. Shen, M. Y.; Li, B. R.; Li, Y. K. *Biosens. Bioelectron.* **2014**, *60*, 101–111.
150. Noor, M. O.; Krull, U. J. *Anal. Chim. Acta* **2014**, *825*, 1–25.
151. Matsumoto, A.; Sato, N.; Cabral, H.; Kataoka, K.; Miyahara, Y. *Procedia Eng.* **2010**, *5*, 926–929.
152. Stefansson, S.; Kwon, H. H.; Ahn, S. N. *J. Nanotechnol.* **2012**, *2012*, 1–8.
153. Sharma, B. K.; Ahn, J. H. *Solid State Electron.* **2013**, *89*, 177–188.
154. Chen, K. I.; Li, B. R.; Chen, Y. T. *Nano Today* **2011**, *6*, 131–154.
155. Wang, X. R.; Shi, Y.; Zhang, R. *Chin. Phys. B* **2013**, *22*, 098505.
156. Patolsky, F.; Zheng, G. F.; Lieber, C. M. *Anal. Chem.* **2006**, *78*, 4260–4269.
157. Janata, J. *ECS Solid State Lett.* **2012**, *1*, M29–M31.
158. Ali, M.; Nasir, S.; Ramirez, P.; Cervera, J.; Mafe, S.; Ensinger, W. *J. Phys. Chem. C* **2013**, *117*, 18234–18242.
159. Whited, A. M.; Park, P. S. H. *Biochim. Biophys. Acta (BBA): Biomembr.* **2014**, *1838*, 56–68.
160. Carvalho, F. A.; Martins, I. C.; Santos, N. C. *Arch. Biochem. Biophys.* **2013**, *531*, 116–127.
161. Pillet, F.; Chopinet, L.; Formosa, C.; Dague, É. *BBA: Gen. Subj.* **2014**, *1840*, 1028–1050.
162. Morita, S. *J. Vac. Sci. Technol. A* **2013**, *31*, 050802.
163. Boisen, A.; Dohn, S.; Keller, S. S.; Schmid, S.; Tenje, M. *Rep. Prog. Phys.* **2011**, *74*, 036101.
164. Pandey, B.; Tan, Y. H.; Fujikawa, K.; Demchenko, A. V.; Stine, K. J. *J. Carbohydr. Chem.* **2012**, *31*, 466–503.
165. Tan, Y. H.; Fujikawa, K.; Pornsuriyasak, P.; Alla, A. J.; Vijaya Ganesh, N.; Demchenko, A. V.; Stine, K. J. *New J. Chem.* **2013**, *37*, 2150–2165.
166. Yang, J.; Chazalviel, J.-N.; Siriwardena, A.; Boukherroub, R.; Ozanam, F.; Szunerits, S.; Gouget-Laemmel, A. C. *Anal. Chem.* **2014**, *86*, 10340–10349.
167. Svedhem, S.; Öhberg, L.; Borrelli, S.; Valiokas, R.; Andersson, M.; Oscarson, S.; Svensson, S. C. T.; Liedberg, B.; Konradsson, P. *Langmuir* **2002**, *18*, 2848–2858.
168. Godula, K.; Bertozzi, C. R. *J. Am. Chem. Soc.* **2012**, *134*, 15732–15742.
169. Wittmann, V.; Pieters, R. J. *Chem. Soc. Rev.* **2013**, *42*, 4492–4503.

170. Sato, Y.; Yoshioka, K.; Murakami, T.; Yoshimoto, S.; Niwa, O. *Langmuir* **2011**, *28*, 1846–1851.

171. Cecioni, S.; Lalor, R.; Blanchard, B.; Praly, J.-P.; Imberty, A.; Matthews, S. E.; Vidal, S. *Chem. Eur. J.* **2009**, *15*, 13232–13240.

172. Cairo, C. W.; Gestwicki, J. E.; Kanai, M.; Kiessling, L. L. *J. Am. Chem. Soc.* **2002**, *124*, 1615–1619.

173. Horan, N.; Yan, L.; Isobe, H.; Whitesides, G. M.; Kahne, D. *Proc. Acad. Sci. Natl. U.S.A.* **1999**, *96*, 11782–11786.

174. Padler-Karavani, V.; Song, X.; Yu, H.; Hurtado-Ziola, N.; Huang, S.; Muthana, S.; Chokhawala, H. A.; Cheng, J.; Verhagen, A.; Langereis, M. A.; Kleene, R.; Schachner, M.; de Groot, R. J.; Lasanajak, Y.; Matsuda, H.; Schwab, R.; Chen, X.; Smith, D. F.; Cummings, R. D.; Varki, A. *J. Biol. Chem.* **2012**, *287*, 22593–22608.

175. Grant, O. C.; Smith, H. M.; Firsova, D.; Fadda, E.; Woods, R. J. *Glycobiology* **2014**, *24*, 17–25.

176. Almant, M.; Mastouri, A.; Gallego-Yerga, L.; García Fernandez, J. M.; Ortiz Mellet, C.; Kovensky, J.; Morandat, S.; El Kirat, K.; Gouin, S. G. *Chem. Eur. J.* **2013**, *19*, 729–738.

177. Tantakitti, F.; Burk-Rafel, J.; Cheng, F.; Egnatchik, R.; Owen, T.; Hoffman, M.; Weiss, D. N.; Ratner, D. M. *Langmuir* **2012**, *28*, 6950–6959.

178. Liu, Y.; Feizi, T.; Campanero-Rhodes, M. A.; Childs, R. A.; Zhang, Y.; Mulloy, B.; Evans, P. G.; Osborn, H. M. I.; Otto, D.; Crocker, P. R.; Chai, W. *Chem. Biol.* **2007**, *14*, 847–859.

179. Palma, A. S.; Feizi, T.; Childs, R. A.; Chai, W.; Liu, Y. *Curr. Opin. Chem. Biol.* **2014**, *18*, 87–94.

180. Rinaldi, S.; Brennan, K. M.; Goodyear, C. S.; O'Leary, C.; Schiavo, G.; Crocker, P. R.; Willison, H. J. *Glycobiology* **2009**, *19*, 789–796.

181. Noble, G. T.; Craven, F. L.; Voglmeir, J.; Šardzík, R.; Flitsch, S. L.; Webb, S. J. *J. Am. Chem. Soc.* **2012**, *134*, 13010–13017.

182. Ma, Y.; Sobkiv, I.; Gruzdys, V.; Zhang, H.; Sun, X.-L. *Anal. Bioanal. Chem.* **2012**, *404*, 51–58.

183. Chevolot, Y.; Laurenceau, E.; Phaner-Goutorbe, M.; Monnier, V.; Souteyrand, E.; Meyer, A.; Géhin, T.; Vasseur, J.-J.; Morvan, F. *Curr. Opin. Chem. Biol.* **2014**, *18*, 46–54.

184. Spinelli, N.; Defrancq, E.; Morvan, F. *Chem. Soc. Rev.* **2013**, *42*, 4557–4573.

185. Gouin, S. G. *Chem. Eur. J.* **2014**, *20*, 11616–11628.

186. Shivatare, S. S.; Chang, S.-H.; Tsai, T.-I.; Ren, C.-T.; Chuang, H.-Y.; Hsu, L.; Lin, C.-W.; Li, S.-T.; Wu, C.-Y.; Wong, C.-H. J. *Am. Chem. Soc.* **2013**, *135*, 15382–15391.

187. Rillahan, C. D.; Schwartz, E.; Rademacher, C.; McBride, R.; Rangarajan, J.; Fokin, V. V.; Paulson, J. C. *ACS Chem. Biol.* **2013**, *8*, 1417–1422.

188. Gray, C. J.; Weissenborn, M. J.; Eyers, C. E.; Flitsch, S. L. *Chem. Soc. Rev.* **2013**, *42*, 6378–6405.

189. Thobhani, S.; Ember, B.; Siriwardena, A.; Boons, G.-J. *J. Am. Chem. Soc.* **2003**, *125*, 7154–7155.

190. Houseman, B. T.; Mrksich, M. *Angew. Chem. Int. Ed.* **1999**, *38*, 782–785.

191. Fang, Y.; Xu, W.; Wu, J.; Xu, Z.-K. *Chem. Commun.* **2012**, *48*, 11208–11210.

192. Horiya, S.; MacPherson, I. S.; Krauss, I. J. *Nat. Chem. Biol.* **2014**, *10*, 990–999.

193. Kadam, R. U.; Bergmann, M.; Hurley, M.; Garg, D.; Cacciarini, M.; Swiderska, M. A.; Nativi, C.; Sattler, M.; Smyth, A. R.; Williams, P.; Cámara, M.; Stocker, A.; Darbre, T.; Reymond, J.-L. *Angew. Chem. Int. Ed.* **2011**, *50*, 10631–10635.

194. Kaplan, J. M.; Shang, J.; Gobbo, P.; Antonello, S.; Armelao, L.; Chatare, V.; Ratner, D. M.; Andrade, R. B.; Maran, F. *Langmuir* **2013**, *29*, 8187–8192.

195. Pedersen, J. W.; Blixt, O.; Bennett, E. P.; Tarp, M. A.; Dar, I.; Mandel, U.; Poulsen, S. S.; Pedersen, A. E.; Rasmussen, S.; Jess, P.; Clausen, H.; Wandall, H. H. *Int. J. Cancer* **2011**, *128*, 1860–1871.

196. Dyukova, V. I.; Dementieva, E. I.; Zubtsov, D. A.; Galanina, O. E.; Bovin, N. V.; Rubina, A. Y. *Anal. Biochem.* **2005**, *347*, 94–105.

197. Compagno, D.; Gentilini, L. D.; Jaworski, F. M.; Pérez, I. G.; Contrufo, G.; Laderach, D. J. *Glycobiology* **2014**, *24*, 899–906.

198. Laderach, D. J.; Gentilini, L. D.; Giribaldi, L.; Delgado, V. C.; Nugnes, L.; Croci, D. O.; Al Nakouzi, N.; Sacca, P.; Casas, G.; Mazza, O.; Shipp, M. A.; Vazquez, E.; Chauchereau, A.; Kutok, J. L.; Rodig, S. J.; Elola, M. T.; Compagno, D.; Rabinovich, G. A. *Cancer Res.* **2013**, *73*, 86–96.

199. van Bueren, J. J. L.; Rispens, T.; Verploegen, S.; van der Palen-Merkus, T.; Stapel, S.; Workman, L. J.; James, H.; van Berkel, P. H. C.; van de Winkel, J. G. J.; Platts-Mills, T. A. E.; Parren, P. W. H. I. *Nat. Biotechnol.* **2011**, *29*, 574–576.

200. Pochechueva, T.; Jacob, F.; Goldstein, D.; Huflejt, M.; Chinarev, A.; Caduff, R.; Fink, D.; Hacker, N.; Bovin, N.; Heinzelmann-Schwarz, V. *Glycoconjugate J.* **2011**, *28*, 507–517.

201. Jacob, F.; Goldstein, D. R.; Bovin, N. V.; Pochechueva, T.; Spengler, M.; Caduff, R.; Fink, D.; Vuskovic, M. I.; Huflejt, M. E.; Heinzelmann-Schwarz, V. *Int. J. Cancer* **2012**, *130*, 138–146.

202. Samraj, A.; Läubli, H.; Varki, N.; Varki, A. *Front. Oncol.* **2014**, *4*, 33.

203. Padler-Karavani, V.; Tremoulet, A. H.; Yu, H.; Chen, X.; Burns, J. C.; Varki, A. *PLoS One* **2013**, *8*, e58443.

204. Chikae, M.; Fukuda, T.; Kerman, K.; Idegami, K.; Miura, Y.; Tamiya, E. *Bioelectrochemistry* **2008**, *74*, 118–123.

205. Hideshima, S.; Hinou, H.; Ebihara, D.; Sato, R.; Kuroiwa, S.; Nakanishi, T.; Nishimura, S.-I.; Osaka, T. *Anal. Chem.* **2013**, *85*, 5641–5644.

206. Nagatsuka, T.; Uzawa, H.; Sato, K.; Kondo, S.; Izumi, M.; Yokoyama, K.; Ohsawa, I.; Seto, Y.; Neri, P.; Mori, H.; Nishida, Y.; Saito, M.; Tamiya, E. *ACS Appl. Mater. Interf.* **2013**, *5*, 4173–4180.

207. Seo, J. H.; Kim, C. S.; Cha, H. J. *Analyst* **2013**, *138*, 6924–6929.

208. O'Neill, E. C.; Rashid, A. M.; Stevenson, C. E. M.; Hetru, A.-C.; Gunning, A. P.; Rejzek, M.; Nepogodiev, S. A.; Bornemann, S.; Lawson, D. M.; Field, R. A. *Chemical Science* **2014**, *5*, 341–350.

209. Shen, Z.; Huang, M.; Xiao, C.; Zhang, Y.; Zeng, X.; Wang, P. G. *Anal. chem.* **2007**, *79*, 2312–2319.

210. Mader, A.; Gruber, K.; Castelli, R.; Hermann, B. A.; Seeberger, P. H.; Rädler, J. O.; Leisner, M. *Nano Lett.* **2011**, *12*, 420–423.

211. Wangchareansak, T.; Sangma, C.; Ngernmeesri, P.; Thitithanyanont, A.; Lieberzeit, P. A. *Anal. Bioanal. Chem.* **2013**, *405*, 6471–6478.

212. Li, Z.; Deng, S. S.; Zang, Y.; Gu, Z.; He, X. P.; Chen, G. R.; Chen, K.; James, T. D.; Li, J.; Long, Y. T. *Sci. Rep.* **2013**, *3*, 2293.

213. Defaus, S.; Gupta, P.; Andreu, D.; Gutierrez-Gallego, R. *Analyst* **2014**, *139*, 2944–2967.

16

NANOTOXICOLOGY ASPECTS OF CARBOHYDRATE NANOSTRUCTURES

Yinfa Ma and Qingbo Yang

Department of Chemistry, Center for Single Nanoparticle, Single Cell, and Single Molecule Monitoring (CS^{3}M), Missouri University of Science and Technology, Rolla, MO, USA

16.1 TOXICOLOGY OF NANOMATERIALS

16.1.1 Applications of Nanomaterials

The unprecedented physicochemical properties of nanomaterials highly improved or created fundamental research as well as industrial field ranging from structural/strength enhancement, fluorescent dye/probe development, nanoscale device fabrication and manipulation to energy collection and conservation, microlaser construction, and antimicrobial/sterilization properties. Recently, manufactured nanomaterials (MNMs) have been widely used in infrastructure construction and agriculture-related products that used in plant germination and growth, plant protection and production, pesticide residue detection as well as plant pathogen detection, etc. [1]. It is crucial to acquire sufficient safety confidence when dealing with these nanomaterials and be more responsible to any possible negative environmental impact to both human beings and surrounding plants and animals or to, say, the whole ecosystem.

Here, we reviewed the state-of-the-art application of MNMs within a variety of research and industrial fields, to overview possible environmental leakage and releasing examples and to better draw a conclusion on how to rationally evaluate adverse effect on a toxicology point of view. Aligned with interdisciplinary judgment of the research and application implications of emerging nanotechnologies, this chapter seeks to promote the awareness of possible negative effect of MNMs in areas

Carbohydrate Nanotechnology, First Edition. Edited by Keith J. Stine.

of materials development and related construction, to support the idea of developing regulatory guidelines for the fabrication, transportation, application, disposal, and even recycling, in order to better protect our society and environment health [2].

Technological advancement has enabled us to manufacture nanomaterials at different shapes, sizes, compositions, etc., and such capabilities are the basis for future modification and engineering for specific aims. Previously, nanomaterials have been used for medical/clinical purposes and were also seen applied in semiconductor as well as food processing, which were successful and had no apparent safety issue appeared. However, lately, applications, for instance, in personal products and agriculture, especially when used for skin contact or crop protection and production, draw a lot of public concerns, and intensive research efforts have been put into these areas. For example, substantially large groups of studies have shown that some nanomaterials composed of semiconductor metal oxides or nonmetal oxides, such as SiO_2, CeO_2, ZnO, Al_2O_3, etc., can pose high amount of cytotoxic risks on different *in vitro* cultured cell lines [3–7]. Moreover, recently published study found out that irradiation (not only UV spectrum but also visible light) greatly enhanced the cytotoxic effect caused by ZnO nanoparticles [8]. Another example showed that the nanomaterials have potential to improve seed germination and thus plant growth. Further, they could provide multiple protections for plants in detecting pathogens, as well as pesticide and herbicide residues [1].

Moreover, the successful use of nanomaterials in the abovementioned fields leads to their intersection with biology and biomedical application and research. Nanobiotechnology (or bionanotechnology) has triggered many applications related to biological cells including cell imaging and cancer treatment tracking. Besides, nanoscale materials, such as nanoparticles, nanowires, nanofibers, nanotubes, etc., have also been tested for varied biological applications, like biosensing, biosample detection and separation, molecular imaging, as well as anticancer therapy. A representative instance can be seen in Figure 16.1, where gold (Au) nanocages were successfully used in cancer cell photothermal treatment.

All these nanoscale fabrications and applications are feasible because of the unique properties, which their bulk formation cannot achieve. These properties, including their high volume/surface ratio, surface tailorability, improved solubility, and multifunctionality, open many new possibilities for biomedicine research and application. Further, the intrinsic properties of nanomaterials, such as optical, magnetic, and biological, provide splendid opportunities in the future, studying and regulating complicated biological process in an unprecedented manner. After all, life itself is fundamentally a handful of processes down to a nanoscale level within each cell [10]. Thus it is necessary to understand the impacts of the presence of nanomaterials inside the cells whenever a comprehensive evaluation is needed for the application of nanomaterials in the biomedical field.

16.1.2 Toxicity of Nanomaterials and Nanostructures

Biocompatibility, cellular/subcellular distribution/localization, induced inflammation, and functional interferences with cells and tissues caused by nanomaterials are some key factors, among others, that determine the toxicity of such engineered

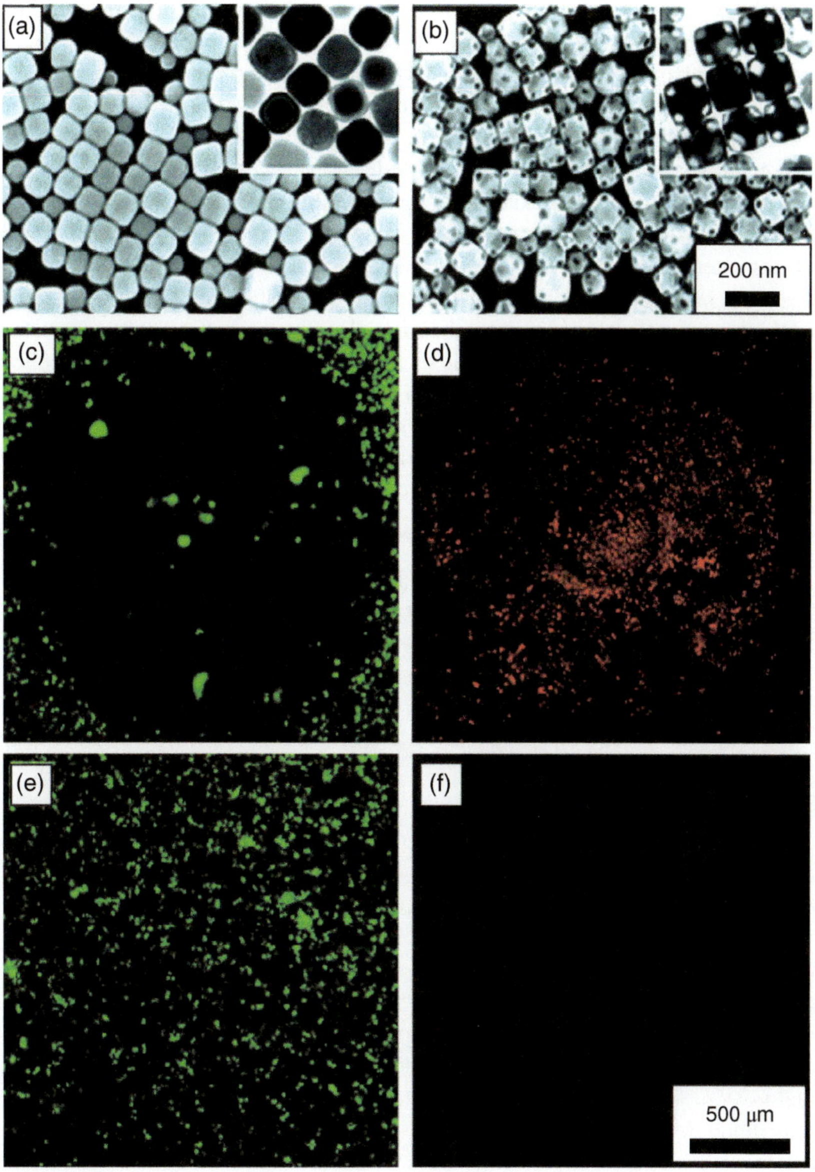

FIGURE 16.1 Cancer treatments through photothermal active gold nanocages. Insets showed TEM images of each SEM image of the gold nanocubes (a) and nanocages (b). Live (c, calcein AM stained) and dead (d, EthD-1 stained) SK-BR-3 breast cancer cells were shown after treatment by immuno(anti-HER2)-Au nanocages combined with 810 nm laser irradiation. Control cells that receive the same laser eradication without nanocage cotreatment showed no sign of cell death (e, f, stained as c and d). *Source*: Reprint with permission from Ref. 9, Copyright 2007, American Chemical Society.

nanoparticles and nanostructures and, therefore, the extent to which they could be used. From a cytotoxic point of view, recent researches reported that engineered inorganic nanoparticles and carbon-based nanostructures may enter into or get in contact with living organisms at realistic doses, but do not cause acute toxic effects. However, their prolonged interaction with living organisms may ultimately disrupt normal activities and lead to malfunction or diseases, especially those nanomaterials that could be used for healing damaged organs. In this case, possible long-term environmental and human health hazards need to be strictly monitored; how those nanostructures enter and are distributed inside the cell and body is critical [11]. An example of nanoparticles entering cells was shown here to establish a concept of how deep and broad the nanomaterials can go into subcellular localizations (Fig. 16.2).

The interactions between nanoparticles and the cells seemingly have some same principles as those between colloidal particles and cells. The van der Waals (VDW) force, electrostatic, solvation, solvophobic, as well as depletion forces are still applicable, but they need special requirements when considering events that occurred at nanoscale

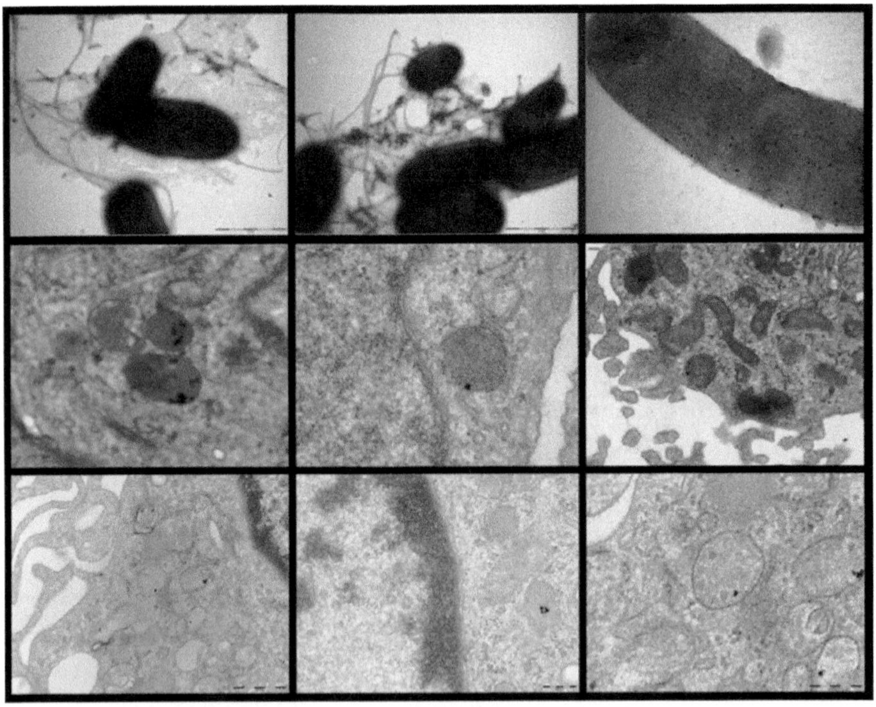

FIGURE 16.2 TEM images of nanoparticles internalized by different cell lines. Top: Bacillus + b-amyloid fibers + gold nanoparticles (AuNPs); middle: AuNP + Hela cells; bottom: Macrophages + AuNP. The two first cases showed passive internalization of NPs through cell membrane recycling, while the third one showed active medication by membrane receptors. The NP internalization, macrophage activation time, as well as the temperature dependence of the endocytosis can serve together for discrimination between active and passive mechanisms. The AuNPs used are 10 nm in diameter. (Middle pictures in collaboration with Silvia Pujals and Ernest Giralt (PCB-UB)). *Source*: Reprinted from Ref. 11, Copyright 2008 with permission from Elsevier Ltd.

[12–14]. For instance, since nanoparticles contain relatively much less atoms, their VDW forces are highly dependent on the positioning of their surface atoms and their standard bulk permittivity functions. This sophisticated phenomenon may greatly influence the interactions between nanomaterials and biological systems. A simple example would be the interaction between two silica particles, while compared with that between a SiO_2 particle and a fibroblast cell [15]. A table of main forces that govern such interactions was provided for better understanding of the complexity (Table 16.1).

TABLE 16.1 Major Forces That Regulate interactions Between Nanomaterials and Biological Systems[a]

Force	Origin and Nature	Range (nm)	Possible Impact on the Interface
Hydrodynamic interactions	Convective drag, shear, lift, and Brownian diffusion are often hindered or enhanced at nanoscale separations between interacting interfaces	10^2–10^6	Increase the frequency of collisions between nanoparticles and other surfaces responsible for transport
Electrodynamic interactions	VDW interactions arising from each of the interacting materials and the intervening media	1–100	Universally attractive in aqueous media; substantially smaller for biological media and cells owing to high water content
Electrostatic interactions	Charged interfaces attract counter ions and repel co ions through Coulombic forces, giving rise to the formation of an electrostatic double layer	1–100	Overlapping double layers are generally repulsive as most materials acquire negative charge in aqueous media but can be attractive for oppositely charged materials
Solvent interactions	Lyophilic materials interact favorably with solvent molecules Lyophobic materials interact unfavorably with solvent molecules	1–10	Lyophilic materials are thermodynamically stable in the solvent and do not aggregate Lyophobic materials are spontaneously expelled from the bulk of the solvent and forced to aggregate or accumulate at an interface
Steric interactions	Polymeric species adsorbed to inorganic particles or biopolymers expressed at the surfaces of cells give rise to spring-like repulsive interactions with other interfaces	1–100[b]	Generally increase stability of individual particles but can interfere in cellular uptake, especially when surface polymers are highly water soluble
Polymer bridging interactions	Polymeric species adsorbed to inorganic particles or biopolymers expressed at the surfaces of cells containing charged functional groups can be attracted by oppositely charged moieties on a substrate surface	1–100	Generally promote aggregation or deposition, particularly when charge functionality is carboxylic acid and dispersed in aqueous media containing calcium ions

[a] Reprint with Permission from Ref. 15, Copyright 2009, Nature Publishing Group.

[b] Depending on the length of adsorbed or expressed polymeric species.

16.1.3 Current Understanding of Nanotoxicity Mechanism

Engineered nanomaterials mostly have structures with at least one dimension of 100 nanometers or less. Along with their increasing application in products like fillers, catalysts, semiconductors, cosmetics, microelectronics, and drug carriers for commercial purposes, material size ranges may approach the length scale at which some specific physical or chemical interactions can occur. Consequently, substantially changed properties will discriminate them from their bulk materials, and this would allow them to perform exceptional feats of conductivity, reactivity, and optical sensitivity. A possible undesirable circumstance of these capabilities is harmful interactions with biological systems and the environment, with the potential to generate toxicity. A possible general mechanism of nanomaterials that interact with biological tissues have been shown in Figure 16.3, which listed several key findings of nanotoxicity that related with physicochemical properties of the materials, Fenton chemistry, dissolution, as well as irradiation-induced radical formation. In this case, the establishment of principles and test procedures for safe manufacturing and application of nanomaterials in the marketplace should be achieved as soon as possible [16].

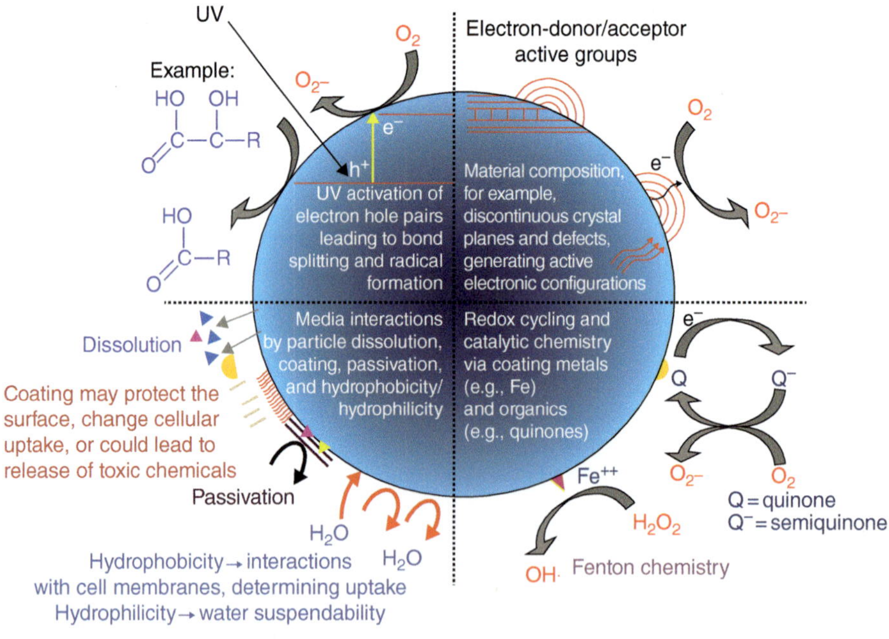

FIGURE 16.3 Main possible mechanisms of interaction between nanomaterial and biological system. Schemes illustrate key properties such as material composition, electronic structure, bonded surface species (e.g., metal containing), surface coatings (active or passive), and solubility, which includes the contribution of surface species and coatings and interactions with other environmental factors (e.g., UV activation). *Source*: Reprint with permission from Ref. 16. Copyright (2006) American Association for the Advancement of Science.

TABLE 16.2 NM Effects as the Basis for Pathophysiology and Toxicity[a]

Experimental NM Effects	Possible Pathophysiological Outcomes
ROS generation[b]	Protein, DNA and membrane injury,[b] oxidative stress[c]
Oxidative stress[b]	Phase II enzyme induction, inflammation,[c] mitochondrial perturbation[b]
Mitochondrial perturbation[b]	Inner membrane damage,[b] permeability transition (PT) pore opening,[b] energy failure,[b] apoptosis,[b] aponecrosis, cytotoxicity
Inflammation[b]	Tissue infiltration with inflammatory cells,[c] fibrosis,[c] granulomas,[c] atherogenesis,[c] acute phase protein expression (e.g., C-reactive protein)
Uptake by reticulo-endothelial system[b]	Asymptomatic sequestration and storage in liver,[b] spleen, lymph nodes,[c] possible organ enlargement, and dysfunction
Protein denaturation, degradation[b]	Loss of enzyme activity,[c] autoantigenicity
Nuclear uptake[b]	DNA damage, nucleoprotein clumping,[c] autoantigens
Uptake in neuronal tissue[b]	Brain and peripheral nervous system injury
Perturbation of phagocytic function,[b] "particle overload," mediator release[b]	Chronic inflammation,[c] fibrosis,[c] granulomas,[c] interference in clearance of infectious agents[c]
Endothelial dysfunction, effects on blood clotting[b]	Atherogenesis,[b] thrombosis,[b] stroke, myocardial infarction
Generation of neoantigens, breakdown in immune tolerance	Autoimmunity, adjuvant effects
Altered cell cycle regulation	Proliferation, cell cycle arrest, senescence
DNA damage	Mutagenesis, metaplasia, carcinogenesis

[a] Reprint with Permission from Ref. 16, Copyright 2006, American Association for the Advancement of Science.
[b] Effects supported by limited experimental evidence.
[c] Effects supported by limited clinical evidence.

Given the increasingly accumulated research data, it is possible that engineered nanomaterials have toxic effect on environment and human health, yet there are currently no conclusive data or scenarios that indicate that these effects will become a major problem or that they cannot be addressed by a rational scientific approach. Meanwhile, complete and comprehensive safety evaluations of nanomaterials should be established immediately. In addition, full understanding of nanotoxicity could also have a positive sequel. Several key nanomaterial effects as well as possible correlations to the pathophysiological outcomes were summed up in the above table (Table 16.2).

A major problem in assessing the hazards of engineered nanomaterials is currently limited by the lack of empirical data. Moreover, more questionable data were from toxicity studies that did not use realistic environmental exposure conditions for determining dose–response parameters. The latter will greatly affect the accurate estimation of health risks associated with the exposure to nanomaterials. Here, a number of open questions have been suggested, including: (i) "developing a combination of different analytical methods for determining engineered nanomaterials concentration, size, shape, surface properties, and morphology in different environmental media, (ii) conducting toxicity studies using environmentally relevant exposure conditions and obtaining data relevant to developing quantitative nanostructure-toxicity relationships (QNTR), and (iii) developing guidelines for regulating exposure of engineered nanomaterials in the environment" [17].

For reducing the considerable knowledge gap between the fabrication and application of nanomaterials and *in vivo* toxicity of NPs, a considerable effort is also needed to study the physiological effects of acute and chronic exposure to NPs. A profound illustration and full understanding of the biological interactions between NPs and proteins and cells and tissues are vital for the future design of safe nanotechnologies. Prior to their wider adoption in everyday products and their clinical use, NP products must be shown to have a high degree of biocompatibility, with minimal negative effects on blood components, genetic material, and cell viability. Table 16.3 lists the most prominent experimental mechanisms of nanomaterial toxicity.

16.2 CYTOTOXICITY OF CARBOHYDRATE-MODIFIED NANOMATERIALS

16.2.1 Carbohydrate-Functionalized Nanomaterials

In recent years, nanomaterials, given their possible cytotoxicity as discussed previously, were further bioactively modified, to improve their solubility, biocompatibility, and biofunctionalities while preserving their desired properties. Among several candidates for bioactive modification, carbohydrate modification seems to have strong interests, and the studies on the usage of carbohydrate-functionalized nanomaterials in detecting bacteria, binding to specific lectins; in delivering glycol-mimetic drug molecules into cells; and in probing cellular activities as biosensors have been developed accordingly. Improved biocompatibility and introduction of biofunctionalities through integration of carbohydrate with nanomaterials are actually broadening the way to the so-called glyconanotechnology and may provide new tools for nanomaterial–cell interaction studies [19]. Liu et al. have fabricated a thin-film network of single-walled nanotubes (SWNTs) modified with monosaccharides (Fig. 16.4) [20]. The PC12 cells were firstly cultured on top of the glycosylated SWNT net, and then when getting in contact with high concentration calcium or potassium solution, a train of electric current spikes will be well detected. Theoretically, the influx of cations triggers the actions of several sets of proteins. This drives immobilization, and final fusion of core vesicles takes place with the plasma membrane and subsequently results in discharge of catecholamine molecules (dopamine, norepinephrine, and epinephrine) into the extracellular space. After being released in between a narrow interface gap of cell and SWNT net, catecholamines immediately diffused onto the nanotubes and interact with them by p–p stacking

TABLE 16.3 Mechanisms of Nanomaterial Cytotoxicity and Potentially Useful Safe Design Features[a]

Nanomaterial	Cytotoxicity Mechanism	Possible Design Features to Mitigate Toxicity
TiO_2	ROS production mediated by electron hole-pairs Glutathione depletion and toxic oxidative stress as a result of photoactivity and redox properties Nanoparticle-mediated cell membrane disruption leads to cell death; protein fibrillation	Cap with surfactants, polymers, or complexing ligands Coat with low molecular weight antioxidants (e.g., ascorbate, glutathione, thiols) and enzymatic scavengers (e.g., superoxide dismutase or its mimics) Alter the chemical composition that, for example, eliminates crystalline states and material defects
ZnO	ROS production Dissolution and release of toxic cations Lysosomal damage Inflammation	
Ag	Dissolution and Ag^+ release inhibits respiratory enzymes and ATP production ROS production Disruption of membrane integrity and transport processes	
Au nanoparticles and nanorods	Disruption of protein conformation	Modify charge, size, and hydrophobicity Induce steric hindrance
CdSe	Dissolution and release of toxic Cd and Se ions	Coat the toxic core with biocompatible polymers or inorganic shells
SiO_2	ROS production by surface defects and impurities Protein unfolding Membrane disruption	Induce particle aggregation Modify the size and/or surface charge Modify the surface with aluminum lactate or polyvinylpyridine *N*-oxide
Fe_3O_4	ROS production and oxidative stress Liberation of toxic Fe^{2+} Disturbance of the electronic and/or ion transport activity in the cell membrane	Functionalize with biocompatible shell Coat with a polymer (e.g., alginate, chitosan)
CeO_2	Protein aggregation and fibrillation	Modify the material to exert antioxidant effects through altered electronic states
MWCNT	Frustrated phagocytosis causes chronic tissue inflammation and DN A oxidative injury	Modify MWCNT structure to reduce stiffness Reduce aspect ratio to yield fiber strands <5 μm Make tubes easier to phagocytose (e.g., γ-irradiation may reduce the stiffness) Use alternative structures (such as multiwalled carbon nano-onions)

(*continued*)

TABLE 16.3 (Continued)

Nanomaterial	Cytotoxicity Mechanism	Possible Design Features to Mitigate Toxicity
SWCNT, MWCNT	Generation of ROS due to the metal impurities trapped inside CNTs Proinflammatory effects due to oxidant injury	Functionalize the surface with antioxidant or metal-complexing agent Coat with a polymer or shell to prevent metal ion release Purify to remove metal contamination Attach ligands to promote receptor-mediated phagocytosis N-dope
	Granulomatous inflammation due to hydrophobic CNT aggregation Interstitial pulmonary fibrosis due to fibroblast-mediated collagen production	Improve the dispersion of CNTs using steric hindrance or electrostatic repulsion through appropriate surface coating/functionalization
Fullerenes	ROS production (spontaneous or photoactivated) Hydrophobic surface increases aggregation but promotes intramembranous localization	Decrease hydrophobicity by coating/functionalizing fullerene surface with anionic head groups
Cationic nanospheres and dendrimers	Membrane damage, thinning, and leakage Damage to the acidifying endosomal compartment by the proton sponge effect that allows entry into the cytosol (Fig. 16.6)	Reduce cationic charge density Replace cationic head groups with amphiphilic head groups Reduce nonspecific cellular interactions through conjugation with appropriate moieties
Co/Ni ferrite nanoparticles, magnetic metallic nanoparticles	Liberation of toxic cations	Coat with end-grafted polymer, carbon, difunctional phosphonic and hydroxamic acids Encapsulate in a polymer coat or liposome Create a core–shell (gold or silica) structure
Al_2O_3	ROS production Proinflammatory response	Coat or functionalize the surface
Cu/CuO	DNA damage and oxidative stress	
MoO_3	Membrane disruption	

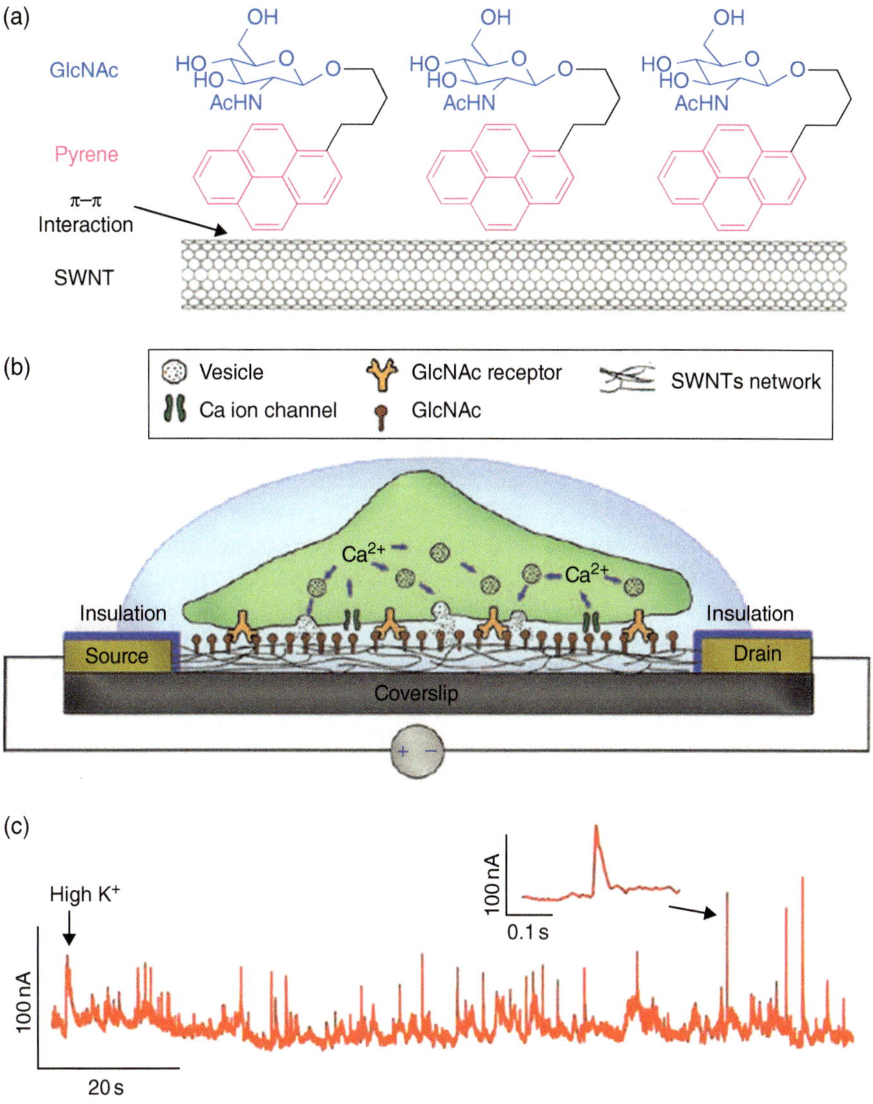

FIGURE 16.4 Cellular exocytosis detection by GlcNAc functionalized SWNTs. (a) SWNTs functionalized by pyrene-tethered GlcNAc. (b) PC12 cell exocytosis triggering and detection by SWNT net. (c) Nanotube responses of high K^+ stimulation of exocytosis by PC12 cells. *Source*: Reprinted with permission from Ref. 20, Copyright 2009, Wiley-VCH Verlag GmbH & Co.

between the aromatic ring of the catecholamine molecule and the sidewalls of CNTs (Fig. 16.4a). The conductance of the nanotube is highly sensitive to the electrochemical perturbations at the surface induced by the interacting molecules. Thus, catecholamines that are released from PC12 cells can easily alter the nanotube's current [20].

16.2.2 Glyconanoparticles, as a Multifunctional Nanocarbohydrate System

As a matter of fact, many carbohydrate-based modifications have been experimented for better quality and biocompatibility not only in nanomaterial manufactures, but some other industrial fields. The pursuing of construction of multivalent carbohydrate systems, with precise geometries that are highly efficient in communicating or interacting with specific proteins, has been a goal of synthetic chemists, since the multivalent nature of carbohydrate-mediated interactions has been discovered. However, challenges are still coming from the control of the spatially or topologically systematic requirements.

Recently, a type of carbohydrate-modified nanoparticles emerged with a variety of modification capability, namely, glyconanoparticles (GNPs) (Fig. 16.5). GNPs are sugar-coated gold, iron oxide, or semiconductor nanoparticles with defined thiol-ending glycosides, which combine the multivalent presentation of carbohydrates (glycol clusters), together with the special chemicophysical properties of the embedded metallic nanocore. The capability of connecting all kinds of carbohydrate chains and other molecules (e.g., luminescent/fluorescent probes, peptides, magnetic chelates) onto the same core nanoparticle in a controlled way (multifunctional GNPs) is crucial and a fundamental principle of such techniques. Moreover, the modification of the core in order to have magnetic or fluorescence properties (multimodal GNPs) made this multivalent glyco-scaffold a suitable platform for so many studies related with carbohydrate-mediated interactions and applications in nanomaterial biocompatibility as well as molecular imaging [21].

Recent researches and new developments highly broadened our vision in carbohydrate nanotechnology; a good example is the quick development and application of chitosan-based nanomaterials. Chitosan, as a biocompatible and biodegradable polymer, has immense structural possibilities for chemical modifications to provide novel properties for the fabricated materials in diverse fields, especially the biomedical field. In recent decades, all sorts of research and development in

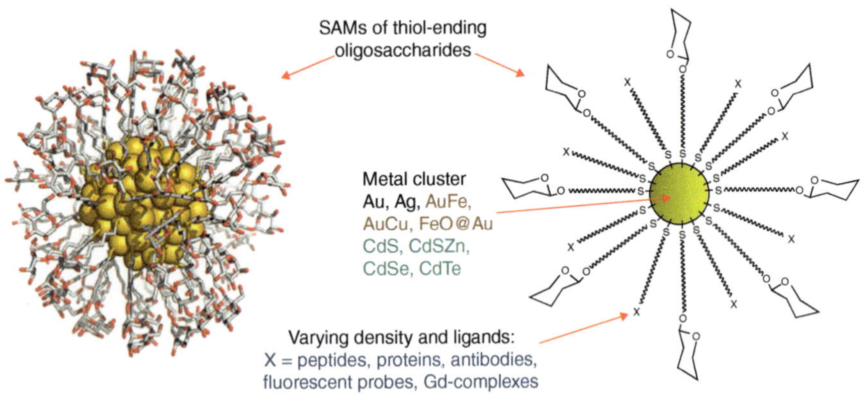

FIGURE 16.5 Calculated gold glyconanoparticle bearing 5-mercaptopentyl-a-D-mannopyranoside and possible modifications to obtain multifunctional and multimodal glyconanoparticles. *Source*: Reprinted with permission from Ref. 21, Copyright 2013, Royal Society of Chemistry.

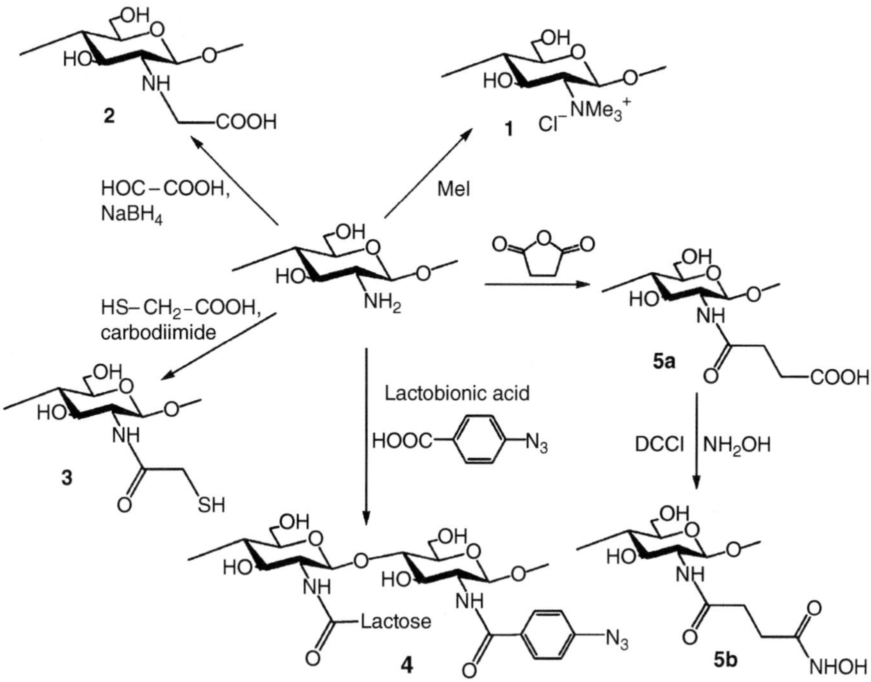

FIGURE 16.6 Chitosan modifications for different applications: (a) methylation, (b) thiolation, (c) azylation, (d) copolymerization, and (e) N-succinylation. *Source*: Adapted with permission from Ref. 18, Copyright 2004, American Chemical Society.

functional biomaterials such as chitosan has actually led to a new branch of industrial field dedicated in providing drug delivery system with superior biocompatibility and high delivery accuracy.

Chitosan is originally a biomaterial; because of its biocompatibility, biodegradability, and nontoxic properties that has recently been found, chitosan started to be applied in much larger fields of science, namely, drug/gene delivery, cell transportation/imaging, biosensors, as well as disease/cancer treatment and diagnosis [22,23]. Moreover, compared with ordinary nanomaterials that are mostly derived from inorganic composition, chitosan-based nanomaterials possess superior physical and chemical properties like high surface area, porosity, tensile strength, conductivity, photoluminescence, and increased mechanical properties. Figure 16.6 highlighted some recent research on different aspect of chitosan-based nanomaterial; some major chemical modifications, like methylation, thiolation, azylation, copolymerization, N-succinylation, were shown in varied applications [22].

Still, given the good biocompatibility and easy fabrication, applications of chitosan-based nanomaterials also found its wide application potential in biosensors [22]. In recent years, this area has gained increasing curiosity for their advantageous properties as analytical tools, specially the potential miniaturization, portability, low cost, and friendly use, compared to other well-established lab-based methods [24].

16.2.3 Nanocarbohydrate–Cell Interaction

Carbohydrate-functionalized nanomaterials can provide different ways for specific cell surface interactions through multivalent binding, plus the benefit of its low cytotoxicity and high solubility in cell culture medium [19]. A recent cytotoxic report on glycopolymer-coated carbon nanotubes done using CHO and Jurkat cells showed good biocompatibility and nontoxicity of the modified CNTs [25]. Bertozzi group also developed a new approach to fabricate CNTs with biomimetics [25,26]. These biomimetic CNTs didn't show any cytotoxicity and proved a successful integration of CNTs into aqueous biological system. Furthermore, CNTs functionalized with sugar chains could be used as new efficient carrier systems for therapeutic molecule transportation due to their nonimmunogenicity, biocompatibility, and low toxicity [19]. Comparatively, nanomaterials composed of pure carbohydrate macromolecules, like chitin and chitosan biopolymers, have also been used as an attractive and effective sorbents because of their nontoxicity, hydrophilicity, biocompatibility, biodegradability, and biofunctionality [27].

16.2.4 Nanocarbohydrate–Virus Interaction

Multivalent GNPs with high affinity to antiviral lectin cyanovirin-N low-mannose (LM) structures find its superior application in a study toward the CV-N variant structures [28]. The authors found that through combining with gold nanoparticles (Au NPs), the affinity of LMs to cyanovirin-N (CV-N) could be greatly amplified. In detail, they found that these GNPs rendered fundamentally higher affinity with CV-N lectins than free glycan ligands, demonstrating that Au NPs within the structure also served as an efficient multivalent scaffold for the enhancement of affinity, which compared well with other sources of multivalent ligands. Thus, the author pointed out that through simple glycans, rather than large and complex sugars, it'll be an easier strategy to graft onto a multivalent scaffold for affinity amplification, which will further aid the future development of new glyconanomaterials for diagnostic and therapeutic applications.

16.2.5 Carbohydrate-Modified Nanomaterial Toxicity In Vitro

Polymeric nanoparticle-based carriers are a type of promising scaffolds for intracellular drug delivery; the efficacy of delivery depends highly on the nanomaterials biocompatibility and cytotoxicity. Physicochemical properties of these materials could be well managed through modification of particle size, shape, surface charge lipophilicity, etc. Recent studies have shown efficient tumor cell targeting through hydrophobically modification of glycol chitosan (HGC) nanoparticles with enhanced membrane permeability and cellular retention. The following study showed that partial free amino derivatizated glycol chitosan nanoparticles with 5β-cholanic acid can enhance its subcellular localization throughout the whole cells, when compared with the parent hydrophilic GC polymers. A closer look with the endocytic inhibitors suggested several distinct uptake pathways that are actually involved during the HGC internalization, including clathrin-mediated endocytosis, caveolae-mediated endocytosis, and micropinocytosis. Even though some HGC nanoparticles were found entrapped on the lysosome membrane, they nevertheless provide a putative way for a versatile drug delivery system for intracellular accurate therapeutic transportation [29].

GNPs have also been fabricated with both carbohydrate and amino group motifs presented at the surface, which embrace great stability and solubility in aqueous solution. Eaton et al. studied the binding of model DNA and plasmid by these nanoparticles through transmission electron microscopy and gel electrophoresis. Varied DNA affinity was correlated with different nanoparticle surface modification; the hybrid amino/Gal nanoparticles had high efficiency, rendered as highly water-soluble DNA binders that can condense the DNA molecules into a compact globular shape. This result indicated a potential benefit of such type of GNPs for nonviral-based gene-transfection reagent [30,31].

Another effective gene delivery system with low cytotoxicity has been proposed recently as well, where a cationic glycopolymer-stabilized gold nanoparticles was used [32]. Under physiological conditions, these GNPs were proved to be biocompatible and are able to resist aggregation. Gel electrophoresis results showed good complexation formation between DNA molecules and the cationic GNPs. A final transfection efficiency was found to be well comparable to the commercialized reagent Lipofectamine 2000 [32].

16.2.6 Carbohydrate-Modified Nanomaterial Toxicity In Vivo

Due to the remarkable biocompatibility and physicochemical properties, carbohydrate-based nanomaterials (such as chitosan-based modification) have become one of the most promising polymers for biomedical application. However, in order to take advantage of chitosan not only on an *in vitro* basis but also moving forward to *in vivo* tests as well as clinical trials, an inherited problem has to be well solved. The cationic nature of chitosan molecules would have poor hemocompatibility when entering into the circulation system and thus would possibly induce thrombosis, making it unsuitable as blood-contacting material [33]. However, for the past decades, researchers have worked thoroughly to attempt to modulate the biopolymer–blood interactions to develop hemocompatible chitosan derivatives and, thus, broaden the application of these biopolymers.

Currently, the strategies used for modifying chitosan hemocompatibility are focusing on two aspects: (i) chemical modifications of chitosan and (ii) association of this biopolymer with compounds that exhibit complementary properties. Recent progress in designing and fabricating hemocompatible, functionalized chitosan-based reagents has been made in areas like drug delivery, central nervous system disease treatment, theragnosis, as well as cardiovascular applications [33]. In sum, either chemical or physical modifications to chitosan can properly make chitosan-related derivatives with higher hemocompatibility. Here, Table 16.4 summarized some significant results that related with chitosan-based system hemocompatibility as potential formulations [33].

TABLE 16.4 The Assessment of Blood Compatibility of Chitosan-Based Systems[a]

Blood Compatibility Test	Chitosan-Based Formulation	Characteristics
Hemolysis	Chitosan emulsion (nanoparticles)	Negligible hemolysis (less than 2%)
	Chitosan–alginate membrane (gel)	No significant hemolysis occurred during 4 h
	Chitosan thermosensitive hydrogel	Injected gel was safe and had no hemolytic effect
	Chitosan membranes	A nonhemolytic result and a low level of cytotoxicity were demonstrated
	Chitosan and chitosan-modified (N-SCs and N.O-SCS) films	Chitosan and its derivatives showed no obvious hemolytic ability; Hemolytic ratio decreased with increasing of DS
Blood coagulation assay	Chitosan and chitosan modified (N-SCs) solution	A prolonged APTT No significant effect on PT and TT
	Chitosan and PC chitosan	PC chitosan exhibits better antiprotein adsorption capability and improved hemocompatibility compared with native chitosan
	Chitosan–biotinylated magnetic particles	Chitosan–biotinylated magnetic particles did not inhibit the activity of prothrombin
Blood coagulation index (BCI)	Chitosan and chitosan-modified films	The BCI value of both derivatives increased with the raising DS No obvious distinction between N-SCs and N, O-SCs
Platelet adhesion and activation	Chitosan–alginate membrane (gel)	Low platelet attachment; platelet activation did not occur on surfaces up to 90 min of exposure
	Chitosan-based surfactant polymers	Charge had a notable effect on platelet adhesion; chitosan surface containing the negatively charged SO_3 exhibited the fewest number of adherent platelets
	Chitosan and LSCs	Absence of platelet aggregation on LSCs Hemostatic effect of chitosan unmodified
	Chitosan and chitin	Increased platelet aggregation and cytokine release from platelets
Rheological measurements	Chitosan films	N-Acyl chitosan is hemocompatible
Thromboelastography (TEG)	N-Acylated chitosan nanoparticles	N-Acyl chitosan nanoparticles are hemocompatible
C3 protein depletion assay	Chitosan and LSCs	Complement activation but in an acceptable region Hemostatic effect of chitosan unmodified

[a] Reprinted from Ref. 33, Copyright 2014 with Permission from Elsevier Ltd.

Finally, in the fields of tumor imaging and cancer therapy, much attention has been drawn about the antiangiogneic reagent or drugs to specifically kill tumor/cancer cells. The antiangiogenic peptide drugs have shown promising targeting capability onto integrins such as avb3 on angiogenic endothelial cells [34]. However, traditional antiangiogenic peptide drugs suffered short half-lives especially *in vivo*, where a fast serum clearance via kidney would quickly invalidate the reagent. Kim et al. prepared self-assembled glycol chitosan nanoparticles as prolonged/sustained drug delivery system for the Arg–Gly–Asp (RGD) peptide in cancer therapy. After being hydrophobically modified using 5b-cholanic acid (HGC), the glycol chitosan nanoparticles can easily encapsulate the RGD peptide with pretty high loading efficiency (>85%). The preliminary *in vitro* test demonstrated prolonged and stabilized RGD releasing for one week. The formed RGD-HGC complex can also inhibit human umbilical vein endothelial cell (HUVEC) adhesion to a βig-h3 protein-coated surface, which indicates the antiangiogenic effect of the RGD peptide in the HGC nanostructures. During the following *in vivo* study, the RGD-HGC nanoparticle–peptide complexes highly inhibited bFGF-induced angiogenesis and thus decreased the hemoglobin content in the matrigel plug assay. Furthermore, intratumoral application of such complexes can also remarkably reduce the tumor growth speed and the amount of microvessel density. Thus, the RGD-HGC method can highly enhance its antiangiogenic activity because of the prolonged and sustained on-site RGD peptide delivery capability [34]. The following figure (Fig. 16.7) showed a representative scheme on how such RGD-HGC complex works.

16.3 TOXICITY-RELATED APPLICATION OF CARBOHYDRATE NANOMATERIALS

16.3.1 Application of Nanocarbohydrate Cytotoxicity for Environmental Utility

As a matter of fact, the characteristic feature of carbohydrates, and its involvement within carbohydrate–protein as well as carbohydrate–carbohydrate interactions under a whole bunch of biological circumstances, leads to a broad spectrum of physiological and pathological processes with regard to phenomenon such as metastasis, inflammation, infection, and so on [35]. A recent report on functionalization of naturally occurring amino sugar, glucosamine, onto silver nanoparticles surfaces (GlcN-AgNPs, size of ~30±5 nm) showed higher antibacterial activity than the raw silver nanoparticles (AgNPs, size ~20±2 nm), against either negative (Klebsiella pneumoniae (ATCC 700603)) or positive (Bacillus cereus) bacteria [36], respectively. The field emission scanning electron microscopy (FE-SEM) images (Fig. 16.8) clearly showed the detailed surface conditions of both AgNPs and GlcN-AgNPs. A possible reason for the higher toxicity of GlcN-AgNPs might attribute to the surface functionalization of glucosamine, the molecule that helped glycol-nanoparticles penetrate into bacterial cells. The highly increased cytotoxicity may be also due to the highly enlarged surface area and roughened particle surface and thus increased

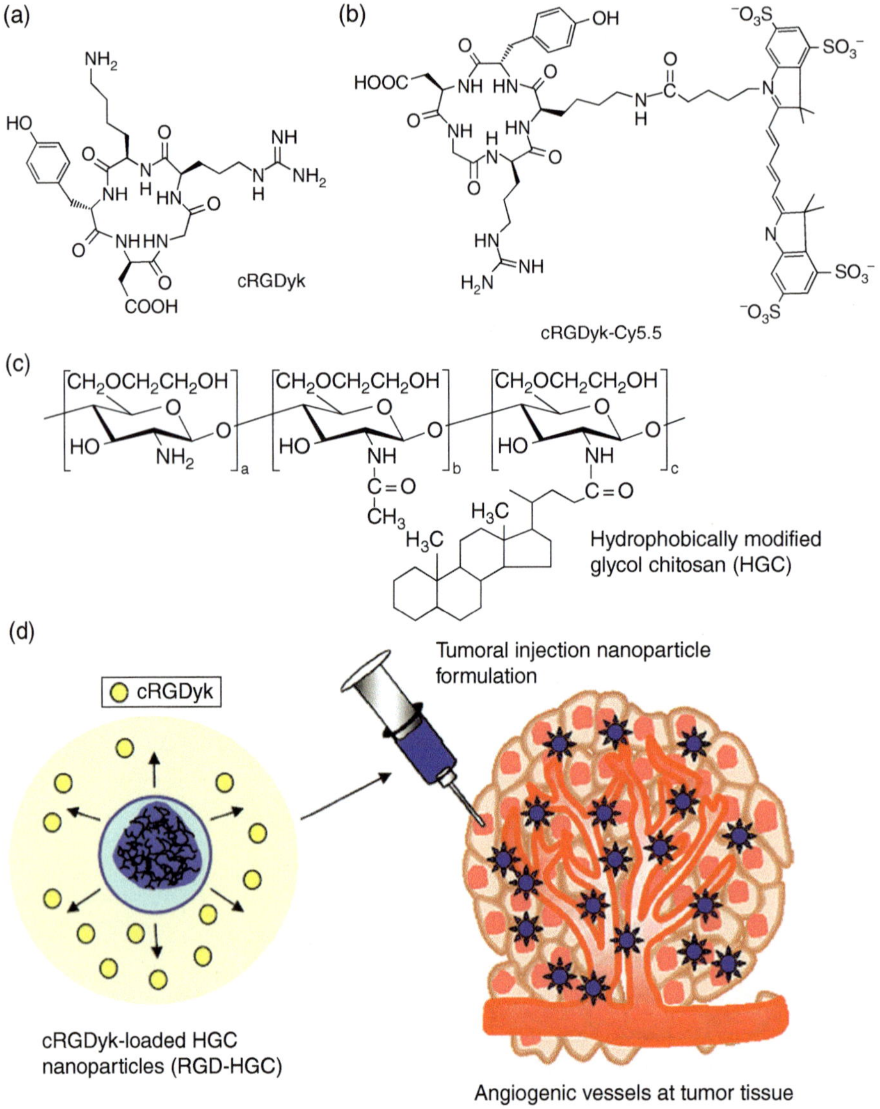

FIGURE 16.7 Representative chemical structures of (a) cRGDyK peptide, (b) Cy5.5-labeled cRGDyK peptide, and (c) HGC nanoparticles (prepared by coupling of glycol chitosan with 5b-cholanic acid). (d) The as synthesized RGD-HGC nanoparticles will be directly administered in solid tumor to release RGD peptides to tumor endothelial cells; thus, the tumor growth could be suppressed through prevention of the formation of microvessels in tumors. *Source*: Reprinted with the permission from Ref. 34, Copyright 2008, Elsevier Ltd.

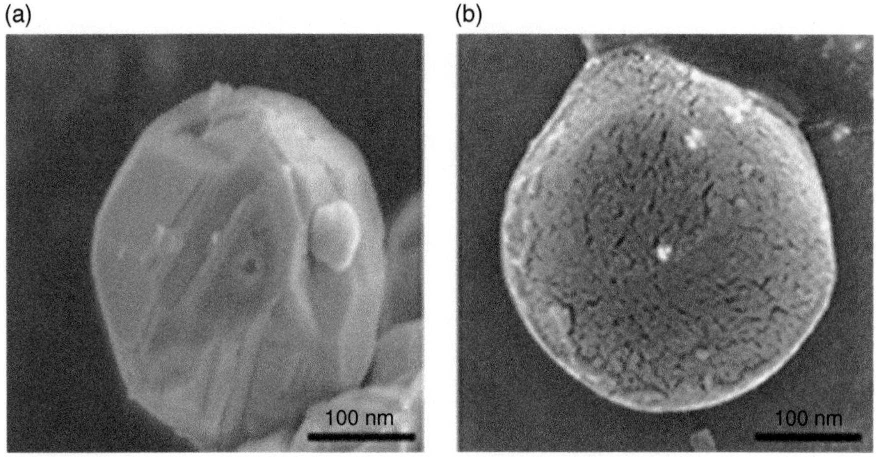

FIGURE 16.8 Surface morphology of (a) AgNPs and (b) GlcN-AgNPs. (b) shows a successful surface functionalization of glucosamine; a smooth spherical layer has been deposited over the nanoparticle (b). *Source*: Reprinted with permission from Ref. 36, Copyright 2010, Springer-Verlag.

contact and interaction area. The author also proposed that it may be thirdly due to the unique activity of the formation of second amide bond [36].

The antibacterial activity of natural polysaccharide has previously been extensively studied. Results showed excellent biocompatibility, biodegradability, and metal complexation [37] Earlier reports also showed that such antibacterial capability could be greatly enhanced when attached to nanoparticles through metal ions such as Cu^{2+} and Zn^{2+}. Interestingly, it was further found out that the increased activity was closely correlated with zeta potential [38]. In another report, a cross-linked chitosan layer (CCTS) coated Ag nano-SiO_2 composite (CCTS-SLS) showed to have high antibacterial activity against Escherichia coli and S. aureus, and the author attributed it to the coordinated action of CCTS and SLS [39]. Besides, aminoglycosides, type of amino-modified sugars, have also been reported to have antibiotic properties. A key issue in antibacterial reagent development is the bacterial resistance to antibiotics. However, from a previous study using aminoglycosides functionalized Au nanoparticles, it showed that such resistance could be compromised, and the activity of applied antibiotics (such as streptomycin, gentamicin, and neomycin) were well protected [40].

The threats coming from waterborne diseases are one of the long-lasting worldwide problems and remain the major cause of death in many developing countries. According to a WHO report published in 2004, at least one-sixth of the world population (1.1 billion people) lack access to safe water [41]. Even in the United States, unexpected outbreaks of waterborne diseases constantly kept at high level. According to the Center of Disease Control Morbidity and Mortality Weekly Report, there were 155 waterborne disease outbreaks and 431,846 cases of illness in public and individual US water systems from 1991 to 2000 [42]. Although later disinfection methods that have been used acquired acceptable water treatment effectiveness in controlling microbial pathogens, researchers in the past decade debated on a dilemma

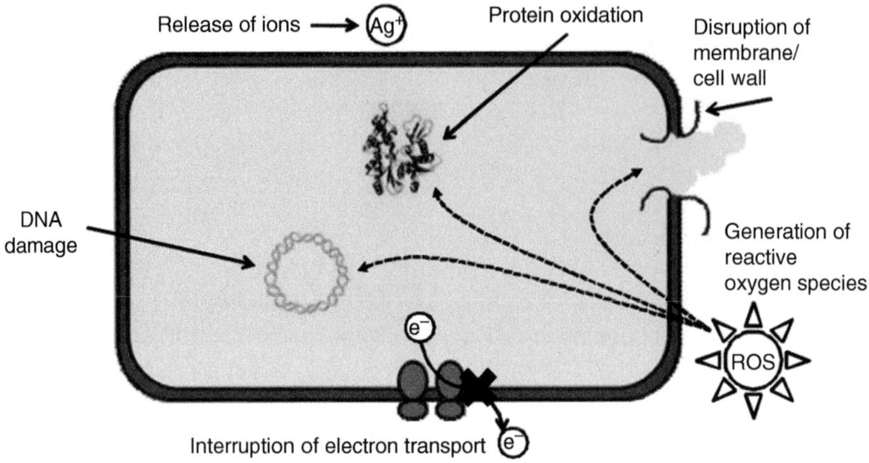

FIGURE 16.9 Proposed mechanisms of antimicrobial capability by nanomaterials. *Source*: Reprinted with the permission from Ref. 42, Copyright 2008, Elsevier Ltd.

between effective disinfection and the formation of harmful disinfection by-products (DBPs). Normally used chemicals for drinking water disinfection like free chlorine, chloramines, and ozone can actually react with many natural water constituents to form DBPs that has carcinogenesis. Given that more than six hundreds of such DBPs are considered as carcinogenic [43], an urgent need to reevaluate conventional disinfection chemicals and methods and employ novel disinfection reagents to avoid DBP formation has been proposed intensely.

For the excellent ability of being used as adsorbents, catalysts, and sensors due to large specific surface area and high reactivity, both natural and engineered nanomaterials have been shown to have strong antimicrobial activities, such as chitosan [44], silver nanoparticles (nAg) [45], photocatalytic TiO_2 [46], fullerol [47], aqueous fullerene nanoparticles (nC_{60}) [48], and carbon nanotubes (CNT) [49]. Different from conventional chemical disinfectants, none of these nanomaterials are strong oxidants so they are relatively inert in aqueous circumstance. Therefore, few DBPs will be produced during exposure.

The nanoparticle's antibacterial abilities are mainly due to three major reasons: naturally acquired antibacterial substances, metal or metal oxides, and novel engineered nanostructures. Several disinfection mechanisms have been proposed and were briefly demonstrated in Figure 16.9. On one hand, the nanoparticles can directly interrupt the microbial cell transmembrane electron transfer, penetrate the cell envelope, react (e.g., oxidize) cell components, or produce secondary products (e.g., reactive oxygen species (ROS) or dissolved heavy metal ions) to cause cellular damages.

On the other hand, naturally occurring chitin and certain peptides have been known for their antimicrobial properties, and it is only recently that these materials were engineered as nanoparticles [44,50]. They are low-cost and low-tech disinfection applications, which are particularly suitable for expanded use and application in developing countries. The antibacterial mechanism of natural peptides, like silk moth cecropins, is the formation of nanoscale channels in bacterial cell membranes, which causes osmotic collapse [50]. Actually, the synthesized antimicrobial peptides were firstly based on known protein structures [51], but step-forward, such synthesize ability can further be used to construct tailored antimicrobial nanostructures [50]. The tailoring processes include many types of optimization: size, morphology, coatings, derivatization, and other properties for specific antimicrobial applications.

Among all above-mentioned nanomaterials, chitosan, which was obtained from chitin in arthropod shells, has long been noticed for the antibacterial activity, but it has just been made into nanostructures recently [44,52]. Astonishingly, nano-chitosan and its derivatives exhibit wide spectrum of antimicrobial effects toward bacteria, viruses, and fungi [44,53–56], though lately was found to be more effective for fungi and viruses than bacteria [54]; as for bacteria, chitosan is more active toward Gram-positive bacteria than Gram-negative ones [55,57]. A wide range of minimum inhibitory concentrations (MICs) from 18 to 5000 ppm were reported and varied upon the target organism, environmental pH, chitosan molecular structure and weight, degree of polymerization, and the presence of lipids and proteins [54,55]. However, the disinfection capability depends upon the material preparation methods and the presence of organics. Chitosan's ability also highly depends on the following: solubility (acidic pH environment) and availability of charged amino groups [54,55].

Several mechanisms have been proposed for chitosan's antimicrobial activity. The major one ascribes such ability to a positively charged particle surface that can interact well with negatively charged cell membranes, and the accumulation of these attachment will finally induce reversed membrane permeability so that permanent damage to the cell membrane could occur [44]. This is in accordance with observation that neither natural chitosan nor engineered chitosan but in a pH environment above 6 will cause little destructive effect to cell membranes, because of the lack of protonated amino groups [44]. In this case, engineered chitosan macromolecules that contain quaternary ammonium groups (e.g., *N,N,N*-trimethyl chitosan, *N*-propyl-*N*, *N*-dimethyl chitosan, and *N*-furfuryl-*N*, *N*-dimethyl chitosan) showed high toxicity than the natural ones, and their activity would increase with decreasing pH [58]. Another proposed mechanism indicated that chitosan may chelate surrounding trace metal ions or atoms, thus destroying cells through inhibition of key enzyme activities [54]. While with a more proactive hypothesis, the chitosan molecules can invade into cell nucleus and bind with DNA to inhibit the normal transcription process [54].

Nevertheless, several important challenges still hinder our way of widely using engineered nanoparticles for drinking water treatment and disinfection, such as the dispersion and retention problem of nanoparticles as well as the sustainability of their antimicrobial activity. Also, even though nanoparticles can originally provide high specific surface area when they are freshly made, most of them would suffer intensive aggregation or agglomeration during the exposure or precipitate processes, which will definitely compromise the disinfection ability.

TiO_2 nanoparticles are known to have aggregation problem when added into water, and even if the nanoparticles are stable in pure water, coagulation will be induced when salt is added [59].

Moreover, a well-dispersed nanoparticle suspension contains particles of extremely small sizes. In this case, a dilemma would be seen when larger amounts of nanoparticles are in need, while fewer of these particles should be leaked out of the disinfection reactors. On one hand, slurry mode may highly improve the disinfection efficiency but leave a huge problem to retain all particles within the reactor; on the other hand, taking advantage of nanoparticle immobilization into more compact layer-shaped reactor will greatly solve the particle leakage problem, but on the contrary, leave very limited reactive surface for the reactor, plus the hardness to equip irradiation capability.

To sum up, several significant limitations exist to fully functionalize these promising nanomaterials for environmental utility and microbial control. It needs more research and fundamental approach to assess the influences of nanoparticle leakage to both environmental and human health; it is also important to acquire better technologies to finely retain nanomaterials for its full functionalization [42].

16.3.2 Applications in Cancer Therapy

There are four major types of organic molecules in biosystems, namely, proteins, lipids, nucleic acids, and carbohydrates. Carbohydrates, by far, are the most abundant organic molecules that are found in all organisms for structure synthesis, cell singling, and metabolizing [60]. As shown in Figure 16.10, the complexity of carbohydrates in the living system is an intrinsic result of some key characteristics of carbohydrate: multiple residue binding (glycosidic bonds) formats, structural heterogeneity of these molecules, varied types of anomeric linkage, and the absence, presence, or position of the branching point [61,62].

Particularly, the complexity and variety of cell surface glycans enable themselves to communicate and attach with the surrounding environment [63–66]. Thus, those glycan molecules are included within many important physiological functions, such as membrane rigidity, embryonic development, cell fate and differentiation, cell trafficking and localization, proliferation, contact inhibition, cell talking and signaling, host–pathogen interaction, immune responses, disease development, metastasis, and so on [61,63,67–77]. More importantly, it was reported that nearly all types of malignant cells and tissues would show apparent alterations in their surface glycosylation patterns when compared to their normal counter parts [78–83].

GNPs, or carbohydrate-functionalized nanoparticles, were thus believed to have the ability to enhance the selective cellular interactions that are used for biosensing application [84,85]. Although the carbohydrate–protein interaction is relatively weak than biotin–streptavidin that has been commercially used, they can be successfully enhanced through multivalent interactions, which is the so-called "cluster glycoside effect" [86]. Multivalent carbohydrate–ligand/protein interactions have been successfully used by Tsai et al. in a rapid and quantitative

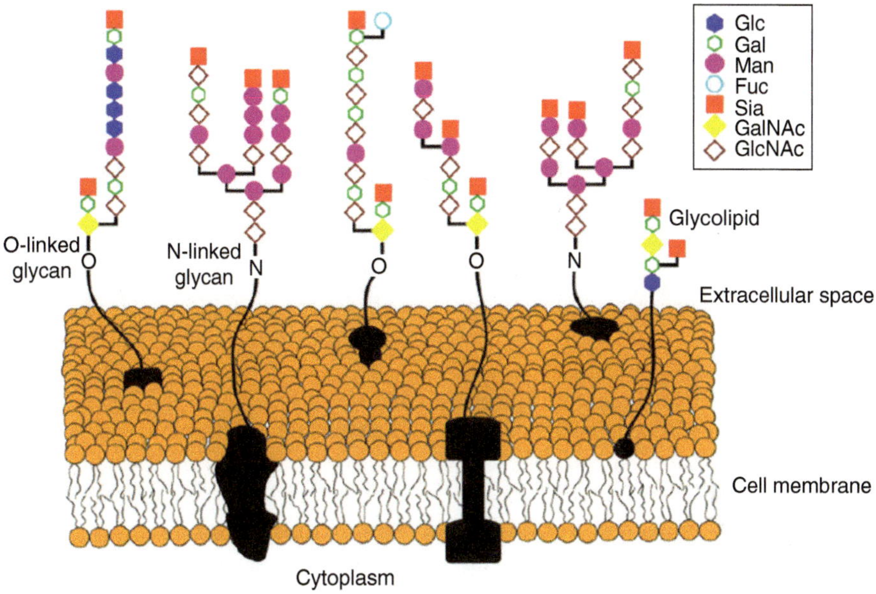

FIGURE 16.10 Carbohydrate heterogeneity found in glycoproteins and glycolipids on cell membrane. Sialic acids (Sia) are normally seen at the terminal residue of O-linked and N-linked glycans of glycoproteins and glycolipids. Glc, glucose; Gal, galactose; Man, mannose; Fuc, fucose; GalNAc, *N*-acetylgalactosamine; GlcNAc, *N*-acetylglucosamine. Illustration prepared by Haike Ghazarian using Corel Draw X4. *Source*: Reprinted with permission from Ref. 60, Copyright 2011, Elsevier GmbH.

measurement for the carbohydrate-binding protein concanavalin A [85]. Furthermore, Penadés et al. have also successfully established a glyconanotechnology for carbohydrate–carbohydrate and carbohydrate–protein interaction study and evaluation, which provide a good model for carbohydrate-mediated biological processes studying [87]. A multifunctional GNPs were then prepared with lactose displayed on the outer surface and were used as a potential platform for carbohydrate-based anticancer vaccines. This platform was reported to have possible polyvalent tailoring capability for the anticancer vaccines and thus would be a good candidate for drug delivery [88].

Another drug delivery system reported recently using hydrophobically modified glycol chitosan nanoparticles as platform, which provides more in-depth progression in the area (Fig. 16.11) [89]. By chemical conjugation of hydrophobic 5β-cholanic acid moieties to the hydrophilic glycol chitosan backbones, the researchers tried to fabricate a water-insoluble camptothecin (CPT) delivery carrier, for better encapsulation of the hydrophobic CPT (above 80%) as well as longer sustainability (one week). This encapsulation would also provide protections for the active lactone ring of CPT, to against possible hydrolysis/degradation when entering the body. Results showed high tumor selectivity and significant antitumor effects toward MDA-MB231 human breast cancer xenografts that are

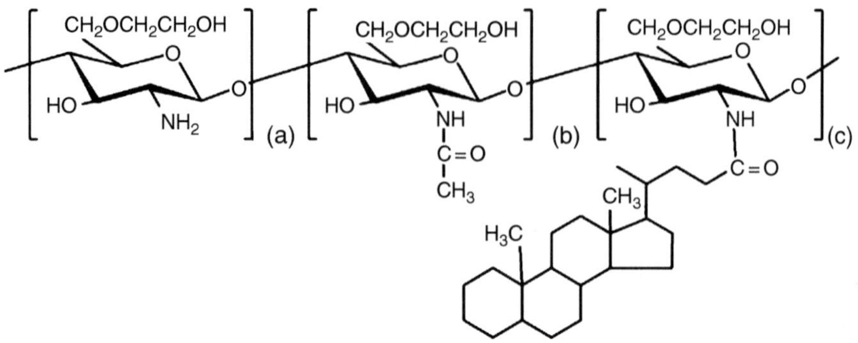

FIGURE 16.11 Hydrophobically modified glycol chitosan (HGC) nanoparticles, prepared through hydrophilic glycol chitosan coupling with hydrophobic 5β-cholanic acids. *Source*: Reprinted with permission from Ref. 89, Copyright 2008, Elsevier B.V.

subcutaneously implanted in nude mice. This promising result indicated the importance of both prolonged blood circulation and high accumulation in tumors, which was assisted by the HGC nanoparticles. And these results, again, revealed the high potentiality of HGC nanoparticles to serve as drug encapsulation and delivery system in future cancer therapy [89].

Similarly, a type of tumor-homing chitosan-based nanoparticles (CNPs), loaded with an anticancer drug paclitaxel (PTX) and labeled with Cy5.5 near-infrared fluorescent dye, was established and tested on its *in vivo* anticancer selectivity and efficiency with SCC7 tumor-bearing mice (Fig. 16.12). Theragnostic nanoparticles, Cy5.5-labeled PTX-CNPs, are highly useful for simultaneous diagnosis of early-stage cancer and drug delivery.

As results, the Cy5.5-CNPs were selectively localized in flank tumor tissue than other organs. Real-time intravascular tracking of CNPs indicated enhanced permeation and retention of the nanoparticles within tumor vasculature. Importantly, such drug delivery platform showed tumor selective localization with different tumor models, such as brain tumors, liver tumors, and metastasis tumor models. Combined with noninvasive near-infrared imaging capability, this technique showed great potential in both cancer therapy and cancer imaging.

In conclusion, the emerging carbohydrate-modified nanomaterials highly extended our view of nanoscale material fabrication and application, especially in a biomedical point of view. Although these materials are still suffering realistic problems like aggregation, leakage, and health and environmental contamination risks, but step by step, studies have showed a lot more excellent properties of these materials, such as lower cost, faster fabrication process, modifiable affinity to several key types of biological molecules, compatibility with both inorganic metal NPs and biological systems, prolonged *in vivo* sustainability, as well as controllable cytotoxicity, which in all make this type of nanoscale materials a great candidate for future biomedical research and application.

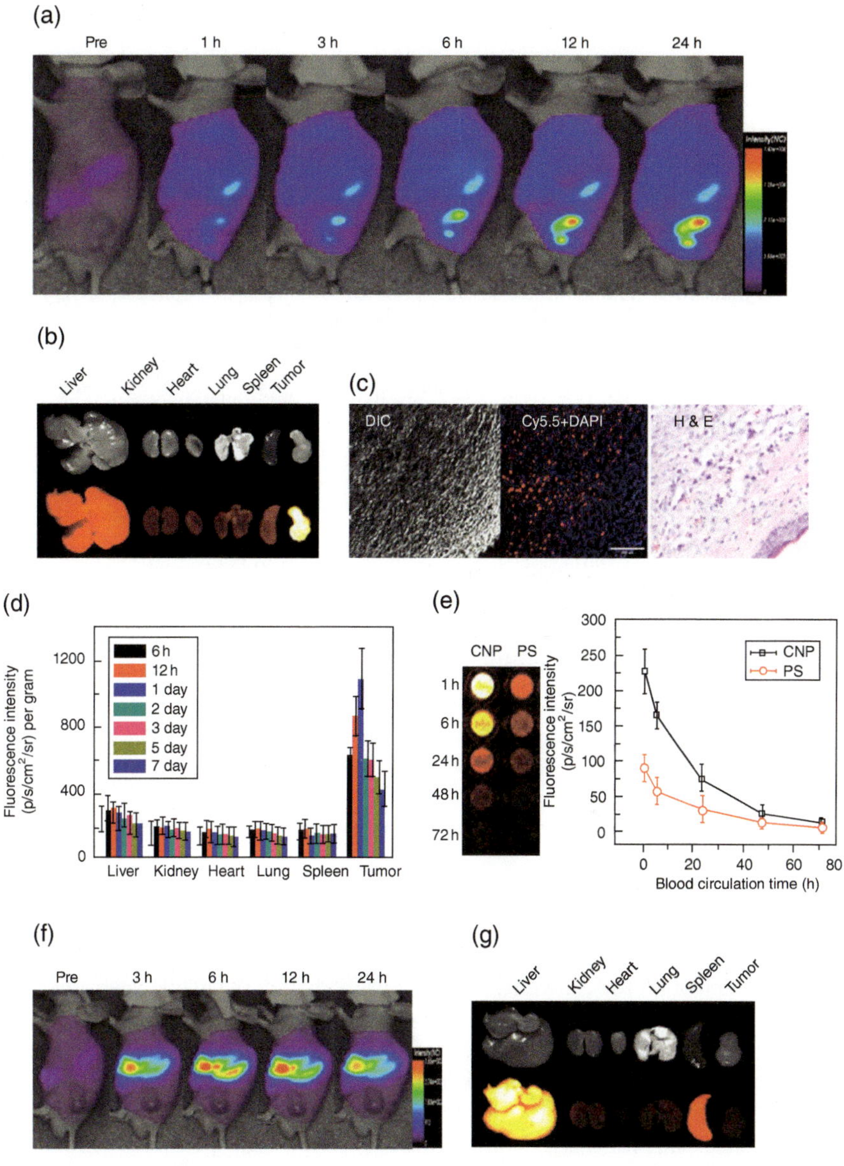

FIGURE 16.12 Flank tumor targeting by Cy5.5-CNPs. (a) A time-dependent whole body distribution with Cy5.5-CNPs. (b) ex vivo NIRF imaging of Cy5.5-CNPs in excised organs after a 24 h injection. (c) Histological analysis of tumor after 24 h postinjection. (d) Quantification of NIRF signal intensities of excised organs that injected with Cy5.5-CNPs. Results were recorded as total photons per centimeter squared per steradian (p/s/cm^2/sr) per gram of each organ at all time points and are presented as mean ± SD (n = 5 per group). (e) NIRF images and intensity graph of blood samples drawn from mice injected with CNPs and PSs. Data are presented as mean ± SD (n = 5). *, statistically significant (P < 0.05). (f) Time-dependent whole body distribution of PSs. (g) Ex vivo organ distribution of PSs after 1 day postinjection. *Source*: Reprinted with permission from Ref. 90, Copyright 2011, Elsevier Ltd.

REFERENCES

1. Khot LR, Sankaran S, Maja JM, Ehsani R, Schuster EW. Applications of nanomaterials in agricultural production and crop protection: a review. *Crop Protection* **2012**;*35*: 64–70.
2. Lee J, Mahendra S, Alvarez PJ. Nanomaterials in the construction industry: a review of their applications and environmental health and safety considerations. *ACS Nano* **2010**;*4*:3580–90.
3. Lin W, Stayton I, Huang Y-w, Zhou X-D, Ma Y. Cytotoxicity and cell membrane depolarization induced by aluminum oxide nanoparticles in human lung epithelial cells A549. *Toxicological and Environmental Chemistry* **2008**;*90*:983–96.
4. Lin W, Xu Y, Huang C-C, Ma Y, Shannon KB, Chen D-R, et al. Toxicity of nano-and micro-sized ZnO particles in human lung epithelial cells. *Journal of Nanoparticle Research* **2009**;*11*:25–39.
5. Lin W, Huang Y-w, Zhou X-D, Ma Y. In vitro toxicity of silica nanoparticles in human lung cancer cells. *Toxicology and Applied Pharmacology* **2006**;*217*:252–9.
6. Lin W, Huang Y-w, Zhou X-D, Ma Y. Toxicity of cerium oxide nanoparticles in human lung cancer cells. *International Journal of Toxicology* **2006**;*25*:451–7.
7. Adams M, Lin W, Ma Y. Quantitative determination of ZnO nano-particles in human bronchoalveolar carcinoma-derived lung cancer cells. *Journal of Undergraduate Chemical Research* **2007**;*6*:19–22.
8. Yang Q, Ma Y. Irradiation-Enhanced Cytotoxicity of Zinc Oxide Nanoparticles. *International Journal of Toxicology* **2014**;*33*:187–203.
9. Chen J, Wang D, Xi J, Au L, Siekkinen A, Warsen A, et al. Immuno gold nanocages with tailored optical properties for targeted photothermal destruction of cancer cells. *Nano Letters* **2007**;*7*:1318–22.
10. Mann S. Life as a nanoscale phenomenon. *Angewandte Chemie International Edition* **2008**;*47*:5306–20.
11. Casals E, Vázquez-Campos S, Bastús NG, Puntes V. Distribution and potential toxicity of engineered inorganic nanoparticles and carbon nanostructures in biological systems. *TrAC Trends in Analytical Chemistry* **2008**;*27*:672–83.
12. Min Y, Akbulut M, Kristiansen K, Golan Y, Israelachvili J. The role of interparticle and external forces in nanoparticle assembly. *Nature Materials* **2008**;*7*:527–38.
13. Dagastine RR, Manica R, Carnie SL, Chan D, Stevens GW, Grieser F. Dynamic forces between two deformable oil droplets in water. *Science* **2006**;*313*:210–3.
14. Kim H-Y, Sofo JO, Velegol D, Cole MW, Lucas AA. van der Waals dispersion forces between dielectric nanoclusters. *Langmuir: The ACS Journal of Surfaces and Colloids* **2007**;*23*:1735–40.
15. Nel AE, Mädler L, Velegol D, Xia T, Hoek EM, Somasundaran P, et al. Understanding biophysicochemical interactions at the nano–bio interface. *Nature Materials* **2009**;*8*: 543–57.
16. Nel A, Xia T, Mädler L, Li N. Toxic potential of materials at the nanolevel. *Science* **2006**;*311*:622–7.
17. Kumar A, Kumar P, Anandan A, Fernandes TF, Ayoko GA, Biskos G. Engineered Nanomaterials: Knowledge Gaps in Fate, Exposure, Toxicity, and Future Directions. *Journal of Nanomaterials* **2014**;*2014*:1–16.

18. Kumar MR, Muzzarelli RA, Muzzarelli C, Sashiwa H, Domb A. Chitosan chemistry and pharmaceutical perspectives. *Chemical Reviews* **2004**;*104*:6017–84.
19. Gorityala BK, Ma J, Wang X, Chen P, Liu X-W. Carbohydrate functionalized carbon nanotubes and their applications. *Chemical Society Reviews* **2010**;*39*:2925–34.
20. Sudibya HG, Ma J, Dong X, Ng S, Li LJ, Liu XW, et al. Interfacing Glycosylated Carbon-Nanotube-Network Devices with Living Cells to Detect Dynamic Secretion of Biomolecules. *Angewandte Chemie International Edition* **2009**;*48*:2723–6.
21. Marradi M, Chiodo F, Garcia I, Penadés S. Glyconanoparticles as multifunctional and multimodal carbohydrate systems. *Chemical Society Reviews* **2013**;*42*:4728–45.
22. Shukla SK, Mishra AK, Arotiba OA, Mamba BB. Chitosan-based nanomaterials: A state-of-the-art review. *International Journal of Biological Macromolecules* **2013**;*59*:46–58.
23. Jayakumar R, Menon D, Manzoor K, Nair S, Tamura H. Biomedical applications of chitin and chitosan based nanomaterials—A short review. *Carbohydrate Polymers* **2010**;*82*:227–32.
24. Teles F, Fonseca L. Applications of polymers for biomolecule immobilization in electrochemical biosensors. *Materials Science and Engineering: C* **2008**;*28*:1530–43.
25. Chen X, Tam UC, Czlapinski JL, Lee GS, Rabuka D, Zettl A, et al. Interfacing carbon nanotubes with living cells. *Journal of the American Chemical Society* **2006**;*128*:6292–3.
26. Chen X, Lee GS, Zettl A, Bertozzi CR. Biomimetic engineering of carbon nanotubes by using cell surface mucin mimics. *Angewandte Chemie International Edition* **2004**;*43*:6111–6.
27. Okuyama K, Noguchi K, Hanafusa Y, Osawa K, Ogawa K. Structural study of anhydrous tendon chitosan obtained via chitosan/acetic acid complex. *International Journal of Biological Macromolecules* **1999**;*26*:285–93.
28. Wang X, Matei E, Deng L, Ramström O, Gronenborn AM, Yan M. Multivalent glyconanoparticles with enhanced affinity to the anti-viral lectin Cyanovirin-N. *Chemical Communications* **2011**;*47*:8620–2.
29. Nam HY, Kwon SM, Chung H, Lee S-Y, Kwon S-H, Jeon H, et al. Cellular uptake mechanism and intracellular fate of hydrophobically modified glycol chitosan nanoparticles. *Journal of Controlled Release* **2009**;*135*:259–67.
30. Eaton P, Ragusa A, Clavel C, Rojas CT, Graham P, Durán RV, et al. Glyconanoparticle–DNA interactions: an atomic force microscopy study. *NanoBioscience, IEEE Transactions On***2007**;*6*:309–18.
31. Ren T, Zhang G, Liu D. Synthesis of galactosyl compounds for targeted gene delivery. *Bioorganic & Medicinal Chemistry* **2001**;*9*:2969–78.
32. Ahmed M, Deng Z, Liu S, Lafrenie R, Kumar A, Narain R. Cationic glyconanoparticles: their complexation with DNA, cellular uptake, and transfection efficiencies. *Bioconjugate Chemistry* **2009**;*20*:2169–76.
33. Balan V, Verestiuc L. Strategies to improve chitosan hemocompatibility: A review. *European Polymer Journal* **2014**;*53*:171–88.
34. Kim J-H, Kim Y-S, Park K, Kang E, Lee S, Nam HY, et al. Self-assembled glycol chitosan nanoparticles for the sustained and prolonged delivery of antiangiogenic small peptide drugs in cancer therapy. *Biomaterials* **2008**;*29*:1920–30.
35. Rojo J, Díaz V, de la Fuente JM, Segura I, Barrientos AG, Riese HH, et al. Gold glyconanoparticles as new tools in antiadhesive therapy. *ChemBioChem* **2004**;*5*:291–7.
36. Veerapandian M, Lim SK, Nam HM, Kuppannan G, Yun KS. Glucosamine-functionalized silver glyconanoparticles: characterization and antibacterial activity. *Analytical and Bioanalytical Chemistry* **2010**;*398*:867–76.

37. Mao H-Q, Roy K, Troung-Le VL, Janes KA, Lin KY, Wang Y, et al. Chitosan-DNA nanoparticles as gene carriers: synthesis, characterization and transfection efficiency. *Journal of Controlled Release* **2001**;*70*:399–421.

38. Du W-L, Niu S-S, Xu Y-L, Xu Z-R, Fan C-L. Antibacterial activity of chitosan tripolyphosphate nanoparticles loaded with various metal ions. *Carbohydrate Polymers* **2009**;*75*: 385–9.

39. Mei N, Xuguang L, Jinming D, Husheng J, Liqiao W, Bingshe X. Antibacterial activity of chitosan coated Ag-loaded nano-SiO_2 composites. *Carbohydrate Polymers* **2009**;*78*:54–9.

40. Nirmala Grace A, Pandian K. Antibacterial efficacy of aminoglycosidic antibiotics protected gold nanoparticles—A brief study. *Colloids and Surfaces A: Physicochemical and Engineering Aspects* **2007**;*297*:63–70.

41. World Health Organization. Water, Sanitation and Hygiene Links to Health, Facts and Figures–updated November 2004. Geneva: WHO **2004**.

42. Li Q, Mahendra S, Lyon DY, Brunet L, Liga MV, Li D, et al. Antimicrobial nanomaterials for water disinfection and microbial control: potential applications and implications. *Water Research* **2008**;*42*:4591–602.

43. Krasner SW, Weinberg HS, Richardson SD, Pastor SJ, Chinn R, Sclimenti MJ, et al. Occurrence of a new generation of disinfection byproducts. *Environmental Science & Technology* **2006**;*40*:7175–85.

44. Qi L, Xu Z, Jiang X, Hu C, Zou X. Preparation and antibacterial activity of chitosan nanoparticles. *Carbohydrate Research* **2004**;*339*:2693–700.

45. Morones JR, Elechiguerra JL, Camacho A, Holt K, Kouri JB, Ramírez JT, et al. The bactericidal effect of silver nanoparticles. *Nanotechnology* **2005**;*16*:2346–53.

46. Cho M, Chung H, Choi W, Yoon J. Different inactivation behaviors of MS-2 phage and Escherichia coli in TiO2 photocatalytic disinfection. *Applied and Environmental Microbiology* **2005**;*71*:270–5.

47. Badireddy AR, Hotze EM, Chellam S, Alvarez P, Wiesner MR. Inactivation of bacteriophages via photosensitization of fullerol nanoparticles. *Environmental Science & Technology* **2007**;*41*:6627–32.

48. Lyon DY, Adams LK, Falkner JC, Alvarez PJ. Antibacterial activity of fullerene water suspensions: effects of preparation method and particle size. *Environmental Science & Technology* **2006**;*40*:4360–6.

49. Kang S, Pinault M, Pfefferle LD, Elimelech M. Single-walled carbon nanotubes exhibit strong antimicrobial activity. *Langmuir: The ACS Journal of Surfaces and Colloids* **2007**;*23*:8670–3.

50. Gazit E. Self-assembled peptide nanostructures: the design of molecular building blocks and their technological utilization. *Chemical Society Reviews* **2007**;*36*:1263–9.

51. Ghadiri MR, Granja JR, Milligan RA, McRee DE, Khazanovich N. Self-assembling organic nanotubes based on a cyclic peptide architecture. *Nature* **1993**;*366*:324–7.

52. Ye W, Xin JH, Li P, Lee KLD, Kwong TL. Durable antibacterial finish on cotton fabric by using chitosan-based polymeric core-shell particles. *Journal of Applied Polymer Science* **2006**;*102*:1787–93.

53. Badawy ME, Rabea EI, Rogge TM, Stevens CV, Steurbaut W, Höfte M, et al. Fungicidal and insecticidal activity of O-acyl chitosan derivatives. *Polymer Bulletin* **2005**;*54*: 279–89.

54. Rabea EI, Badawy ME-T, Stevens CV, Smagghe G, Steurbaut W. Chitosan as antimicrobial agent: applications and mode of action. *Biomacromolecules* **2003**;*4*:1457–65.

55. No HK, Young Park N, Ho Lee S, Meyers SP. Antibacterial activity of chitosans and chitosan oligomers with different molecular weights. *International Journal of Food Microbiology* **2002**;*74*:65–72.

56. Chirkov S. The antiviral activity of chitosan (review). *Applied Biochemistry and Microbiology* **2002**;*38*:1–8.

57. Don T-M, Chen C-C, Lee C-K, Cheng W-Y, Cheng L-P. Preparation and antibacterial test of chitosan/PAA/PEGDA bi-layer composite membranes. *Journal of Biomaterials Science, Polymer Edition* **2005**;*16*:1503–19.

58. Jia Z, Xu W. Synthesis and antibacterial activities of quaternary ammonium salt of chitosan. *Carbohydrate Research* **2001**;*333*:1–6.

59. Lyon D, Brown D, Alvarez P. Implications and potential applications of bactericidal fullerene water suspensions: effect of nC60 concentration, exposure conditions and shelf life. *Water Science and Technology* **2008**;*57*:1533–8.

60. Ghazarian H, Idoni B, Oppenheimer SB. A glycobiology review: carbohydrates, lectins and implications in cancer therapeutics. *Acta Histochemica* **2011**;*113*:236–47.

61. Mody R, Joshi Sa, Chaney W. Use of lectins as diagnostic and therapeutic tools for cancer. *Journal of Pharmacological and Toxicological Methods* **1995**;*33*:1–10.

62. Gorelik E, Galili U, Raz A. On the role of cell surface carbohydrates and their binding proteins (lectins) in tumor metastasis. *Cancer and Metastasis Reviews* **2001**;*20*:245–77.

63. Sharon N, Lis H. Lectins as cell recognition molecules. *Science* **1989**;*246*:227–34.

64. Sharon N, Lis H. Carbohydrates in cell recognition. *Scientific American* **1993**;*268*:82–9.

65. Ofek I, Hasty DL, Doyle RJ. *Bacterial adhesion to animal cells and tissues*. Washington, DC:ASM Press **2003**.

66. Ofek I, Hasty DL, Sharon N. Anti-adhesion therapy of bacterial diseases: prospects and problems. *FEMS Immunology & Medical Microbiology* **2003**;*38*:181–91.

67. Sharon N. Carbohydrates as future anti-adhesion drugs for infectious diseases. *Biochimica et Biophysica Acta (BBA)-General Subjects* **2006**;*1760*:527–37.

68. Akiyama S, Yamada S, Yamada K. Analysis of the role of glycosylation of the human fibronectin receptor. *Journal of Biological Chemistry* **1989**;*264*:18011–8.

69. Sharon N, Lis H. History of lectins: from hemagglutinins to biological recognition molecules. *Glycobiology* **2004**;*14*:53R–62.

70. Chammas R, Veiga S, Line S, Potocnjak P, Brentani R. Asn-linked oligosaccharide-dependent interaction between laminin and gp120/140. An alpha 6/beta 1 integrin. *Journal of Biological Chemistry* **1991**;*266*:3349–55.

71. Opdenakker G, Rudd PM, Ponting CP, Dwek R. Concepts and principles of glycobiology. *The FASEB Journal* **1993**;*7*:1330–7.

72. Zheng M, Fang H, Tsuruoka T, Tsuji T, Sasaki T, Hakomori S. Regulatory role of GM3 ganglioside in alpha 5 beta 1 integrin receptor for fibronectin-mediated adhesion of FUA169 cells. *Journal of Biological Chemistry* **1993**;*268*:2217–22.

73. Disney MD, Seeberger PH. The use of carbohydrate microarrays to study carbohydrate-cell interactions and to detect pathogens. *Chemistry & Biology* **2004**;*11*:1701–7.

74. Nimrichter L, Gargir A, Gortler M, Altstock RT, Shtevi A, Weisshaus O, et al. Intact cell adhesion to glycan microarrays. *Glycobiology* **2004**;*14*:197–203.

75. Zachara NE, Hart GW. Cell signaling, the essential role of O-GlcNAc! *Biochimica et Biophysica Acta (BBA): Molecular and Cell Biology of Lipids* **2006**;1761:599–617.
76. Blomme B, Van Steenkiste C, Callewaert N, Van Vlierberghe H. Alteration of protein glycosylation in liver diseases. *Journal of Hepatology* **2009**;*50*:592–603.
77. Mukhopadhyay B, Martins MB, Karamanska R, Russell DA, Field RA. Bacterial detection using carbohydrate-functionalised CdS quantum dots: a model study exploiting *E. coli* recognition of mannosides. *Tetrahedron Letters* **2009**;*50*:886–9.
78. Orntoft TF, Vestergaard EM. Clinical aspects of altered glycosylation of glycoproteins in cancer. *Electrophoresis* **1999**;*20*:362–71.
79. Oppenheimer SB. Cellular basis of cancer metastasis: a review of fundamentals and new advances. *Acta Histochemica* **2006**;*108*:327–34.
80. Danussi C, Coslovi A, Campa C, Mucignat MT, Spessotto P, Uggeri F, et al. A newly generated functional antibody identifies Tn antigen as a novel determinant in the cancer cell–lymphatic endothelium interaction. *Glycobiology* **2009**;*19*:1056–67.
81. Patsos G, Hebbe-Viton V, Robbe-Masselot C, Masselot D, San Martin R, Greenwood R, et al. O-Glycan inhibitors generate aryl-glycans, induce apoptosis and lead to growth inhibition in colorectal cancer cell lines. *Glycobiology* **2009**;*19*:382–98.
82. Powlesland AS, Hitchen PG, Parry S, Graham SA, Barrio MM, Elola MT, et al. Targeted glycoproteomic identification of cancer cell glycosylation. *Glycobiology* **2009**;*19*:899–909.
83. Shida K, Misonou Y, Korekane H, Seki Y, Noura S, Ohue M, et al. Unusual accumulation of sulfated glycosphingolipids in colon cancer cells. *Glycobiology* **2009**;*19*:1018–33.
84. Schofield CL, Field RA, Russell DA. Glyconanoparticles for the colorimetric detection of cholera toxin. *Analytical Chemistry* **2007**;*79*:1356–61.
85. Tsai C-S, Yu T-B, Chen C-T. Gold nanoparticle-based competitive colorimetric assay for detection of protein–protein interactions. *Chemical Communications* **2005**;*34*: 4273–4275.
86. Lundquist JJ, Toone EJ. The cluster glycoside effect. *Chemical Reviews* **2002**;*102*:555–78.
87. de la Fuente JM, Penadés S. Glyconanoparticles: types, synthesis and applications in glycoscience, biomedicine and material science. *Biochimica et Biophysica Acta (BBA): General Subjects* **2006**;*1760*:636–51.
88. Ojeda R, de Paz JL, Barrientos AG, Martín-Lomas M, Penadés S. Preparation of multifunctional glyconanoparticles as a platform for potential carbohydrate-based anticancer vaccines. *Carbohydrate Research* **2007**;*342*:448–59.
89. Min KH, Park K, Kim Y-S, Bae SM, Lee S, Jo HG, et al. Hydrophobically modified glycol chitosan nanoparticles-encapsulated camptothecin enhance the drug stability and tumor targeting in cancer therapy. *Journal of Controlled Release* **2008**;*127*:208–18.
90. Na JH, Koo H, Lee S, Min KH, Park K, Yoo H, et al. Real-time and non-invasive optical imaging of tumor-targeting glycol chitosan nanoparticles in various tumor models. *Biomaterials* **2011**;*32*:5252–61.

INDEX

Carbohydrate Nanotechnology, First Edition. Edited by Keith J. Stine.